£82. 25
1/ 1990

GLASS
SCIENCE AND TECHNOLOGY

VOLUME 4B
Advances in Structural Analysis

Contributors

Philip J. Bray
G. N. Greaves
D. L. Griscom
C. R. Masson
Phillip E. Stallworth
G. Tomandl
J. Zarzycki

GLASS
SCIENCE AND TECHNOLOGY

Edited by   *D. R. UHLMANN*

DEPARTMENT OF MATERIALS SCIENCE
AND ENGINEERING
UNIVERSITY OF ARIZONA
TUCSON, ARIZONA

N. J. KREIDL

SANTE FE, NEW MEXICO

VOLUME 4B
Advances in Structural Analysis

 1990

ACADEMIC PRESS, INC.
Harcourt Brace Jovanovich, Publishers

Boston   San Diego   New York   Berkeley   London   Sydney
Tokyo   Toronto

This book is printed on acid-free paper. ∞

ACADEMIC PRESS, INC.
1250 Sixth Avenue, San Diego, CA 92101

United Kingdom Edition published by
ACADEMIC PRESS LIMITED
24–28 Oval Road, London NW1 7DX

Library of Congress Cataloging-in-Publication Data
(Revised for vol. 4A–4B)

Glass—science and technology.

Includes bibliographies and indexes.
Contents: 1. Glass-forming systems—v. 2, pt. 1.
Processing—[etc.]—v. 4A. Structure,
microstructure, and properties. v. 4B. Advances in
structural analysis.
1. Glass—Collected works. I. Uhlmann, D. R.
(Donald Robert) II. Kreidl, N. J.
TP848.G56   666'.1   80–51
ISBN 0-12-706704-3 (vol. 4A)
ISBN 0-12-706707-8 (vol. 4B)

To the memory of
Dr. C. R. Masson

Contents

Chapter 5   Mössbauer Effect in Glasses
G. Tomandl

Chapter 6   Chromatography
C. R. Masson

Contents of Volume 4A

List of Contributors

Numbers in parentheses indicate the pages on which the authors' contributions begin.

PHILIP J. BRAY (77), *Department of Physics, Brown University, Providence, Rhode Island 02912*

G. N. GREAVES (1), *SERC Daresbury Laboratory, Warrington WA4 4AD, United Kingdom*

D. L. GRISCOM (151), *Optical Sciences Division, Naval Research Laboratory, Washington, DC 20375*

C. R. MASSON (313), *Atlantic Research Laboratory, National Research Council of Canada, Halifax, N. S. B3H 3Z1 Canada*

PHILLIP E. STALLWORTH, (77), *Department of Physics, Brown University, Providence, Rhode Island 02912*

G. TOMANDL, (273), *Institut fur Werkstoffwissenschaften III (Glas und Keramik), Universitat Erlangen, West Germany*

J. ZARZYCKI (253), *Laboratory of Science of Vitreous Materials, University of Montpellier, France*

Preface

Contemporary Concepts in Structure

The present volumes offer the thoughtful comments of leading workers in the field of glass structural science. The respective chapters cover, with two notable exceptions, the principal methods used to obtain experimental data on glass structure.

The first of these exceptions, wide angle x-ray scattering (WAXS), has been well summarized in a recent work by Wright (1989) and is embraced to differing extents in several of the present chapters. While WAXS may well be regarded as the cornerstone of structural investigations, the work by Wright is so cogent and comprehensive that a presentation at this time would likely represent in large measure a restatement or a brief update.

The second group of important structural characterization methods, which will not be found in the present volumes, is that of infrared and Raman spectroscopy. These methods yield information about the optical characteristics of materials as well as their structure. After lengthy consideration, it was decided that they would best be included in a forthcoming volume on optical properties (Uhlmann, to be published).

As a framework for any consideration of glass structure, it is useful to recall an underlying principle that was enunciated long ago by Warren. Specifically, it is unreasonable to demand that an amorphous structure be uniquely established by a particular structural investigation. Rather, such investigations should be viewed as providing data with which any proposed structural model must be consistent. This perspective is shared by other distinguished contributors to the science of amorphous structure, such as Professor Porai-Koshits, and underlies the recent comment of Wright (1989) to the effect that, "Frequently, however, comparisons between models and experiments are at a very superficial level and hence the literature contains a wide range of structural models all claiming to be in 'good agreement' with the experiments with the result that the structures of even the simplest amorphous solids are not well established beyond the first few co-ordination shells." Again, "the greatest barrier to progress in understanding the structure of amorphous solids lies with the development of adequate modeling techniques rather than in the improvement of the diffraction data themselves."

From Zarchariasen's celebrated remark of 1932, "It must be frankly admitted that we know practically nothing about the atomic arrangement in glasses," it is not a great leap to Professor Bishay's comment at the Kreidl Symposium of 1984 that, "It is a pleasure to return to a meeting on glass science after a multiyear absence and see many familiar faces discussing many of the same problems, such as 'What is the structure of glass'? (Bishay, oral comm., 1984)." Yet definite progress *has* been made. The data of today are much more detailed and more extensive than those of few decades ago, and, even more important, data are available based on a wider range of methods.

In the opinion of the present authors, one of the principal challenges of the coming decade will be the integration of data provided by a variety of techniques combined with more critical evaluations of the consistency of such data with structural models. The effective response to this challenge will require scientists who are conversant in detail with the capabilities and limitations of a range of experimental methods, the likely accuracy (or inaccuracy) of data provided by each method, the techniques of data manipulation and refinement used with each method, and the limits within which the various types of data can be used to distinguish among structural models.

A second notable challenge for the coming decade is the development of models and data that provide improved and more detailed descriptions of structural features on various scales of structure. Of particular importance in this regard is the elucidation of features in the intermediate range of structure—i.e., from 5–8 Å out to 20–30 Å. This represents a formidable challenge for theoreticians as well as experimentalists, and *must* be accompanied by critical assessments of the utility of various methods for providing the desired insight (hard insight rather than "pie in the sky"). It seems likely that meeting this challenge will require the development of novel methods/approaches coupled with novel extensions of existing methods.

A third notable challenge is the use of structural models to predict the properties or responses of materials. Most individuals feel that it *does* matter whether a glass is a random network (at some level) or an array of paracrystals, but they would be hard-pressed to specify how such structural differences would influence properties or processing. To see how effective such predictions of properties based on structural models can be, one need only turn to the field of polymeric materials. Thanks to the pioneering work of Flory and his successors, the implications of the random coil model for a variety of properties of amorphous polymers have been explored. In fact, when that model was challenged by reports of nodular features in a number of amorphous polymers, the agreement between predicted and measured properties was used to support the random coil model and to cast doubt on the reports of the nodular features (a conclusion which was later supported by direct structural investigations). The time is long overdue for developing predictive structure-property relations for amorphous materials other than polymers.

A fourth notable challenge for the coming decade is the elucidation of the dependence of glass structure on the mode of formation of the glass. Even for melt-derived glasses, relatively little detailed information exists on the effects of cooling rate on structure; and the situation is much worse when one includes quite different methods of forming glasses (e.g., sol-gel methods, vapor deposition methods, and electrochemical methods). Structural differences between amorphous coatings and bulk glasses provide further examples of this type of investigation. In several instances, notable changes in structure and defects with mode or conditions of glass formation have been suggested, but the issue requires much more extensive and critical exploration. Such exploration has been the subject of recent conferences organized by Weeks (1985).

A further notable challenge involves the characterization of defects, including but not restricted to point defects and composition-related defects, in glass structure. Thanks to the initiative of Weeks (1980) and Griscom (1980), this area has begun to receive increased attention in recent years. Such attention has served to clarify a number of outstanding questions, but it has also led to the posing of new questions, and has emphasized the need for improved structural models. Improved understanding of intermediate-range structure will undoubtedly lead to important new questions concerning defects at this level of structure. In addition, there exists a particular need for improved models that relate defects to chemistry, melting conditions, heat treatment conditions, radiation exposures, etc.—as well as models that represent the relationships between defects of different types and the properties of the glasses.

An additional notable challenge involves the elucidation of structural features, including defect structures, in a broader range of composition including complex compositions. Even in the case of simple glasses, there remain important unresolved questions concerning structure, but these are amplified when complex multicomponent compositions are considered. Systematic investigations employing combinations of experimental techniques will almost surely yield critical new insight into structural issues, including insight into the structure of end-member glasses and the generality/validity of structural models.

The issue of models for glass structure seems deserving of particular comment. Such models have considerable value as guides for thought, even when they are wrong in detail. The early disputes between proponents of random network and crystallite models applied to the classic oxide glasses were effectively resolved in favor of the random network concept, based largely on the arguments of Warren concerning the size of the crystallites (estimated from line broadening) and the absence of intense small-angle scattering. This does not exclude deviations from an overall random arrangement in the form of either disordered local variations in concentration and/or structure as well as defects. On the contrary, such deviations have been suggested even in simple, single-component glasses such as SiO_2.

Subsequent small angle x-ray scattering (SAXS) studies of a number of simple oxide glasses indicated an intensity of asymptotic scattering that is very similar to that expected from thermodynamic fluctuation theory, assuming that the fluctuations present at the glass transition temperature are frozen-in as the glass is formed. The observations suggest that the glasses are substantially homogeneous and seem inconsistent with models that are based on distinct heterogeneities that differ to any considerable extent in electron density from the intermediate "glue."

SAXS studies on simple oxide glasses, carried out as a function of temperature, have indicated a level of fluctuation scattering that is substantially independent of temperature up to the glass transition (T_g), and then increases with increasing temperature above T_g in accordance with predictions of fluctuation theory. In the case of amorphous polymers, the observed asymptotic SAXS intensity is again consistent with the predictions of thermodynamic fluctuation theory, and again the intensity of such scattering increases with increasing temperature above T_g. With polymers, however, the intensity of the scattering—and hence the magnitude of density fluctuations—decreases with falling temperature over a range of temperature below T_g (down to the secondary transition temperature). The temperature dependence of the fluctuation scattering below T_g is markedly smaller than that observed above T_g. The results suggest that portions of the chains retain mobility over a range of temperature below that at which large-scale chain mobility is frozen-in.

The picture that emerges from such studies is one of glasses as fluctuated systems, with a level of fluctuations characteristic of T_g (in the case of oxide glasses) and of a temperature somewhat below T_g (in the case of polymer glasses). The observed fluctuation scattering is of a type that would be expected on the basis of thermodynamic fluctuation theory for even the simplest fluids.

While SAXS studies of glasses of different types consistently indicate a level of asymptotic scattering consistent with thermodynamic fluctuations, a number of electron microscope studies of both oxide glasses and polymer glasses have indicated the presence of structural heterogeneities on a scale of tens to hundreds of angstroms. These heterogeneities have been designated as micelles or nodules and have been interpreted in various ways. All of the interpretations involve the suggestion that the glass contains regions of locally high order, in some cases approaching that of a crystal. To be consistent with the observed widths of the wide-angle x-ray scattering peaks, the degree of order is typically suggested to be sufficiently high to permit the features to be seen in the electron microscope (as by diffraction contrast), but sufficiently small as to give rise to the broad Bragg peaks. That is, the degree of order in these local regions is suggested to be like that of a recipe—not too hot and not too cold, but just right.

Subsequent electron microscope observations of both oxide and polymeric glasses have indicated the absence of observable heterogeneities on the scale on which the nodules or micelles were reported. In the case of polymer glasses, it has even been suggested that radiation damage from the electron beam of the electron microscope would destroy the order and would substantially preclude the observation of regions of local order even if they were present. To avoid this problem, electron microscope studies have been carried out on a number of glassy polymers stained with heavy metal ions, using the technique of Z contrast electron microscopy. In this case, the radiation damage from the electron beam, even if it destroyed the local order, would not destroy the compositional contrast. Even with such methods, the studies have failed to reproduce the observations of the nodular features. Taken as a whole, then, the electron microscope observations of glasses— whether organic or inorganic—are consistent with the structure being highly homogeneous and with models such as that of a random network or random coil. These observations seem inconsistent, however, with structural models based on distinct heterogeneities differing in order and density (and hence structure) from the intermediate "glue."

The overall results of wide-angle x-ray scattering studies, as discussed by Wright for example, are generally consistent with structural models of a random array type (whether random network or random coil, etc.). At least two qualifications to the statement should, however, be noted: (1) the scale of structure probed by the technique is principally that of short-range structure (as distinguished from the intermediate-range structure discussed above); and (2) the random array may not be that of the smallest possible structural units. Recall, for example, the case of B_2O_3 and borate glasses, whose structures may be described as a random array, but not of simple BO_3 triangles. Rather, the scattering data are consistent with a random array of boroxyl rings + BO_3 triangles, as well as BO_4 tetrahedral groups for borates containing modifying cations.

In the case of borate glasses, the technique of NMR spectroscopy has provided particularly valuable insight into structural features on a scale of near-neighbor distances, and the results have also been employed to infer structural characteristics on the intermediate scale of structure (see Chapter 2, Volume 4B, by Bray for details). Recent developments in this area, such as magic angle spinning with solid samples and high field spectrometers, offer promise for introducing a new era of NMR investigations of amorphous materials.

By random array models, the present authors denote structural models in which no unit of the structure is repeated at regular intervals in three dimensions. Such structures need not be uniform on a molecular level. Indeed, the density fluctuations discussed above—as well as the compositional fluctuations in multicomponent glass compositions—imply variations in structure from place to place within a material. Even in the simplest single component glasses, the local structures are often similar to those of corresponding crystals (a sufficiently near-sighted fly might experience some difficulty in determining whether he is sitting on an atom in a crystal or in a glass). Point defects and grouping of point defects are present in variable quantities depending on the material. Higher order groups such as rings (typically with different numbers of atoms in the rings) and chains as well as defects in ring and chain structures are also present.

The similarity in very local structure between glass and corresponding crystal is not surprising in light of the usual similarity in bonding and density between the two forms (recall that the typical glass differs in density from the corresponding crystal by about 10% or less, although larger differences are known, as are cases where the glass is more dense than the corresponding crystal).

In the case of multicomponent glasses, the situation is more complicated. In many systems, the phenomenon of phase separation (liquid–liquid immiscibility) is widespread, and even in materials that do not exhibit phase separation, compositional fluctuations of considerable intensity can be anticipated. Even in single-phase multicomponent glasses, the distributions of the respective species are typically not random. For example, the alkali and alkaline earth ions in simple silicate glasses appear to occur frequently in pairs or clusters containing significant numbers of the respective modifying cations.

The situation is even more complicated when one includes the occurrence of ring and chain structures of different types, such as are characteristic of many glasses. In these cases, the ring and chain distributions in glass and crystal can be quite different (with crystallization involving a notable change in such distributions). Defects such as chain ends can have a considerable influence on properties. Recall, for example, the technologically important role of hydrogen doping in affecting the properties of amorphous silicon. Similarly, in the case of cross-linked organic polymers, one anticipates significant variations in the local concentrations of cross links and hence in the overall structure.

The picture of glass structure that emerges from the myriad of data obtained using a variety of techniques is one in which a random array picture seems to represent a useful first-order approximation to the structure. This term, random array, seems notably preferable to that of a random network. (What does a random network mean in the case of, for example, glassy lithium niobate or glassy polycarbonate?)

As we have seen, while the random array picture provides a useful guide to thought about glass structure, the present state of random array modeling does not permit prediction of many important or potentially important features—e.g., clustering, point defects, rings, chains, and even the detailed descriptions of rotation angles between structural units or distributions of constituent units.

The developments of the past few decades have considerably expanded the depths of questions being raised with respect to glass structure. The discussion over random network vs. crystallite models has been extended to include paracrystals or paracrystalline arrays, regions of locally high order, and more recently, quasicrystals and a variety of supercell structures. Further modeling based on molecular dynamics simulations, but using more realistic potentials and larger arrays, combined with more detailed treatments of spatially chaotic configurations and consideration of the impact of local chemical bonding on structure, will undoubtedly yield important insight into the general question of glass structure.

In the chapters that follow, it will be demonstrated how defects as well as short-range order in glasses have been considerably elucidated by the techniques of NMR, ESR, Mössbauer, SAXS, and XAFS. It will also be shown how rather unconventional methods (from the perspective of inorganic glasses), such as chromatography, can also provide valuable structural information. It is recognized also that structure is strongly related to electronic states and electrical properties. These are addressed in the present volume and will be considered at length in the chapter by Feltz in a companion volume on *Electrical Properties of Glasses.*

Overall, the field of glass structure has seen important advances during these decades. Our concepts of structure are growing increasingly refined, and discussions of issues such as defects have opened up entirely new vistas for speculation and investigation. The issue of glass structure seems fundamental to discussions of the properties and processing of glasses, but a critical need exists, as discussed above, for more extensive modeling of these relations. The authors feel that the present chapters provide a useful perspective on the present state of knowledge of glass structure. They are confident that the coming decades will continue to see notable programs in the area, but they are also confident that similar volumes produced 20 or even

100 years from now will still be replete with unsolved problems and outstanding challenges for future generations of scientists to elucidate.

Even in cases where the techniques are well established, recent advances in methodology pose new challenges and opportunities. As an example, the advent of magic-angle spinning in NMR spectroscopy has led to refinements and revisions in previous interpretations of structure. In the coming years, these interpretations will undoubtedly be extended to more complex glasses than those for which data are presently available, including compositions of commercial importance.

D. R. Uhlmann

N. J. Kreidl

References

Griscom, D. L. (1980). "Point Defects in Amorphous SiO_2: What Have We Learned from 30 Years of Experimentation" Defects in Glasses, *Mat. Res. Soc.*, pp. 213–225. Pittsburgh, Pennsylvania.

Uhlmann, D. R. "Optical Properties," to be published in *A Cer. Soc.* Westerville, Ohio.

Weeks, R. A., Kinser, D. L., Kordas, G., eds. (1985). "Effects of Modes of Formation on the Structure of Glasses." North Holland, Amsterdam.

Weeks, R. A. and Nelson, C. M. (1980). "Point Defects in Amorphous SiO_2," *J. Appl. Phys.* **31** 1555.

Wright, A. C. (1989). "Diffraction Studies," to be published in *Proc. International Congress on Glass.* Leningrad.

Wright, A. C., Clare, A., Grimley, D., Hulme, R., and Sinclair, R. (1989a). "A Neutron Diffraction Study of Network Glasses with a Structural Unit Connectivity of Three." Glass Melting, to be published in *A. Cer. Soc.*

Wright, A., Price, D., and Sinclair, N. (1989b). "The Interpretation of the First Diffraction Peak for a Network Glass: Fact and Fantasy," Ibid.

GLASS
SCIENCE AND TECHNOLOGY

VOLUME 4B
Advances in Structural Analysis

CHAPTER 1

X-Ray Absorption Spectroscopy

G. N. Greaves

SCIENCE AND ENGINEERING RESEARCH COUNCIL
DARESBURY LABORATORY
WARRINGTON, UNITED KINGDOM

I. Introduction

Since the early 1980s, X-ray absorption spectroscopy (XAS) has added considerable color to our picture of the structure of glass. From the 1930s up

1

until the early 1980s, we have had to be content with the average environment of atoms in glasses which can be obtained from diffuse scattering of X-rays and neutrons. However, most glasses are multicomponent and the mean local structure offers only a monochrome version of the structural chemistry. This is a serious limitation in elucidating the complex structure of oxide glasses, for instance, where the environment of modifiers is masked by the much stronger pair-correlation functions associated with the network. However, by tuning to the absorption edges of the different elements in a glass, XAS techniques can be used to extract individual pair-correlation functions.

X-ray absorption fine structure, like many other X-ray phenomena, has its origins in the 1920s and 1930s. In some respects it is a rediscovered technique. What was Kossel structure (Kossel, 1920) is now called X-ray absorption near edge structure (Durham et al., 1981) or XANES. Kronig structure (Kronig, 1931) is referred to as EXAFS or extended X-ray absorption fine structure (Sayers et al., 1972). There have been two major causes for this resurgence of interest. One has been the establishment of a viable theory for the scattering of photoelectrons in condensed matter (Lee and Pendry, 1975). Second, the advent of synchrotron radiation sources has enabled X-ray absorption spectra to be measured with sufficient energy resolution and signal-to-noise quality for the atomic environments of separate elements to be accurately determined.

The same partial radial distribution functions obtainable from XAS can also be extracted in principle from X-ray or neutron diffraction using the complementary techniques of anomalous dispersion or isotopic substitution. In practice, however, these are difficult techniques which have enjoyed only a restricted application as of the late 1980s. By contrast, EXAFS and XANES measuring in transmission, which is the simplest XAS geometry, is a relatively trivial experiment to perform, taking only 30 minutes or thereabouts to execute, and requiring only a few mgms of material. Accordingly, XAS has found wide application in the study of glass structure. Extracting the pair-distribution functions of modifiers such as Na in oxide glasses is a striking example (e.g., Greaves et al., 1981). The various environments of inter-mediates have also been conclusively determined—e.g., Ti (Sandstrom et al., 1980) and Fe (Calas et al., 1984)—in many cases for the first time. In chalcogenide glasses, thermal (Neminich et al., 1978) and photostructural (Lowe et al., 1986) annealing can be detected and followed using XAS. By employing glancing angles of incidence, XAS becomes surface-sensitive and is proving to be a powerful technique for exploring the structural chemistry of glass corrosion (Thornley et al., 1986, Greaves et al., 1989). Another quite different application of EXAFS has been in the area of amorphous biological systems. Here the environments of Ca and P are naturally of interest, but also the corrosion effects of toxic cations such as Mn can be revealed through XAS

(Taylor et al., 1988). Although outside the scope of this chapter, XAS techniques have also been applied, with considerable success, to metallic glasses in the identification of chemical order. The reader is referred, for instance, to the reviews by Gurman (1982) and Wong (1983).

This chapter begins by outlining the basic principles of XAS—the data reduction and analysis required to extract pair distribution functions. This is juxtaposed with a brief description of models of glass structure to emphasize the ways the technique can benefit the structural physics and chemistry of the glassy state. Although most of the XAS studies I shall describe were accomplished using the conventional transmission geometry, other modes of detecting X-ray absorption offer exciting possibilities. These will be discussed next.

The burden of this chapter, however, concerns the widened perspective XAS techniques have added to our perception of glass structure. The two principle areas of network glasses and modified glasses will be reviewed by a series of examples. The increased information derived from atomic specific radial distribution functions in some cases helps to discriminate between various models — for example, in identifying "wrong bonds" in chalcogenide glasses—and in other cases to confirm previous ideas or lead to a more comprehensive model such as the modified random network for oxide glasses. The chapter concludes with a few examples of XAS applied to diffusion-related problems. XAS techniques are particularly powerful in this context by virtue of their ability to monitor the atomic environment of the mobile species.

II. Basic Principles

A. INFORMATION CONTENT OF XAS

The Golden Rule or dipole approximation equates photoelectron absorption, α, to the matrix elements of the initial $(\psi_{l_0 m_0})$ and final states (ψ_{lm}):

$$\alpha \propto \sum_{lm} \sum_{m_0} |\langle \psi_{lm} | \varepsilon \cdot \mathbf{r} | \psi_{l_0 m_0} \rangle|^2,$$

ε being the electric field polarization vector. The fine structure, χ, in an XAS spectrum results from perturbations in the final states from the free electron values. This originates in the interference that occurs between the photoelectron emanating from the excited atom and that fraction that is reflected back from the potential wells of neighboring atoms. XAS therefore is a signature in k-space of the immediate environment of the element whose absorption edge is being probed. The free electron wave vector, **k**, is given by

$$E = E_{XR} - E_o = \frac{\hbar^2 k^2}{2m} \tag{1}$$

where E_{XR} is the x-ray photon energy. Note that the origin of the free electron energy, E_o, does not coincide with the edge of the conduction band but falls in the occupied valence levels, i.e. below the absorption threshold. Accordingly, the minimum wavevector that can be measured is always finite. k_{min} is typically around 2Å^{-1} at threshold so there is no XAS analogue to X-ray or neutron small-angle scattering. On the other hand, the momentum transfer for a back-scattered photoelectron is $2k$ compared to k in conventional diffraction. The effective wave-vector limit then for an EXAFS spectrum running out to 15 or 20Å^{-1} is 30 to 40Å^{-1}, adding considerable precision to the determination of interatomic distances. By the same token, XAS is sensitive to static distortion and anharmonicity in the local environment— particularly nearest neighbors. We shall see that this is of relevance for analyzing modifiers in oxide glasses.

Because the scattering is complex, the momentum transfer of the photo-electron is modified by a phase shift $2\delta_i + \psi_j$, where δ_i is the change in phase encountered passing forward and backward through the central atom potential and the change resulting from back reflection by a neighboring atom, ψ_j. The interference phenomenon and the phase shift contribution to $\chi(k)$ can be most easily identified in the Plane Wave expression for EXAFS (Lee and Pendry, 1975):

$$k\chi_i(\mathbf{k}) = -\sum_j \frac{N_{ij} f_j(\mathbf{k}, \pi) \sin(2kr_{ij} + 2\delta_i + \psi_j)}{r_{ij}^2}$$
$$\times \exp(-2\mathbf{k}^2 \mu_{ij}^2) \cdot \exp(-r_{ij}/\lambda). \tag{2}$$

In Eq. (2), $f_j(\mathbf{k}, \pi)$ is the back-scattering amplitude and N_{ij} is the number of atoms of type j at distance r_{ij}. These together govern the amplitude of the oscillations in χ. On the other hand, the two exponential terms relate, first, to the damping of the fine structure caused by static and thermal atomic displacements of type j atoms relative to atoms of type i, μ_{ij}^2 (Debye-Waller factor), and, second, to the broadening effects due to the finite mean free path of the photoelectron, λ.

The local atomic structure information available from EXAFS then is the coordination number, N_{ij}, the interatomic distance, r_{ij}, and the variance in r_{ij}, μ_{ij}^2. All of these are specific to the element whose absorption edge is being measured. It will be obvious from what has been said that translational invariance is not a prerequisite and that XAS is equally sensitive to environments in amorphous solids and liquids as it is to crystalline materials. Finally, it goes without saying that there are no silent elements for XAS. The scope for studying the structural physics and chemistry of amorphous materials therefore is fairly comprehensive. (See Table I.)

The information content of an EXAFS spectrum is illustrated in Fig. 1. CaO has a rock-salt structure, and the presence of many well-defined shells of atoms contribute to the fine structure at the Ca K-edge. The fine-structure function $\chi(\mathbf{k})$ normalized per electron is obtained from the measured X-ray absorption, α, in the following way:

$$\chi(\mathbf{k}) = \frac{\alpha - \alpha_a}{\alpha_a - \alpha_v}. \tag{3}$$

α_v is the residual X-ray absorption underpinning the edge in question and is obtained by extrapolating the pre-edge absorption above threshold. α_a is the atomic absorption—simply the measured absorption with the fine structure removed. α_a-α_v is therefore the step height corresponding to the particular core level transition (in this case, Ca 1s → O 2p). The wavevector scale is given by Eq. (1) corrected for E_o. In this respect, as the empirical definition of the absorption threshold is the first major turning point in α, this then needs to be corrected in analysis by -5 to -20 eV, depending on material.

It should be evident from Eq. (2) that the Fourier transform of $\mathbf{k}\chi(\mathbf{k})$ will yield a radial distribution function or RDF. This is much the same situation as in diffuse scattering except that in EXAFS the RDF is specific to the excited atom. It is convenient to correct for the nearest neighbor phase shift and back-scattering amplitude as well as the dampening due to thermal and static disorder (Gurman and Pendry, 1976):

$$F(r) = \frac{1}{2\pi} \int_{k_{min}}^{k_{max}} \frac{W(\mathbf{k})\mathbf{k}^n\chi(\mathbf{k})\exp[-i(2kr + 2\delta_i + \psi_j)]}{f(\mathbf{k})\exp(-2\mathbf{k}^2\mu_{ij}^2)} \, d\mathbf{k}.$$

$W(\mathbf{k})$ is a suitable window function and $\chi(\mathbf{k})$ is weighted ($n = 2$ or 3) to provide a reasonably symmetrical function over the $\mathbf{k}$-range available. The

TABLE I

UTILITY OF SOFT AND HARD X-RAYS FOR XAS

Energy range	Edge type	Elements	Comments
500 eV–4 keV	K	O → K	Vacuum required. Limited use for multi-
	L	V → Sn	component systems.
4 keV–30 keV	K	K → Sn	Measurements can be made in air.
	L	Sn → U	Suitable for most multicomponent systems.

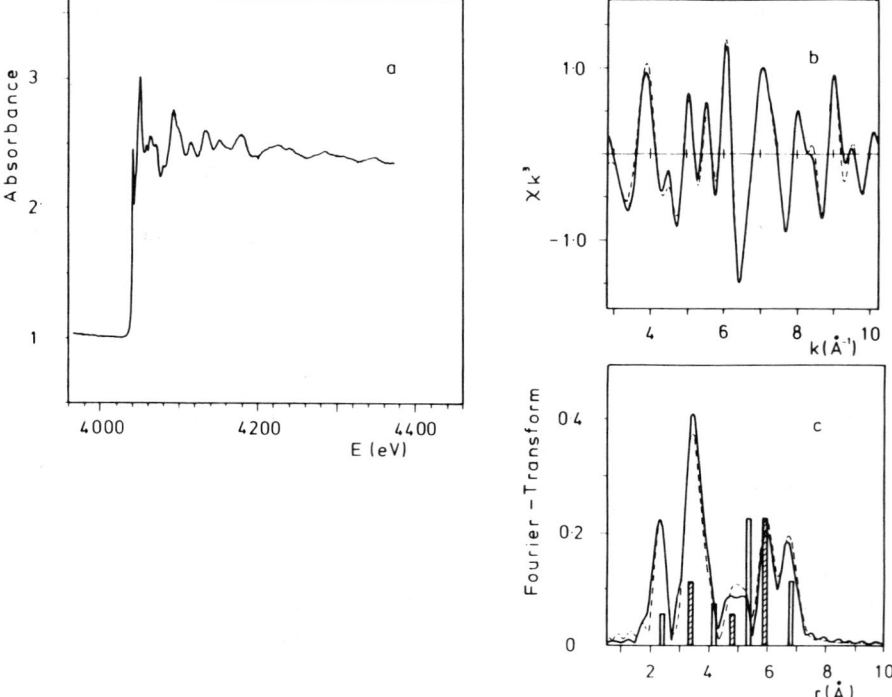

FIG. 1. Ca K-edge EXAFS for CaO (a) measured absorption spectrum; (b) normalized EXAFS, χ, weighted by k^3; (c) Fourier transform of χk^3. Solid lines in (b) and (c) refer to the experimentally determined structure while the dashed lines refer to CWT calculations using seven shells of atoms. The histogram in (c) relates to the crystallographic positions of shells of oxygen and calcium around a central calcium atom. (Binsted et al., 1985).

peaks in the RDF shown in Fig. 1 correspond to the shells of O and Ca atoms surrounding each Ca atom. CaO is a useful model compound for studying Ca local structure in oxide glasses. The Ca and O phase shifts can be refined, E_o obtained, the mean free path determined for a calcium oxide environment, and the single-photon quantum efficiency estimated for Ca 1s excitations.

While the EXAFS radial-distribution function (Fig. 1c) is a helpful guide for assessing local atomic structure, it contains various artifacts. For example, there is always some structure because of the finite **k** range available and from the particular form of W(**k**) used (e.g., Gaussian "bell," Hanning window). Secondly, peak positions and amplitudes may be distorted; F(r) can only be phase-shift–corrected for particular near-neighbors. More importantly, there are inaccuracies due to inadequacies in the plane wave approximation (PWA) implicit in F(r). (See Eq. 2.) The most accurate description of EXAFS is embodied in the curved wave theory (CWT) of Lee and Pendry (1975). This employs spherical waves rather than plane waves to describe the final state wave functions $\psi_{l,m}$. The dotted line in Fig. 1b and c is the result of least squares fitting the measured EXAFS of CaO using CWT. In this case, as in other crystalline model systems, the agreement with diffraction results defines the accuracy attainable for the absorption edge in question. Usually for shells of atoms out to 3 or 4Å and a wavevector range of $3-10\text{Å}^{-1}$, the precision of EXAFS analysis is

$$\Delta r \lesssim 0.02\text{Å}; \qquad \frac{\Delta N}{N} \lesssim 0.1; \qquad \frac{\Delta \mu^2}{\mu^2} \lesssim 0.2.$$

Although in principle EXAFS can be employed as an ab initio technique, the inclusion of crystalline models in experiments is advisable in order to refine the photoelectron parameters previously listed. Muffin-tin calculations are usually sufficient to obtain the photoelectron-scattering factors to reasonable accuracy, but these may be refined against crystallographic rs, Ns, and μ^2s. For the case of simple alloys and elements where there may be close similarities between the crystalline and amorphous states, comparative study adds considerably to the precision to which r, N, and μ^2 can be determined. This will be demonstrated in Section IV.A for amorphous semiconductors.

Most of the discussion so far has concerned the EXAFS part of XAS. In general this extends from around 60 eV above threshold onward. At these energies the photoelectron is only weakly scattered, principally by pairs of atoms. Close to the absorption edge itself, scattering is much stronger, often involving three, four, five, and six atom groups—more in the case of metals. This is the XANES region and multiple scattering makes an important addition to the underlying scattering between pairs of atoms. As a result, XANES is sensitive to local symmetry in addition to pair correlations. Also, at these low energies, λ is large, so scattering from distant neighbors can be more pronounced in the XANES than in the EXAFS part of the spectrum.

The absolute position of the absorption threshold reflects the chemical state of the atom. There are two contributions: core level screening and modifications in the electronic structure of unoccupied states. The former usually dominates and the threshold moves to higher energies with oxidation state. In transition metals, chemical shifts are usually of the order of 5 or 10 eV but can be larger. Of course, oxidation state is also mirrored in the nearest-neighbor bond length obtainable from EXAFS. In this way, XANES and EXAFS can be used to complement one another, as will be seen in Section VI.C in the study of intermediates.

Finally, it should be noted that fine structure is sometimes observed below the principle absorption edge. This is generally attributable to transitions not strictly allowed by the dipole selection rules ($\Delta l = \pm 1$), e.g., $1s \rightarrow 3d$. Pre-edge features occur when hybridization confers final states with some p-like symmetry. The strength of pre-edge features are influenced by the occupancy of these levels and hence the chemical state of the atom. Pre-edge features are pronounced in transition-metal spectra (e.g., Figs. 17 and 18)—increasing with oxidation state and, for metals, decreasing with increasing atomic number.

Line widths in XAS spectra increase as the lifetime of the excited state decreases. (This can be appreciated by appealing to the uncertainty principle.) The lifetime of a particular core hole (e.g., 1s) decreases with increasing Z. A rough empirical rule relating the full line width ΔE_τ to the binding energy E_{BE} is

$$\Delta E_\tau \sim 2.10^{-4} E_{BE}. \tag{4}$$

This is the intrinsic width of features in XANES spectra. The fine structure in EXAFS spectra are usually broader than this, being governed by the maximum r_{ij} detected. This is also dependent on λ and μ_{ij}^2 and will obviously vary from one material to the next. For nearest neighbors at 2Å, the minimum EXAFS line width is around 12 eV. ΔE_τ becomes a significant fraction of this for binding energies in excess of 30 keV which is why for heavy atoms such as Cs, Pb, and U, the L_{III}-edge is used in preference to the K-edge.

EXAFS and XANES experiments that are sufficiently accurate for quantitative structural analysis require a combination of high spectral resolution and excellent signal to noise ratio (S/N). Synchrotron radiation brings these two requirements together. In measuring dilute systems, some trade-off between spectral resolution for increased S/N is generally desirable. Spectral resolution effects the number of frequencies (i.e., interatomic distances) that can be separated when data is analyzed, whereas the signal to noise ratio achieved primarily dictates the precision with which the amplitude-related functions N and μ^2 (Eq. 2) can be extracted. In order to obtain these parameters reliably, a S/N in χ of at least 50 is desirable. To achieve this, more than 10^{10} photons are required per experimental point. Synchrotron sources in conjunction with perfect crystal monochromators can deliver these quantities of photons on a second-by-second basis, which, as we have seen, is part of the reason for the renaissance in XAS during the 1980s.

B. MODELS FOR GLASS STRUCTURE

Models for glass structure are as numerous as radial-distribution functions from diffuse scattering are underdetermined. The availability of pair-distribution functions from XAS measurements, though, have sharpened the criteria by which models can be judged. This chapter is concerned with the way EXAFS and XANES experiments are leading to improved models for glass structure. Before I describe these, it is worthwhile to briefly review the principal models proposed for network and modified glasses.

1. Continuous Random Network

Many of the original ideas for simple covalently bonded glasses put forward by Zachariasen in 1932 still hold good—in particular, the notion that local order, long-range disorder, and connectivity are reconciled by allowing the network to consist of rings of different sizes. The continuous random network (CRN) is the embodiment of these concepts. Specific models have been built for many elementary glasses. These reveal in three dimensions how the dominant peaks in experimental RDFs develop. More importantly, modeling exercises have resulted in a better understanding of the topology of disordered networks. CRNs for glasses such as SiO_2 (Bell and Dean, 1972), Ge (Polk and Boudreaux, 1973), As (Greaves and Davis, 1974), and Se (Long et al., 1976) have demonstrated how rings of different size can be accommodated with rigid interatomic interactions by bond angle and dihedral angle disorder. Indeed, the ring statistics can be used to predict these angular distributions (Davis et al., 1977). CRNs have also proved versatile in predicting the structurally sensitive electronic and phonon properties (Mott and Davis, 1979; Elliott, 1983). For elemental glasses, XAS reveals just how perfect the nearest-neighbor coordination sphere can be in the amorphous state (Section IV.A).

CRNs are generally built on the assumption of chemical order. This is undoubtedly the case for stoichiometric materials such as SiO_2, $GeSe_2$, and As_2S_3, with the 8-N rule applying. However nonstoichiometric derivatives (usually prepared as thin films) imply the presence of homopolar or "wrong bonds" (i.e., bonds between like atoms) in the structure if connectivity is to be maintained. Wrong bonds may be widespread for SiO_x as the random bonding model of Phillipp (1971, 1972) proposed. Alternatively, some degree of phase separation (viz. Si and SiO_2) may occur (Temkin, 1975). For germanium chalcogenides, changes in coordination have been established making heteropolar bonds more likely (Fuoss et al., 1981). In arsenic chalcogenides, on the other hand, there is evidence for As-As bonds indicating the possible incorporation of As_4S_4 molecules in the glass structure (Nemanich et al., 1978). In all of these cases, an experiment such as EXAFS that isolates individual pair-distribution functions has an important role to play in identifying the existence or otherwise of chemical order (Sections IV.B and V.B).

2. Modified Random Network

Although model builders of nontetrahedral networks such as As or Se have simulated Van der Waals interactions, ionic bonds have yet to be incorporated into physical models. Real multicomponent glasses such as soda-lime-silica glass comprise a mixture of covalent and ionically bonded atoms. In the early Zachariasen picture of glass structure, the local environments of the modifiers and, indeed, the nonbridging oxygens were underspecified. Cations such as Na and Ca were predicted to occupy a "statistical distribution" of holes or voids in the covalent network. X-ray and neutron-scattering studies of modified oxide glasses provide no strong evidence to the contrary (e.g., Misawa et al., 1980). EXAFS experiments, on the other hand, have contradicted this premise, revealing well-defined sites for modifiers such as Na (Section VI.A). Figure 2 reproduces the modified random network (MRN) model that was proposed to demonstrate how regular modifier ligands could be incorporated within the context of a CRN (Greaves, 1985). The structure is partly ionic and partly covalent and, in the proportions

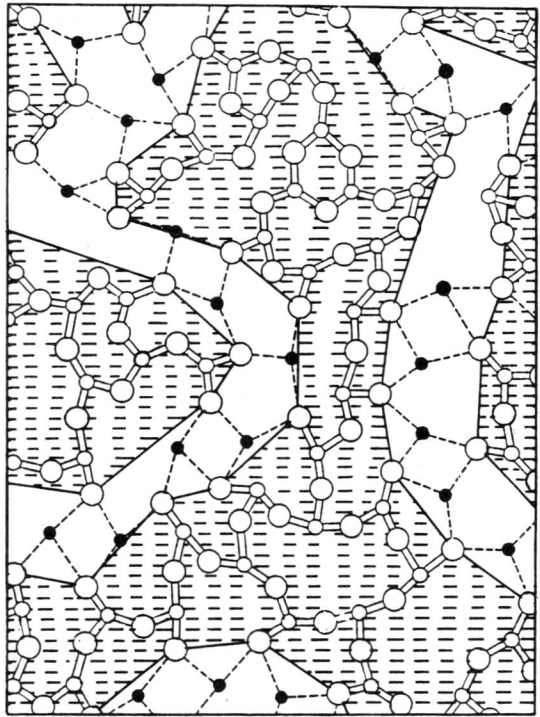

FIG. 2. Modified random network model for an oxide glass. Covalent bonds are shown by solid lines and ionic bonds by dotted lines. Modifying cations and glass-forming cations are shown threefold coordinated. Bridging oxygens are twofold coordinated, while nonbridging oxygens are threefold coordinated. The shaded regions are defined by nonbridging oxygens and highlight the percolation channels of the modifier that run through the network. (Greaves, 1985).

chosen (which are roughly 1:2), the structure is characterized by modifier regions within which cations will be mobile. Whether these regions in the MRN will extend to percolation pathways for ionic transport will depend on the composition—more particularly the volume fraction of the modifier. In three dimensions the percolation threshold occurs at 16% (Zallen, 1983), which will usually be exceeded for stable glass compositions.

It is implicit in the concept of the MRN model that, if the local environment of the modifying cation is specific, so too will be the surroundings of the nonbridging oxygens to which it is bonded. In Fig. 2, nonbridging oxygens are coordinated to two modifying cations and one network cation. It is clear that nonbridging oxygens in the MRN model link ionic to covalent regions and as such will constitute lines of easy shear in the structure. In this respect, the fracto-emission results of Dickinson and colleagues (1988) are pertinent. They report copious emission of Na atoms from the fracture of $Na_2Si_3O_7$ glass. Complementary to this result are the SIMS measurements of Richter and coworkers (1985) on freshly fractured surfaces, which they found to be Na-rich and Si-deficient. It would appear that glasses indeed fracture along modifier pathways. By the same token, the line density of nonbridging oxygens will govern the overall viscosity. The more copious these lines are, the lower the viscosity will be. This is certainly the case for simple binary melts.

The MRN model provides a useful picture of the way ionic conductivity changes with composition. In particular, transport in alkali silicates is perceived to change from a vacancy to an interstitialcy mechanism as the concentration of modifier increases. The activation energy falls and the correlation factor—the ratio between self-diffusion and ionic conductivity—falls. Both level out at between 10 and 20% modifier content (Frischat, 1975; Wegener and Frischat, 1983; Kelly et al., 1980). At low concentrations of modifier in an MRN structure, the ionic mobility will be governed by the most difficult hop, but as the concentration increases toward the percolation threshold, the activation energy should decrease as conducting channels of modifier are established. Likewise, correlated transport will become increasingly prevalent. Needless to say, the replacement of an alkali by another modifier of different size and/or coordination is likely to have a deleterious effect on melt viscosity and on ionic conductivity in the glass. We will discuss these matters in more detail in the review of the EXAFS of modifiers in Section VI.

3. Molecular Dynamics Models

Finally, brief mention should be made of molecular-dynamics (MD) calculations. It is generally recognized that MD structures are no better than the interatomic potentials that are employed to generate them. Early work involved the use of central forces (Angell et al., 1977; Soules, 1979; Mitra et al., 1981). This combined with limited cluster size and unrealistic cooling rates while adding something to our knowledge of amorphous condensation did little to provide realistic models for glass structure. However, improvements involving three-body potentials have resulted in far more specific structures (Vessal et al., 1988) furnishing well-coordinated networks with realistic bond angles and bond-angle distributions. These developments offer a most promising route forward, particularly in relation to realizing specific MRNs. Indeed, attempts at condensing binary silicates have revealed vestiges of modifier channels (Garofalini, 1988).

III. XAS Techniques

X-ray absorption spectroscopy attracts a variety of techniques as does the measurement of photoelectron absorption cross sections at other energies. Figure 3 illustrates these schematically. In the transmission geometry that detects the primary process, the whole X-ray absorption is measured. However the cross section of the element of interest might only be a small fraction of this. A particular partial cross section is more prominent in the excitation spectrum of a secondary process such as X-ray fluorescence or Auger emission. In particular, where photoelectron detection is concerned,

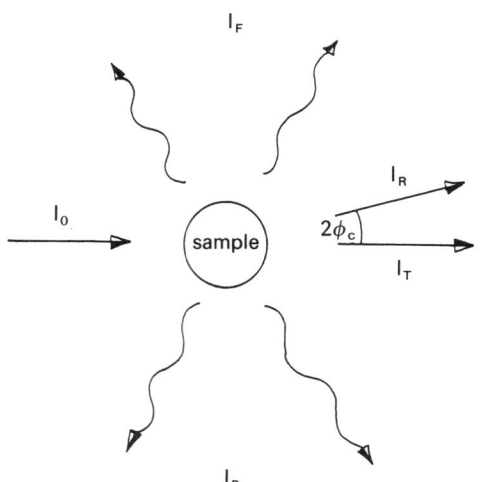

FIG. 3.   Modes for measuring XAS. I_0 is the intensity of the monochromatic X-rays incident on the sample and I_T the transmitted intensity. I_R is the reflected signal if I_0 is incident at the critical angle for total external reflection $2\phi_c$. Secondary emission is indicated by I_F and I_P (X-ray fluorescence and photoelectron emission, respectively).

the average escape depth of electrons (which is typically 500Å) adds considerable surface specificity to this technique compared to the much longer X-ray attenuation length of several microns which governs X-ray detection. However, improved surface sensitivity can be obtained from flat polished specimens using the geometry of total external reflection. In this case, surfaces and interfaces can be studied without the need for an ultrahigh vacuum sample environment. The various techniques shown in Fig. 3 will now be described in more detail.

A. TRANSMISSION

The standard arrangement used at synchrotron sources for transmission XAS is shown in Fig. 4. Monochromatic radiation with a band pass at least as small as ΔE_r (Eq. 4) is required and this is readily obtained using two-bounce crystal monochromators. Silicon is used for X-ray energies down to a few keV. At soft X-ray energies, InSb, Ge, or KAP are preferable (McDowell et al., 1988). Total X-ray absorption, α, is obtained from the ratio of transmitted (I_T) to incident (I_0) beam intensity:

$$I_T = I_0 \exp(-\alpha t). \qquad (5)$$

t is the sample thickness and signal-to-noise ratio in χ is maximized when

$$\alpha t = 2.6.$$

(For this, incidentally, the reference ion chamber should be set to absorb 22% of the incident X-rays.) The optimum thickness of a multicomponent sample for a transmission measurement depends of course on the average atomic density. For the same absorption edge, an oxide glass sample, for example, will need to be up to three or four times thicker than a chalcogenide glass sample. At the same time, the optimum sample thickness for the same material also depends strongly on the energy of the absorption edge of the element being measured. It is convenient to take 4 keV as the division between hard and soft X-rays. At hard X-ray energies, running typically from around 4 up to 30 keV, t for an EXAFS measurement will range from 10 to 100 μm. For a synchrotron beam measuring a few mm², this represents around a mgm of material active in the measurement. Most EXAFS and XANES experiments of glasses are made with hard X-rays in an environment of air. The particular problems of working in the soft X-ray region (500 eV– 4 keV) are considered by Greaves and Raoux (1983). Briefly, samples need to be measured under vacuum and optimum specimen thicknesses for transmission are usually less than a μm. Photoyield is often the more convenient detection mode for soft X-ray edges as thicker specimens can be employed. (See Section III.C.) For the structural chemist, however, the chief distinction

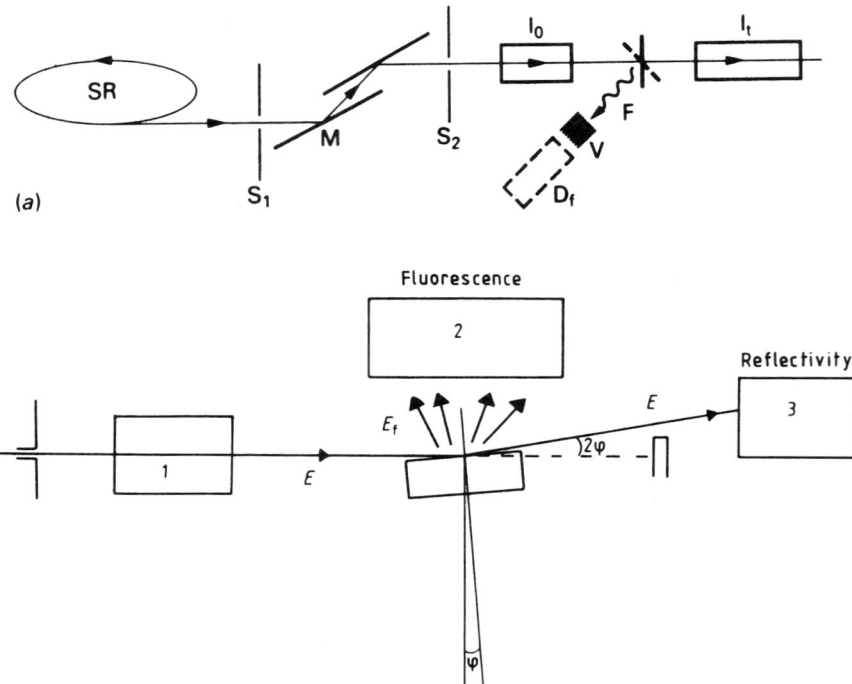

Fig. 4.   Experimental arrangements for XAS studies. (a) Transmission and fluorescence. SR: synchrotron radiation source. S_1, S_2: vertical slits. M: two-crystal monochromator. V: soller slits. I_0, I_T: ion chambers. D_F: scintillation counter or solid-state detector. (b) Glancing-angle geometry. ϕ angle of incidence, detectors 1 and 3 are ion chambers, detector 2 may be an ion chamber or a solid-state detector.

between hard and soft X-ray energies lies in the elements whose absorption edges can be measured. The demarcation in Table I should be of value.

Measurements of elements lighter than 0 are possible but the separation between edges of different elements is usually only sufficient for XANES studies. Note the switch to L-edges for heavy elements. This is due to the electron-hole lifetime broadening (ΔE_τ), which washes out the fine structure for the corresponding K-edge.

Fabrications of thin-film specimens of amorphous materials several μms thick is straightforward if this can be done by depositing onto Al or Be substrates. Self-supporting samples of oxide glasses can often be prepared by blowing thin films. Thicker specimens are usually prepared by powdering the glass or pelletizing with an X-ray transparent binder—BN or graphite. Alternatively, tapes can be made up using glue or varnish. Effective specimen thicknesses can of course be increased by "doubling over" and also by inclining at small angles to the incident beam.

The relevant signal, S, in an XAS experiment is the normalized fine structure (Eq. 2), which is usually only a few percentage points of the X-ray absorption cross section, α_i, of the element being measured. This in turn will generally only be a fraction of the total cross section of the specimen α. If the concentration of the element in question is c_i,

$$S \sim c_i \alpha_i \chi_i.$$

The photon limited noise, N, depends on the geometry employed. For transmission through a sample of optimum thickness

$$N_T \sim \frac{2\alpha}{\sqrt{I_0}}$$

giving a signal-to-noise ratio of

$$\frac{S}{N_T} \sim \frac{\chi_i}{2} \left\{ \frac{c_i \alpha_i}{\alpha} \right\} I_0^{1/2}. \tag{6}$$

This clearly falls away linearly with concentration. A value of around 50 is needed for full EXAFS analysis and so for 10^{10} photons per experimental point, Eq. 6 gives a lower limit for c_i of a few percent if the transmission mode is used.

B. FLUORESCENCE

Returning to Fig. 4, characteristic X-ray emission is best collected in a synchrotron experiment at right angles to the beam, where polarization effects minimize the Compton scattering contribution to the background. Raleigh scattering is generally reduced using X-ray filters or an energy-discriminating detector. For a dilute system, the characteristic absorption α_i is related to the fluorescence signal I_F by

$$I_F \sim I_0 \varepsilon \frac{c_i \alpha_i}{2\alpha}. \tag{7}$$

Eq. 7 assumes that the specimen is opaque (i.e. $\alpha t \gg 1$). This clearly removes the necessity to fabricate samples with an optimum thickness t. ε is the overall efficiency (quantum x solid-angle x detector) which can seldom be greater than 5%—the primary limitation being the solid angle. The photon-limited noise

$$N_F \sim 2 \sqrt{\alpha \frac{c_i \alpha_i}{\varepsilon I_0}}$$

is yielding a signal-to-noise ratio for fluorescence detection of XAS

$$\frac{S}{N_F} \sim \frac{\chi \varepsilon^{1/2}}{2} \left\{ \frac{c_i \alpha_i}{\alpha} \right\}^{1/2} I_0^{1/2}. \tag{8}$$

Comparing Eq. 8 with Eq. 6, clearly improvements in S/N for the same I_0 accrue for fluorescence compared to transmission modes, as the sample concentration c_i falls. The two modes of measurement have comparable sensitivity as:

$$\frac{c_i \alpha_i}{\alpha} \to \varepsilon,$$

i.e., for concentrations of around 5%. Lower concentrations can be achieved by using multiple scans and increasing I_0. With present synchrotron sources, the minimum concentration is about 0.1%—equivalent to an impurity level of around 10^{19} atoms cm^{-3}. This is for an environment of medium atomic number. For heavy or light atom matrices, the minimum concentration may be up to five times greater or less respectively.

Now, for concentrated thick specimens Eq. 7 indicates the fluorescence signal I_F is nonlinear in α_i. However, for thin specimens $\alpha t \ll 1$,

$$I_F \sim I_0 \varepsilon c_i \alpha_i t$$

and the problem of nonlinearity disappears. However, for specimens measured at normal incidence to the X-ray beam, so does the signal. The appropriate geometry for measuring concentrated systems in fluorescence is to employ small angles of incidence (Heald et al., 1984). In this case, t is replaced by z, the X-ray penetration depth. (See Section III.D.) The use of glancing angles also makes more efficient use of beam height (Eq. 11).

Fluorescence techniques are customarily exploited for hard X-ray absorption edges where the X-ray emission efficiency is high. In the soft X-ray range where efficiencies are usually a few percent or less, however, Compton scattering cross sections are almost negligible and energy discrimination of the detector is less important. Accordingly, the overall efficiency of detection, ε, can be improved, as Stöhr and coworkers (1985) have shown.

C.. PHOTOYIELD

In the soft X-ray region, XAS is most conveniently measured by detecting photoemission from the specimen. After all, a vacuum environment that is clearly essential for electron detection is also necessary for soft X-rays because the attenuation of air is so great. At the same time, at low photon energies, Auger emission dominates the relaxation processes. Unless a characteristic Auger line is windowed for detection, however, photoyield measures the total absorption coefficient, α, i.e.

$$I_{PY} \sim I_0 \varepsilon \alpha L \qquad (9)$$

where L is the electron escape depth. This increases from around 100Å at soft X-ray energies to as much as 1000Å in the hard X-ray region. For most glasses—provided there are no leaching effects (Section VIII.C)—the structure at depths even as small as 100Å is usually equivalent to the bulk. However, because only a small fraction of the X-ray attenuation depth is sampled, photoyield is inherently less sensitive compared to transmission, which also detects the total X-ray cross section (Eqs. 5 and 9). Nevertheless photoyield, particularly for soft X-ray absorption edges, is simpler to measure and sample fabrication is less of a problem. At hard X-ray energies, the photoyield technique can also be advantageous if samples cannot be prepared as thin films or powders. A vacuum environment is not necessary if an ion-chamber geometry is employed (Guo and denBoer, 1985).

D. REFLECTIVITY

Finally, attention is drawn to the advantages offered by the use of glancing angles in achieving surface sensitivity. Applications, particularly in the study of glass corrosion, will be given later in Section VII.C.

The geometry for total external reflection is shown schematically in Fig. 4. At glancing angles of incidence, an evanescent wave is generated at the surface or interface, penetrating a few 10's Å into the sample. If ρ (gm cm^{-3}) is the physical density, the critical angle for total external reflection ϕ_c is given by

$$\phi_c \sim \frac{20}{E_{XR}} \sqrt{\rho} \text{ mrad.}$$

The minimum penetration depth is also a function of density but is energy independent.

$$z_{min} \sim \frac{48\text{Å}}{\sqrt{\rho}}.$$

At 10 keV, for instance, the critical angle for silica is 3.1 mrad compared to 9.0 mrad for gold. The respective minimum penetration depths for SiO_2 and Au are 31 and 11Å.

In Fig. 5, the reflectivity, R, and the X-ray penetration depth, z, are plotted as a function of the angle of incidence. At ϕ_c, R falls rapidly toward zero but z increases approaching

$$z = \frac{\sin \phi}{\alpha}.$$

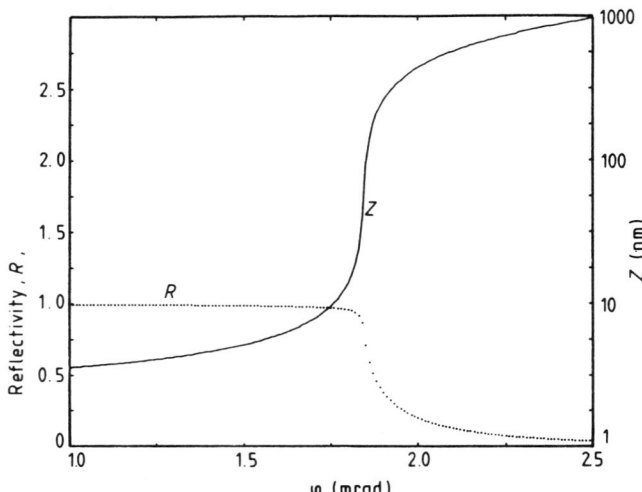

FIG. 5. Dependence of reflectivity R and penetration depth z on angle of incidence ϕ. Calculations are made for a borosilicate glass of density 2.6 gm cm^{-3} at a wavelength of 0.7Å (17.7 keV). (Thornley et al., 1986).

Hence, by suitable choice of the angular position of the incident beam surface, specific structural information can be obtained. This will come from the first atomic layers for an angle of incidence less than the critical angle or from a surface region which may be over a μm thick in the case of an angle above the critical angle.

Referring back to Fig. 4, both the reflected beam, I_R, and the fluorescence signal, I_F, carry information about the X-ray absorption at the surface. In particular, the reflectivity measures the total X-ray absorption, α, for $\phi \leqslant \phi_c$:

$$\frac{\alpha z}{\phi} \propto 1 - \frac{I_R}{I_0} \tag{10}$$

and the fluorescence is given by

$$I_F \sim I_0 \varepsilon \frac{c_i \alpha_i z}{\phi}. \tag{11}$$

(Note that when $\phi = \phi_c$, the intensity of the evanscent wave quadruples due to the standing wave resulting from constructive interference between the incident and reflected waves.) Equation 10 ceases to hold for $\phi \geqslant \phi_c$ because of an increasing contribution to R from the real part of the refractive index. Accordingly, where reflectivity is useful for measuring coatings on flat surfaces, fluorescence enables the structure, particularly of dilute components, to be measured as a function of depth.

IV. Network Glasses

Most nonmetallic glasses are comprised, more or less, of a network structure. I shall begin by examining the local structure viewed from the network-forming cations, first in elemental glasses such as a-Ge going on to simple oxide glasses such as a-SiO$_2$. XAS, with its large effective $\mathbf{k}$ range, brings considerable precision to defining the immediate environment and the degree to which this differs with respect to crystalline isomorphs or is retained when glass formers are alloyed with one another or when modifiers are introduced. Although low $\mathbf{k}$ information is limited in XAS, the technique can still be used to identify neighbors outside the first coordination sphere. This facility is increasingly important in compositionally complicated glasses where element-specific correlations cannot be deconvoluted easily from the total RDF obtained from diffraction experiments. It can also be used very effectively to detect the onset of devitrification.

A. ELEMENTAL AMORPHOUS SEMICONDUCTORS

Although elemental amorphous semiconductors such as a-Ge or a-As have the simplest structures and therefore do not appear to be obvious candidates for XAS experiments, many of the systematic uncertainties in data analysis can be circumvented by direct comparison with crystalline allotropes that share the same or have very similar photoelectron parameters. As a result, precision in the structural parameters r, N, and μ^2 can be greatly improved over what can usually be achieved for multicomponent systems, provided of course the crystallographic values are accurately known. In favorable cases, the limits given in Section II.A can typically be improved by as much as a decade. Notably N and μ^2, which are highly correlated in least squares fitting a single spectrum, can be deconvoluted if crystalline and amorphous spectra are ratioed (Sayers et al., 1975). Returning to the PWA Eq. (2), if χ_g and χ_c are the fine structure of the glass and crystal, respectively, Fourier filtered to isolate a particular shell of common radius,

$$\ln\left(\frac{\chi_g}{\chi_c}\right) = \ln\left(\frac{N_g}{N_c}\right) - 2(\mu_g^2 - \mu_c^2)\mathbf{k}^2. \tag{12}$$

Plotting $\ln(\chi_g/\chi_c)$ versus $\mathbf{k}^2$ enables differences in coordination number (ordinate intercept) and Debye-Waller factor (slope) between the crystalline and glassy state to be measured to an accuracy of a few percentage points. Improvements in the precision with which interatomic distances can be determined can also be made by comparative analysis (e.g., Stegemann and Lengeler, 1986, estimate $\Delta r \sim \pm 0.003$Å).

If the crystalline model is highly ordered, the Debye-Waller factor will be entirely thermal in origin and so $(\mu_g^2 - \mu_c^2)$ relates directly to the static disorder present in the glass. For instance, bond-angle distortion can be estimated from the spread in second-neighbor distances. It should be stressed, though, that static disorder in nearest-neighbor distances is usually extremely small in network glasses; rather, μ_g^2 relates to the strength of the covalent bond as it does in ordered crystals. In particular the Debye-Waller factor recorded in an EXAFS experiment is a measure of the relative motion between the absorbing atom and its back-scattering neighbor along the bond direction. The short-range correlation in this motion is dominated by the optical phonons and can be approximated by an Einstein model (Sevillano et al., 1979)

$$\mu^2 = \frac{\hbar}{2m\omega_E} \coth \frac{\hbar\omega_E}{2kT} \tag{13}$$

where m is the reduced mass, k is Boltzmann's constant, and ω_E is the Einstein frequency. ω_E is closely related to the appropriate stretching frequency. For a-Ge, As or Si optic modes peak around 300 cm^{-1} and Eq. (13) gives Debye-Waller factors, μ^2, of around 0.004Å^2 at room temperature (Yang et al., 1986). For stronger bonds such as those in SiO$_2$ for which $\omega_E \sim 1100$ cm^{-2}, μ^2 drops to 0.001Å^2 (Greaves et al., 1981).

The approach just outlined can also be extended to the study of binary glasses, such as As$_2$S$_3$ and GeSe$_2$, provided the atom type for a given shell can be unambiguously identified. In particular, by comparing different specimens of the same or similar composition, subtle changes in structure resulting from different modes of preparation, different thermal treatments, or photostructural effects can be investigated (Section V.B).

1. Amorphous Ge, Si, and Ge$_x$Si$_{1-x}$

Although a-Ge was one of the earliest materials to be studied using XAS (Sayers et al., 1972), it is only latterly with improvements in measurement (harmonic-free X-rays, uniform samples of optimum thickness, avoidance of specimen pin holes, etc.) that the benefits in precision just outlined have been achieved. EXAFS studies of a-Ge have revealed a first coordination shell almost identical to crystalline Ge. Stegemann and Lengeler (1986) report $N = 4 \pm 0.04$, $R = 2.44(8) \pm 0.003$Å, and $\mu_g^2 - \mu_c^2 = 0.001(8) \pm 0.00015$Å^2 for both evaporated and glow discharge deposited films. The structure of a-Ge is therefore almost fully coordinated and point defects cannot be responsible for the density deficit, which is typically 9% for films prepared at room temperature.

Stegemann and Lengeler estimate that voids of minimum radius 14Å are needed to account for the lower density of a-Ge. This is qualitatively in agreement with the results of small-angle scattering data. As I have already pointed out (Section II.A), XAS yields information about more distant shells. Although fine structure extends only as far down as a $\mathbf{k}_{min}$ of around 2Å^{-1} (the effective wavevector minimum being 4Å^{-1}), measurements of crystalline materials, for instance, reveal atomic shells more than 8Å away from the excited atom. (For example, see results for CaO shown in Fig. 1.) This is particularly true of c-Ge recorded at 77 K (Crozier and Seary, 1981; Paesler et al., 1983; Stegemann and Lengeler, 1986). In a-Ge, the second shell due to two-bond neighbors is readily resolved, enabling bond-angle distortion to be estimated. RMS values, $\sigma(\theta)$, of between 7 and $9°$ have been reported, compared to $10°$ obtained from X-ray diffraction (Paul et al., 1973). Combined with the highly coordinated first shell, significant bond-angle distortion is well described by the CRNs of Polk and others (Section II.B).

Differences in $\sigma(\theta)$ for films of a-Ge can be related to material preparation and posttreatment. In particular, the onset of crystallization in evaporated Ge films can be detected using EXAFS. Stegemann and Lengeler (1986) find this begins for substrate temperatures in excess of 220°C and is complete by 290°C. Evangelisti and coworkers (1981) claim "good" films are produced for substrate temperatures below 130°C. On the other hand, Stegemann and Lengeler report that a-Ge prepared by glow discharge deposition is completely amorphous even for substrate temperatures as high as 350°C. Indeed, compared to other techniques sensitive to structural ordering, such as X-ray diffraction and Raman scattering, EXAFS has proved to be an extremely precise method, able to detect crystallinity on the scale of λ (i.e., $\leq 10\text{Å}$). Stern and colleagues (1983), for instance, examined a-Ge sputtered with hydrogen at different substrate temperatures and employed the ratioing technique (Eq. 12) on the second-neighbor shell to search for incipient crystallization. For a sample showing no evidence of crystallization by X-ray diffraction or optical absorption EXAFS analysis pointed to microcrystallinity on the 10Å scale affecting 20% of the total volume. Indeed, the authors find evidence for reversible conversion between the microcrystallites and the amorphous phase with temperature. Paeseler and coworkers pursued a lowest–free-energy state for a-Ge by monitoring the temperature-dependence of the Raman optical phonon peak. The smallest $\sigma(\theta)$ they found from EXAFS was $7°$. Further temperature treatment resulted in an abrupt transition to crystallinity where $\sigma(\theta) = 0°$. Complementary changes associated with the relaxed amorphous state and the crystalline state in Ge have also been detected in the K and L_{III} near-edge structure (Morrison et al., 1985; Stegemann and Lengeler, 1986).

Despite the great technological interest in amorphous Si, this material has been studied only in the 1980s by EXAFS (Bellissent et al., 1983; Filipponi et al., 1986; Menelle et al., 1986). The delay in tackling this most important of amorphous tetrahedrals is mainly the difficulty of measuring fine structure from soft X-ray edges (Section III.A), a measurement that only became routine in the 1980s. Because Si is a much weaker backscatterer than Ge, the k-range is shorter and the precision with which ΔN, $\Delta \mu^2$, and ΔR can be measured with respect to crystalline Si is not as great. Nevertheless, several interesting findings have emerged.

As with evaporated a-Ge, sputtered a-Si is found to be almost fully coordinated ($N = 3.9(8)$; Filipponi et al., 1986a), despite its low density compared to c-Si. Once again, this points to the presence of large voids in films of a-Si. In both hydrogen-free and hydrogenated a-Si, the primary bond length is the same as in c-Si (i.e., 2.35Å). However, unlike a-Ge, hydrogenated Si—both glow discharge and sputtered films—have a coordination number significantly lower than 4. Filipponi and colleagues find this decreases with increasing hydrogen content, falling to 3.7(2) for a film containing 21% hydrogen. Menelle et al. (1986) quote a similar reduction. Interestingly, the amorphous film density, ρ_g, judged from the height of the Si K-edge, also decreases compared to c-Si. The fall in N_g/N_c and ρ_g/ρ_c is also accompanied by some static disorder, $\Delta \mu^2 \sim 2$ to $5 \ 10^{-3} \text{Å}^2$, which, as we have seen, is comparable with the thermal disorder in c-Si at room temperature. a-Si(H) films appear to be less disordered than hydrogen-free films.

These results point to the following model for the effect of increasing hydrogen content. Initially hydrogen is homogeneously distributed through a-Si(H) films, relieving the strain that would otherwise be present in the absence of hydrogen and thereby allowing the material to approach its lowest free-energy state. Little deviation from $N = 4$ is expected. As more hydrogen is added, however, at some stage polymerized silane chains form, resulting in a lowering of N without a significant change in static disorder. EXAFS experiments would suggest this heterogeneous hydrogen distribution is established when the hydrogen content reaches around 20%.

Alloying amorphous Ge with Si provides an interesting variable band gap semiconductor but is also of fundamental interest in so far as it mixes structural disorder with compositional disorder (Connell and Lukovsky, 1978). In the crystalline state, the lattice parameter of Ge_xSi_{1-x} alloys changes linearly with composition (Johnson and Christian, 1954). EXAFS studies have provided direct evidence that in the amorphous state these alloys are randomly mixed. In particular, both the Ge and Si K-edges have been examined in $a\text{-}Ge_xSi_{1-x}$; H (Incoccia et al., 1985; Mobilio and Filipponi, 1987). End members were used to establish the appropriate phase shifts. The coordination was assumed to be tetrahedral. Because of the similarity in covalent radii (Ge 1.23Å, Si 1.17Å) the individual Ge-Ge, Si-Si, and Ge-Si interatomic distances cannot be easily identified in the Fourier transforms of the Ge and Si EXAFS because of the restricted band width (Section VI.B, Eq. 14). However, careful analysis of the fine-structure amplitudes reveals that

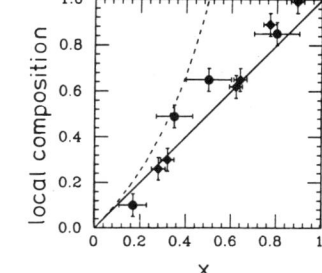

FIG. 6. Local composition obtained from EXAFS compared to bulk composition for a-$Si_{1-x}Ge_x$:H films. Nearest-neighbor environments for Si (circles) and Ge (diamonds) taken from analysis of the respective K-edge EXAFS. (Filipponi et al., 1986).

changes in the Si and Ge coordination numbers do occur with composition. In particular, the $N_{Si}/N_{Si} + N_{Ge}$ and $N_{Ge}/N_{Si} + N_{Ge}$ ratios follow the profile expected for random bonding, i.e., a linear dependence with composition. This is the solid line shown in Fig. 6. The circles are taken from the Si K-edge and the diamonds from the Ge K-edge spectra. The dotted line corresponds to chemical ordering for the minority component. At the midcomposition ($x = 0.5$), Si would be surrounded by Si and Ge by Ge. This is clearly not the case. Another interesting outcome of this study of Ge_xSi_{1-x}:H films is that Ge-Ge, Si-Si and Ge-Si bond lengths do not change with composition; neither do the corresponding static Debye-Waller factors. All of this points to the interatomic potential for Ge-Si being similar to those for Ge-Ge and Si-Si. As these are known to be almost equal (Keating, 1966), random bonding in the amorphous state is not unexpected.

2. Amorphous As and P

Unlike amorphous tetrahedrals, amorphous Group Vs are threefold coordinated with shorter bond lengths and lower densities than their crystalline allotropes. The latter are layered structures exhibiting some degree of mesomeric or back bonding between the layers. Specifically this is an attractive interaction between p-like lone-pair orbitals and antibonding p-σ* orbitals on a neighboring atom. In the amorphous state this extra coordination is largely lost and the primary covalent bonds are strengthened. Secondary bonding has a marked effect on the electronic properties: rhombohedral As is semimetallic, orthorhombic As is a narrow-gap semiconductor, while amorphous As is a semiconductor with energy gap of 1.2 eV, comparable in size to that of a-Si.

Bordas and coworkers were the first to apply EXAFS to these materials and their results are presented in the review of Greaves et al. (1979). They examined a-As alongside the rhombohedral and orthorhombic forms. The intra- and interlayer distances are readily distinguished and the foreshortening of the primary bond—from 2.52 (rhombohedral) to 2.47 (orthorhombic) to 2.42 (amorphous)—is well reproduced by **k**-space fitting. Thermal and static disorder have been investigated by Fontana and coworkers (1982) and by Yang et al. (1986). In common with a-Ge and a-Si, there is little evidence for static disorder ($\Delta\mu^2 \leqslant 0.002\text{Å}^2$). The temperature dependence can be fitted to the Einstein model (Eq. 13) yielding a characteristic frequency of $234\,\text{cm}^{-1}$—almost identical to the observed Raman symmetrical stretching frequency. Moreover, the same treatment for rhombohedral As results in a lower frequency ($216\,\text{cm}^{-1}$), in line with the lower Raman mode and the longer bond length associated with the extensive mesomeric bonding.

Second neighbors can be identified in the EXAFS for amorphous As. Taken together with the nearest-neighbor distance, this yields a bond angle of 98°. The coordination number however is significantly greater than 6, the number expected for two-bond neighbors. The extra coordination derives from the residual mesomeric bonding in the amorphous structure. CRN studies have demonstrated that there is a wide variety of neighbors generated around the two-bond distance as a result of a broad dehedral-angle distribution (Greaves and Davis, 1974). Because of the mixture of neighbors contributing to the second shell of atoms, bond-angle distortion is difficult to isolate accurately. Bordas and colleagues estimate a spread of $\pm 5°$, which is less than what is observed in tetrahedral networks.

There are significant differences between the structure of a-P and a-As (Davis et al., 1977; Elliott et al., 1985). Apart from the shorter bond length, the P bond angle of 103° is larger. More interestingly, there is some structure in the second shell of the RDF and the arrangement of shells of atoms beyond this is more complicated than the simple zigzag signature found in a-As (and also a-Si and a-Ge). K-edge EXAFS of sputtered a-P is shown in Fig. 7. There are three nearest neighbors at 2.2(0)Å with a room-temperature Debye-Waller factor of 0.005Å^{-2} (peak A). The mean position of the second shell (peak B) yields a bond angle of 10(4)°. However there is more than one contribution to this feature suggesting either a smaller bond angle on some sites or a significant number of eclipsed or semistaggered dihedral-angle configurations. Elliott and colleagues (1985) came to similar conclusions in their neutron-scattering study of amorphous red P. In particular, the semistaggered ($\phi = 70°$) and the eclipsed ($\phi = 0°$) geometries would throw up three-bond neighbors at 2.8 as well as 4.3Å. The EXAFS Fourier transform of a-P (Fig. 7) exhibits features close to their distances. Among the crystalline allotropes, eclipsed and semistaggered configurations only occur in monoclinic P which is based on cagelike P_8 and P_9 clusters. Accordingly, both neutron diffraction and EXAFS point to these structural units being prevalent in a-P and affecting the short- and intermediate-range order.

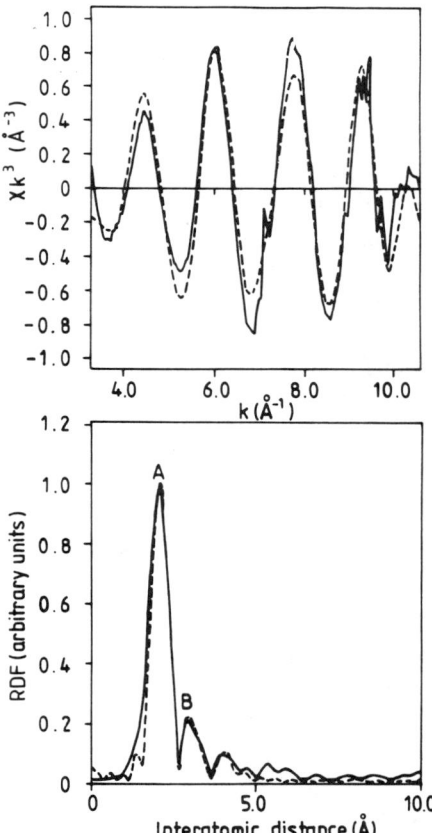

FIG. 7. Normalized EXAFS and Fourier transform measured from the K-edge of *a*-P (McDowell et al., 1988). Dotted lines indicate least squares CWT fit.

B. SIMPLE OXIDE GLASSES

The immediate atomic environment of glass-forming cations such as Si and Ge has been well studied using X-ray and neutron diffraction. Indeed, the first two peaks in the total RDF, which relate to nearest neighbors and to O-O correlations, are the most reliable structural features to emerge from diffuse-scattering measurements of glasses (e.g., reviews by Wright and Leadbetter, 1976, and Porai-Koshits, 1977). XAS measurements on germania and germanate glasses (Sayers et al., 1975; Cox and McMilland, 1981; Lapeyre et al., 1983), silica and silicate glasses (Greaves et al., 1981) and also As_2O_3 glass (Gurman and Pettifer, 1979) have demonstrated that equivalent and in some cases more precise structural information can be obtained using XAS techniques. In nonstoichiomatic oxides such as SiO_x (Greaves et al., 1986), XAS has a unique role to play in differentiating between chemical ordering and random bonding in what is already a structurally disordered solid.

1. Glass Former Environments in a-GeO₂, a-SiO₂, and Binary Glasses

The EXAFS Fourier transforms of the K-edge of Ge are shown in Fig. 8 for vitreous GeO_2(3) and two of its crystalline polymorphs: α-quartz form (2) and rutile form (1). These are the results of Lapeyre and coworkers (1983) and demonstrate how the first and second shells of atoms in the glass closely match in position the tetrahedral Ge sites in the α-quartz structure. The rutile structure, which contains octahedral Ge sites, has a longer bond length of 1.88Å and a split second shell. The Ge-O bond length in $a\text{-}GeO_2$ measured from EXAFS is 1.74Å in, equal to the bond length in α-GeO_2. At the same time the Ge-Ge distance of 3.14Å in the glass matches that in α-GeO_2,

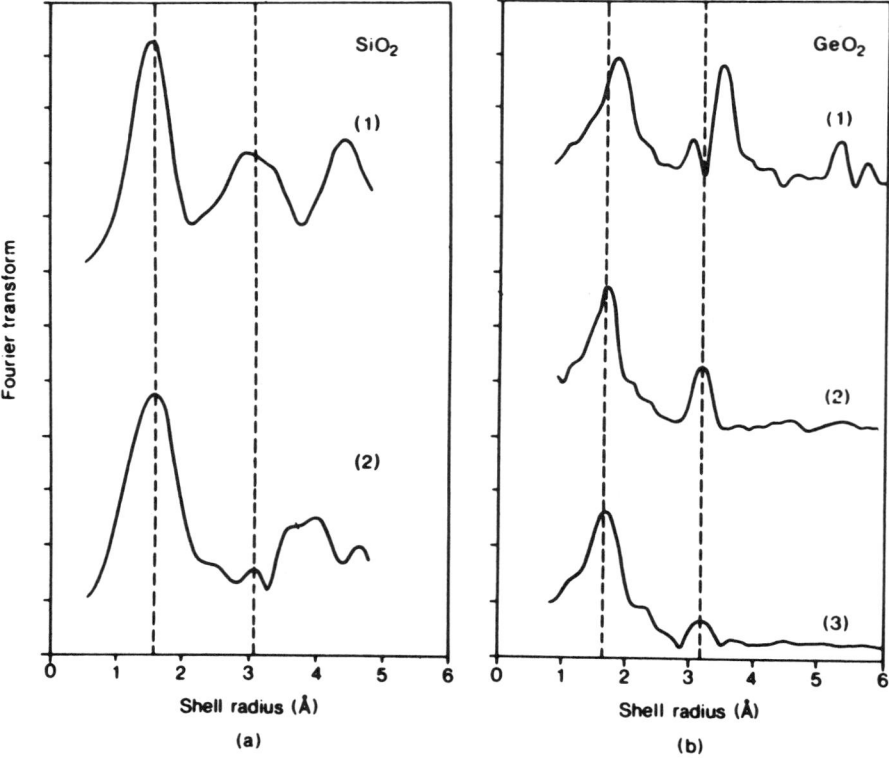

Fig. 8. Phase-shift–corrected Fourier transforms of K-edge EXAFS of crystalline and vitreous silica (Greaves et al., 1981) and germania (Lapeyre et al., 1983). Dotted lines mark the first and second shells for the glasses.

indicating that the crystalline oxygen bond angle of 130° is replicated in the amorphous network.

However, there are considerable differences in the EXAFS Debye-Waller factors associated with the first two shells in amorphous and crystalline GeO_2. In particular, the mean square variation in bond length in the glass is far less than in the crystal and only changes slightly with temperature (Wong and Lytle, 1980). At the same time, there is significant bond-angle distortion in a-GeO_2, estimated at 6° (Lapeyre et al., 1983), leaving the Ge-Ge shell in the glass much diminished compared to the crystal (Fig. 8). These important observations point to the tetrahedral GeO_2 unit being far more rigid in a-GeO_2 than in the crystal, but less rigidly connected. Sayers et al. (1975) suggest this stems from the loss of mesomeric bonding in the glass. Regular mesomeric configurations in the crystalline structure will be reduced in a continuous random network, as we have seen in a-As. However, because of the symmetrical tetrahedral geometry in a-GeO_2, a shortening of the primary covalent bond is neither expected nor observed.

Figure 8 also includes the EXAFS Fourier transforms of the Si K-edge for silica (2) and α-quartz (1) obtained from soft X-ray transmission measurements (from Greaves et al., 1981). Once again, the similarity between bond lengths and bond angles in the amorphous and crystalline state can be seen. Curved wave theory analysis reveals no change in the measured Si-O distance of 1.61Å, but the Si-Si distances differ by 0.1Å with the result that the bond angle in the glass is estimated to be 160° compared to 144° in α-SiO_2. The precise spread in bond lengths and bond angles that can be deduced is more difficult to judge for the Si K-edge where only a limited k-range is available. Analysis shows there is no measurable difference in bond-length disorder between a-SiO_2 and αSiO_2—the observed μ^2 of 0.001Å^2 being close to that calculated from Eq. (13) for an Einstein frequency of $1100\,cm^{-1}$. The spread in Si-O-Si bond angles, however, is found to be 15–20° greater in the glass than the crystal. This is comparable with the σ oxygen bond-angle widths reported in the modelling studies of Mozzi and Warren (1969) (XRD 16°), Mitra (1982) (MD 14°), and Dupree and Pettifer (1984) (MAS-NMR 18°). It should be emphasized, though, that none of these distributions are symmetrical. The peak positions, which EXAFS analysis is sensitive to unless anharmonicity is included, vary considerably from 144° (Mozzi and Warren), 157° (Mitra), and 147° (Dupree and Pettifer). The general consensus is that the oxygen bond angle is larger in a-SiO_2 than in α-SiO_2 but the precise mode, mean and full width half maximum (FWHM) have yet to be unambiguously identified.

Mention should also be made of O K-edge EXAFS from a-SiO_2 grown on silicon measured using photoyield detection and reported by Stöhr and coworkers (1979). The first shell at 1.6(1)Å is due to O-Si correlations and complements the equivalent feature in the Si K-edge EXAFS, but with a coordination of 2 Si's rather than 4 O's. Stöhr attributes a second shell in the O K-edge EXAFS to O-O correlations, obtaining a silicon bond angle of $10(8)°$—lying close to the expected tetrahedral angle. Given the measurement difficulties at 530 eV X-ray energies, no figure is put on the local disorder as judged from oxygens in a-SiO_2. In any case, the definition of SiO_4 tetrahedra in a-SiO_2 has been well studied by diffuse scattering techniques (Wright and Leadbetter, 1976). Indeed, the principle features in the X-ray diffraction pattern QI(Q) can be generated from EXAFS by adding together the Si and O $k\chi(k)$'s (Greaves and Raoux, 1983).

Germania-silica glasses (of relevance to optical fibre cores) have been studied by Lapeyre's group, who monitored the Ge K-edge EXAFS as a function of composition (1983). They observed a small contraction in the Ge-O bond length from 1.74 to 1.73Å but otherwise perfect tetrahedral coordination was maintained. However, the second peak found in a-GeO_2 (Fig. 8) and attributed to cation–cation correlations was found to be absent in GeO_2-SiO_2 glasses. This is a particularly interesting finding, being a strong indicator of random mixing of GeO_4 and SiO_4 tetrahedral units. (I discussed an analogous situation in respect to a-Ge_xSi_{1-x} films in Section IV.A.1.) If there were substantial clustering of GeO_2, then Ge-Ge distances would be resolved. Alternatively, the differences in oxygen bond angles for GeO_2 and SiO_2 discussed previously together with the weaker back-scattering afforded by Si neighbors will suppress the cation-cation frequency in Ge fine structure, which is how things appear to be. Added to this is the slight but nevertheless measurable shortening of Ge-O bonds which is expected with the inevitable modification to the charge on neighboring oxygens brought about by juxtaposition with more strongly electronegative Si's. As of the late 1980s, the K-edge EXAFS of Si has yet to be measured in GeO_2-SiO_2 glasses, but one would expect a complementary small increase in Si-O bond lengths.

The Ge environment in alkali germanate glasses has been measured by several groups (Cox and McMillan, 1981; Sakka and Kamiya, 1982; Lapeyre et al., 1983). If these studies were inspired by an attempt to elucidate the structure of silicate glasses by choosing a glass-forming cation with a more amenable X-ray K-edge, this analogy was mistaken. EXAFS reveals significant changes in the oxygen shell surrounding Ge, so that as alkali is added, the average Ge-O distance in the glass increases. Cox and McMillan (1981), who chose the Li_2O-GeO_2 system, applied CWT k-space fitting and found the Ge oxygen shell was split in the presence of lithia. Returning to Fig. 8, rutile GeO_2 has a bond length 0.14Å longer than α-GeO_2. Accordingly, the binary glasses were modelled by a mixture of fourfold and sixfold Ge sites with the respective bond lengths 1.74 and 1.88Å. Cox and McMillan found the sixfold sites increased with Li_2O to saturate at around 25%. Complementary changes were also observed in the white-line intensity at the absorption edge. The μ^2 values for the two sites were found to mirror the oxygen ligands in the two crystalline polymorphs, indicating the network in the binary glass is composed of rigid α-quartz and rutile-like units. Lapeyre et al. (1983) as well as Sakka and Kamiya (1982) found the same to be true for Na_2O and K_2O germanates. Interestingly, Lapeyre observed Ge-Ge correlations in Na_2O-GeO_2 glasses at the same distance and with the same prominence as in a-GeO_2 (Fig. 8). This was true for all glasses, including those with the highest alkali content (viz. $Na_2O(GeO_2)_2$). The natural conclusion is that, although the random tetrahedral network is clearly depolymerized as alkali is added with the generation of some sixfold Ge sites, the local tetrahedral structure is relatively undisturbed. It is reminiscent of the MRN model for which the network and modifier components are independent of each other but for the linking nonbridging oxygens (Fig. 2).

The effect of modifiers on the network in binary silicate glasses is different. The present author found that the immediate environment of Si was like that in a-SiO_2 and essentially stayed unchanged for the addition of modifiers such as Na and Ca. Both the bond length and the bond angle remained intact, indicating tetrahedral SiO_4 units predominate and a silica-like network is retained in these modified glasses (Greaves et al., 1981). Clearly depolymerization does take place in these glasses. The generation of nonbridging oxygens has been detected in neutron-diffraction experiments (Misawa et al., 1980) and MASNMR (Dupree et al. 1984b). Changes in the oxygen environment should be directly detectable in EXAFS from the oxygen K-edge. Such experiments have yet to be made as of the late 1980s.

2. Arsenates and Phosphates

Arsenate and phosphate glasses have been little studied by XAS, but the technique offers some potential in detecting molecular structure. Arsenolite, As_2O_3, is a molecular crystal composed of As_4O_6 units arranged on a diamond lattice. Gurman and Pettifer (1979) examined this alongside As_2O_3 glass using As K-edge EXAFS. Intra- and intermolecular distances out to 4Å in the mineral were detected and successfully analyzed using CWT k-space fitting. Unlike a-SiO_2 and a-GeO_2, they found the local environment in a-As_2O_3 was far more disordered. The first two principal shells in arsenolite at 1.80Å and 3.24Å are well reproduced in the glass, indicating the same As-O-As bond angle. However, the mean square variation in the As-O bond length, μ^2, was found to be much larger in the glass than in the mineral—410^{-3} compared to 10^{-3}Å^2—pointing to substantial bond straining. On the other hand, deconvoluting this from the second shell width yielded a static bond-angle mean square variation in a-As_2O_3 of only 3°, i.e., much smaller than in the tetrahedral oxides. Although a CRN interpretation cannot be ruled out, this combination of strained bonds and reasonably well defined oxygen bond angles is strongly indicative of a distorted molecular structure. In the CRN picture, which so well describes a-SiO_2 and a-GeO_2, loss of long-range order is primarily componded from oxygen bond-angle disorder. In a-As_2O_3, long-range disorder would appear to stem principally from variations in the As-O bond length.

The P K-edge has been recorded for a series of iron-lead phosphate glasses by the present author and coworkers. The iron and lead EXAFS will be discussed in Section VI.C.3. The phosporous XANES are shown in Fig. 9, together with spectra for a polyphosphate $Pb_2P_2O_7$ and an orthophosphate $Fe_3(PO_4)_2.8H_2O$. The two crystalline models exhibit different near-edge signatures: the white line is followed by a double peak for the pyrophosphate, whereas for the orthophosphate the white line is stronger but is followed by a broad single peak. All the glasses reveal a triplet situation similar to $Pb_2P_2O_7$, suggesting the substantial presence of polyphosphate groups in the glass.

3. Lower Oxides of Silicon—a-SiO_x

The structural properties of lower oxides of silicon, like amorphous SiO, have attracted attention every since the optical absorption edges of thin films were reported by Phillipp (1971, 1972). The continuous opening up of the energy gap in a-SiO_x from 1 to 10 eV as x increased from 0 to 2, Phillipp suggested was due to random bonding arrangements. In addition to the chemically ordered environments of silicon and silica, viz. $Si(Si_4)$ and $Si(O_4)$, the following configurations were proposed: $Si(Si_3O)$, $Si(Si_2O_2)$, and $Si(SiO_3)$. The proportions of each type of configuration would depend statistically on the oxygen content, x. The optical properties of a-SiO_x films were confirmed by Holzenkämpfer and coworkers (1979), who studied these in conjunction with ESR measurements, finding both were consistent with a random bonding interpretation. In contrast to work on thin films, bulk SiO has been studied by X-ray diffraction (Temkin, 1975), neutron diffraction (Etherington et al., 1983), and magic angle spinning (MAS) NMR (Dupree et al., 1984a). These authors have interpreted the structure of SiO in terms of a mixture model (Temkin, 1975) that chiefly consists of Si and SiO_2 regions interspersed with a random bonding component.

Si K-edge EXAFS measurements were made on thin films of a-SiO_x:H across the series by the present author and colleagues (Greaves et al., 1986) and results are shown in Fig. 10. The band pass of the EXAFS spectra (Eq. 14) are sufficient to clearly resolve Si-O (1.6Å) from Si-Si (2.4Å) bonds. Of particular interest is the way these distances change with composition. The behavior portrayed in Fig. 10 is precisely what is expected if the structure of these films were randomly bonded. On the other hand, if the structure were

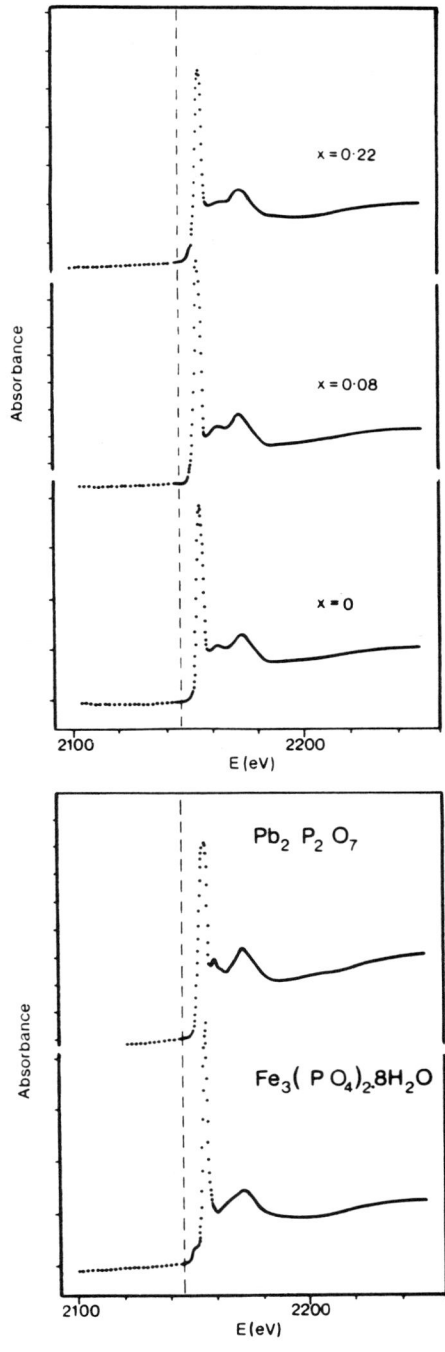

FIG. 9. P K-edge XANES for $(Fe_2O_3)_x(Pb(PO_3)_2)_{1-x}$ glasses compared to crystalline $Pb_2P_2O_7$ and $Fe_3(PO_4)_2\cdot8H_2O$. (Greaves et al., 1988a).

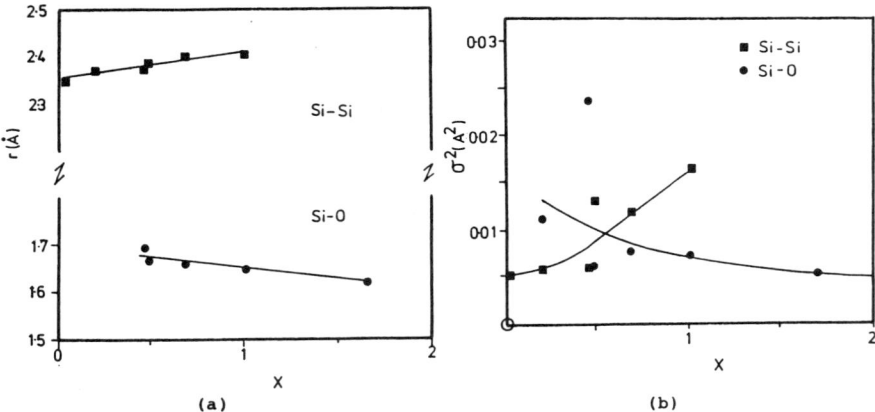

FIG. 10. Compositional changes in Si-O and Si-Si bond lengths. (a) As measured from Si K-edge EXAFS in a-SiO$_x$:H films. Debye-Waller factor variations are shown in (b). (Greaves et al., 1986).

composed simply of mixtures of silicon and silica, no change would be expected. Rather, in a chemically disordered structure, the gradual incorporation of oxygen into the nearest-neighbor environment of silicon will increase the charge on silicons, causing them to repel one another, but increasing their attraction to oxygens. The qualitative behavior illustrated in Fig. 10 has also been reported by Forty et al. (1986) for evaporated figures of SiO$_x$.

Interestingly, the amplitudes of EXAFS oscillations for intermediate oxides are found to be considerably reduced compared to those for silicon and silica. While a fall in total coordination number cannot be ruled out, the most likely explanation for this is an increase in static disorder. Indeed, we would expect site variations to be particularly pronounced in randomly bonded structures where configurations accommodating differing proportions of oxygen and silicon neighbors were juxtaposed. The Si-O and Si-Si EXAFS Debye-Waller factors for a-SiO$_x$:H films are also shown in Fig. 10. Note the incongruent behavior as the oxygen content is varied. With increasing oxygen content in a randomly bonded structure, disorder should decrease for oxygen ligands and increase for silicon ligands as the configurations Si(Si$_3$O), Si(Si$_2$O$_2$), and Si(SiO$_3$) became successively dominant.

One consequence of the increased disorder of intermediate configurations in SiO$_x$ is that the core levels will be distributed giving site variations in the chemical shifts. However, this will be less true for the chemically ordered configurations Si(Si$_4$) and Si(O$_4$). Accordingly, at the SiO composition, a splitting of XPS lines should be detected. Similarly, the XANES will be dominated by the well-ordered $Si(Si_4)$ and $Si(O_4)$ configurations. This splitting should weaken for other compositions where intermediate strained configurations are more likely. Just this behavior is reported by Forty et al. (1986), who measured Si 2p XPS and is also reflected in the Si K-edge XANES (Greaves et al., 1986). Similar effects are expected in MAS NMR spectra and have been observed by Dupree and coworkers (1984) in bulk SiO. It should be remembered that NMR and also ESR spectra cannot be relied on to quantify the proportions of various configurations if not all the Si centers in the specimen are active. XAS and XPS on the other hand probe all Si sites.

In a related system, a-$(SiO_2)_x(CeO_2)_{1-x}$, EXAFS has been used to probe the environments of silicon and cerium. Gurman (1988) report Si K-edge and Ce L_{III} edge data for sputtered thin films. These reveal that the coordination of Ce as well as Si is tetrahedral, although the oxygen shell is more disordered for the heavier cation. Interestingly, a shell of Si atoms at 3.1Å is resolved around Ce, confirming that the structure is not a simple mixture of the two oxides. Moreover, oxygen-deficient films mirror the local structure found in a-SiO_x films. In particular, cerium coordinates to oxygen at the expense of silicon, whose environment incorporates silicon as well as oxygen nearest neighbors. The same monotonic decrease in the Si-O bond length as illustrated in Fig. 10 is also observed in these amorphous silica-ceria films, indicating random bonding is occurring.

V. Chalcogenide Glasses and Photostructural Effects

Chalcogenide glasses contain Group VI elements such as S, Se, and Te, usually alloyed with As or Ge. Like simple oxide glasses they form random networks. However, chalcogenide glasses are less stable than oxide glasses. The cation-anion electronegativity differences are not so marked in chalcogenides and compositional disorder is always a possibility. In particular, photostructural changes are often exhibited. These are configurational modifications taking place at temperatures well below the glass transition temperature, T_g, by the action of band gap light (Tanaka, 1980).

A. STOICHIOMETRIC GLASSES

Glasses such as As_2S_3 and $GeSe_2$ have K absorption edges which (apart from sulphur) occur around 12 keV and, accordingly, XAS measurements are comparatively easy to make using the standard transmission geometry. Not surprisingly, chalcogenides were some of the earliest glasses to be studied using EXAFS (Sayers et al., 1974; Pettifer, 1979). In some respects these

studies yielded rather straightforward answers—confirming Pauling covalent radii and the 8-N rule for coordination numbers. Having said that, the
ability of EXAFS to determine the coordination numbers of specific elements
is a distinct advantage compared to conventional diffraction methods where
nearest-neighbor coordination numbers can only be deconvoluted rather
imprecisely.

Stoichiometric arsenic chalcogenide glasses have been studied by Sayers et
al. (1974), Nemanich and coworkers (1977), and Pettifer (1979). In a-As_2S_3
arsenic was reported to be coordinated to three sulphurs at 2.24Å with a
room-temperature Debye-Waller factor of $3 \ 10^{-3}$Å^2 (slightly smaller than for
a-As). Two-bond As neighbors were recorded at 3.5Å giving an As-S-As bond
angle of 101°. These contributions to the partial RDF can be seen in Fig. 11.
A similar picture emerges from studies of a-As_2Se_3 where arsenic is rigidly
threefold coordinated with an As-Se bond length of 2.40Å (Sayers et al., 1974;
Pettifer, 1979). In this case, however, two-bond neighbors are scarcely
detectable, demonstrating greater bond-angle distortion in the selenide than
in the sulphide. Compositional disorder has been detected in mixed chalcogenides. In a-$As_2S_3Se_3$ Pettifer found arsenic had both selenium and sulphur
nearest neighbors. These were in the ratio 1:2, but at the expected covalent
distances found in the binary glasses (Pettifer, 1979).

The heteropolar bonding reported in arsenic sulphides and selenides
contrasts with the situation found in a-As_2Te_3, where evidence for homopolar bonding has been found (Gurman, 1988). Gurman has reanalyzed the
earlier work of Pettifer and coworkers (1979) and finds arsenic coordinated to
2.5 telluriums at 2.68Å and 0.5 arsenics at 2.53Å. While the local environment
displays some chemical disorder, the network is essentially threefold coordinated in marked contrast to crystalline As_2Te_3 where arsenic occupies
three- and sixfold sites. There is considerable mesomeric bonding and c-
As_2Te_3 is semimetallic compared to a-As_2Te_3 which is semiconducting. It
should be noted that crystalline arsenic sulphides and selenides are semiconducting. Their structures also exhibit back-bonding configurations but not so
extreme as in c-As_2Te_3. We saw the relationship between mesomeric bonding
and electronic characteristics previously in the case of allotropes of arsenic
(Section IV.A.2). In chalcogenide glasses, EXAFS has an important role to
play in untangling partial RDFs and, while the results for a-As_2S_3 and a-
As_2Se_3 are not unexpected, chemical disorder in a-As_2Te_3 is an interesting
result. We shall see later that some homopolar bonding is also observed in a-
As_2S_3 for only small departures from stoichiometry. Indeed, in the case of the
thin films, wrong bonds are almost impossible to avoid and are undoubtedly
linked to photostructural behavior (Greaves et al., 1987).

Turning to germanium chalcogenides, stoichiometric a-Ge_2Se_3 has been
studied by EXAFS at both Ge and Se K-edges (Sayer et al., 1974; Nemanich

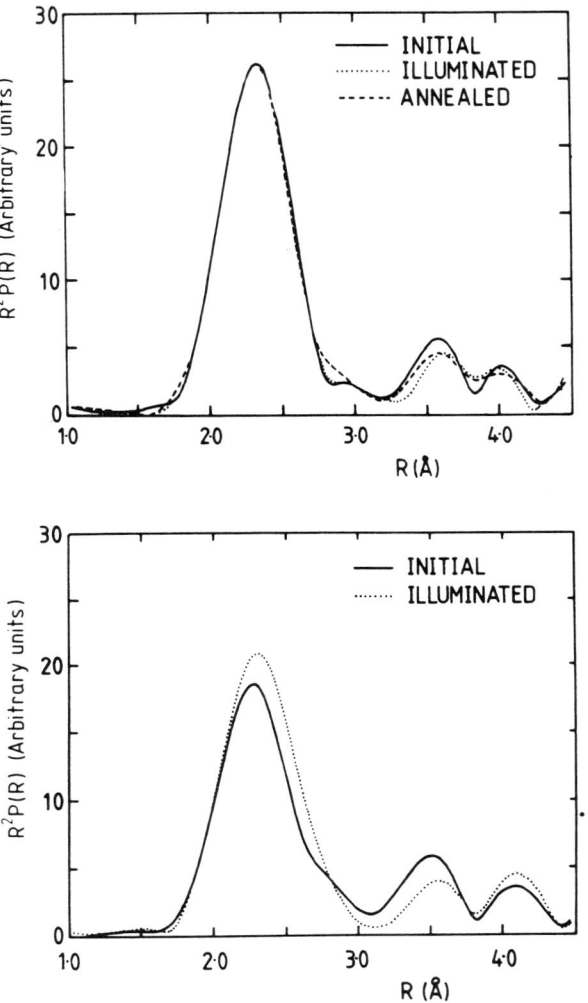

FIG. 11.   Arsenic RDFs for amorphous arsenic sulphide obtained from K-edge EXAFS. (a) Bulk As_2S_3 glass in the annealed, illuminated, and partially reannealed states. (b) Evaporated film before and after illumination. (Lowe et al., 1986).

et al., 1978). The local structure is reported to be purely covalent, with germanium fourfold coordinated and selenium twofold coordinated. The Ge–Se bond length is 2.36Å, which is close to that found in crystalline β-GeSe$_2$ where germanium occupies tetrahedral sites and selenium twofold sites. The local structures in the amorphous and crystalline states are therefore very similar. Because both germanium and selenium K-edges EXAFS are easily measured in the same glass, the presence or absence of homopolar bonds can be checked out. a-$GeSe_2$ is found to be free from wrong bonds. The same conclusions have been corroborated in anomalous dispersion experiments using both K-edges by Fuoss and coworkers (1981). Raman measurements have indicated that the same 4:2 geometry persists in chalcogen-rich glasses such as $GeSe_3$—the extra selenium being accommodated in Se-Se chains (Tronc et al., 1973).

Crystalline and amorphous $GeSe_2$ and GeS have been studied by Oyanagi and coworkers (1981) using germanium K-edge EXAFS. While a-GeS_2 appears to have a 4:2 coordination arrangement like a-$GeSe_2$, germanium is reported to be threefold coordinated in a-GeS—realizing a prediction made by Bienenstock on the basis of conventional X-ray diffraction measurements (1973). Fuoss and colleagues came to the same conclusions for a-GeSe in their anomalous dispersion studies. Interestingly, c-GeS and c-GeSe have distorted rock-salt structures, characterized by extensive mesomeric bonding arrangements. The similarity between these structures and that of black phosphorous has been made (Bienenstock, 1983). Now Oyanagi et al. (1981) report that the Ge-S bond length decreases in going from the crystalline to the amorphous state, falling from 2.45 to 2.37Å, which strongly indicates a breaking of back-bonds in a-GeS as we saw earlier for arsenic (Section IV.A.2).

Note that the structural behavior of germanium chalcogenides contrasts with that of the oxides of silicon discussed previously (Section IV.B.3). In a-SiO_x, mesomeric interactions are inhibited by the silicon environment, which is always tetrahedral. At the same time, anion deficiency ($x \leqslant 2$) is accommodated, not by a triply coordinated structure, but by a 4:2 coordination arrangement accompanied by random bonding configurations.

B. PHOTOSTRUCTURAL EFFECTS

Chalcogenide glasses have optical gaps in the visible and near UV and, although they are sensitive to a wide variety of ionizing and particle radiation, pronounced effects are produced with band gap light (Tanaka, 1980). Since photostructural changes take place well below T_g, it is not surprising these are strongly dependent on the way the glass is prepared. Indeed, photostructural changes cannot be induced at the glass transition, but generally increase in size as the temperature falls. In some cases the original configuration can be restored by annealing at T_g (reversible changes), and in other cases the new configuration is permanent (irreversible changes).

Photodarkening and photoexpansion usually accompany photostructural modifications. Depending on the method of preparation, photoinduced phenomena may be inverted—photobleaching and photodensification are sometimes observed. As chalcogenide glasses can be obtained by quenching from the melt and by thin-film deposition (e.g., evaporation), glasses and films offer contrasting photostructural properties. In the case of annealed glass, these are small and reversible but for thin films—particularly if these are deposited obliquely to the substrate—much larger effects are seen and these are usually irreversible. One intriguing irreversible effect that also displays technological promise is photodiffusion. I shall describe this in more detail at the end of this chapter (Section VII.A).

Compared to tetrahedral networks such as a-Si or a-Ge where all valence electrons are incorporated into hybridized bonds, in chalcogenides—like pnictides discussed earlier—the average coordination number is lower than four, which results in a fraction of electrons occupying lone-pair states at the top of the valence band. It is the interaction between lone-pair orbitals on one atom and antibonding orbitals on a neighboring atom that constitutes the mesomeric or back-bond. Crystalline chalcogenides have layerlike structures (As_2S_3, $GeSe_2$, etc.) or are chainlike (S, Se, Te). Indeed, evidence for molecular-, layer, or chainlike groups in the amorphous state is strong, particularly in evaporated films (Moss and Price, 1985). It is convenient to refer to primary and secondary bonds in chalcogenides as intra- and intermolecular interactions, respectively. However, in contrast to the crystalline state, local steric constraints are less severe in the glass so the breaking of both intra- and intermolecular bonds by absorbing UV radiation is likely to promote subtle configurational changes because of the variety of meta stable arrangements available at the local level. It is worth noting that the broad matrix of photostructural phenomena exhibited by chalcogenide glasses is not manifest in the crystalline state.

The EXAFS technique enables direct short-range structural information to be obtained about photo-induced changes. Changes in medium-range order can be monitored in the shape, height, and position of the first diffraction peak occurring around 1Å (Tanaka, 1975). As we have seen, these lie outside the k-range accessible to EXAFS (Section II.A). However, we shall see that, in general, photostructural changes are promoted by the breaking of homopolar bonds and the generation of heteropolar bonds. The detection of changes in the immediate atomic environment, therefore, provides essential clues as to configurational changes taking place in intermediate-range order.

1. Reversible and Irreversible Changes in a-As_2S_3

The different photostructural changes in amorphous chalcogenides are best compared by contrasting the behavior of an annealed glass with that of a vapor-deposited film. Figure 11 is taken from the work of Lowe and coworkers (1986) where the As environments in vitreous and evaporated As_2S_3 obtained from the As K-edge EXAFS are compared. Differences resulting from illumination are most marked in the vicinity of the nearest-neighbor peak in the RDF centred at 2.25Å. The glass displays no perceptible change in area or position on illumination, whereas the equivalent features for the film increases substantially in height. There is clearly scarcely any evidence for intramolecular bond breaking when the annealed glass is illuminated. However, detailed k-space analysis reveals that the nearest-neighbor peak contains contributions from As-As as well as As-S correlations. This is particularly marked for the film where, before illumination, As atoms are surrounded by 2.(5) S's at 2.25Å and 1.(1) As's at 2.48Å. This gives an overall coordination of approximately three atoms, but the network is not entirely chemically ordered. Similar results for evaporated a-As$_2$S$_3$ have been reported by Nemanich et al. (1978). The presence of homopolar bonds (As-As) can be explained by assuming that As$_4$S$_4$ groups are incorporated in the structure. This is the dominant molecule in the vapor phase and its structure is characterized by arsenic atoms coordinated to two sulphurs at 2.23Å and one arsenic at 2.59Å. After illumination, the balance of homopolar and heteropolar interactions in evaporated films alters: the As-As coordination falls, whereas the As-S coordination is found to increase (Sayers et al., 1987). For instance, in Fig. 11 the As-S subshell increases from 2.(5) to 2.(8) S's. EXAFS measurements then suggest that irreversible photostructural effects in a-As$_2$S$_3$ films involve the breaking of homopolar (As-As) bonds and the generation of heteropolar (As-S) bonds. Similar observations have been reported by Nemanich on thermal annealing of a-As$_2$S$_3$ films and both processes are envisaged in terms of the depolymerization of As$_4$S$_4$ molecular units.

Reversible photostructural effects in a-As$_2$S$_3$ are more subtle. They can be detected in the annealed glass beyond the first coordination sphere. The small changes evident in Fig. 11 are too small to be quantified but qualitative trends can clearly be seen. Illumination causes a broadening of the 3.5Å peak with a shift to larger r. This peak is assigned to As next-near neighbors and the photoinduced change signifies either an increase in the As-S-As bond angle (101°) or a weakening of intermolecular interactions. In either event, it is clearly secondary mesomeric bonds rather than the primary covalent bonds that are being affected by the absorption of band-gap light. The changes seen in the glass (Fig. 11) are consistent with the photoexpansion of As$_2$S$_3$ glass discussed by Tanaka (1980). There is some indication too from the EXAFS of the glass that these changes are indeed reversible. The behavior of the structure of the evaporated film for $r \geqslant 3$Å is similar to the glass, although some of the photoinduced changes are bound to be coupled to irreversible effects involving the redistribution of primary bonds.

2. Reversible Photostructural Changes in a-GeS$_2$ and a-GeSe$_2$

I have already outlined how the local structure in amorphous germanium chalcogenides changes with composition. The lower chalcogenides such as GeSe are threefold coordinated. In the dichalcogenides and trichalcogenides, a 4:2 coordination arrangement prevails. With this variety of local structure, it is not unreasonable to presume that for most compositions the glass structure will include some traces of each type of environment. Accordingly, for GeS_2, for instance, while the dominant coordination geometry will be 4:2 and the majority of bonds heteropolar, a small fraction of threefold-coordinated atoms and some S-S linkages are to be expected—particularly in unannealed material. Furthermore, it will be the primary and secondary bonds associated with these chemically disordered configurations that will be most vulnerable to change when UV light is absorbed.

Using neutron diffraction, Gladden (1986) has measured a small decrease of the average coordination number in a-GeS_2 of 0.04 atom upon illumination. The Ge K-edge EXAFS from the same batch of annealed and illuminated glasses was also measured. The coordination number of germanium analyzed from the annealed glass was $4(\pm 0.4)$, but a significant change in the EXAFS amplitudes occurred on illumination, indicating an increase of approximately 0.5 atom in the coordination shell of germanium. Now the change in the occupancy of the coordination shell of sulphur consistent with this change and the neutron-diffraction decrease amounts to a fall of 0.3 atoms. The individual coordinations of germanium and sulphur in GeS_2 glass then are altered in opposite senses by illumination. If the structure were perfectly chemically ordered, for instance, the breaking of heteropolar bonds would initially affect both atom types in the same way. The observed incongruent behavior lends support to the model in which homopolar S-S bonds already exist in the annealed glass compensated by undercoordinated germaniums. The action of light will preferentially break these S-S connections allowing the undercoordinated germaniums to relax to fully coordinated tetrahedral configurations. Such a switch from homopolar to heteropolar interactions is consistent with the photodarkening observed in annealed glasses.

Turning now to a-$GeSe_2$, evidence for Se-Se bonds comes from both Raman (Phillips, 1981) and Mössbauer (Boochland et al., 1982) results. K-edge EXAFS spectra, for both Ge and Se in $GeSe_2$ glass, are reproduced in Fig. 12 from the work of Gladden et al. (1985). Qualitatively similar behavior to GeS_2 glass is observed: illumination causes the amplitudes of the Ge fine structure to increase. Moreover, the amplitudes of the Se EXAFS decrease. Note too how the photoinduced coordination numbers analyzed for germanium and selenium in the annealed glass are $4(\pm 0.4)$ and $2(\pm 0.2)$, respectively, so it is tempting to infer the same photostructural behavior in a-$GeSe_2$ as proposed above for a-GeS_2. Changes in EXAFS amplitudes, however, do not necessarily derive from changes in coordination number but may also occur when there are modifications in the local order. As we have seen, N and μ^2 are strongly correlated in any least-squares fitting procedure but can be deconvoluted when the bond length is invariant by using Eq. (12).

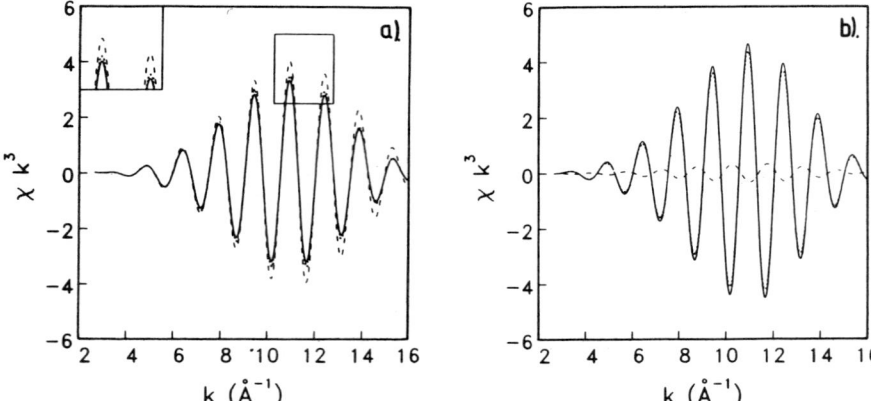

FIG. 12. Weighted normalized EXAFS spectra for GeSe$_2$ glass. (a) Ge K-edge and (b) Se K-edge in the annealed (——) and illuminated (−−−) states. The reversibility on reannealing ($\cdots$) is shown in the inset to (a). (Gladden et al., 1985.)

This procedure was used to measure the 0.5 atom increase in the germanium environment in a-GeS$_2$ previously quoted; the change in μ^2 was insignificant ($\Delta\mu^2 \leqslant 1\%$). In a-GeSe$_2$, the situation was found to be different with a less-than-significant change in the germanium coordination number but a 13% decrease in μ^2 in comparing the annealed to the illuminated glass. This reflects a photoinduced improvement in the definition of the first coordination sphere. Interestingly, amplitude-ratioing analysis for the selenium EXAFS revealed a fall in the selenium coordination number of 0.2 atoms with little or no change in μ^2. This is a smaller change than in a-GeS$_2$ glass, but it is in the same sense and reinforces the view that photostructural effects in amorphous germanium chalcogenides involve the breaking of homopolar chalcogen bonds and the improved configuration of germanium sites. Compared with As$_2$S$_3$ glass, the EXAFS spectra of germanium chalcogenide glasses do not resolve shells of atoms beyond nearest neighbors. However, it is clear from the neutron-scattering data reported by Gladden (1986) that subtle changes are occurring at these distances and it is these that will be primarily responsible for the photoexpansion observed in annealed glasses (Tanaka, 1980). Nearest-neighbor distances are unaffected by illumination as the invariance of the fine structure frequency in Fig. 12 makes clear.

VI. Modified Glasses

Oxides such as soda and lime break up the network of simple oxide glasses such as silica or germania. They dramatically affect the viscosity of the melt and the optical gap of the resultant glass shrinks from 9 eV in silica, for instance, to 5 or 6 eV in sodium disilicate. Modified glasses are ionic conductors and, although the Raman and infrared modes associated with the network are retained from the simple glass, broad modes appear around 200 to 300 cm^{-1}. All these facts are well known and their qualitative explanation lies in the fact a significant fraction of the glass structure is composed of oxygen ionically bonded to modifying cations. The interfacing of these regions to the network will lower the viscosity of the melt and modify the elasticity of the resultant glass. The presence of ionically bonded oxygens will alter the distribution of 0 2p states at the top of the valence band narrowing the energy gap of the glass. The modified ionic regions will provide numerous sites supporting cationic transport and, like ionic crystals, this component of the glass structure will absorb in the far infrared.

Since the early 1980s, EXAFS studies have produced a wealth of quantitative structural information about the role of modifiers in glasses that has previously been lacking. This work confirms that modifying cations by and large are chemically ordered in glasses and that their environments are often similar to the corresponding sites in crystalline isomorphs. Coordination numbers are generally higher than for the more covalently bonded network-forming cations. In some cases, the oxygen shell distribution around the modifier is found to be nongaussian and there is significant static disorder. This is particularly true of alkaline earths such as Ca, for instance. In other cases (Fe^{3+} sites in silicates, for example), metal environments may be better ordered locally than in the crystalline state.

The difficulties associated with measuring oxygen K-edge EXAFS (see Section III.A) have precluded the direct probing of oxygen environments in modified glasses to data. However, these can be inferred once the cation environments have been determined. Glass-forming cations mainly coordinate to bridging oxygens and much is known about their geometry from conventionally obtained total RDFs. Modifying cations, on the other hand, chiefly bond to nonbridging oxygens. These are manifest in XPS by split oxygen core levels (Jen and Kalinowski, 1980). Nonbridging oxygens associate one-to-one with alkali cations. This correspondence means that if the coordination number of the modifying cation is greater than the network cation, then nonbridging oxygens are also likely to have a greater coordination number—certainly compared to bridging oxygens. Precisely this situation occurs in the crystalline state as Fig. 13 illustrates. In oxide glasses, EXAFS studies have revealed well-defined environments for many modifying cations, and this implies that the environments of nonbridging oxygens will also be well defined. Firmly prescribed local order around modifiers and nonbridging oxygens forms the basis of the MRN model introduced in Section II.B.2 and illustrated in Fig. 2.

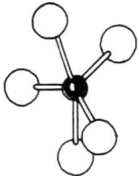

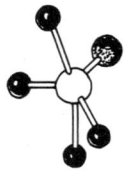

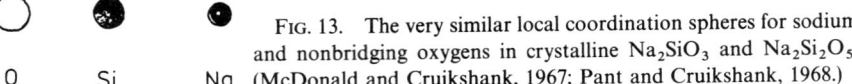

O   Si   Na

FIG. 13.   The very similar local coordination spheres for sodium and nonbridging oxygens in crystalline Na_2SiO_3 and $Na_2Si_2O_5$. (McDonald and Cruikshank, 1967; Pant and Cruikshank, 1968.)

Where the environment of a particular modifier metal is more or less fixed for a particular glass, we can expect this will vary from one glass system to another as occurs in the minerals. This chemical variability is in marked contrast to network cations whose immediate local structure remains almost unchanged. It has to be that the structure of the modified component in a glass holds the key to the material's particular physical, optical, and transport properties. XAS techniques are especially suited to exploring the structure of modified glasses, as partial RDFs of modifying cations can be obtained without confusion from the strong peaks associated with the network that characterize the total RDF (e.g., Si-O and O-O correlations in silicates). Moreover, where several modifiers are present in the same glass, individual environments can be selected, which is a clear advantage in examining cooperative phenomena such as the mixed alkali effect. This section will review results for alkalis in oxide glasses, alkaline earths, and also transition metals. The latter play an intermediate structural role between that of a glass modifier and that of a glass former.

A. ALKALI ENVIRONMENTS

Sodium was the first modifier to be examined using EXAFS (Greaves et al., 1981). More recently, potassium (Jackson et al., 1987) and cesium EXAFS have been measured (Greaves 1989). I shall describe these cation environments separately and then in relation to the mixed-alkali effect.

1. Sodium

Results for three glasses ($Na_2Si_2O_5$, $Na_2CaSi_5O_{12}$, and $Na_2B_2O_7$) are shown in Fig. 14. These were obtained from thin-blown films using the transmission geometry (Section II.A). A KAP monochromator was used to reach the Na K-edge at 1.08 KeV. Note that the sodium environments of all three glasses are quite different. In sodium disilicate glass, five oxygens are analyzed at a distance of 2.30Å. The Debye-Waller factor, μ^2, is 0.006Å^2, which corresponds to an Einstein frequency of around 200 cm^{-1} (Eq. 13). This is typical of the far-infrared absorption in silicate glasses and suggests

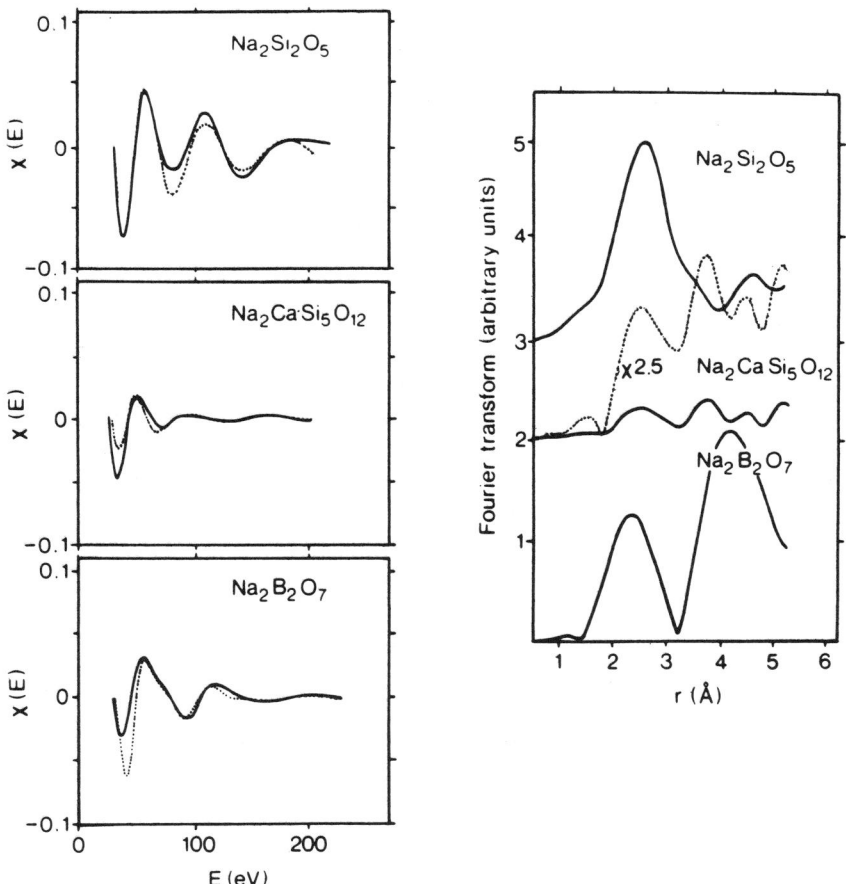

FIG. 14. The changing environment of sodium in various oxide glasses measured from Na K-edge EXAFS. The normalized fine structure is shown on the left and the phase-shift–corrected Fourier transforms on the right. (Greaves et al., 1981.)

there is little static disorder associated with these sodium sites. The fivefold coordination is reminiscent of the environments of sodium in crystalline Na_2SiO_3 (McDonald and Cruikshank, 1967) and $Na_2Si_2O_5$ (Pant and Cruikshank, 1968), which consist of the trigonal biprisms illustrated in Fig. 13. The mean Na-O bond length is 2.40Å, slightly longer than that measured in the glass.

Interestingly, in their early X-ray diffuse-scattering studies of oxide glasses close to the disilicate composition, Warren and Biscoe (1938) also reported Na-O distances near 2.35Å and a coordination number of around 5. In the total RDF for a sodium silicate glass, Na-O correlations only comprise a small outcrop at the base of the second strong peak at 2.63Å resulting from O-O correlations. Deconvoluting these contributions is difficult, and subsequent workers have been less confident in extracting modifier environments. For instance, in analyzing their accurate neutron data, Misawa and coworkers (1980) preferred to concentrate on the integrated weight of O-O pairs demonstrating that these decreased in proportion to alkali content, underlining the one-to-one correspondence between alkalis and nonbridging oxygens. This familiar relationship has given rise to the popular misconception that alkalis and nonbridging oxygens are singly attached and linked to the glass network rather like hydroxyl groups might be configured. However, such a picture is completely foreign to the structural chemistry of most of the minerals. It is clearly inconsistent with the EXAFS results illustrated in Fig. 14 and elsewhere in this section and indeed to the original interpretation of Warren and Biscoe (1938).

Turning now to the EXAFS of the calcium-containing glass $Na_2CaSi_5O_{12}$ shown in Fig. 14, here the EXAFS is much weaker with an inner shell of two or three oxygens at 2.43Å and an outer shell of three atoms at 3.3Å that are possibly oxygens. In crystalline Na_2CaSiO_4 (Wyckoff, 1964) the oxygen shell is split, the site geometry consisting of a trigonal antiprism, with three oxygens at a bonding distance of 2.3Å and a further three oxygens at nonbonding distances of around 3Å. The Debye-Waller factors for the sodium EXAFS are two or three times those in $Na_2Si_2O_5$, pointing to significant static disorder in this ternary glass.

The environment of sodium in $Na_2B_2O_7$ glass is again different from the other two glasses shown in Fig. 14, with six oxygens analyzed at 2.25Å and a Debye-Waller factor of 0.03Å². In crystalline sodium borates, sodium is six- or 7fold coordinated with an average Na-O bond length of 2.3Å which, however, often exhibits considerable variance (Wyckoff, 1964).

In all three glasses, EXAFS partial RDFs give an indication of more distant shells of atoms beyond the nearest-neighbor oxygens. These are likely to comprise chiefly Na-Si and Na-Na correlations and are most apparent for sodium diborate glass. Here a second shell possibly due to sodiums at 3.8Å creates the asymmetry in the EXAFS oscillations, beating with the inner shell of oxygens. No Na-B separations are detected because of the weak back-scattering factor, f_j $(\mathbf{k}, \pi)$, of boron.

Sodium EXAFS in sodium aluminosilicate glasses has been reported by McKeown et al. (1985b) who also analyze a coordination number of 5 or 6 but longer Na-O distances of around 2.6. Their investigation also included $Na_2Si_2O_5$ glass where they reported a similar environment for sodium. However, the group were unable to provide an explanation for a Na-O bond length more typical of alumino silicates such as albite ($NaAlSi_3O_8$). Their measurements were made using photoyield detection (Section III.C) which is surface-sensitive, in which case the discrepancy might result from compositional differences with respect to the bulk glass. Another explanation may lie in the fact that beryl monochromator crystals were employed. Sodium impurities in the beryl can add a significant EXAFS signature to the flux incident on the sample, which is often difficult to normalize out.

Brown's group (McKeown et al., 1985a, b) also measured the aluminum K-edge in the sodium aluminosilicate glasses, this time using a quartz monochromator. Striking differences in the XANES were reported for minerals in which aluminum was four- or sixfold coordinated. Octahedral configurations result in a split "white line," while a tetrahedral arrangement gives a single resonance at the edge. All the glasses exhibited a tetrahedral-like XANES. The EXAFS $\mathbf{k}$-range was limited but yielded Al-O distances of around 1.8Å, contrasting with 1.9Å bond lengths expected for an octahedral Al environment. Both XANES and EXAFS confirm that aluminum enters these glasses as a network former alongside silicon. These results are in contrast to germanate glasses where, as we have seen (Section IV.B.1), six- as well as fourfold sites are present, depending on the alkali content.

2. Potassium and Cesium

Experiments by Greaves and colleagues (1990) on $K_2Si_2O_5$ (potassium K-edge) and $Cs_2Si_2O_5$ (cesium L_{III}-edge) glasses reveal environments for potassium and cesium bearing similarities to that of sodium in $Na_2Si_2O_5$ (Figs. 14 and 15). The average coordination number found is approximately 5 and the alkali-oxygen distances scale with cationic radius: 2.64Å for the K-O separation and 3.00Å for the Cs-O separation. μ^2 values are larger than for Na-O interactions, lying between 0.02 and 0.03Å^2. While the room-temperature thermal contribution will be larger than for Na-O correlations, there is clearly more static disorder around these larger alkali cations.

Jackson et al. (1987) report potassium K-edge EXAFS measurements on orthoclase glass ($KAlSi_3O_8$) and other aluminosilicates, finding K-O distances of 3.0Å and high coordination numbers of 9 or 10. However, they also found that the near-edge structure for orthoclase glass exhibited a split white line similar to that of leucite ($KAlSi_2O_6$) where potassium has a more regular

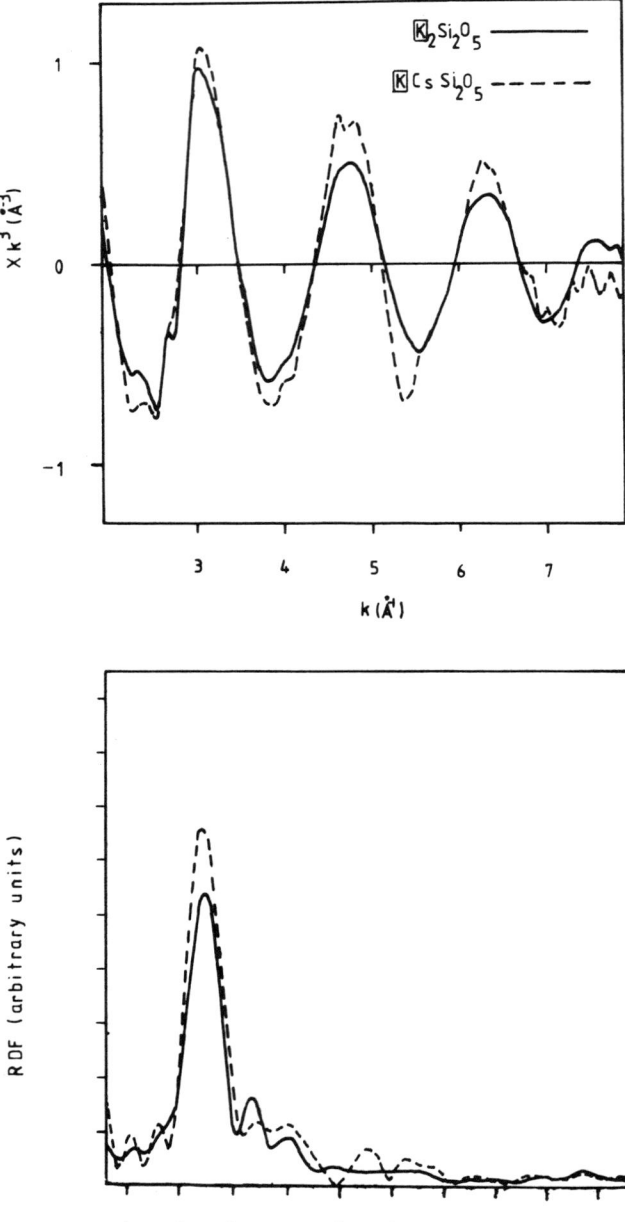

FIG. 15(a).

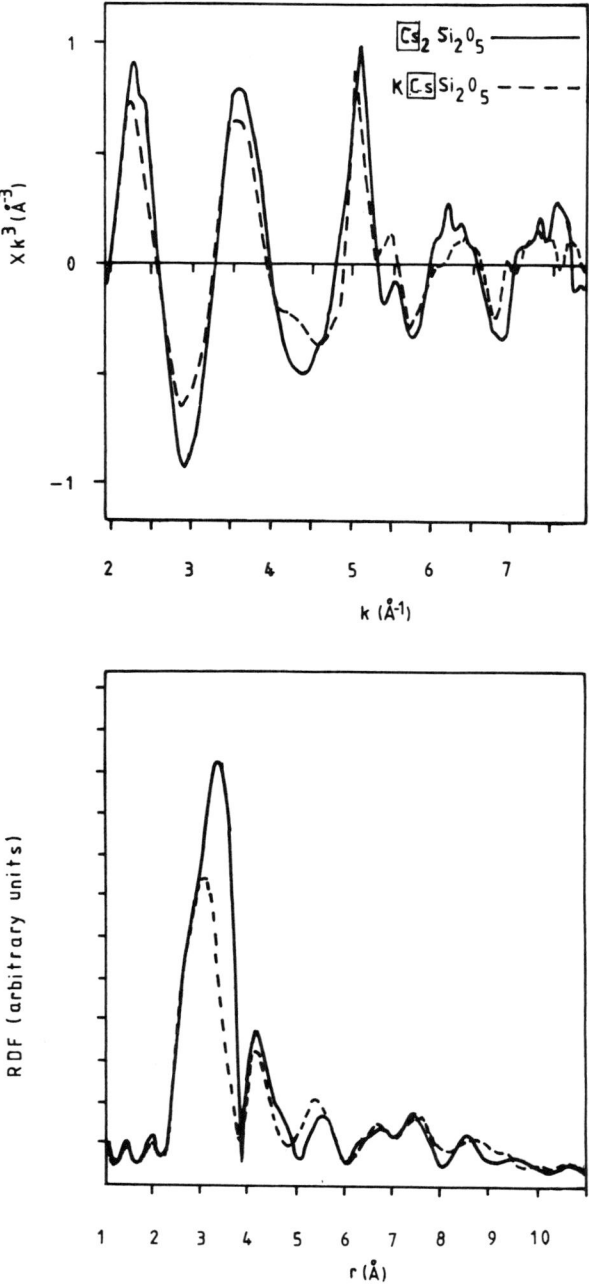

FIG. 15. Local structure of cesium and potassium in the mixed-alkali glass $CsKSi_2O_5$. Potassium K-edge EXAFS and its Fourier transform are shown on the left (a) with cesium L_{III}-edge data on the right (b) (Greaves, 1990). EXAFS for the binary glasses $K_2Si_2O_5$ and $Cs_2Si_2O_5$ are also shown for comparison.

sixfold geometry. While at face value there is some inconsistency between EXAFS and XANES, there seems to be little doubt that K-O distances are longer in aluminosilicate glasses than in silicates.

3. Mixed-Alkali Effect

When more than one alkali is introduced into the same glass (e.g., $[K_xCs_{1-x}]_2Si_2O_5$), although the density varies linearly with composition, there are striking departures from Vegard's law for ionic conductivity and viscosity (Day, 1976, and references therein). The mixed-alkali effect is greatest at $x \sim 0.5$ where the ionic conductivity can be four or five decades lower than in the single-alkali glass ($x = 0, 1$). Interestingly the diffusivity of the majority of alkali is lowered by increasing amounts of the second alkali, suggesting one cation interferes with the transport of the other. The mixed-alkali effect is maximized when the difference in size between the two alkalis is greatest. Because of its element specificity, EXAFS offers a unique tool to establish if this dynamic effect has any structural constraints.

In addition to studying the potassium and aluminum EXAFS of potassium alumino silicate glasses, Jackson and colleagues (1987) also examined mixed-alkali glasses made from mixtures of orthoclase ($KAlSi_3O_8$) and albite ($NaAlSi_3O_8$). As with single-alkali glasses, they found the aluminum to be tetrahedrally coordinated and network-forming, independent of the Na/K ratio. At the same time, they found the coordination number of potassium and the K-O bond length were correlated to the mixed-alkali ratio, both increasing slightly at $x = 0.5$. However, as sodium EXAFS was not reported, it is not clear whether this behavior was common to both alkalis.

In a study of $[K_xCs_{1-x}]_2Si_2O_5$ glasses, (Greaves et al. (1989)) the local structure of each alkali was recorded. Results for one of the glasses ($x = 0.5$) are shown in Fig. 15. The partial RDF of cesium comprises several peaks but the main feature is due to five nearest-neighbor oxygens at 3.00Å. While the detailed Ns and μ^2s of the remaining peaks cannot be unambiguously analyzed, they can be assigned with reasonable confidence to a shell of Si's at 3.6Å followed by an alkali shell near 4.8Å. Figure 15 also illustrates the partial RDF of potassium which comprises a major peak due to five oxygens at 2.56Å. More distant features can be attributed to silicons and alkalis as in the cesium RDF. Figure 15 includes data for $K_2Si_2O_5$ glass. There are clearly differences both in the oxygen shell and in the cation shells. Taken together there is no doubt that the local structure of potassium is affected by the presence of cesium and vice versa. The two alkalis coordinate into the network and possibly with one another.

In the MRN model introduced earlier (Section II.B, Fig. 2), it was proposed that ionic transport takes place along percolation pathways comprising ionically bonded modifiers. At the disilicate composition, this is believed to occur by an interstitialcy mechanism, the geometry of the sites being governed by the geometry of the modifying cations. For a mixed-alkali glass, there is no reason to suppose percolation pathways will be anything other than a random mixture of alkalis and nonbridging oxygens. Transport of alkalis of different sizes will mutually interfere if they share the same percolation pathways. Sites involved in transport are unlikely to be matched to both alkalis which will lower individual mobilities. In particular, the activation energy for a given alkali is expected to increase as its concentration decreases and bad hops become more likely. Statistically, the overall ionic conductivity should fall to a minimum at the 50 : 50 composition, as indeed is observed (Day, 1976).

It is clear from these measurements of potassium and cesium disilicate glasses that there is substantial structural basis to the mixed-alkali effect.

B. CALCIUM ENVIRONMENTS

Of the alkaline-earth modifiers in oxide glasses, chiefly calcium has been studied as of the late 1980s (Geere et al., 1983; Binsted et al., 1985; Hardwick et al., 1985). Taking the calcium local structure in the minerals as a guide, environments of this modifying cation are often more disordered than those of an alkali modifier such as sodium, for instance. Moreover oxygen distributions are not always symmetrical.

Approaches to EXAFS analysis for materials where the partial RDF is asymmetric have been developed by Hayes et al. (1978) and Eisenberger and Brown (1979). These are geared toward r-space analysis and involve parameterizing anharmonicity in the Coulomb potential seen by a neighboring atom. In glasses, site anharmonicity is not known a priori and in any case may be masked by site variability. An alternative approach, particularly suited to k-space fitting, is to employ subshell analysis. In the case of an oxide glass, for example, the oxygen shell is divided into a number of subshells, each is then set initially to the same bond length and coordination number—the Debye-Waller factor being given by the thermal μ^2 obtained from model compounds. The Fourier filtered oxygen shell component of the EXAFS is obtained by windowing the first major r-space peak and back transforming from r into k-space. The subshell bond lengths and coordination numbers are allowed to float until a best fit to experiment is obtained.

Examples of asymmetric Ca environments in crystalline and glassy silicates obtained by EXAFS subshell analysis are shown in Fig. 16. It should be stressed that these are not Fourier transforms, but Gaussian broadened histograms of k-space analysis—each subshell being weighted by its coordination number and broadened by μ^2. Accordingly, the integrated area under the distribution equals the total coordination number of the oxygen shell. There is of course a limit to the number of subshells that are meaningful

G. N. GREAVES

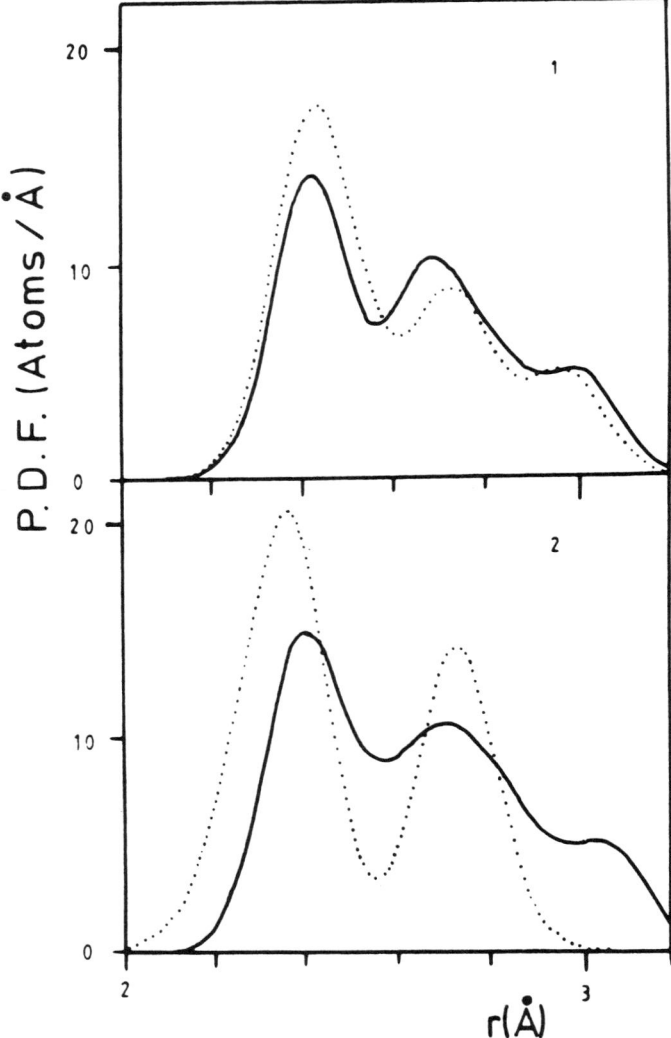

FIG. 16. Pair-distribution functions for Ca-O interactions in anorthite glass (1) and diopside glass (2) (Binsted et al., 1985). These are Gaussian broadened histograms of least squares subshell fitting of the nearest-neighbor oxygens. (See text for details.) Dotted curves are the results for crystalline anorthite (1) and crystalline diopside (2) obtained in the same way.

in this type of analysis. This is governed by the band pass of the EXAFS experiment, $k_{max} - k_{min}$. It is easy to show from Eq. (2) that the minimum separation between subshells, Δr, is given by

$$\Delta r \gtrsim \frac{\Pi}{k_{max} - k_{min}}. \tag{14}$$

Taking a typical EXAFS band pass of 10Å^{-1}, $\Delta r \gtrsim 0.3\text{Å}$. This subshell resolution can be seen in the residual structure of the distributions presented in Fig. 16. In certain cases, it is worthwhile analyzing whole EXAFS distributions comprising several shells of atoms. An example of this will be found in Fig. 27 (Section VII.C).

The oxygen distributions surrounding Ca illustrated in Fig. 16 are for anorthite ($CaAl_2Si_2O_8$) and diopside ($CaMgSi_2O_6$) and the respective glasses prepared by rapid quenching from the melt. Samples were powdered and measured in transmission. The structures of the two minerals are crystallographically quite different. Anorthite is a three-dimensional network in which Ca^{2+} ions occupy irregular sevenfold positions balancing the negatively charged feldspar rings $(Al,Si)_2O_4^{2-}$ (Wainwright and Starkey, 1971). By contrast, diopside has a pyroxene structure characterized by infinite chains of SiO_4 units. These are cross-linked by Mg and Ca in six- and sevenfold sites (Clarke et al., 1969). Accordingly, although the coordination number of Ca is similar in the two materials, the detailed local environment is very different. This can be seen in Fig. 16, the EXAFS oxygen distributions agreeing closely with X-ray crystallography results (Binsted et al., 1985). In particular, the mean Ca-O bond length in anorthite (2.6Å) is longer than in diopside (2.50Å) and this is reflected in the densities—anorthite being 14% less dense than diopside.

It is noteworthy that the densities of the two glasses are both very similar to the density of anorthite. Figure 16 demonstrates that the Ca environments in the two glasses are also almost identical, sharing the same coordination number of seven oxygens and the same triangular asymmetry and spread. The mean Ca-O bond length in the glasses is 2.63Å, with the maximum occurring at 2.41Å. Static disorder in the glasses amounts to a spread in bond lengths about the mean of 0.23Å, which is more than twice the Ca-O thermal disorder of 0.09Å measured at room temperature in CaO (Fig. 1). However, there is a significant difference between the calcium environment in diopside and in diopside glass which qualitatively explains the difference in density. Evidently the regular silicate chain structure in the crystalline state is replaced in the glass by a network that is more three-dimensional in character. Figure 16, therefore, illustrates the strength of EXAFS to explore internetwork structure. By contrast, X-ray diffraction and Raman spectroscopy, being mainly sensitive to infranetwork structure, reveal little difference between the crystalline and amorphous state in these materials.

Other studies of Ca EXAFS in silicate glasses include Wollastonite ($CaSiO_3$) glass (Geere et al., 1983) where the peak in the oxygen distribution was found at 2.38Å with a spread of 0.2Å. In a basaltic glass $Ca_3Mg_4Al_2Si_7O_{24}$ Hardwick et al. (1985) report an average Ca-O distance of 2.48Å and a coordination number of 9.

C. LOCAL STRUCTURE OF INTERMEDIATES

While sodium and calcium have ionic radii close to 1Å, those of transition metals such as iron, for instance, are significantly shorter. The corresponding metal–oxygen bonds range from 1.8 to 2Å in length and as such are intermediate between Na-O or Ca-O which are around 2.4–2.6Å in length and those of Si-O and Al-O bonds which are 1.6–1.7Å in length. Accordingly, transition metals are commonly classed as intermediates: they coordinate with oxygens sometimes octahedrally and sometimes tetrahedrally depending on oxidation state and glass composition. XAS techniques are particularly suited to differentiate between modifying and network environments, especially as intermediates are commonly a minority component in the glass. Measurements are conveniently made in transmission, although fluorescence detection is needed for dilute quantities. Transition metals were some of the first metals in oxide glasses to be investigated by EXAFS (Brown et al., 1978). Octahedral and tetrahedral bond lengths are readily differentiated in the period of the EXAFS oscillations. Near-edge structure is also valuable in this context. (See Section II.A.) The position of the absorption threshold indicates the oxidation state of the metal. The initial XANES provides clues as to the local symmetry; this is particularly true of the pre-edge feature characteristic of 3d metal oxide K-edges. This feature should increase in intensity in the configurational order: regular octahedron → distorted octahedron → tetrahedron. Figure 17 illustrates these points for the case of ferric iron in aegerine where Fe^{3+} is octahedrally coordinated and its glass where it is tetrahedrally coordinated.

1. Titanium

The addition of TiO_2 to silica and silicate glasses serves to reduce the expansion coefficient to close to zero. It also acts as a nucleating agent. Sandstrom and colleagues in a series of interesting studies have examined the K-edge XANES and EXAFS of Ti in TiO_2-SiO_2 glasses prepared by flame hydrolysis (Sandstrom et al., 1980; Gregor et al., 1983). TiO_2 and Ba_2TiO_4 were used to reference octahedral and tetrahedral Ti sites, respectively. Sandstrom's group found that the oxidation state in SiO_2 was always Ti^{4+}. Nevertheless, changes in coordination occurred with concentration so that

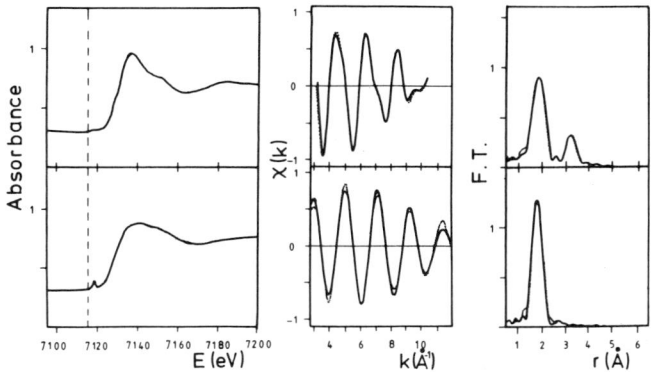

FIG. 17. The K-edge XANES, EXAFS, and Fourier transform of Fe^{3+} in aegerine ($NaFeSi_2O_6$) crystal (top) and glass (bottom). (From Binsted et al., 1985.)

below 7 wt % TiO_2 all of the Ti was fourfold coordinated with a Ti-O bond length of 1.81Å. At higher concentrations, some sixfold coordinated Ti occurred with a Ti-O bond length of 2.15Å—the proportion increasing roughly linearly with concentration to 14 wt % where upon TiO_2 began to crystallize out. At the onset of crystallization, 30% of the Ti sites were found to be sixfold coordinated. This behavior is truly intermediate with fourfold Ti adding to the SiO_2 network and sixfold Ti acting as a traditional modifier. Incidentally, at very low concentrations of TiO_2 ($\leqslant 0.5$ wt %), Sandstrom's group found that only sixfold Ti sites were present in SiO_2. At this level, Ti presumably acts as a chemical defect, occupying voids in the silica network. Figure 18 beautifully shows the changes occurring in the Ti K-edge XANES as the titanium concentration is increased. Tetrahedral titaniums enhance the pre-edge feature (Gregor et al., 1983).

Dumas and Petiau (1986) have studied Ti environments in aluminosilicate glasses, observing similar intermediate behavior. In particular, they investigated the nucleating properties of Ti and found that with progressive heat treatment, the local environment of Ti changes gradually from tetrahedral to octahedral symmetry, with Al_2TiO_5 being the first titanium-containing phase to devitrify. The ambivalent structural role of titanium evidently aids nucleation in oxide glasses.

2. Vanadium

Unlike titanium, the oxidation state of vanadium in oxide glasses is affected by the redox state of the melt. A mixture of V^{4+} and V^{5+} centers is generally formed. At sufficient concentration these will support small polaron transport, the electron hopping from V^{4+} to V^{5+} sites (Greaves, 1973).

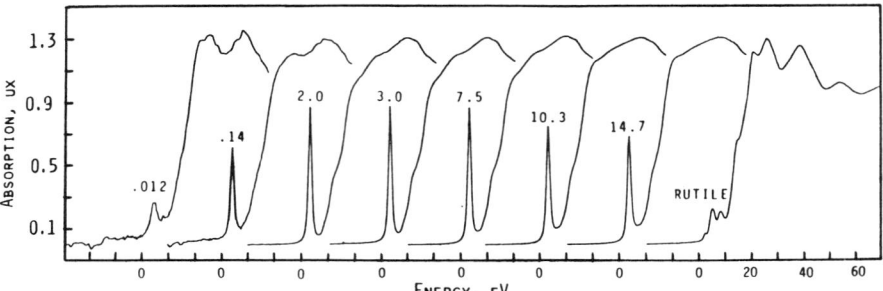

FIG. 18.   Titanium K-edge XANES in TiO_2-SiO_2 glasses. Note the changes in the pre-edge feature reflecting the balance between tetrahedral and octahedral Ti^{4+}. (Gregor et al., 1983.)

Bianconi and colleagues have made a series of XAS studies of vanadium in various glasses, including V_2O_5-P_2O_5 glasses (Stizza et al., 1984; Benfatto et al., 1985). In this system, the V^{4+}/V^{5+} ratio changes with composition. At low concentrations of V_2O_5, V^{4+} predominates and the near-edge structure of vanadium approaches that of VO_2 for which vanadium is sixfold coordinated. As the V_2O_5 content is increased, the vanadium XANES becomes a broadened version of V_2O_5 which contains fivefold coordinated vanadium. This suggests a gradual change for vanadium from a modifying to a network-forming role.

3. Iron

Iron, like vanadium, exists in different oxidation states—chiefly Fe^{2+} and Fe^{3+}—dependent on the glass composition and on the melting conditions. For example, Calas and Petiau (1983) conclude from XANES measurements of pyroxenes that in reduced glasses ferrous iron generally occupies fairly regular octahedral sites, evidently behaving as a modifier. The same conclusion is reached from the study of optical absorption and Mössbauer spectra.

In oxidized glasses, all the indications are that role reversal takes place and that ferric iron adopts tetrahedral symmetry and contributes to the network (Brown et al., 1978). This behavior is quite distinct from the crystalline state where Fe^{3+} is usually sixfold coordinated. Binsted et al. (1985) have looked in detail at a series of iron silicate glasses and minerals. Metallic iron, fayalite (Fe_2SiO_4), and aegerine ($NaFeSi_2O_6$) were used as model compounds for Fe^0, Fe^{2+}, and Fe^{3+}, respectively: viz., to establish chemical shifts, XANES fingerprints, and EXAFS parameters. For instance, the chemical shifts for Fe^{2+} and Fe^{3+} are 8 eV and 13 eV compared to Fe^0. Mean Fe^{2+}-O and Fe^{3+}-O distances for octahedrally coordinated iron are 2.22 and 2.03Å, respectively. Fe K-edge XANES and EXAFS for aegerine and its glass are contrasted in Fig. 17. The XANES for the glass is rather featureless apart from the prominent pre-edge feature indicative of hybridized bonding characteristic of tetrahedral coordination. This feature is absent in aegerine where Fe^{3+} occupies octahedral sites. Corroborative evidence for tetrahedral coordination in aegerine glass comes from the EXAFS spectra also included in Fig. 18. Notably the mean Fe-O distance analyzed for the glass is 1.86Å compared to 2.03Å for the crystal. Moreover, the spread in Fe^{3+}-O distance for aegerine glass is much narrower: 0.06Å compared to 0.11Å for the crystal. This can be seen in the sharper peak of the Fourier transform. There is no doubt the majority of iron sites in the glass are tetrahedral. The absence of any other peaks in the partial RDF of the glass confirm that iron does not form clusters in the structure but integrates into the silicate network.

A similar picture has emerged for Ca-bearing glasses such as $CaFeSiO_4$ and $CaFeSi_2O_6$. However, Fe-O distances are slightly longer than for the Na-bearing glass (1.92Å compared to 1.86Å), suggesting that some of the iron may be present in octahedral coordination—possibly as a minor fraction of Fe^{2+}.

The environment of ferric iron in lead phosphate glasses has been examined by Greaves et al., 1988A). The structural role of Fe^{3+} is quite different from its behavior in silicate glasses and this has important implications for the chemical resistance of these glasses (Sales and Boatner, 1984). In particular, when 5 to 10 wt % Fe_2O_3 is added to lead metaphosphate glass, the resistance to aqueous attack improves dramatically. EXAFS measurements at the Fe K-edge and the Pb L_{III}-edge demonstrate that both fulfill glass-modifying roles but their individual environments are strikingly different. This is a nice example of the power of the technique to differentiate between cations of similar function. Iron is found to be octahedrally coordinated with a mean Fe-O bond length of 1.95Å. The root mean square spread of oxygen distances is 0.09Å, which is narrower than in crystalline Fe_2O_3, for instance. Lead, on the other hand, has a coordination number of 8 and a mean Pb-O bond length of 2.49Å, while the spread in Pb-O bond length of 0.24Å is considerably broader than in crystalline PbO_2. Gaussian broadened histograms of Fe-O and Pb-O distances are shown in Fig. 19a. The network architecture in phosphate glasses is beautifully revealed through liquid chromatography (Sales et al., 1986). Iron-lead-phosphate glasses are comprised of long polyphosphate chains which shorten in length as iron is added. The EXAFS results demonstrate that Fe and Pb serve to cross-link the phosphate anions as illustrated in Fig. 19b. This is a modified random network, and in this context cations will migrate along the channels between the polyphosphate chains. Clearly the presence of iron centers in the glass will create "knots" in the percolation pathways impeding mobility as I established earlier for mixed-alkali silicates (Section VI.A.3). While aqueous corrosion in

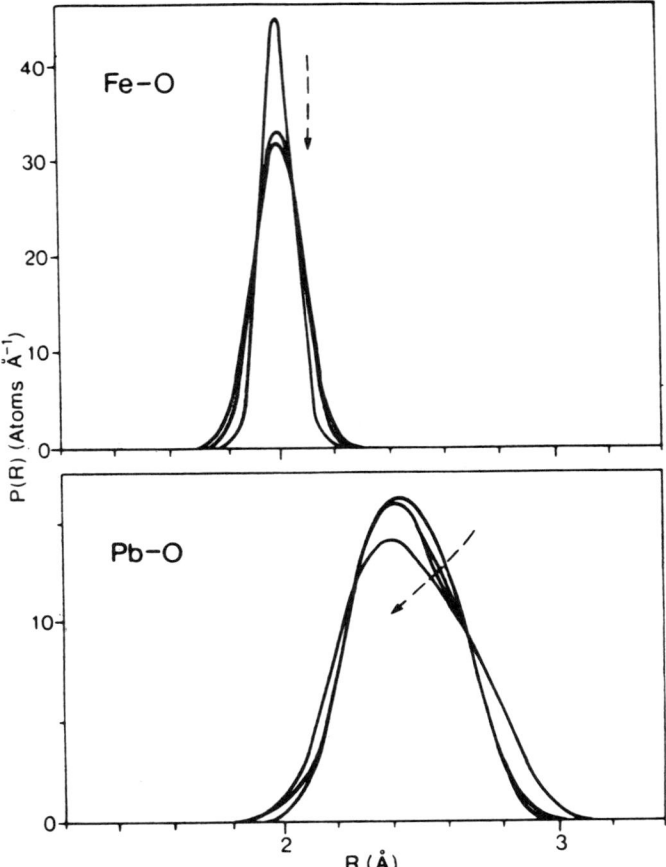

FIG. 19. (a) Gaussian broadened histograms for the oxygen shells of iron and lead in iron-lead-phosphate glasses. Arrows indicate the small trends evident with increasing iron content. (Greaves et al., 1988a.)

glasses terminates in the dissolution of the network, it is initiated by ion exchange of modifiers for protonated water diffusing from the surface into the bulk. In the MRN model, this will occur along the modifier channels and, clearly if these are constricted by modifiers of different size being present, the hydration process will be restricted and corrosion resistance should improve, as observed.

VII. Applications

The bulk of this chapter has been concerned with the contribution XAS techniques are making to solving the classical problems of the structural

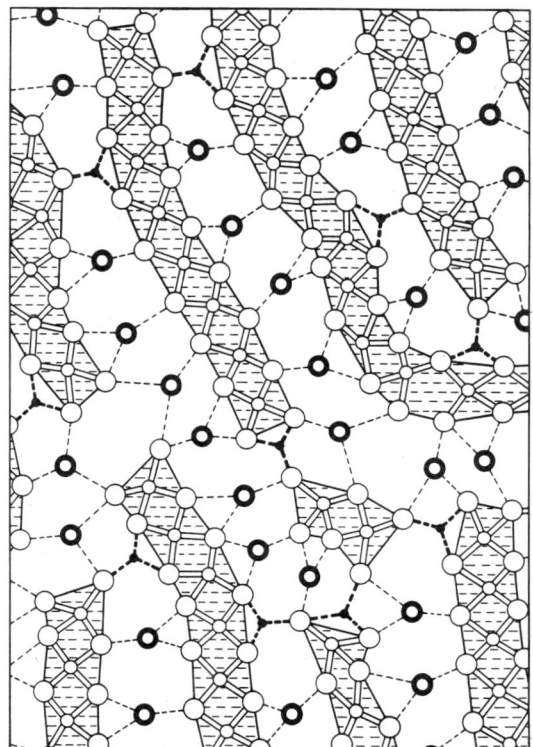

Fig. 19. (b) Modified-random-network model for iron-lead-phosphate glass. Polyphosphate chains are shown edge sharing for convenience. These are cross-linked by lead atoms (open circles) and iron atoms (filled circles), the latter forming "knots" in the percolation pathways which will inhibit diffusion and, by the same token, the onset of aqueous corrosion.

chemistry of glasses: the geometry of network formers and the degree of structural and chemical order, modifier environments and their relationship to the overall structure and properties of oxide glasses, the structural role of the intermediates, and so on. These findings have been interpreted in the context of continuous and modified random-network models for glass structure. This section will show how these ideas can be applied to three rather different diffusion related problems: photodiffusion of silver in chalcogenide glasses, corrosion of biogenic concretions, and leaching of uranium in nuclear-waste glasses.

A. PHOTODIFFUSION

Transition metals alloyed with chalcogenide glasses are frequently mobile and dramatically affect the electronic properties. Early EXAFS measure-

ments on the alloying of copper with As_2Se_3 glass at the level of 5 to 10%
revealed that the metal was covalently bonded to four seleniums (Hunter and
coworkers, 1977). The most likely way for this bonding configuration to be
realized would be through utilization of lone-pair electrons from the
neighboring seleniums. In this sense copper could be envisaged as a glass
former, although the seleniums it is bonded to would also be shared by arsenic
atoms. Mott (1976) and Kastner (1979) have argued that the increase in
coordination of the chalcogen implied by this arrangement should affect the
balance of charged intrinsic defects postulated for these glasses (D^+D^-
centers or VAPs), thereby modifying the position of the Fermi level. This
would then explain the huge increase in electrical conductivity at room
temperature exhibited by chalcogenide glass doped in this way. Transition
metals, however, are often extremely mobile and will readily diffuse from the
glass surface into the bulk either thermally or under the action of band gap
light. There is very little lateral diffusion in the photochemical change and, as
the metal alters the chemical etch rate of the glass, photodiffusion in
chalcogenide glasses offers potential in the area of high-resolution lit-
hography. Ag, Cu, Zn, In, and other metals exhibit this phenomenon, out of
which Ag has been studied the most.

The EXAFS of silver measured at the K-edge (25.5 keV) for an annealed
glass ($AgAs_4S_6$) and a photodoped film are compared in Fig. 20—taken from
Steel et al. (1988). Several glasses and photodoped films were measured and
the results were found to be practically the same. The compositions of the
glasses were chosen to be close to those of the photodiffused films. These were
made by overlaying 5000–6000Å films of As_2S_3 with 1000Å of Ag and then

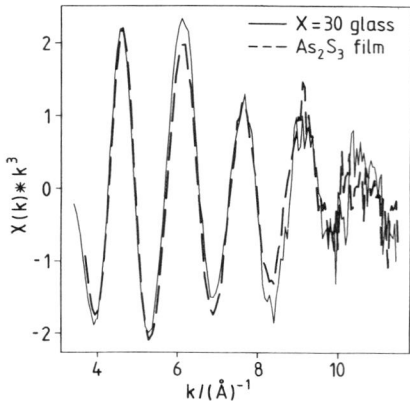

FIG. 20. Silver K-edge EXAFS for the glass $AgAs_4S_6$ compared to that of a photodif-fused As_2S_3 film. (Steel et al., 1988.)

irradiating with an IR filtered mercury lamp. From a comparison of the edge positions of the glasses and films with crystalline silver sulphides, silver was established to be monovalent in all cases. Analysis of the data illustrated in Fig. 20 reveal a single shell of sulphur atoms at 2.50Å with a Debye-Waller factor μ^2 of 0.01Å^2 (Steel and coworkers, 1988). Clearly, silver is covalently bonded in amorphous As_2S_3, whether it is alloyed or photodiffused. Evidently it does not behave as a glass modifier in the strict sense of the term, although (like alkalis in oxide glasses) silver readily diffuses. It is also important to realize that, while the overall coordination number of silver in amorphous As_2S_3 is similar to crystalline sulphides such as Ag_2S and Ag_3AsS_3 (Wyckoff, 1964), the latter materials exhibit considerable disorder and asymmetry in the chalcogen shell. (FWHMs are typically four times as great as the glass.) The regular sulphur coordination sphere of silver in amorphous As_2S_3 suggests that silver is located in between the molecular-like units prevalent in the evaporated films (Section V.B.1). As we have seen, significant reordering occurs in films of As_2S_3 when band gap light is absorbed (Fig. 11b). If these photostructural changes occur in the presence of silver, diffusion is likely to be aided by what will amount to an "opening up" of intermolecular pathways.

The photodiffusion of silver has also been studied in Ge chalcogenide glasses and films (Rennie, 1986). As with As_2S_3, the silver K-edge EXAFS for annealed $GeSe_2$ glasses and photodoped films are almost the same. The measured Ag-Se bond length in $GeSe_2$ is 2.60Å and the coordination number is approximately 3. The same is true for $GeSe_2$ films alloyed with silver by thermal diffusion. If the environment is almost the same when silver is introduced into the film, the diffusion mechanism will most likely be similar. In amorphous $GeSe_2$, as in As_2S_3, it would appear silver is mobile in the regions between molecular units in the base glass. In view of the EXAFS results on photostructural effects in germanium chalcogenides (Section V.B.2), it would appear that photodiffusion of silver may well be promoted by the conversion of mesomeric Se-Se bonds into primary Ag-Se bonds.

Figure 21 demonstrates the potential of glancing-angle geometry (Section III.D) for studying photodiffusion. The Fourier transform for the silver K-edge EXAFS before and after illumination is shown for a 300Å silver film evaporated onto 0.5 μm of a-$GeSe_2$. Measurements were made close to the critical angle for total external reflection. Before illumination there are two principal features in the silver RDF: a peak close to 2Å due to Ag-O correlations characteristic of Ag oxide and a peak at 2.9Å due to the Ag-Ag distance in metallic silver. It is well known that silver metal carries an oxide coating of several atomic layers. After illumination with band gap light, silver should diffuse approximately 2000Å into the $GeSe_2$ film. In the glancing-

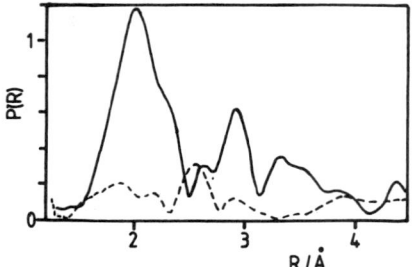

FIG. 21. Phase-shift–corrected Fourier transform for silver K-edge EXAFS before (——) and after (– – –) photodiffusing into a 5000Å GeSe₂ film. The measurements were made at glancing angles. Illumination was in situ at room temperature.

angle experiment, the silver K-edge threshold is observed to drop significantly, resulting in increased noise in the fine structure, which is responsible for the lower amplitudes in Fig. 21. Nevertheless the qualitative change in the silver RDF is striking. While "ghosts" of the Ag-O and Ag-Ag peaks persist indicating photodissolution is not entirely complete at the surface, the major feature is now at 2.6Å, which clearly signifies the presence of the Ag-Se bonds associated with dissolved silver.

Silver is of course also highly conducting in many oxide glasses. Dalba and coworkers (1986) have reported silver K-edge EXAFS from silver borate glasses around the composition $Ag_2O(B_2O_3)_5$. They find a low coordination number for silver of around 2 and a bond length of 2.27Å, indicating a fair degree of covalency. The analogy with the alloying of silver with chalcogenide glasses is tempting. Certainly silver cannot be considered as a conventional oxide modifier; the low coordination number rules this out. At the same time, the Ag-O bond length (which is longer than in Ag_2O) suggests that the metal is bonding to the borate network probably by utilizing the lone-pair orbitals of neighboring bridging oxygens.

B. Corrosion of Intracellular Granules

For many organisms, part of the system for regulating intracellular ions appears to involve the deposition of minerals within membrane bound vesicles in the cytoplasm of particular cells. In the context of the present chapter, the fact these deposits are both inorganic and amorphous makes them especially relevant. The best studied of these systems has been the $CaMgP_2O_7$ granule, which occurs in a number of organisms but which has been characterized in the hepatopancreas of garden snails (Howard et al., 1981). However, when a toxic cation such as Mn^{2+} is presented in the diet or injected into the blood, this becomes rapidly assimilated into the gut and is incorporated into the granules, modifying their morphology, composition, and structure. This can be appreciated from Fig. 22, which shows electron micrographs of doped and undoped deposits. If the toxicity loading is severe, the cells deteriorate, appearing in the faeces, after which the animal normally

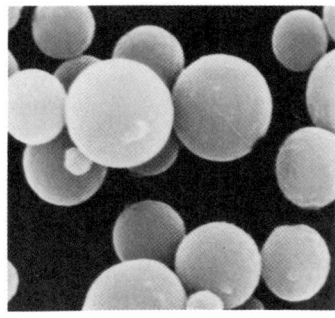

a

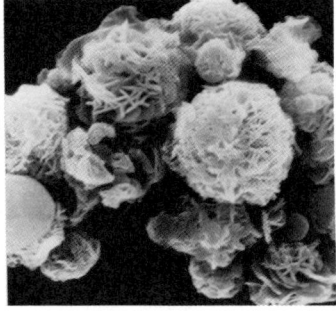

FIG. 22. Scanning-electron micrographs of amorphous intracellular granules isolated from the hepatopancreas of garden snails. (a) Granules from normally fed animals. (b) Granules from animals whose diet was sprinkled with manganese carbonate. (Taylor et al., 1988.)

b

survives and rapidly recovers. The structure and stability of the intracellular granules are clearly pivotal in this detoxification cycle.

The K-edge EXAFS of both Ca and Mn have been studied in deposits from untreated and treated snails. Gaussian broadened histograms of the respective oxygen shells obtained by subshell fitting (Section VI.B) are presented in Fig. 23. The mean Ca-O and Mn-O distances are 2.30Å and 2.17Å, respectively. There is also evidence in the complete EXAFS for a shell of phosphorous atoms at 3.2–3.3Å. The overall microscopic structure of these amorphous deposits is likely to have much in common with the phosphate MRN shown in Fig. 19b, metal ions cross-linking the polyphosphate chains. From infrared measurements and also phosphorous K-edge XANES, the dominant chain length is 2, viz. $P_2O_7^{2-}$. The presence of some water in the granules is evident in the cation EXAFS from the lack of cation–cation correlations. It is possible water also contributes to the interchain structure.

The difference between the Ca and Mn sites suggests toxic doping does not take place by a simple ion-exchange process. The striking changes in the morphology of the granules evident in Fig. 22 confirm this. The new precipitation on the doped granule surfaces is rich in manganese, but the

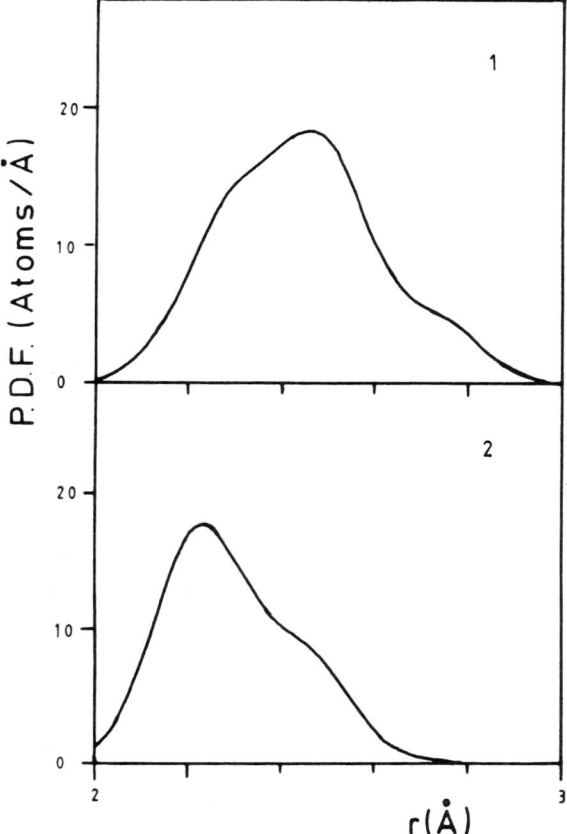

FIG. 23. Gaussian broadened histograms for the oxygen shells of calcium (1) and manganese (2) taken from undoped and doped amorphous intracellular granules illustrated in Fig. 22. (Greaves et al., 1984.)

granule centers are heavily depleted of calcium. Interestingly, these changes are not found in abiological pyrophosphates and this underlines the way the membrane surrounding intracellular granules controls the passageway of metal ions.

An attractive interpretation of these results is that they are associated with a corrosion process (Taylor et al., 1988). Suppose that manganese passes through the granule membrane as a hydrated ion and that it is capable of dissociating to liberate a hydrated proton. (Evidence for this comes from in vitro experiments indicating that manganese induces an increase in pH.) The released proton could then ion exchange with calcium and magnesium ions which should diffuse via percolation channels to the surface where reprecipitation could occur. Here manganese would be more soluble than magnesium but less soluble than calcium—manganese should reprecipitate and calcium should leave. This compositional redistribution is confirmed by electron microprobe measurements (Howard et al., 1981). These reveal reprecipitation of amorphous manganese magnesium phosphate occurs all of which point to a rapid efflux of calcium from the membrane into the cytoplasm. It is well known that calcium can be pharmacologically very toxic to cells when its intracellular concentration exceeds certain levels. The tissue becomes soft; the cell dies off and is discharged from the animal. This model of metal-induced aqueous corrosion then also provides an explanation for the biochemical detoxification function of the intracellular granules.

Little is known about the role of the amorphous state in biological mineralization. XAS techniques have an important contribution to play in elucidating the physical and chemical properties of these intriguing materials.

C. LEACHING OF URANIUM AT THE SURFACE OF BOROSILICATE GLASSES

As of 1989, aqueous corrosion of borosilicate glasses represents the major limitation on the use of these materials for the safe long-term storage of nuclear waste. During corrosion a surface region is formed whose properties can be significantly different from those of the bulk glass. Qualitatively it is now reasonably certain that corrosion is initiated by the depletion of alkali ions from the surface and the influx of water, culminating in the hydrolysis and subsequent dissolution of the glass network (Douglas and El-Shamy, 1967; Clark and Hench, 1983). For a simple oxide glass the depletion layer can extend several microns into the glass surface after an hour's leaching in water at 70 or 80°C (Frischat, 1975). By this stage, dissolution of the network is the dominant leaching mechanism. Extensive corrosion leads to the establishment of a gel layer which is usually followed by the formation of an insoluble—possibly crystalline—layer at the surface, as Schreiber's group has demonstrated (Schreiber et al., 1985). This may result from the precipitation of the least soluble species at the surface—which is the case for iron-lead-phosphate glasses (Section VI.C.3)—or alternatively it may recondense out of the gel layer. The corrosion of uranium in borosilicate glass appears to follow the latter course.

Although sodium is depleted from the glass surface, the uranium content is enhanced. This can be measured quite easily under conditions of total external reflection of X-rays (Section III.D) by monitoring the size of the fluorescence signal at the absorption edge of uranium. Figure 24 shows the rise in the relative step height of the L_{III}-edge with leaching. The main purpose of measuring XAS at glancing angles, however, lies in its ability to probe the local environments of metals in the glass surface (Thornley et al., 1986). These will alter with diffusion, hydration, and recondensation. As of the late 1980s,

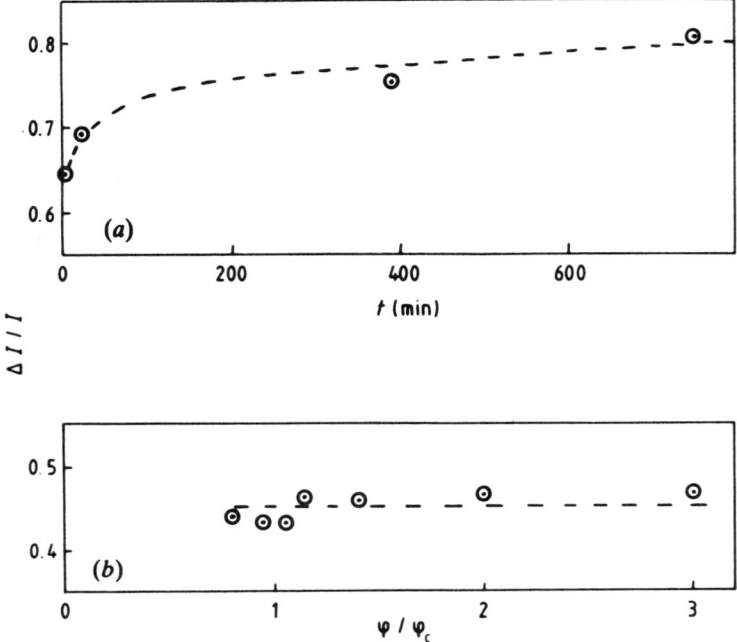

FIG. 24. Normalized step height $\Delta I/I$ of the U L_{III} edge for a borosilicate glass (Thornley et al., 1986). Close to the critical angle ϕ_c, this increases with leaching time indicating that uranium is accumulating at the surface. For an unleached glass, $\Delta I/I$ is constant as a function of ϕ/ϕ_c, demonstrating uranium is uniformly distributed in depth.

the structural chemistry underlying corrosion processes in glasses has been studied using analytical methods which, for the most part, give no direct structural information. Moreover many of the techniques require a high vacuum sample environment, precluding in situ experiments. At hard X-ray energies XAS can be measured in air under natural conditions. This is a particular advantage for studying aqueous corrosion because water is adsorbed into the surface and any subsequent evacuation of the specimen necessarily alters the composition and structure of the leached layer. Furthermore, without X-ray reflection techniques, depth profiling of compositional variations can often only be achieved destructively by successive etching or sputtering of the glass surface. Using glancing-angle X-rays, changes in structure can be recorded nondestructively as a function of depth by varying the angle of incidence, ϕ, of the incoming X-rays. Figure 5 shows the variation of X-ray penetration depth, z, with ϕ for borosilicate glass at the energy of the uranium L_{III}-edge, 17.2 keV.

The surface-sensitivity of the glancing-angle geometry can be appreciated in Fig. 25, which is taken from the author's work (Greaves et al. 1989). Here the uranium L_{III} EXAFS spectra for 0.2 at % U in sodium borosilicate glass are compared. Spectrum (c) was measured at $\phi = \pi/4$ registering the bulk structure of the glass, while spectrum (d) was measured at $\phi = 3.6$ mrad from a leached glass. For this angle of incidence, $\phi/\phi_c = 2$, making the depth of penetration of X-rays 1.4 μm. The difference compared to the bulk structure is substantial. Both spectra were measured using fluorescence detection (Section III.B) as other XAS detection modes are insensitive to this level of dilution. Also included in Fig. 25 are spectra for uranium metal (a) and UO_3 (b). Structure in $f(\mathbf{k}, \pi)$ for uranium neighbors can clearly be seen in the metal fine structure. U-U correlations in UO_3 can be seen in the EXAFS,

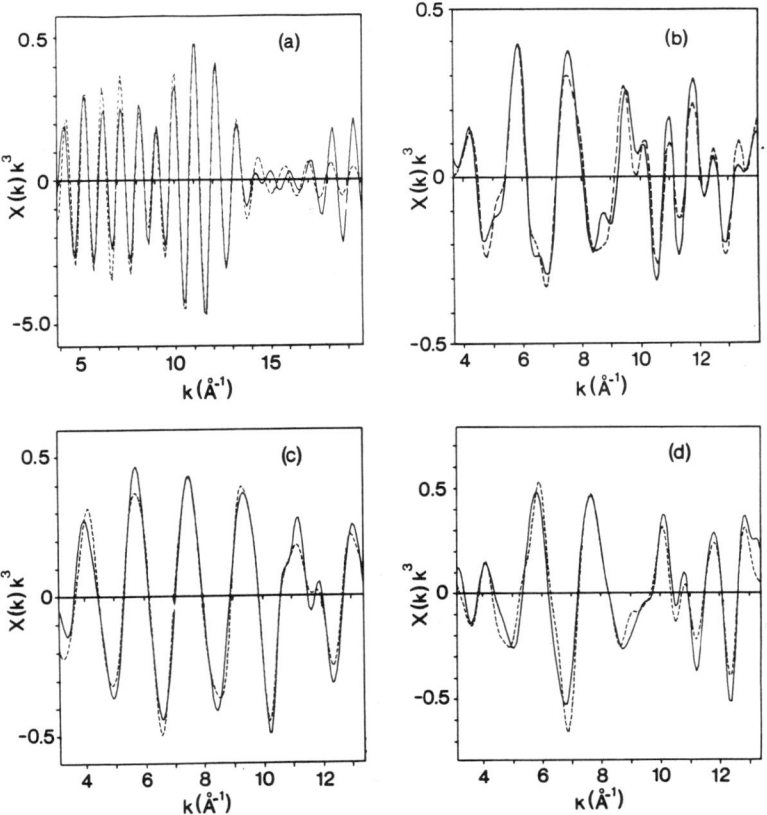

FIG. 25.   Weighted normalized U L_{III} EXAFS for (a) uranium metal, (b) UO_3, and uranium containing borosilicate glass measured (c) at $\phi = \pi/4$ and (d) at $\phi/\phi_c = 2$ for a leached glass (Greaves et al., 1989). The differences between (c) and (d) demonstrate the surface-sensitivity of the glancing-angle geometry. Dotted lines are the result of CWT least squares fitting.

building up around 12Å^{-1}. The success with which EXAFS can reproduce the crystallographic RDFs can be seen in Fig. 26 for uranium metal and UO_3 data. Both structures exhibit considerable disorder and the subshell fitting routine described earlier (Section VI.B) has been used. In analyzing the uranium L_{III} EXAFS for the glasses, a simple modelling scheme was introduced: distances from 1.6 to 2.7Å were ascribed to oxygen shells, structure between 2.7 and 3.3Å was assumed to be due to U-Si correlations, while all distances out to 5Å were taken as uranium shells.

The uranium RDF analyzed from the EXAFS of the bulk glass (Fig. 25c) is shown in Fig. 27a where the modelling scheme is indicated. The most obvious feature of this RDF and those at the surface (b–f) collated in Fig. 27 is the split shell of nearest-neighbor oxygens. In the bulk glass structure there are just two distinct oxygen distances: a subsiduary subshell at 1.95Å with the majority of oxygen neighbors centered at 2.29Å. The total oxygen coordination number is 7.(4). Uranium in oxidized borosilicate glasses is present as U^{6+}, judged from the L_{III}-edge position. The oxidation state and the

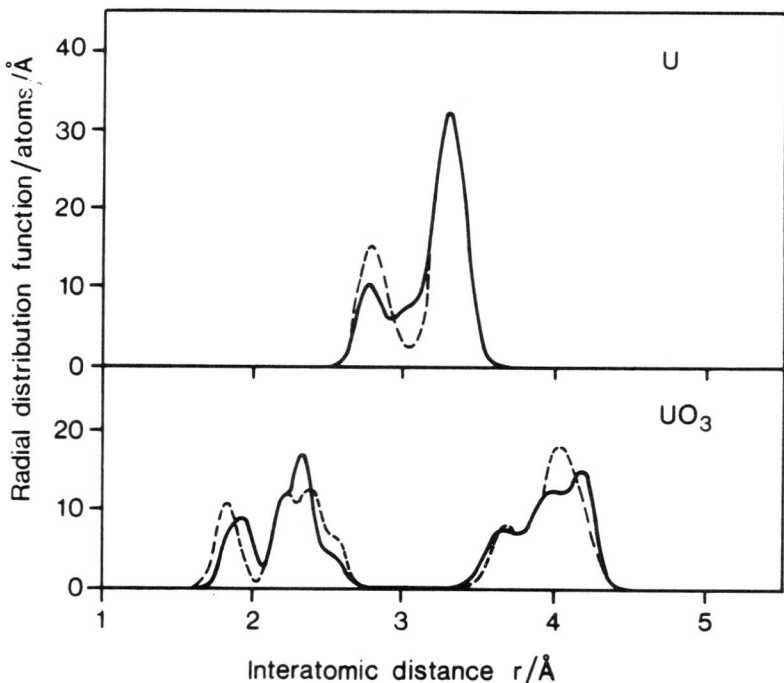

FIG. 26.   Gaussian broadened histograms for the uranium environment in uranium metal and UO_3 taken from fitting spectra (a) and (b) in Fig. 25. Dashed curves are the crystallographically determined RDFs.

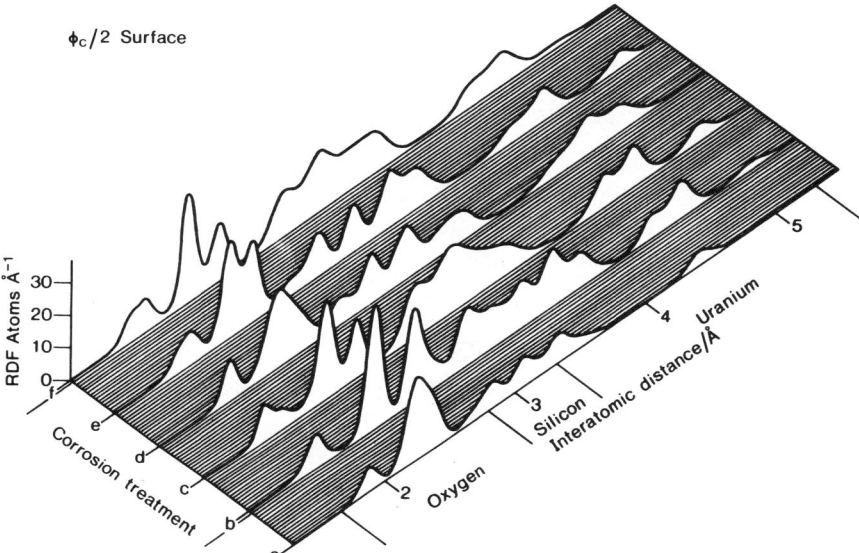

φc/2 Surface

FIG. 27. Gaussian broadened histograms for the uranium environment in the surface of borosilicate glasses (Greaves et al., 1989). These were obtained by CWT least squares fitting U L_{III} EXAFS. The bulk structure is shown in (a) which was measured at $\phi = \pi/4$. The remaining distributions were obtained at $\phi/\phi_c = 0.5$ where the X-ray penetration depth is a few 10sÅ (Fig. 5). Histograms (b)–(f) correspond to different amounts of leaching at 100 C: (b) polished surface, (c) 15 min. leaching, (d) 15 min. leaching, followed by 12 hours drying in vacuum, (e) 30 min. leaching, and (f) 90 min. leaching. The modelling scheme used to identify near neighbors is shown alongside the y axis.

oxygen shell splitting are the hallmark of crystalline uranyl compounds, suggesting that uranium in borosilicate glasses has a similar distorted octahedral oxygen coordination sphere. In the crystalline state, uranyl bonding promotes layered structures (Weigel, 1986). The axial and equatorial bond lengths are influenced by the outer environment: the types of cation neighboring uranium and the degree of interlayer bonding. The distances just quoted for the bulk glass (Fig. 27a) correspond closely to the uranium environment in alkali uranates such as Na_2UO_4 (Kovba, 1971).

The shape of the oxygen shell and also the detection of uranium shells have previously been reported in transmission EXAFS studies of glasses containing rather more uranium ($\gtrsim 5$ at %) by the groups of Knapp (1984) and Petiau (1986). However, with the dilution of uranium at the 0.2 at % level, if the metal were evenly dispersed throughout the glass, the average U–U separation would be between 15 and 20Å. Accordingly, with spacings detected in the 3.3 to 5Å range, some clustering must be taking place in the bulk glass. A striking feature of Fig. 27, however, is that clustering of uranium is exaggerated at the surface of the glass by corrosion treatment. This observation matches the increase in uranium at the surface detected by the rise in the fluorescence signal (Fig. 24). The microscopic aggregation of uranyl-like groups in oxide glasses can be understood by reference to the modified random-network model shown in Fig. 2. Uranium, by virtue of its oxidation state and coordination number, will behave as a modifier and along with sodium should congregate into islands or filaments within the covalently bonded random network. Clearly, if uranium is restricted to these areas, U–U correlations will be more in evidence. As a modifying cation U^{6+} should also bond to the network and a silicon shell around 3Å in the RDF is expected as we observed for cesium and potassium in silicate glasses (Section VI.A.2). Because uranium in this borosilicate glass is far less concentrated than sodium, we expect it will diffuse less readily, which explains why initially sodium is preferentially leached from the glass.

The uranium RDFs presented in Fig. 27 labelled b–f were analyzed from data obtained at glancing angles where $\phi/\phi_c = 0.5$ and the penetration of X-rays is reduced, in principle, to 40Å (Fig. 5). Surface-sensitivity, however, is also related to surface quality, and for glasses this visibly deteriorates with extended leaching. Nevertheless, provided the surface figure is retained, an increase in roughness is not likely to seriously impair the surface-sensitivity. Indeed, this study demonstrated considerable distinction between $\phi/\phi_c = 0.5$ and $\phi/\phi_c = 2$ data while at the same time revealing vestiges of the same corrosion processes at 1.4-μm penetration as at 40Å but in a slower cycle (Greaves et al., 1989).

Apart from the increase in the U–U contribution to the RDF compared to the bulk, the most notable distinction at the surface is the extra structure that develops in the oxygen subshells at 1.9 and 2.3Å, suggesting that different types of uranyl-like units are being generated as a result of aqueous corrosion. In particular there is a new configuration with a shorter "axial" oxygen bond at 1.8Å and a longer "equatorial" bond at 2.4Å. This environment closely matches the uranium local structure in $UO_2(OH)_2$—a layered material morphologically similar to sodium uranates (Weigel, 1986). The existence of a hydrated oxide environment in the leached glass is consistent with the MRN interpretation just given where, in the initial stages of leaching, Na^+ ion exchanges for H_3O^+. Attention is drawn to RDFs c and d in Fig. 27 which both correspond to glasses leached for 15 minutes. They differ, in that glass d was held under vacuum for 12 hours rather than being measured undried as was the case for glass c. The $UO_2(OH)_2$ configuration disappears and the alkali uranate environment found in the bulk is restored. It is also interesting to note that the "untreated" polished surface structure (RDF b) is different from all the rest, exhibiting an oxygen coordination sphere that is split into three subshells. This distinctive environment is reminiscent of hydrated uranyl silicates such as Weeksite (Stohl and Smith,

1981) where uranyl groups bond to SiO_4 units to form urano-silicate layers—the alkali and water occupying interlayer regions. As in $UO_2(OH)_2$, the interlayer bonding is weak leaving the axial oxygens shorter than in uranates. The equatorial bonds, on the other hand, tend to be split leaving several around 2.2Å and the remainder at 2.6Å. The as-polished specimen RDF (Fig. 27b) conforms well to this geometry, confirming that surface corrosion is inevitable in conventional polishing using water. We have found that this coating can be etched off in the initial leaching, but it underscores the difficulty, alluded to in Section II.B, of creating glass surfaces with an equivalent chemistry to the bulk.

Finally, attention is drawn to the U–U correlations particularly evident in the surface RDFs. These fall into two groups whose shell radii are approximately $\sqrt{2}R$ and $2R$, where R is the equatorial oxygen bond length. This suggests the clustering of uranyl groups into square-planar island structures. Indeed, semiquantitative agreement can be achieved in predicting the uranium shells from the equatorial oxygen shell distributions. In particular, the relative weights of oxygen and uranium shells can be used to estimate the size and shape of the uranyl clusters. Uranyl clusters derived in this way for the bulk and at the surface are contrasted in Fig. 28. The perimeter oxygens in these square planar islands of modifier are nonbridging, linking the uranyl groups at the edges to the borosilicate network or to adjacent Na_2O components that comprise the majority of the conducting channels in the modified random network.

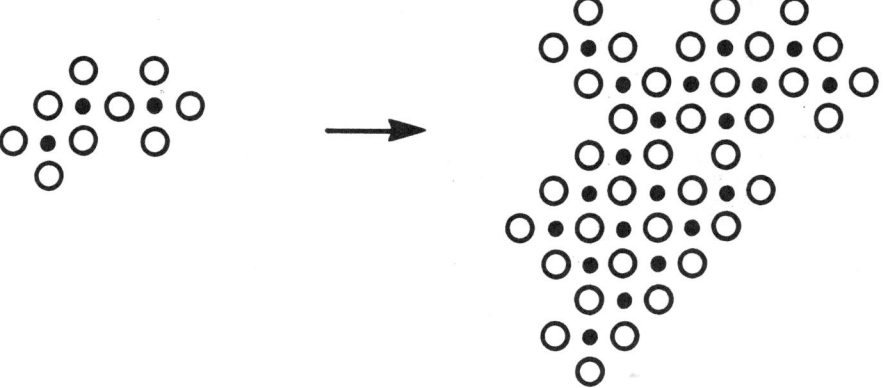

FIG. 28.   Uranyl square planar islands in borosilicate glasses. These are modelled from the U–U coordination numbers at $\sqrt{2}R$ and $2R$ distances from Fig. 27. Solid circles represent uranium atoms, and open circles represent oxygens. The cluster on the left corresponds to the bulk structure and that on the right to the surface structure following leaching.

VIII. Conclusion

In reviewing the X-ray absorption fine structure of glasses, I have stressed the unique versatility of the technique to study order and disorder at both the structural and chemical levels. Table II lists the glasses considered in this chapter, catalogued according to atom type and types of neighboring elements.

There are, of course, drawbacks in the scope of the application of XAS. These come mainly from the lack of low k information which necessarily restricts XAS to short-range order (1–5Å), leaving medium-range order (5–15Å) obscure. In principle, XANES contains specific information about more distant shells of atoms as well as local symmetry. Unfortunately the complexity of calculations of near-edge fine structure has inhibited its general application to date. However, there will be considerable payoff as increased computer power becomes more readily available and analytical techniques can be directed at what is often the richest part of the measured fine structure.

Since the early 1980s, new modes for measuring XAS have been introduced, notably fluorescence detection and glancing-angle geometry. The exploitation of these developments to solve problems in glass science is only just beginning. With fluorescence detection the structures of impurities down to 100 ppm are already being studied. Problems in nucleation and growth of minor phases, coloring agents, semiconductor dopants, and so on can now be tackled. Surface-sensitivity achieved by total external reflection has a wealth of applications. Aqueous corrosion has already been mentioned. Other areas that spring to mind include ion exchange from molten salts, ion implantation, coatings, multi layers, and the surfaces of glass melts. Some of these will be difficult from the standpoint of the sample environment, but empirical

TABLE II

ENVIRONMENTS OF ELEMENTS IN GLASSES DISCUSSED IN THE TEXT OBTAINED USING XAS TECHNIQUES. BOND LENGTHS, R AND COORDINATION NUMBERS, N, ARE GIVEN. WHERE DISTRIBUTIONS ARE ASYMMETRIC THE MEAN BOND LENGTH IS GIVEN, FOLLOWED BY THE MODE. WHERE TWO-BOND NEIGHBORS ARE DETECTED, DISTANCES ARE GIVEN FOLLOWED BY THE BRIDGING BOND ANGLE

Element	Absorption edge	Threshold energy (keV)	Glass	Bond	R(Å)	N
As	K	11.87	As	As–As	2.4(2)	3
				As–As	3.6[98°]	
			As_2O_3	As–O	1.8(0)	3
				As–As	3.2[148°]	
			As_2S_3	As–S	2.2(5)	3
				As–As	3.5[101°]	
			As_2Se_3	As–Se	2.4(0)	3
				As–As	3.6[97°]	
			$As_2S_3Se_3$	As–S	2.2(5)	2
				As–Se	2.4(0)	1

TABLE II (*cont.*)

Element	Absorption edge	Threshold energy (keV)	Glass	Bond	R(Å)	N
			As_2Te_3	As–Te	2.6(8)	2.(5)
				As–As	2.5(3)	0.(5)
Ca	K	4.039	$CaSiO_3$	Ca–O	2.3(8)	
			$Ca_3Mg_3Al_2Si_7O_{24}$	Ca–O	2.4(8)	9
			$CaAl_2Si_2O_3$	Ca–O	$\begin{cases} 2.6(4) \\ (2.4(2)) \end{cases}$	7
			$CaMgSi_2O_6$	Ca–O	$\begin{cases} 2.6(3) \\ (2.4(0)) \end{cases}$	8
			Snail granules	Ca–O	2.3(0)	8
Ce	L_{III}	5.723	CeO_2–SiO_2	Ce–O	2.3(7)	4
				Ce–Si	3.1[99°]	
				Ce–Ce	4.0[112°]	
Cs	L_{III}	5.012	$CsSi_2O_5$	Cs–O	3.0(0)	5
			$KCsSi_2O_5$			
Cu	K	8.979	Cu–As_2Se_3	Cu–Se	2.0(9)	4
Ge	K	11.103	Ge	Ge–Ge	2.44(8)	4
			GeO_2	Ge–O	1.7(4)	4
				Ge–Ge	3.2[130°]	
			M_2O–GeO_2	Ge–O	$\begin{cases} 1.7(1) \\ 1.8(4) \end{cases}$	4 / 6
			GeS	Ge–S	2.3(7)	3
			GeSe	Ge–Se	2.3(8)	3
Fe	K	7.112	$NaFeSi_2O_6$	Fe^{3+}–O	1.8(6)	4
			Fe_2O_3–$Pb(PO_4)_2$	Fe^{3+}–O	1.9(5)	6
Pb	L_{III}	13.055	$Pb(PO_4)_2$	Pb^{4+}–O	2.4(9)	8
Mn	K	6.539	Snail granules	Mn–O	2.1(7)	6
O	K	0.543	SiO_2	O–Si	1.6(1)	2
				O–O	2.6[108°]	
P	K	2.149	P	P–P	2.2(0)	3
				P–P	3.5[104°]	
K	K	3.608	$K_2Si_2O_5$	K–O	2.7(9)	5
			$KCsSi_2O_5$	K–O	2.7(9)	5
			$KAlSi_3O_8$	K–O	3.0	9
Si	K	1.839	Si	Si–Si	2.3(5)	4
			Si:H	Si–Si	2.3(5)	3.(7)
			SiO_2 }	$\begin{cases} Si–O \\ Si–Si \end{cases}$	1.6(1) / 3.1[160°]	4
			$Na_2Si_2O_5$ }			
			SiO_x:H	Si–O	1.6(2)–1.6(9)	4
				Si–Si	2.4(4)–2.3(5)	
			CeO_2–SiO_2	Si–O	1.6(0)–1.6(5)	4
				Si–Si	2.3(5)	
				Si–Si	3.1[160°]	
				Si–Ce	3.1[99°]	
Ag	K	25.514	$Ag_2B_{10}O_{16}$	Ag–O	2.2(7)	2
			Ag–As_2S_3	Ag–S	2.5(0)	3
			Ag–As_2Se_3	Ag–Se	2.6(0)	3
Na	K	1.071	$Na_2Si_2O_5$	Na–O	2.3(0)	5
			$Na_2B_2O_2$	Na–O	2.2(5)	6
Ti	K	4.966	TiO_2–SiO_2	Ti^{4+}–O	1.8(1)	4
					2.1(5)	6
U	L_{III}	17.166	$(Na_2O)_4(B_2O_3)_3/$	U–O	$\begin{cases} 1.95 \\ 2.29 \end{cases}$	(2) / (5)
			$(SiO_2)_{11}(Al_2O_3)$	U–U	$\begin{cases} 3.2[90°] \\ 4.6[180°] \end{cases}$	

difficulties should be considerably eased compared to conventional surface techniques since a vacuum environment is not required.

The element-specific local structural information that has become available from XAS studies since the early 1980s has been very timely in respect to new ideas about the structure of glass. In simple stoichiometric glasses, EXAFS has revealed the considerable perfection in the arrangement of first-shell atoms. However, in many ways the most interesting discoveries have concerned departures from the continuous-random-network model. The detection of chemical disorder in chalcogenide glasses, for instance, has provided clues as to the presence of molecular fragments in nonstoichiometric thin films. Photostructural and thermal annealing effects are found to be associated with the reestablishment of chemical order through the removal of wrong bonds. Where photodiffusion is concerned, EXAFS indicates that highly mobile metals bond into the glass matrix utilizing mesomeric interactions with chalcogens. In oxide glasses, the specific environments detected by XAS for alkalis, alkaline earths, and transition metals lead directly to the notion that the structure comprises a network interspersed with a modifying component. The modified-random-network model accords specific environments to nonbridging oxygens as well as to modifying cations and results in an archeplago structure in which network peninsulas and islands are threaded by channels and lakes of modifier (Figs. 2 and 19b). This is not just an imaginative vignette; it also provides a plausible structural basis for understanding the dynamic properties of glasses: ionic transport, chemical resistance, viscosity, etc. Perhaps the most exciting prospect lies in the possibility of modelling the structure of a glass surface. Taken in conjunction with XAS measurements, we should now be able to explore the extent to which the surface differs from the bulk and the ways it is altered by ionic diffusion processes.

While X-ray absorption spectroscopy has limitations as a local structural probe and should not be considered the exclusive technique for glass-structure determination, it is a powerful method that has come of age through advancements in the theory of photoelectron phenomena and in the increased availability of synchrotron radiation facilities. Its ease of application to most elements in the Periodic Table, together with its inherent surface-sensitivity means that it is bound to play a major part in endeavors to elucidate the detailed structure of glasses and glass surfaces for many years to come.

Acknowledgments

Much of the work featured in this chapter was done at the SERC's Synchrotron Radiation Source at Daresbury Laboratory. The director and staff are thanked for provision of X-ray and computing facilities.

References

Angell, C. A., Cheeseman, P. A., Clarke, J. H. R., and Woodcock, L. V. (1977). "Structure of Non-Crystalline Materials I." Taylor & Francis, London, pp. 191–194.

Bell, R. J., and Dean, P. (1972). *Phil. Mag.* **25**, 1381–1398.

Bellissent, R., Chenevas-Paule, A., Lagarde, P., and Bazin, D. (1983). *J. Non-Cryst. Solids* **59–60**, 237–240.

Benfatto, M., Bianconi, A., Davoli, I., Garcia, J., Marcelli, A., Natoli, C. R., and Stizza, S. (1985). *J. Non-Cryst. Solids* **77–78**, 1325–1328.

Bienenstock, A. (1973). *J. Non-Cryst. Solids* **11**, 447–458.

Binsted, N., Greaves, G. N., and Henderson, C. M. B. (1985). *Contrib. to Mineral Petrol.* **89**, 103–109.

Boochland, P., Grothaus, J., Bresser, W. J., and Suranyi, P. (1982). *Phys. Rev.* **B25**, 2975–2978.

Brown, G. E., Keefer, K. D., and Fenn, P. M. (1978), *Abstr. Geol. Soc. Amer.* **10**, 373.

Calas, G., Levitz, P., Petiau, J., Bondat, P., and Loupias, G. (1980). *Rev. Phys. Appl.* **15**, 1161–1167.

Calas, G., Bassett, W. A., Petiau, J., Steinberg, M., Tchoubar, D., and Zarka, A. (1984). *Phys. Chem. Minerals*, **11**, 17–36.

Calas, G., and Petiau, J. (1983). "Structure of Non-Crystalline Materials II." Taylor & Francis, London, pp. 18–28.

Clark, D. E., and Hench, L. L. (1983). "Scientific Basis for Nuclear Waste Management VI." Elsevier Science, Amsterdam.

Clarke, J. R., Appleman, D. E., and Papike, J. J. (1969), *Geol. Soc. Amer. Spec. Paper* 2, 31–51.

Connell, G. A. N., and Lukovsky, G. (1978). *J. Non-Cryst. Solids* **31**, 123–155.

Cox, A. D., and McMillan, P. W. (1981). *J. Non-Cryst. Solids* **44**, 257–264.

Cradwick, M. E., and Taylor, H. F. W. (1972). *Acta Crystallogr.* **B28**, 3583–3587.

Dalba, G., Fornasini, P., Rocca, F., Bernieri, E., Burattini, E., and Mobilio, S. (1986). *J. Non-Cryst. Solids* **91**, 153–164.

Davis, E. A., Elliott, S. R., Greaves, G. N., and Jones, D. P. (1977), "Structure of Non-Crystalline Materials II." Taylor & Francis, London, pp. 205–209.

Day, D. E. (1976). *J. Non-Cryst. Solids* **21**, 343–372.

Dickinson, J. T., Langford, S. C., Jensen, L. C., McVay, G. L., Kelso, J. F., and Pantano, C. G. (1988). *J. Vac. Sc. Technol.* **A6**, 1084–1089.

Douglas, R. W., and El-Shamy, T. M. M. (1967). *J. Am. Ceram. Soc.* **50**, 1–8.

Dumas, T., and Petiau, J. (1986). *J. Non-Cryst. Solids* **81**, 201–210.

Dupree, R., Holland, D., and Williams, D. S. (1984a). *Phil. Mag.* **B50**, L13–L18.

Dupree, R., Holland, D., McMillan, P. W., and Pettifer, R. F. (1984b). *J. Non-Cryst. Solids* **68**, 399–410.

Dupree, R., and Pettifer, R. F. (1984), *Nahre* **308**, 523–526.

Durham, P. J., Pendry, J. B., and Hodges, C. H. (1981). *Solid State Commun.* **38**, 159–162.

Eisenberger, P., and Brown, G. S. (1979). *Solid State Commun.* **29**, 481–484.

Elliott, S. R. (1983). "Physics of Amorphous Materials." Longman, London and New York.

Elliott, S. R., Dore, J. C., and Marseglia, E. (1985). *J. de Phys.* **C8**, 349–353.

Etherington, A. C., Wright, A. C., and Sinclair, R. N. (1983). "Structure of Non-Crystalline Materials II." Taylor & Francis, London, pp. 501–514.

Evangelisti, F., Proietti, M. G., Balzarotti, A., Cousin, F., Incoccia, L., and Mobilio, S. (1981). *Solid State Commun.* **37**, 413–416.

Filipponi, A., Fiorini, P., Evangelisti, F., Balerna, A., and Mobilio, S. (1986a). "EXAFS IV," *J. de Phys.* **C8**, 357–361.

Filipponi, A., della Sala, D., Evangelisti, F., Balerna, A., and Mobilio, S. (1986b). "EXAFS IV," *J. de Phys.* **C8**, 375–377.

Fontana, M. P., Lottici, P. P., Razzetti, C., Bianchi, D., Antonioli, G., and Emiliani, U. (1982). *Solid State Commun.* **43**, 561–565.

Forty, A. J., Kerbar, M., Robinson, J., El-Mastiri, S. M., and Farrabough, E. N. (1986), "EXAFS IV," *J. de Phys.* **C8**, 857–860.

Frischat, G. H. (1975). "Ionic Diffusion in Oxide Glasses." Trans. Tech. Publ., Aedermannsdorf, pp. 89–93.

Fuoss, P. H., Eisenberger, P., Warburton, W. K., and Bienenstock, A. (1981), *Phys. Rev. Lett.* **46**, 1537–1540.

Garofalini, S. H. (1988). Private communication.

Geere, R. G., Gaskell, P. H., Greaves, G. N., Greengrass, J., and Binsted, N. (1983). "EXAFS II." Springer-Verlag, Berlin, pp. 256–260.

Gladden, L. F. (1986), Ph.D. thesis, Cambridge University.

Gladden, L. F., Elliott, S. R., Greaves, G. N., Cummings, S., and Rayment, T. (1985). *J. Non-Cryst. Solids* **77–78**, 1199–1202.

Greaves, G. N. (1973). *J. Non-Cryst. Solids* **11**, 427–446.

Greaves, G. N. (1985). *J. Non-Cryst. Solids* **71**, 203–217.

Greaves, G. N., and Davis, E. A. (1974). *Phil. Mag.* **29**, 1201–1206.

Greaves, G. N., and Raoux, D. (1983). "Structure of Non-Crystalline Materials II." Taylor & Francis, London, pp. 55–80.

Greaves, G. N., Elliott, S. R., and Davis, E. A. (1979). *Adv. Phys.* **28**, 49–141.

Greaves, G. N., Fontaine, A., Lagarde, P., Raoux, D., and Gurman, S. J. (1981). *Nature* **293**, 611–616.

Greaves, G. N., Simkiss, K., and Taylor, M. (1984). *Biochem. J.* **221**, 855–868.

Greaves, G. N., Jiang, X. L., Jenkins, R. N., Holzenkämpfer, E., and Kalbitzer, S. (1986). "EXAFS IV," *J. de Phys.* **C8**, 853–856.

Greaves, G. N., Elliott, S. R., Gladden, L. F. and Spence, C. A. (1987). *Diffusion and Defect Data*, **53–54**. Trans. Tech. Publ., Aedermannsdorf, pp. 155–166.

Greaves, G. N., Barrett, N. T., Antonini, G. M., Thornley, F. R., Willis, B. T. M., and Steel, A. (1989). *J. Am. Chem. Soc.* **111**, 4313–4324.

Greaves, G. N., Catlow, C. R. A., Henderson, C. M. B., Vessal, B. and Zhu, R., unpublished results.

Greaves G. N. (1990). *Phil. Mag.* (in press).

Gregor, R. B., Lytle, F. W., Sandstrom, D. R., Wong, J., and Schultz, P. (1983). *J. Non-Cryst. Solids* **55**, 27–44.

Guo Tie and denBoer, M. L. (1985). *Phys. Rev.* **B31**, 6233–6237.

Gurman, S. J. (1982). *J. Materials Science* **17**, 1541–1570.

Gurman, S. J. (1988). Private communication.

Gurman, S. J., and Pendry, J. B. (1976). *Solid State Commun.* **20**, 287–290.

Gurman, S. J., and Pettifer, R. F. (1979). *Phil. Mag.* **B40**, 345–359.

Hardwick, A., Whittaker, E. J. W., and Diakun, G. P. (1985). *Mineralogical Mag.* **49**, 25–30.

Hayes, T. M., Boyce, J. B., and Beeby, J. L. (1978). *J. Phys.* **C11**, 2931–2937.

Heald, S. M., Keller, E., and Stern, E. A. (1984). *Phys. Lett.* **103A**, 155–158.

Holzenkämpfer, E. Richter, F.-W., Stuke, J., and Voget-Grote, U. (1979). *J. Non-Cryst. Solids* **32**, 327–338.

Howard, B., Mitchell, P. C. M., Ritchie, A., Simkiss, K., and Taylor, M. G. (1981). *Biochem. J.* **194**, 507–511.

Hunter, S. H., Bienenstock, A., and Hayes, T. M. (1977). "Structure of Non-Crystalline Materials I." Taylor & Francis, London, pp. 73–76.

Incoccia, L., Mobilio, S., Proietti, M. G., Fiorini, P., Giovannella, G., and Evangelisti, F. (1985). *Phys. Rev.* **B31**, 1028–1033.

Jackson, W. E., Brown, G. E., and Ponader, C. W. (1987). *J. Non-Cryst. Solids* **93**, 311–322.

Jen, J. S., and Kalinowski, M. R. (1980). *J. Non-Cryst. Solids* **38–39**, 21–26.

Johnson, R., and Christian, S. M. (1954). *Phys. Rev.* **95**, 560–561.

Kastner, M. (1979). *Phil. Mag.* **B37**, 127–133.

Keating, P. N. (1966). *Phys. Rev.* **145**, 637–645.

Kelly, J. E. III, Cordaro, J. F., and Tomozawa, M. (1980). *J. Non-Cryst. Solids* **41**, 47–55.

Knapp, G. S., Veal, B. W., Lam, D. J., Paulikas, A. P., and Pan, H. F. (1984). *Mat. Lett.* **2**, 253–256.

Kossel, W. (1920). *Z. Phys.* **1**, 119–134.

Kovba, L. M. (1971). *Radiokhimiya* **13**, 309–311.

Kronig, R. de L. (1931). *Z. Phys.* **70**, 317–323.

Lapeyre, C., Petiau, J., and Calas, G. (1983). "Structure of Non-Crystalline Materials II." Taylor & Francis, London, pp. 42–51.

Lee, P. A., and Pendry, J. B. (1975). *Phys. Rev.* **B11**, 2795–2811.

Long, M., Galison, P., Alben, R., and Connell, G. A. N. (1976). *Phys. Rev.* **B13**, 1821–1829.

Lowe, A. J., Elliott, S. R., and Greaves, G. N. (1986). *Phil. Mag.* **B54**, 483–490.

McDonald, W. S., and Cruikshank, D. W. J. (1967). *Acta Crystallogr.* **22**, 37–43.

McDowell, R. A., West, J. B., Greaves, G. N., and Van der Laan, G. (1988). *Rev. Sci. Inst.* **59**, 843–852.

McKeown, D. A., Waychunas, G. A., and Brown, G. E. (1985). *J. Non-Cryst. Solids* **74**, 325–348.

McKeown, D. A., Waychunas, G. A., and Brown, G. E. (1985). *J. Non-Cryst. Solids* **74**, 349–371.

Menelle, A., Flank, A. M., Lagarde, P., and Bellissent, R. (1986). "EXAFS IV," *J. de Phys.* **C8**, 379–382.

Misawa, M., Price, D. L., and Suzuki, K. (1980). *J. Non-Cryst. Solids* **37**, 85–97.

Mitra, S. K., Amini, M. N., Fincham, D., and Hockney, R. W. (1981). *Phil. Mag.* **B43**, 365–372.

Mobilio, S., and Filipponi, A. (1987). *J. Non-Cryst. Solids* **97–98**, 365–372.

Morrison, T. I., Paeseler, M. A., Sayers, D. E., Tsu, R., and Gonzales-Hernandez, J. (1985). *Phys. Rev.* **B31**, 5474–5478.

Moss, S. C., and Price, D. L. (1985). "Physics of Disordered Materials," D. Adler, H. Fritzsche, and S. R. Ovshinsky, eds. Plenum, New York and London, pp. 77–95.

Mott, N. F. (1976). *Phil. Mag.* **B34**, 1101–1108.

Mott, N. F., and Davis, E. A. (1979). "Electronic Processes in Non-Crystalline Materials." Clarendon Press, Oxford.

Mozzi, R. L., and Warren, B. E. (1969). *J. Appl. Cryst.* **2**, 164–172.

Nemanich, R. J., Connell, G. A. N., Hayes, T. M., and Street, R. A. (1978). *Phys. Rev.* **B18**, 6900–6914.

Oyanagi, H., Tanuka, K., Hasoya, S., and Minomura, S. (1981). *J. de Phys.* **42**(C4) 221–224.

Paeseler, M. A., Sayers, D. E., Tsu, R., and Gonzalez-Hermandez, J. (1983). *Phys. Rev.* **B28**, 4550–4557.

Pant, A. K., and Cruikshank, D. W. J. (1968). *Acta Crystallogr.* **B24**, 13–19.

Paul, W., Connell, G. A. N., and Temkin, R. J. (1973). *Adv. Phys.* **22**, 529–665.

Petiau, J., Calas, G., Petitmaire, D., Bianconi, A., Benfatto, M., Marcelli, A. (1986). *Phys. Rev.* **B34**, 7350–7361.

Pettifer, R. F. (1979). "Proceedings 4th EPS General Conference." Institute of Physics, London, pp. 522–531.

Pettifer, R. F., McMillan, P. N., and Gurman, S. J. (1977). "Structure of Noncrystalline Materials I." Taylor & Francis, London, pp. 63–67.

Phillipp, H. R. (1971). *J. Phys. Chem. Solids* **32**, 1935–1945.

Phillipp, H. R. (1972). *J. Non-Cryst. Solids* **8–10**, 627–632.

Phillips, J. C. (1981). *J. Non-Cryst. Solids* **43**, 37–78.

Polk, D. E., and Boudreaux, D. S. (1973). *Phys. Rev. Lett.* **31**, 92–95.

Porai-Koshits, E. A. (1977). *J. Non-Cryst. Solids* **25**, 87–129.

Rennie, J. H. S. (1986). Ph.D. thesis. Cambridge University.

Richter, T., Frischat, G. H., and Borchardt, G. (1985). *Phys. Chem. Glasses* **26**, 208–212.

Sakka, S., and Kamiya, K. (1982). *J. Non-Cryst. Solids* **49**, 103–116.

Sales, B. C., and Boatner, L. A. (1984). *Science* **226**, 45–58.

Sales, B. C., Ramsey, R. S., Bates, J. B., and Boatner, L. A. (1986). *J. Non-Cryst. Solids* **87**, 137–158.

Sandstrom, D. R., Lytle, F. W., Wei, P. S. P., Gregor, R. G. Wong, J., and Schultz, P. (1980). *J. Non-Cryst. Solids* **41**, 201–207.

Sayers, D. E., Stern, E. A., and Lytle, F. W. (1971). *Phys. Rev. Lett.* **27**, 1204–1207.

Sayers, D. E., Lytle, F. W., and Stern, E. A. (1972). *J. Non-Cryst. Solids* **8–10**, 401–407.

Sayers, D. E., Lytle, F. W., and Stern, E. A. (1974). "Amorphous and Liquid Semiconductors V." Taylor & Francis, London, pp. 403–412.

Sayers, D. E., Stern, E. A., and Lytle, F. W. (1975). *Phys. Rev. Lett.* **35**, 584–587.

Sayers, D. E., Yang, C. Y., and Paeseler, M. A. (1987). "Disordered Semiconductors," M. A. Kastner, G. A. Thomas, and S. R. Ovshinsky, eds. Plenum, New York and London, pp. 273–282.

Schreiber, H. D., Balazs, G. B., and Solberg, T. N. (1985). *Phys. Chem. Glasses* **26**, 35–45.

Sevillano, E., Meuth, H., and Rehr, J. J. (1979). *Phys. Rev.* **B20**, 4908–4911.

Sharma, S. K., and Yoder, H. S. (1978). "Carnegie Inst. Washington Year Book" **78**, pp. 526–532.

Singh, A., Edwards, A. M., Gurman, S. J., and Davis, E. A. (1988). "EXAFS V" (in press).

Soules, T. F. (1979). *J. Chem. Phys.* **71**, 4570–4578.

Singh, A., Gurman, S. J., and Davis, E. A. (1989). *J. Non-Cryst. Solids* (in press).

Soules, T. F. (1979). *J. Chem. Phys.* **71**, 4570–4578.

Steel, A. T., Greaves, G. N., Firth, A. P., and Owen, A. E. (1988). *J. Non-Cryst. Solids* **107**, 155–162.

Stegemann, G., and Lengeler, B. (1986). "EXAFS IV," *J. de Phys.* **C8**, 407–410.

Stern, E. A., Bouldin, C. E., von Roedern, B., and Azoulay, J. (1983). *Phys. Rev.* **B27**, 6557–6560.

Stizza, S. Davoli, I., Tomellini, M., Marcelli, A., Bianconi, A., Gzowski, A., and Murawski, L. (1984). "EXAFS II," *Springer Proc. Phys.* **2**, 331–334.

Stohl, F. V., and Smith, D. K. (1981). *Amer. Mineralogist* **66**, 610–624.

Stöhr, J., Johansson, L., Linday, I., and Pianetta, P. (1979). *Phys. Rev.* **B20**, 664–680.

Stöhr, J., Kollin, E. B., Fischer, D. A., Hastings, J. B., Zaera, F., and Sette, F. (1985). *Phys. Rev. Lett.* **55**, 1468–1471.

Tanaka Ke (1975). *App. Phys. Lett.* **26**, 243–245.

Tanaka Ke (1980). *J. Non-Cryst. Solids* **35–36**, 1023–1034.

Taylor, M. G., Simkiss, K., Greaves, G. N., and Harries, J. (1988). *Proc. R. Soc. Lond.* **B234**, 463–576.

Temkin, R. J. (1975). *J. Non-Cryst. Solids* **17**, 215–230.

Thornley, F. R., Barrett, N. T., Greaves, G. N., and Antonini, G. M. (1986). *J. Phys.* **C19**, L563–L569.

Tronc, P., Bensoussan, M., Brenac, A., and Sebenne, C. (1973). *Phys. Rev.* **B8**, 5947–5956.

Vessal, B., Leslie, M. and Catlow, C. R. A. (1989). "Molecular Simulation" (in press).

Wegener, W., and Fischat, G. H. (1983). "Structure of Non-Crystalline Materials II." Taylor & Francis, London, pp. 326–334.

Warren, B. E., and Biscoe, J. (1938). *J. Amer. Ceram. Soc.*, **21**, 259–265.

Wainwright, J. E., and Starkey, J. (1971). *Zeit Krist* **133**, 75–84.

Weigel, F. (1986). "Chemistry of the Actinide Elements," **1**. Chapman and Hall, London.

Wong, J. (1983). *Topics in Appl. Phys.* **46**, 45–75.

Wong, J., and Lytle, F. W. (1980). *J. Non-Cryst. Solids* **37**, 273–284.

Wright, A. C., and Leadbetter, A. J. (1976). *Phys. Chem. Glasses* **17**, 122–145.

Yang, C. Y., Paeseler, M. A., and Sayers, D. E. (1986). "EXAFS IV," *J. de Phys.* **C8**, 391–394.

Zachariasen, W. H. (1932). *J. Am. Chem. Soc.* **54**, 3841–3851.

Zallen, R. (1983). "Physics of Amorphous Solids." Wiley, New York, chap. 4.

CHAPTER 2

Nuclear Magnetic Resonance in Glass

Phillip E. Stallworth
Philip J. Bray

DEPARTMENT OF PHYSICS
BROWN UNIVERSITY
PROVIDENCE, RHODE ISLAND

I. Introduction

Nuclear magnetic resonance (NMR) techniques have been used since the 1950s in order to study dynamical and structural behavior of glasses (Silver and Bray, 1958; Bray et al., 1963, 1988; Müller-Warmuth and Eckert, 1982; Kirkpatrick et al., 1986a). Up to the 1970s, work involved principally continuous-wave (CW) spectrometers and low-field (< 2.5 Tesla) electromagnets. Early NMR work focused primarily on dipolar, chemical shift, and quadrupolar interactions of various nuclei in glasses: ^{1}H, ^{7}Li, ^{9}Be, ^{10}B, ^{11}B, ^{19}F, ^{23}Na, ^{51}V, ^{203}Tl, ^{205}Tl, and ^{209}Pb. NMR studies of glasses were performed almost exclusively, upon borate and borosilicate glasses resulting in the analysis of ^{10}B, ^{11}B lineshapes and resonances of certain sensitive "glass-modifier" nuclei—e.g., ^{7}Li (Hendrickson and Bray, 1974), ^{209}Pb (Kim et al., 1976), and ^{203,205}Tl (Baugher and Bray, 1969). Since 1980, high-power pulsed spectrometers with high-field superconducting magnets (4–11 Tesla) have been employed, often using magic-angle sample spinning (MASS) techniques, to study nuclei such as ^{29}Si (Dupree et al., 1984; Dupree and Pettifer, 1984; Grimmer et al., 1984) and ^{27}Al (Müller et al., 1983; Oestrike et al., 1987; Risbud et al., 1987).

Described in the following sections are the fundamental interactions and experimental procedures used by various workers in the study of glasses. The purpose of this chapter is: (1) to achieve a focused understanding of NMR in glass, (2) to present a comprehensive account of past studies, and (3) to provide a summary of current research. Regretfully, it is not possible to discuss and refer to all published information; hence, much work has not been included here. In particular, NMR studies of chalcogenide and metallic glasses have been omitted; the interested reader may refer to the following reviews and discussions: Drain (1967), Alloul (1977), Müller-Warmuth and Eckert (1982), and Bray et al. (1988).

II. NMR Theory in Glasses

A. NMR HAMILTONIAN FOR GLASSES

In nuclear magnetic resonance spectroscopy, photons or quanta are used to stimulate transitions between nuclear energy levels. The energy levels are set up by interactions between the nuclear magnetic dipole moments and an applied magnetic field (Zeeman levels); but those levels can be modified by interactions of the magnetic dipole moments with the local magnetic fields produced by neighboring nuclei (dipolar interaction), by modifications of the applied field by the electron distribution about the nuclei (chemical shift interaction), and by the interaction of nuclear quadrupole moments with electric-field gradients at the nuclear sites (quadrupolar interaction). Each of these interactions can be represented by a Hamiltonian (Andrew, 1955; Abragam, 1961; Slichter, 1978) that yields the relevant energies. The total NMR Hamiltonian is the sum of all the interaction Hamiltonians. For the glass studies mentioned in this chapter, the Zeeman contribution is the dominant interaction; all other interactions are treated perturbatively. The Hamiltonian is expressed as follows:

$$H = H_Z + H_D + H_{CS} + H_Q \tag{1}$$

where H_Z is the Zeeman Hamiltonian, H_D is the dipolar Hamiltonian, H_{CS} is the chemical-shift Hamiltonian and H_Q is the quadrupolar Hamiltonian. These individual contributions to the total NMR interaction in glasses are described shortly.

1. Zeeman Interaction

Nuclear magnetic resonance occurs for nuclei possessing a spin I (an integer or half integer) of $1/2$ or greater. These nuclei exhibit a nuclear magnetic dipole moment given by

$$\mathbf{\mu} = \gamma \hbar \mathbf{I} \tag{2}$$

where $\mathbf{\mu}$ is the nuclear magnetic moment, γ is the gyromagnetic ratio, $\hbar$ is Planck's constant (h) divided by 2π, and $\mathbf{I}$ is the nuclear spin vector. γ is, in general, different for each nucleus and can be expressed in terms of the Bohr nuclear magneton μ_N as

$$\gamma = \frac{g}{\hbar}\left(\frac{e\hbar}{2m_p}\right) = \frac{g\mu_N}{\hbar} \tag{3}$$

where g is the nuclear g-factor, e is the proton Coulombic charge, m_p is the proton mass and μ_N (encompassing the quantity in parentheses) is the Bohr nuclear magneton.

The interaction of the nuclear magnetic moment with an external magnetic field $\mathbf{H}_o$ yields the quantum mechanical interaction energy

$$H_Z = -\mathbf{\mu} \cdot \mathbf{H}_o = -\gamma \hbar \mathbf{I} \cdot \mathbf{H}_o = -\gamma \hbar m H_o \tag{4}$$

where m (the magnetic quantum number) is an integer taking on $2I + 1$ values ranging from $+I$ to $-I$. Hence, in the presence of a magnetic field, $2I + 1$ nondegenerate equally spaced energy levels will be established. The energy difference between adjacent levels is

$$\Delta E_Z = -\gamma \hbar m H_o - [-\hbar\gamma(m + 1)H_o]$$
$$= \gamma \hbar H_o \tag{5}$$
$$= h\nu_o$$

where ν_o is the Larmor resonant frequency which is the required frequency for transitions between Zeeman levels. These transitions are induced by subjecting the nucleus to a radio-frequency electromagnetic wave (RF magnetic field). The RF magnetic field oscillates at ν_o and is transverse to the external field $\mathbf{H}_o$. Classically, ν_o is the frequency of precession of the nuclear magnetic dipole moment about $\mathbf{H}_o$.

2. Magnetic Dipolar Interaction

In solid samples containing many nuclei, one must consider not only how the magnetic moments interact with the external field, but also how each moment interacts with its neighboring moments. In other words, all nuclei possessing magnetic moments act like small magnets and can influence the magnetic environments of the surrounding nuclei. This coupling between spins is called the magnetic dipole–dipole interaction; it is small in magnitude compared to the Zeeman energy (usually less than 50 kHz) and is roughly proportional to r^{-3} (r is the dipole–dipole distance). Like the Zeeman interaction, the magnetic dipolar interaction can be understood by considering the classical expression. The dipolar energy between two moments (1 and 2) separated by a distance r is given as (Abragam, 1961; Slichter, 1978):

$$E_D = \frac{\boldsymbol{\mu}_1 \cdot \boldsymbol{\mu}_2}{r^3} - \frac{3(\boldsymbol{\mu}_1 \cdot \mathbf{r})(\boldsymbol{\mu}_2 \cdot \mathbf{r})}{r^5} \tag{6}$$

where $\mathbf{r}$ is the vector between the interacting spins. By considering all dipole–dipole pairs of a macroscopic system containing many magnetic nuclei, one arrives at the following dipolar Hamiltonian:

$$H_D = \frac{1}{2} \sum_{j=1} \sum_{k=1} \left[\frac{\boldsymbol{\mu}_j \cdot \boldsymbol{\mu}_k}{r_{jk}^3} - \frac{3(\boldsymbol{\mu}_j \cdot \mathbf{r}_{jk})(\boldsymbol{\mu}_k \cdot \mathbf{r}_{jk})}{r_{jk}^5} \right] \tag{7}$$

where $\mathbf{r}_{jk}$ is the vector joining the jth and kth nuclei. The sums are taken over all possible pair combinations. Since the magnetic dipolar interaction is energetically small relative to the Zeeman interaction, it can be greatly simplified through a perturbative approximation, yielding the result for homonuclear spins

$$H_D = \frac{1}{4} \gamma^2 \hbar^2 \sum_{j,k} \frac{(1 - 3 \cos^2 \theta_{jk})}{r_{jk}^3} (3 I_{z_j} I_{z_k} - \mathbf{I}_j \cdot \mathbf{I}_k). \tag{8}$$

θ_{jk} is the angle between the external field $\mathbf{H}_o$ and the vector $\mathbf{r}_{jk}$ and I_z is the z-component of the spin vector (Eq. 2). Quantum mechanically this expression yields a spectrum of perturbations of the Zeeman levels and gives rise to a symmetric distribution about the resonance frequency ν_o. Hence, a width is established for the resonance line (Fig. 1). Although in practice this expression is not used, it is found that the dipolar interaction can be described, in many cases, by a simple symmetric distribution in frequencies about the Larmor frequency. Often for glasses, Gaussian and/or Lorentzian distribution functions can be used to represent the spread in resonant frequencies about ν_o.

In considering only the Zeeman and dipolar energies, the dipolar width increases as the values of r_{jk} decrease. In light of this, analytical techniques have been developed that calculate the average distance between interacting

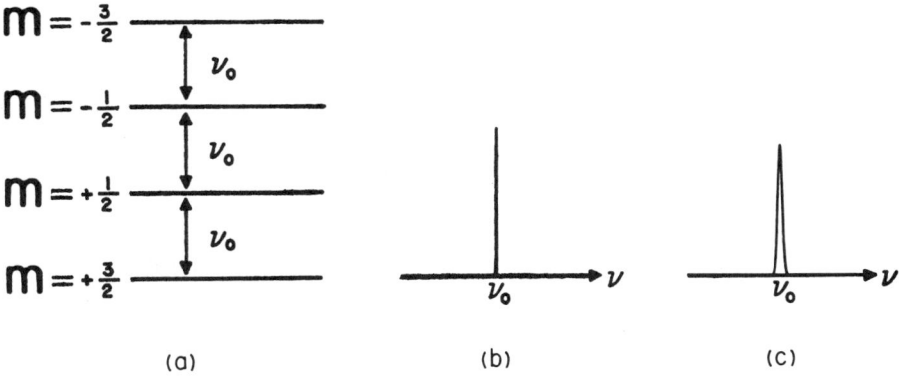

FIG. 1.   (a) Energy levels for a spin-3/2 nucleus in the presence of a magnetic field H_o. (b) The resultant NMR absorption is represented as a single transition occurring at frequency v_o. (c) Dipolar interactions yield a symmetric distribution about the resonance frequency.

spins. The classic method, developed by Van Vleck, makes use of the second-moment of the lineshape function in order to reveal the linewidth dependence upon nuclear spin–spin distances (Andrew, 1955; Abragam, 1961; Slichter, 1978).

3. Quadrupolar Interaction

The energy expression for the interaction of a nucleus in the presence of an external magnetic field and the electronic interaction between the nucleus and its surrounding electrons (without dipolar broadening) can be written (Cohen and Reif, 1957; Abragam, 1961):

$$H = H_Z + H_Q$$

$$= \gamma \hbar I_z H_o + \frac{e^2 qQ}{4I(2I-1)}\left[3I_z^2 - I(I+1) + \frac{1}{2}\eta(I_+^2 + I_-^2)\right]. \tag{9}$$

The first term is the Zeeman interaction; the second term arises as a result of the interaction of the nuclear-charge distribution with the surrounding electric-field gradient (EFG). eq is a component of the electric field gradient, Q is the nuclear quadrupole moment, I_z, I_+ and I_- are the standard spin angular momentum operators and η is the asymmetry parameter which reflects the deviation from cylindrical symmetry of the EFG:

$$\eta = \frac{V_{yy} - V_{xx}}{V_{zz}} \tag{10}$$

where V_{xx}, V_{yy}, and V_{zz} are the components of the EFG in its principal axes system of coordinates labelled so that:

$$|V_{zz}| = eq \quad \text{and} \quad |V_{zz}| \geqslant |V_{yy}| \geqslant |V_{xx}|.$$

Quadrupole coupling constants, $Q_{cc} = e^2 qQ/h$, many of which are quoted in NMR and NQR (nuclear quadrupole resonance) literature, indicate the size of the quadrupolar interaction. Only nuclei with $I \geqslant 1$ have electrical quadrupole moments.

Treatment of H_Q in Eq. (9) as a perturbation on the Zeeman energy levels yields the following expressions correct through second-order effects for the transition energies (Abragam, 1961):

$$hv_{m \leftrightarrow m-1} = hv_o - \frac{hv_Q}{2}(m - 1/2)[3 \cos^2 \theta - 1 - \eta \cos 2\phi \sin^2 \theta]$$

$$+ \frac{hv_Q^2}{12v_o}\left[\frac{3}{2} \sin^2 \theta[(A + B)\cos^2 \theta - B]\right.$$

$$+ \eta \cos 2\phi \sin^2 \theta[(A + B)\cos^2 \theta + B]$$

$$\left. + \frac{\eta^2}{6}[A - (A + 4B)\cos^2 \theta - (A + B)\cos^2 \phi(\cos^2 \theta - 1)^2]\right] (11)$$

where

$$v_Q = \frac{3e^2 qQ/h}{2I(2I - 1)}$$

$$A = 24m(m - 1) - 4I(I + 1) + 9$$

$$B = \frac{1}{4}[6m(m - 1) - 2I(I + 1) + 3].$$

Here θ and ϕ are the polar angles for the orientation of the EFG axes with respect to the magnetic field direction, and only the allowed transitions $\Delta m = \pm 1$ have been considered. The second term in Eq. (11) arises from first-order effects; the third term, involving v_Q^2/v_o, arises from second-order effects. If one considers a collection of nuclei that find themselves in environments for which the EFG takes on all orientations (i.e., noncrystalline solids or powdered crystals), then all polar angles θ and ϕ are realized and the corresponding spectral intensity will be weighted accordingly. By plotting the number of combinations of θ and ϕ values that give a particular resonance frequency, v, as a function of the resonance frequency (v), one obtains the "powder pattern" for the sample (Cohen and Reif, 1957).

From Eq. (11) it is seen that the central transition ($m = 1/2 \leftrightarrow m = -1/2$) for half-integral spins is unshifted by first-order quadrupolar effects; however, it is also seen that for the satellite transitions ($m \leftrightarrow m - 1$ with $m \neq 1/2$), first-

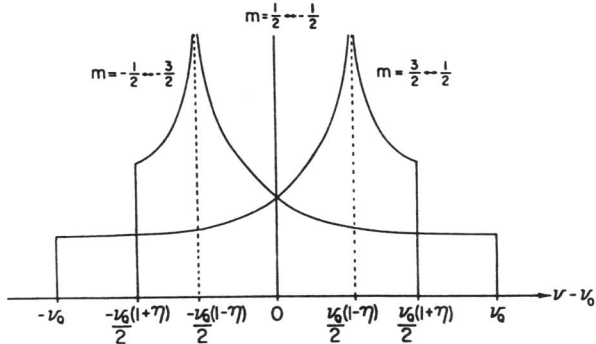

FIG. 2. First-order quadrupolar powder pattern for a randomly oriented ensemble of spin-3/2 nuclei. Here $v_Q = 3Q_{cc}/2I(2I - 1)$.

order shifts do occur. The resultant powder pattern for glasses and polycrystalline powders is displayed in Fig. 2. The second-order expression (third term in Eq. (11)) is only required when the quadrupole interaction becomes sufficiently large as shown in Fig. 3. For a large quadrupole interaction, the satellite transitions will be located at frequencies far distant from v_0 and

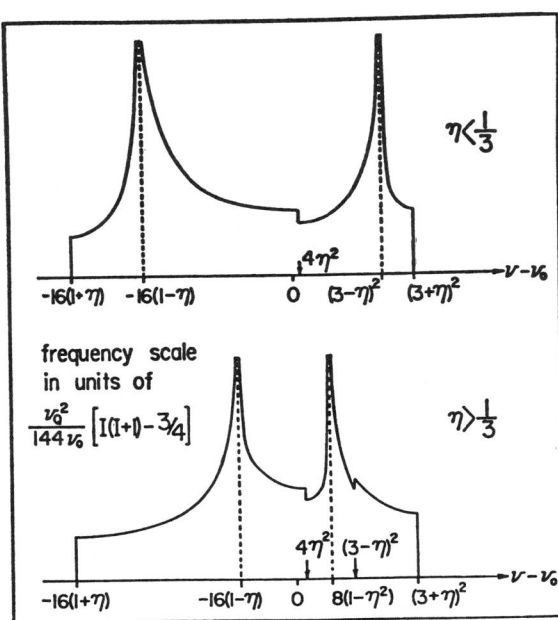

FIG. 3. Second-order quadrupolar powder pattern for the central transition of a half-integral nuclear spin ($m = 1/2 \leftrightarrow m = -1/2$). The two cases shown indicate the general lineshapes for η less than or greater than 1/3. Here $v_Q = 3Q_{cc}/2I(2I - 1)$.

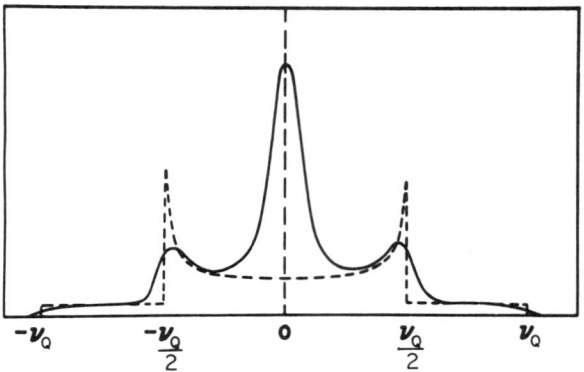

FIG. 4. Dipolar broadened lineshape for the first-order ($I = 3/2$) quadrupolar interaction where $\eta = 0$.

generally are not seen for glass samples at lower fields. The quadrupolar parameters η and e^2qQ can be determined from the positions of the divergences and shoulders.

Real spectra of glass samples yield smoother lineshapes than those shown in the theoretically derived powder patterns. This smoothing is produced through magnetic dipole-dipole interactions as shown in Figs. 4 and 5.

The electron distributions that generate the EFG tensor determine the strength of the quadrupole-coupling constant Q_{cc} and the value of the asymmetry parameter η. Thus, values of Q_{cc} and η determined from the NMR spectra provide information about properties of the electron distribution

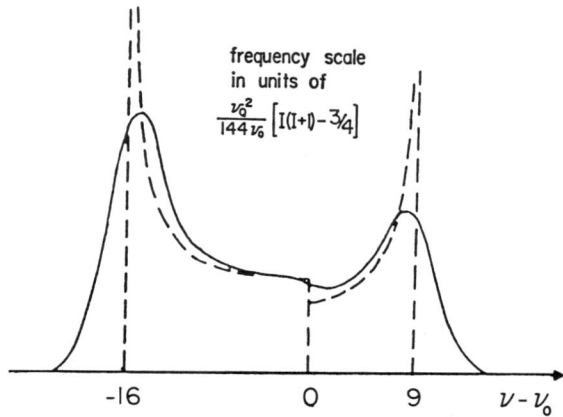

FIG. 5. Dipolar broadened lineshape for the second-order quadrupolar interaction (central transition) where $\eta = 0$.

such as the occupied electron orbitals, bond angles, and type and number of atomic neighbors.

4. Chemical-Shift Interaction

In the presence of an external magnetic field, atomic electrons will undergo induced motions giving rise to currents. These currents in turn generate magnetic fields that can add to or subtract from the applied field at the nuclear site. The combined induced magnetic fields generated by surrounding electrons plus the external field result in a net magnetic field at the nuclear site. The corresponding resonance frequency is thus shifted from the "bare nucleus" Larmor frequency. Chemical-shift magnitudes, which can be as large as 100 kHz, are indicative of the size of the induced electron currents; and the signs ($+$ or $-$) of the shift indicates the degree of paramagnetic and diamagnetic contributions. The difference between the applied field and the field at the nuclear site is referred to as the chemical shift (Abragam, 1961). The resultant induced magnetic field is related to the applied field through the chemical shift tensor $\bar{\sigma}$.

$$\mathbf{H}_{\text{nucleus}} = -\bar{\sigma} \cdot \mathbf{H}_o. \tag{12}$$

Hence the chemical-shift Hamiltonian becomes:

$$H_{CS} = -\boldsymbol{\mu} \cdot \mathbf{H}_{\text{nucleus}}$$

$$= \boldsymbol{\mu} \cdot \bar{\sigma} \cdot \mathbf{H}_o. \tag{13}$$

A first-order perturbation treatment for the $m \leftrightarrow m - 1$ transition yields for the chemical shift:

$$v = v_o(1 - \sigma_{zz}) \tag{14}$$

where σ_{zz} is the chemical-shift tensor component along the direction of $\mathbf{H}_o$. Since the chemical-shift effect is quite small relative to the Zeeman energy, higher perturbative terms can be disregarded. σ_{zz} can be written in terms of the diagonal elements σ_{11}, σ_{22}, and σ_{33} ($\sigma_{11} \leqslant \sigma_{22} \leqslant \sigma_{33}$) of the chemical-shift tensor in its principal axes, and the polar angles θ and ϕ that give the orientation of the principal axes with respect to the external field. Then

$$v = v_o[(1 - \sigma_{11})\sin^2 \theta \sin^2 \phi + (1 - \sigma_{22})\cos^2 \theta \sin^2 \phi + (1 - \sigma_{33})\cos^2 \theta]. \tag{15}$$

Upon rearrangement, this equation becomes:

$$v = v_o[1 - \sigma_{iso} - \sigma_{ax}(3 \cos^2 \theta - 1) - \sigma_{aniso} \sin^2 \theta \cos 2\phi] \tag{16}$$

with

$$\sigma_{iso} = \frac{1}{3}(\sigma_{11} + \sigma_{22} + \sigma_{33})$$

$$\sigma_{ax} = \frac{1}{6}(2\sigma_{33} - \sigma_{11} - \sigma_{22})$$

$$\sigma_{aniso} = \frac{1}{2}(\sigma_{22} - \sigma_{11}).$$

As in the case of the quadrupole interaction, Eqs. (15) or (16) can be used to generate the powder pattern of responses for a polycrystalline powder or glass for which all values of θ and ϕ are present. Figure 6 shows two chemical-shift powder patterns for the asymmetric anisotropy and axially symmetric anisotropy. The measure of axial symmetry is given as:

$$\zeta = 2\left(\frac{\sigma_{22} - \sigma_{11}}{\sigma_{33} - \sigma_{11}}\right). \tag{17}$$

The shoulders and divergences of the powder patterns yield the values for the components of the chemical-shift tensor.

B. Relaxation and Dynamics in Glasses

1. *Spin-Lattice Relaxation*

In solids, the dynamics of how an individual nuclear spin absorbs and releases energy depends upon how that spin interacts with other nuclear spins, paramagnetic ions, electron spins, and phonons. The equilibrium configuration of a system of nuclear spins, which are distributed among a set of energy levels (e.g., the Zeeman energy levels), is governed by a Boltzmann distribution. Hence, the lower-energy spin states have a larger population. Upon absorption of energy, the spins in the lower-energy levels are raised into the higher-energy states. If the spins cannot relax back to the lower levels, the populations of the states become equal and no net power

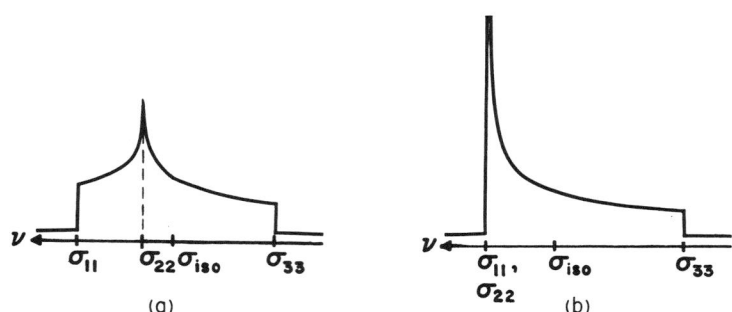

FIG. 6.   Illustrations of chemical-shift powder patterns for (a) asymmetric and (b) axially symmetric anisotropies.

absorption can occur: the system is "saturated" and no NMR signal can be obtained. Through interactions of the spins with the lattice (i.e., other nuclear spins, paramagnetic ions, etc.), energy can be transferred from the spins to the lattice and hence bring the populations of energy levels back to the Boltzmann values.

The spin-lattice relaxation time T_1 is the parameter that characterizes the rate at which the z-axis component M_z of the bulk magnetization (which is generated by the z-axis components of all the nuclear magnetic moments) returns to its equilibrium value (determined by the Boltzmann population) after a disturbance. The equation defining this process is the following (Abragam, 1961):

$$\frac{dM_z}{dt} = -\frac{(M_z - M_o)}{T_1} \tag{18}$$

where M_o is the equilibrium value of M_z.

Ionic motion gives rise to fluctuating magnetic and electric fields at the nuclear sites and can strongly influence T_1. Assuming that ionic motion is random and that T_1 is determined by time-varying interactions arising from ion motion then, as shown by Bloembergen, Pound, and Purcell (Bloembergen et al., 1948; Bloembergen, 1961), the relaxation rate is given by

$$\frac{1}{T_1} = C[J(\omega) + 4J(2\omega)] \tag{19}$$

where $J(\omega)$ is the spectral density or the Fourier transform of the autocorrelation function describing the fluctuating magnetic field and C is a constant determined by the specific time-varying interaction (e.g., dipolar, quadrupolar, or chemical shift). Furthermore, if the motion is determined by a single correlation time, τ_c (which is the average time that a nuclear spin remains at any given site in the case of ion hopping) then Eq. (19) as shown by Abragam (1961) can be simplified to give

$$\frac{1}{T_1} = C\left[\frac{\tau_c}{1 + \omega^2\tau_c^2} + \frac{4\tau_c}{1 + 4\omega^2\tau_c^2}\right] \tag{20}$$

where ω is the angular frequency ($2\pi v$, where v is the resonance frequency). When the ionic motion is thermally activated,

$$\tau_c = \tau_o e^{E_A/k_B T}, \tag{21}$$

where E_A is the activation energy, k_B is the Boltzmann constant, and T is the temperature.

Ionic motion in glasses is generally a complicated matter due to the nonperiodic structure and consequent distributions of ion sites, correlation times, and activation energies. In order to deal with this, distributions $G(\tau)$ of correlation times are used to convolute the expression for the reciprocal of the spin-lattice relaxation time (Noack, 1971).

$$\frac{1}{T_1} = C \left[\int_0^\infty \frac{\tau G(\tau)}{1 + \omega^2 \tau^2} \, d\tau + 4 \int_0^\infty \frac{\tau G(\tau)}{1 + 4\omega^2 \tau^2} \, d\tau \right] \qquad (22)$$

where

$$\int_0^\infty G(\tau) \, d\tau = 1.$$

2. Spin–Spin Relaxation

When the bulk magnetization **M** is perturbed from its equilibrium value by tipping the direction of the magnetization away from the z-axis direction, a component of magnetization will be created in the direction transverse to the z-axis direction (x-y plane). After the perturbation, the system will adjust to reestablish the Boltzmann population among its energy levels and the bulk magnetization will again align itself along the z-axis direction. The rate at which the magnetization recovers along the z-axis is related to T_1. The rate at which the x-y component of the magnetization decays is related to T_2, the spin–spin or transverse relaxation time.

The individual nuclear magnetic moments precess with characteristic frequencies centered about the Larmor frequency. Different nuclei may precess at rates somewhat above or below the Larmor frequency because local fields, arising from neighboring nuclei, add to or subtract from the applied magnetic field H_o. The net effect can be described as a "fanning-out" or "dephasing" of the individual transverse magnetization components such that with time, their vector sum falls to zero. In the absence of magnetic-field inhomogeneity (which will also cause the fanning-out or dephasing), T_2 characterizes the rate of this dephasing process.

For most solid samples, $T_2 \ll T_1$. However, for many materials, T_1 and T_2 merge asymptotically with increasing temperature (Fig. 7). Phenomenological expressions, proposed by Bloch, for the sample magnetization in terms of T_1 and T_2, are described by Abragam (1961) and Slichter (1978). It should be mentioned that there is a direct inverse relationship between linewidths and relaxation times. Larger dipolar linewidths are indicative of small values of T_2, as in solids, and small dipolar linewidths are indicative of larger values of T_2, as in liquids.

Increased motion of a resonating nucleus can average the local rigid lattice interactions (dipolar, quadrupolar, and chemical-shift anisotropy) to reduced values. T_2 is intimately involved with the change in the dipolar width of the

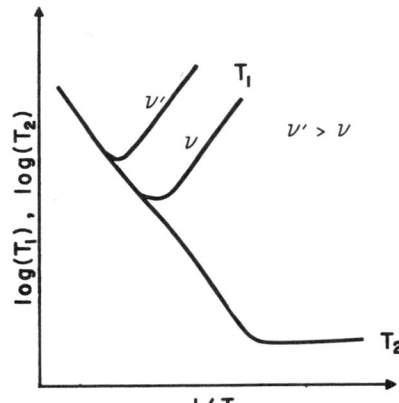

FIG. 7.   Relaxation time T_1 (spin-lattice) and T_2 (spin–spin) behavior, shown here as a log dependence upon reciprocal temperature. The frequency dependence is indicated for two different values v' and v.

response as a function of temperature. In general, atomic motion increases with temperature and has the net effect of reducing the breadth of the NMR linewidth (motional narrowing).

III. Experimental Methods

A. CONTINUOUS-WAVE NMR

Continuous-wave NMR (CW) can be observed when a sample is placed in an applied field, H_o, and subjected to a continuous radio-frequency field (Andrew, 1955). The RF field is a small orthogonal field (relative to H_o) oscillating at or near the Larmor frequency. The RF field is swept through the resonance frequency (v_o), which is determined by the Zeeman interaction and the perturbing interactions. Consequently, the nuclei under study (in the sample) will be observed to resonate (absorb energy) within some frequency width of the Larmor frequency. The lineshape (which for glasses is the NMR absorption powder pattern) is dependent upon the various nuclear interactions mentioned previously and therefore gives information concerning the nuclear environment. Experimentally, it is found that sweeping the large external field, rather than the frequency of the RF power source, is more practical since the spectrometer bandwidth is usually quite narrow. The NMR intensity for solid samples in a single sweep is usually very small so that low signal-to-noise (S/N) ratios are the rule; however, with signal averaging and lock-in detection, acceptable signals can be obtained. Figure 8 shows a typical CW experimental setup.

A convenient probe for CW is the crossed-coil probe (Abragam, 1955; Bucholtz, 1982). Here, the transmitter and receiver coils, being orthogonal to

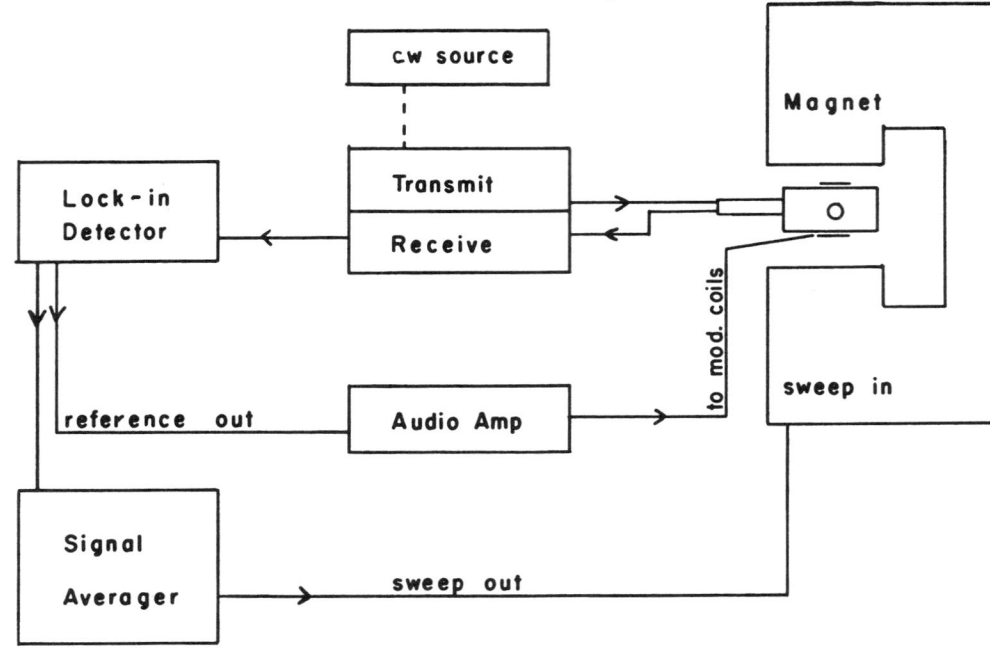

FIG. 8.   Block diagram of the continuous-wave NMR spectrometer.

each other, transmit the RF field and detect the time-varying magnetic flux (due to the sample magnetization), respectively. The NMR signal is proportional to the energy absorbed by the spin system. Care must be taken, however, in increasing the RF power since saturation effects could occur.

Some NMR signals from solids are conveniently narrow and strong so that they can be accumulated much faster than traditional lock-in sweep times. Nonadiabatic superfast passage (NASP) methods have been developed that allow the experimentalist to use higher power and larger sweep rates in obtaining good spectra (Segel et al., 1983; Bray et al., 1988). The external field can be swept quickly through the use of Helmholtz coils. The limitations, outside of saturation considerations, of these techniques hinge upon the magnitude and rate at which one can sweep the Helmholtz coils. Larger sweeps imply the use of larger coils which may affect field homogeneity.

B. Pulsed Methods

Upon placement of a sample into a magnetic field, H_o, a bulk magnetization will proceed to form. The sample magnetization will reach equilibrium at a rate governed by T_1. This occurs as the nuclear spins tend to align themselves with the external field. By subjecting the sample to a powerful transverse RF pulse operating at v_o, the nuclear spins will be tilted away from the magnetic field H_o. Immediately after the pulse, the bulk magnetization, **M**, of the sample will have rotated relative to its equilibrium direction. The longer the pulse duration, the greater the tilt of the magnetization. The angle of the tilt that the magnetization vector develops in a time t_p is given by:

$$\Theta = \gamma H_1 t_p \tag{23}$$

where γ is the gyromagnetic ratio of the nucleus undergoing the excitation, H_1 is the magnitude of the RF field, and t_p is the pulse duration. The pulse that results in a magnetization tilt of $\Theta = \pi/2$ or 90° is called a 90-degree pulse. Similarly, the tilt of $\Theta = \pi$ or 180° is termed a 180-degree pulse. For this case, the magnetization vector **M** is found antiparallel to the external field direction. Pulses can be applied to rotate the magnetization to any desired angle by varying the pulse duration (Fig. 9).

Application of a pulse other than the 180-degree pulse leaves a component of magnetization in the x-y plane. This component will precess about the external field and thus can be detected by a receiver coil oriented orthogonal to H_o. As mentioned earlier, the x-y component will decay in a time T_2. Hence, the output of the detector is a decaying sinusoidal signal called a free induction decay or FID. The FID presents the NMR response as a function of the time after a pulse has been applied.

The precessional frequency of the nucleus is dependent upon the local nuclear environment. Local fields reside within $\pm\Delta H$ of H_o; therefore, precessional frequencies are spread uniformly within $\pm\Delta v$ of v_o. The richness of this distribution in precessional frequencies is reflected in the features of the FID. The usual NMR absorption curve, which presents the power absorption as a function of frequency, can be generated by a Fourier transformation of the FID (Fukushima and Roeder, 1981).

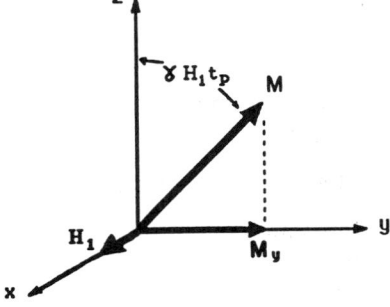

FIG. 9.   Action of an RF pulse of width t_p in rotating the bulk magnetization **M** through the angle $\gamma H_1 t_p$ in the rotating reference frame. The component of **M** in the $x - y$ plane is proportional to the detected signal strength.

Application of a square-wave pulse of duration t_p bathes the sample in a spectrum of frequencies of width $1/t_p$. Hence, a pulse experiment can excite all of the nuclei in a sample if the width in frequency of the absorption curve is less than $1/t_p$. Continued repetition of the pulse and summing each acquired FID with a signal averager can rapidly yield a signal strength exceeding that obtained after longer periods of CW signal averaging. (The appropriate time delay t between pulses must satisfy $t \gg T_1$ in order to avoid sample saturation.)

C. Spin-Lattice Relaxation-Time Measurements (T_1)

Two popular methods for measuring T_1 are the inversion-recovery and the saturation-recovery methods (Fukushima and Roeder, 1981). The inversion-recovery experiment utilizes a $180° - \tau - 90°$ pulse sequence where τ is the time delay between the 180 and 90° pulses. For such a sequence, the sample magnetization, $M(\tau)$, as a function of time (τ) behaves as

$$M(\tau) = M_o(1 - 2e^{-\tau/T_1}). \qquad (24)$$

M_o is the equilibrium magnetization magnitude. By monitoring the magnetization intensity as a function of the delay between pulses (τ) and plotting $1 - M/M_o$ on a semilog graph versus τ, a straight line of slope $(-1/T_1)$ and intercept $\ln 2$ can be obtained. Hence, the spin-lattice relaxation time can be obtained by measuring the slope.

The saturation recovery pulse uses a $90° - \tau - 90°$ sequence to obtain the same information as the inversion-recovery sequence. The advantage of the saturation-recovery sequence is that the magnetization is initially flipped into the x-y plane, not into the complete antiparallel position, so that M_o can be measured easily. Therefore, one does not have to wait as long for the magnetization to recover fully in order to implement the next sequence. After the first 90-degree pulse a time interval, τ, is allowed to pass. In this time, the magnetization begins to recover. After τ, a second 90-degree pulse is applied that rotates the recovering magnetization back down into the x-y plane where it is detected as a FID. The saturation-recovery sequence yields for $M(\tau)$

$$M(\tau) = M_o \, (1 - e^{-\tau/T_1}). \qquad (25)$$

Therefore, a simple semilog plot of $1 - M/M_o$ versus τ will yield the time constant T_1.

D. Spin–Spin Relaxation-Time Measurements (T_2)

For simple cases where there are no quadrupolar or anisotropic chemical-shift interactions (considering perfect field homogeneity), the value for T_2 can be obtained directly from the accumulated FID. The envelope of the FID decays as $\exp(-t/T_2)$. Therefore, through a semilog plot a linear dependence is found for the logarithm of the FID envelope with respect to time (t); the slope of the plot yields the value for T_2.

The presence of magnetic-field inhomogeneity will prevent the direct measurement of T_2 from the FID envelope. Nevertheless, alternative methods have been devised that succeed in overcoming this hindrance by subjecting the sample to echo trains and recording the resultant echo-amplitude maxima as a function of time (Slichter, 1978; Fukushima and Roeder, 1981). The echo train is a pulse sequence which successfully refocuses the FID after dephasing. The envelope of echo maxima traces a decaying, $\exp(-t/T_2)$, behavior from which T_2 can be deduced. As with T_1, T_2 measurements, obtained as functions of temperature and frequency (Fig. 7), are important in understanding thermally activated motional processes in glasses.

E. ACTIVATION-ENERGY MEASUREMENTS

Motional line-narrowing studies can give insight into the behavior of ionic conduction in glass. Ionic motion is described as "hopping from site to site." Therefore, the site is understood as an energy well of depth E_A (the activation energy for the site). The time that a hopping ion will spend in the site is given by the correlation time, τ_c. The Arrhenius expression, Eq. (21), indicates that τ_c and, hence, the ionic motion are temperature-dependent.

Activation energies can be obtained through consideration of T_1 phenomena. The log of measured spin-lattice relaxation times can be plotted as a function of inverse temperature $1/T$ (Fig. 7) as indicated by Eqs. (20,21). Ideally, this should yield symmetric "V" curves where the asymptotic slopes give the activation energy. However, in most situations for glasses as just indicated, there are presumably distributions in activation energies and one does not obtain ideal "V" curves. Distribution functions must be employed in order to adequately describe the behavior (e.g., the Cole-Davidson distribution as used by Göbel et al. (1979) in their analysis of ^{7}Li NMR responses in various silicate glasses).

Another useful method for obtaining activation energies involves the measurement of the NMR linewidth as a function of inverse temperature. These results, when displayed on a semilog plot (Fig. 7), can be fit to an appropriate model, whereby the activation energies are obtained.

One such model involves the BPP liquid-relaxation model derived by Bloembergen, Purcell and Pound (Bloembergen et al., 1948; Bloembergen, 1961; Abragam, 1961). A modified version of the BPP model has been made by Gutowsky and Pake (1950) and Gutowsky and McGarvey (1952) in order to explain motional narrowing in solids. This has had some success but does not account for all the observed behavior of relaxation in glasses. The modified BPP formula is given as:

$$v_j = \alpha W \cot \left[\frac{\pi}{2} \left(\frac{W^2 - D^2}{A^2 - D^2} \right) \right] \tag{26}$$

and

$$v_j = v_o e^{-E_A/k_B T} \tag{27}$$

where v_j is the hopping frequency (inverse of the correlation time), α is a constant of order unity, W is the halfwidth (at half intensity) of a lower temperature (wider) line, D is the halfwidth of the narrowed line at higher temperature, and A is the square toot of the second moment of the rigid lattice line.

Some activation energies obtained through conductivity measurements have been found to be substantially larger than values gathered from NMR linewidth measurements. The accuracy of a model obviously depends upon its compatibility with other measurements. The Hendrickson-Bray (HB) model (Hendrickson and Bray, 1973), has been used successfully for various ionic conducting glass systems. It is employed to describe motional narrowing when the conducting species is understood to be found in both mobile and stationary sites. The HB equation is given as:

$$W = \frac{A}{\left[1 + \left(\frac{A}{B} - 1 \right) e^{-E_A/k_B T} \right]} + D \tag{28}$$

where W is the linewidth measured as a function of temperature, A is the rigid lattice linewidth, B is the high-temperature linewidth, and D is the field inhomogeneity correction. For measurements carried out in homogeneous fields a semilog plot of $(1/W - 1/A)$ versus $(1/T)$ gives a straight line whose slope yields the activation energy.

F. SIGNAL-ENHANCEMENT TECHNIQUES

Brief discussions will be given for three of the more popular techniques used in reducing linewidths. At this time, their application in the study of glasses has not been completely exploited, yet it is expected that general usage of such techniques is imminent.

1. Line-Narrowing through Magic-Angle Sample Spinning

Dipolar interactions are reduced almost to zero by the averaging caused by rapid motions in liquids. Solids, however, display large dipolar broadenings since the rigid lattice does not allow the averaging that occurs in liquids. Often for solids, structural information can be concealed by large dipolar broadenings. For example, isotropic chemical-shift measurements can give valuable information concerning the structural environment, but accurate values and fine chemical-shift detail may be unobtainable due to dipolar broadening. Magic-angle sample-spinning techniques (Andrew and Eades, 1953; Andrew et al., 1958) and variable-angle sample-spinning (VASS) techniques (Ganapathy et al., 1982) can effectively reduce both homo- and heteronuclear dipolar interactions by ultrafast rotation of the entire sample about an axis tilted from the direction of H_o by the magic angle of 54.7° (or the variable angle). At the magic angle, the interaction terms containing the factor $(3 \cos^2 \theta - 1)$ are averaged out. Hence, as can be seen from Eq. (8), sample spinning at 54.7° will reduce the dipolar influence to zero.

This averaging of terms containing $(3 \cos^2 \theta - 1)$ does not occur just for dipolar effects. Axial and anisotropic contributions to the chemical shift (Eq. (16)) can broaden spectral features, but rapid sample rotation at the magic angle can eliminate the axial contribution and reduce the anisotropic angularly dependent terms as well. The effect of averaging out the anisotropic terms increases with the spinning rate. Similarly, MASS reduces the influence of first-order quadrupolar effects. (See Eq. (11).)

Ideally, the spinning rate should be greater than or equal to the static linewidth. Spinning rates of several kilohertz are generally necessary to narrow sufficiently ^{29}Si and ^{31}P resonances in glass samples. However, NMR responses of most other nuclei in glasses exhibit broadening by quadrupolar and anisotropic chemical-shift interactions; hence, their linewidths can be larger than any physically realizable spinning rate. Application of MASS to very wide lines, produced by such broadening mechanisms, can yield an entire set of narrow responses, separated by the spinning frequency; the envelope of this set of narrow responses reproduces the absorption response. If the spinning rate is not too low with respect to the breadth of the static line, the spectrum can be reduced to one narrow response (centered about the center-of-gravity) flanked by the associated narrow responses called "spinning sidebands."

2. Line-Narrowing through Multipulse Sequences

Pulse sequences can be employed to narrow a line sufficiently, without requiring rotation of the sample. They typically employ a repetitive sequence of pulses and are referred to as spin-flip narrowing (Slichter, 1978) or multipulse (Fyfe, 1983) sequences. The first multipulse sequence, developed by Waugh et al. (1968), succeeds in averaging out the dipolar interaction, thereby eliminating this effect in spectra. Even though heteronuclear dipolar interactions and anisotropic chemical-shift interactions cannot be reduced, homonuclear dipolar interactions can be effectively reduced by these techniques.

3. Heteronuclear Dipolar Decoupling and Cross Polarization

In the event that dipolar interactions between unlike spins contribute significantly to the width of the resonance line, hence obscuring spectral detail, MASS or heteronuclear dipolar decoupling schemes can be employed which can reduce the linewidth. MASS will reduce all heteronuclear dipolar effects to zero; however, decoupling techniques have the feature of selectivity. That is, specific nuclei, in principle, can be decoupled from unlike nuclei being observed, thereby reducing the contribution to the linewidth. Upon subjection of the sample to a decoupling RF field centered at the Larmor frequency of the "interfering" nucleus (while also imposing the resonance RF field at v_o for the nuclei under study), one achieves a mixing up of spins inside the interfering spin system. If the decoupling field is powerful enough, the nonobserved interfering spin system will be rapidly flipped and the corresponding heteronuclear dipole contribution to the Hamiltonian will be time averaged to zero. As a result, the observed resonance line will be narrowed.

Sensitivity problems may arise for solids containing a dilute concentration of the NMR nuclei of interest and/or containing nuclei with naturally low sensitivities. In such cases, direct observation of the resonance may prove difficult or impossible if heteronuclear dipolar effects are large enough. One can, under special conditions, transfer magnetization from one spin system to another in order to increase the sensitivity. This technique is called cross-polarization (CP). The weakly abundant or insensitive nuclei become thermally connected to a larger reservoir of abundant polarized spins. Thermal contact is made by maintaining the Hartmann-Hahn condition (Abragam, 1961; Fukushima and Roeder, 1981):

$$\gamma_a H_a = \gamma_b H_b. \tag{29}$$

Either H_a or H_b is the decoupling field, the other field being the RF field of the observed nucleus. The γ values are the gyromagnetic ratios of the decoupled and observed nuclei. The magnetically polarized spin system is maintained through application of a strong pulse of long duration. As indicated by Eq. (29), energy can be transferred between the decoupled and observed nuclear-spin systems such that the polarization of the insensitive nuclear spins becomes enhanced. Therefore, the added polarization of the insensitive nuclei aids in increasing the strength of the signal.

IV. NMR Investigations of Glass Systems

A. STUDIES OF B_2O_3 AND SiO_2 GLASS

B_2O_3 and SiO_2 are perhaps the most important constituents of oxide glasses. It is, therefore, important to understand basic physical properties of these major glass formers. Structural and relaxational information, which may be impossible to gather through any other means, has been obtained through ^{10}B, ^{11}B, ^{17}O, and ^{29}Si NMR.

The ^{11}B isotope, a quadrupolar nucleus having $I = 3/2$, is 81.17% abundant and exhibits a moderately large gyromagnetic ratio. Therefore, sensitive NMR measurements can be performed without isotopic enrichment. Natural abundance ^{11}B NMR structural investigations of B_2O_3 glass were carried out first by Silver and Bray (1958). Through observation of the second-order quadrupolar broadened lineshape, they were able to ascertain that the boron atoms were unequivocally bound to three oxygen atoms. The ^{11}B coupling constant (Q_{cc}) was found to be 2.76 ± 0.05 MHz (assuming $\eta = 0$). Tetrahedrally bound boron atoms, implying greater symmetry at the nuclear site, would reveal much narrower spectra due to a reduced quadrupolar interaction; such spectral features were not observed for ^{11}B in B_2O_3 glass. Further support of these results was indicated by the calculated quadrupole coupling of 2.6 MHz for ^{11}B in planar symmetric $BO_{3/2}$ groups by Bassompierre (1953).

Later studies (Kline et al., 1968) of crystalline B_2O_3 indicated that similar or equivalent trigonal $BO_{3/2}$-type units, which comprised the glass, form the crystal. The trigonal ^{11}B response yields the same value of Q_{cc} for the glass as for the crystal; however, the glass ^{11}B derivative spectrum (Fig. 10) appears more rounded (less intense) in the lower-frequency region and more sharp in the higher-frequency region when compared to the crystalline lineshape. This subtle difference in lineshape indicates a distribution in quadrupolar parameters Q_{cc} and η. According to Zachariasen's proposal of glass structure (Zachariasen, 1932), such distributions among nuclear sites is to be expected. The observation of the same structural unit for the glass as in the crystal for B_2O_3 supports the theory, proposed by Krogh-Moe (1965), that the glass is made of structural units present in the crystal. More comprehensive experiments involving ^{10}B, ^{11}B, and ^{17}O NMR (Kriz and Bray, 1971; Jellison and Bray, 1976; Jellison et al., 1977) were performed in order to gain further

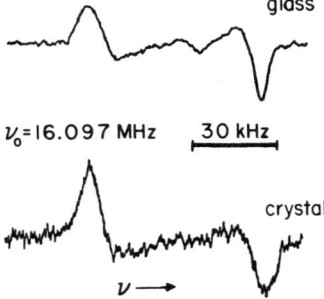

FIG. 10. The ^{11}B NMR spectrum observed for vitreous B_2O_3 (top) and crystalline B_2O_3 (bottom). (Kline et al., 1968.)

insight into the distributions of the quadrupolar parameters and to better understand what these distributions mean in terms of the structure of B_2O_3 glass.

Through observation of the ^{10}B nucleus ($I = 3$) in isotopically enriched B_2O_3 glass, it was found (Jellison et al., 1977) that certain portions of the ^{10}B spectrum are sensitive exclusively to distributions in either Q_{cc} or η (Fig. 11). This is exploited in order to find the widths (assuming Gaussian distributions) of these quadrupolar distributions. Possible deviations from $\eta = 0$ for the boron environment, as mentioned by Silver and Bray (1958), indicate that the electron distributions about boron nuclei are slightly distorted from perfect axial symmetry. This is reflected in the average η value of 0.12 where the distribution in η is represented by the Gaussian width $\sigma_\eta = 0.043$. The proposed structure of B_2O_3 glass is that of randomly oriented six-membered boroxol rings (Krogh-Moe, 1965; however, there remain doubts due to differences in results gathered from infrared, NMR, and more recent NQR studies). Two oxygen sites are expected, one corresponding to oxygen atoms in boroxol rings [O(R)] and one corresponding to oxygen atoms outside, but connected to, the boroxol rings [O(C)]. For a glass consisting entirely of boroxol rings, the ratio of the number of O(R) oxygen atoms to O(C) oxygen atoms is 2:1. ^{17}O NMR can be used to measure the oxygen ratio directly. Since the NMR-sensitive isotope for oxygen (^{17}O: $I = 5/2$) is only 0.032% abundant, isotopic enrichment is necessary for adequate signal to noise. Jellison et al. (1977) used CW NMR to obtain ^{17}O quadrupolar broadened derivative lines (Fig. 12). Their computer simulation reveals two sites: one site is characterized by a zero distribution in η and is identified as O(R); the other site displays a large distribution in η and is identified as O(C). Better simulations were obtained upon a 6:5 weighting of the two sites, O(R) to O(C). In contrast to these earlier findings, ^{11}B NQR spectra of B_2O_3 glass gathered by Gravina (1988) indicate the presence of two distinct boron environments possibly arising from ring $BO_{3/2}$ (boroxol) and linking $BO_{3/2}$ boron sites.

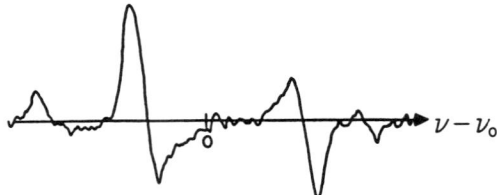

FIG. 11.   The ^{10}B NMR spectrum observed for B_2O_3 glass (derivative of the absorption response). $\nu_0 = 7$ MHz. (Jellison and Bray, 1976.)

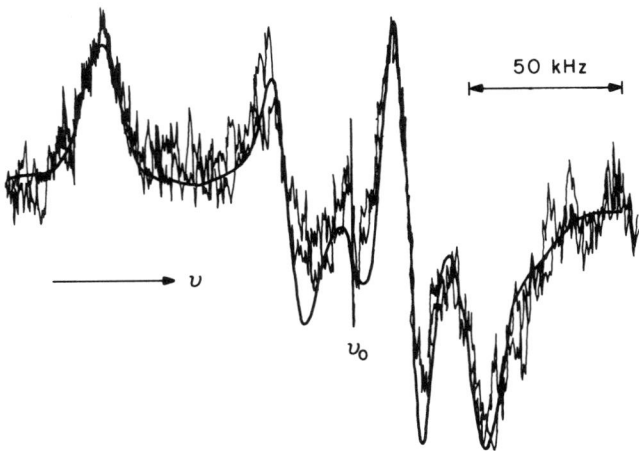

FIG. 12. Experimental ^{17}O derivative response with accompanying two-site computer-generated simulation (solid line). $v_o = 7\,MHz$. (Jellison et al., 1977.)

France and Wadsworth (1982) have carried out additional ^{11}B NMR studies of vitreous B_2O_3. They measured directly the positions of the divergences and shoulders of the first-order quadrupolar powder pattern spectrum, then calculated, from these, the values for Q_{cc} and η. Their results are in good agreement with the values found by Jellison et al. (1977).

^{11}B spin-lattice relaxation times have been measured in B_2O_3 glass ($10\mu\,sec \leqslant T_1 \leqslant 100\,sec$) for temperatures ranging between 1.2 and 1000K. Szeftel and Alloul (1975) found that T_1^{-1} is virtually independent of frequency and varies as $T^{1.3}$ for temperatures between 1.2 and 77K. For temperatures between 150 and 500K, Rubinstein et al. (1975) and Rubinstein and Resing (1976) have indicated that nuclear spin-lattice relaxation occurs via a Raman mechanism (Van Kronendonk, 1954) that involves a two-phonon quadrupolar relaxation process. This process proceeds through the annihilation of an incident phonon at the nuclear site such that its energy is imparted to the flipping of the nuclear spin and creation of an energy-conserving phonon. The more familiar modified BPP theory is used by Rubinstein (1976) in describing T_1^{-1} throughout the temperature range 500–1000K (Fig. 13).

Unlike the ^{10}B and ^{11}B NMR spectra for B_2O_3, the ^{29}Si spectra of SiO_2 are not quadrupolar broadened ($I = 1/2$ for ^{29}Si); therefore, subtle structural information can only be gotten through analysis of the ^{29}Si dipolar and chemical-shift interactions. Since quadrupolar relaxation is not possible, spin-lattice relaxation times for ^{29}Si in SiO_2 glass have been measured to be many hours (Holzman et al., 1955). T_1 can be greatly reduced through the incorporation of paramagnetic impurities into the glass. Presumably, this is

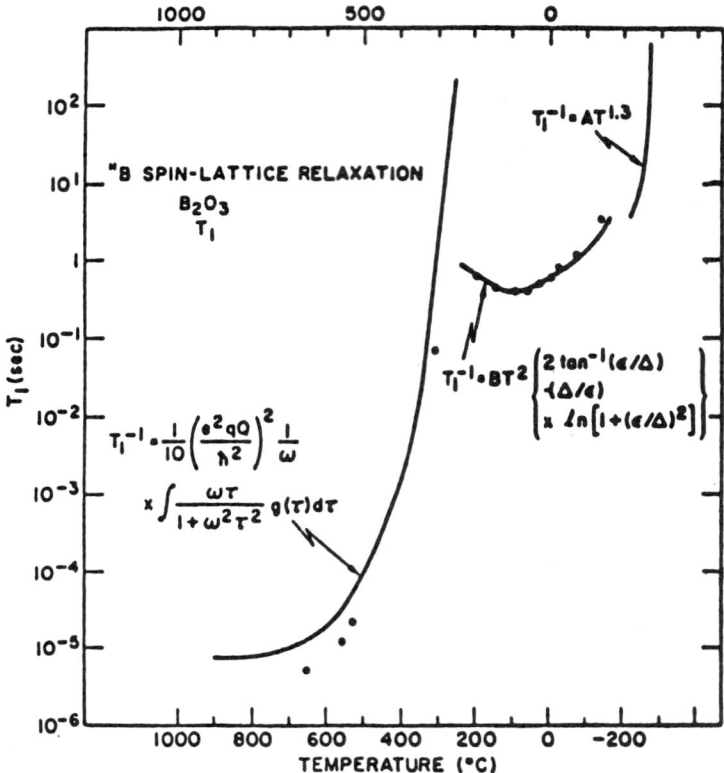

FIG. 13. Temperature variation of ^{11}B spin-lattice relaxation time T_1 in B_2O_3 glass at 15.450 MHz. Data points are black dots. (Rubinstein, 1976.)

the cause of lower T_1 values as reported in other SiO_2 studies. NMR experiments are further hampered by the moderately low sensitivity of the ^{29}Si nucleus. Nevertheless, prior to high-field Fourier transform techniques, important chemical-shift and spin-lattice relaxation-time measurements in silicate glasses were made using high RF power in detecting the silicon dispersion signal (Holzman et al., 1956).

SiO_2 glass is understood to be a random network of $SiO_{4/2}$ tetrahedra, where the silicon is bound to the oxygen atoms. The oxygen atoms, in turn, are bound to two silicon atoms. ^{17}O CW NMR studies by Geissberger and Bray (1983) gave, through evaluation of the quadrupolar parameters, the most probable hybridization of oxygen and the range of bond angles (α) for the Si—O—Si link. Computer simulation of the ^{17}O lineshape yielded most probable values for Q_{cc} and η (Q_{cc}^o and η^o) along with the respective Gaussian variances $\sigma_{Q_{cc}}$ and σ_η. (Refer to Fig. 14.) A Townes-Dailey calculation

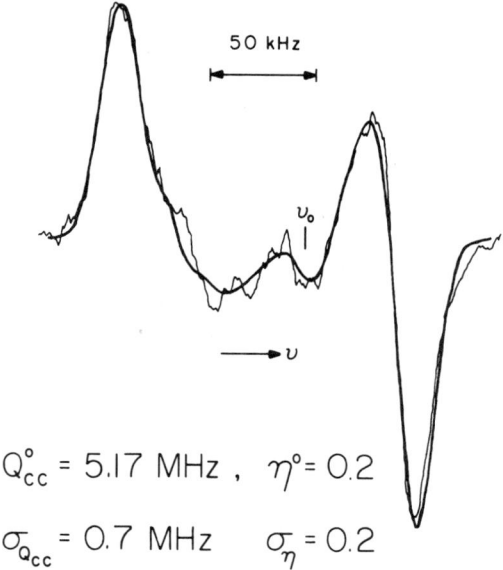

$$Q^o_{cc} = 5.17 \text{ MHz} \, , \quad \eta^o = 0.2$$

$$\sigma_{Q_{cc}} = 0.7 \text{ MHz} \quad \sigma_\eta = 0.2$$

FIG. 14. The one-site ^{17}O NMR experimental spectrum and its computer simulation (smooth line) for SiO$_2$ glass. $\nu_o = 7$ MHz. (Geissberger and Bray, 1983.)

(Townes and Dailey, 1948) employing the ^{17}O quadrupole parameters weakly supports the presence of π bonding between silicon 3d and oxygen 2p orbitals. The Si—O—Si bond angle, α, was found to have an average value of 144° and a range of 130 to 180°.

High-field MASS ^{29}Si NMR investigations of SiO$_2$ have been performed by Dupree and Pettifer (1984). By simulating the ^{29}Si MASS spectrum using a variety of Si—O—Si bond-angle distribution functions, they have found previous models to be inconsistent with the experimental results. The asymmetric MASS ^{29}Si line is best described when simulated using a broad distribution function for α with a nonzero probability inside the range $120° < \alpha < 180°$ and nearly constant probability over the range $140° \leqslant \alpha \leqslant 155°$. Murdoch et al. (1985) have measured ^{29}Si chemical shifts for SiO$_2$ and have calculated Si—O—Si bond angles (Smith and Blackwell, 1983) to range between 135 and 160°. Oestrike et al. (1987) have performed similar measurements but indicate, through a different analysis, an average Si—O—Si bond angle of about 151° and a range of 130 to 170°.

B. STUDIES OF MODIFIED BORATE GLASSES

Alkali oxide is usually incorporated into B$_2$O$_3$ glass ionically. The alkali metal (e.g., Li$^+$) becomes associated with either a negatively charged tetrahedrally bound boron atom ($BO_{4/2}^-$) or with a nonbridging oxygen (which is in turn bound to a trigonally coordinated boron). Furthermore, the glass will be composed of structural groupings (boroxol, tetraborate, diborate, metaborate, pyroborate, and orthoborate) present in the various crystal structures of the alkali borate compounds (Fig. 15). The structural groupings are comprised of uncharged trigonal units like those associated with the boroxol group, $BO_{3/2}$; singly charged tetrahedral units as in the diborate group, $BO_{4/2}^-$; singly charged trigonal units with one NBO (non-bridging oxygen), BO_2^- (as in the metaborate group); doubly charged trigonal units with two NBOs, $BO_{5/2}^{-2}$ (as in the pyroborate group); and triply charged trigonal units with three NBOs, BO_3^{-3} (orthoborate). The abundance and type of structural unit present depend upon the amount of alkali oxide incorporated into the glass.

Second-order quadrupolar broadened ^{11}B NMR spectra (Fig. 16) reveal different responses for the various structural units. The tetrahedral response is characterized by a small EFG at the nuclear site, due to the symmetric charge distribution of the surrounding electrons. Hence, the resulting $BO_{4/2}^-$ response appears narrow and featureless. Typical $BO_{4/2}^-$ quadrupolar values for alkali and alkaline earth borate glasses are $Q_{cc} \simeq 250$ to $700\,kHz$ and $\eta \simeq 0$. Both the trigonal uncharged boroxol $BO_{3/2}$ and the trigonal charged orthoborate BO_3^{-3} groups yield low values for η (0 to 0.15) yet display large values for Q_{cc} (2.4 to 3.0 MHz). The NMR responses of these units are similar, indicating the comparable symmetries of their nuclear environments. The charged trigonal metaborate BO_2^- and pyroborate $BO_{5/2}^{-2}$ units yield similar

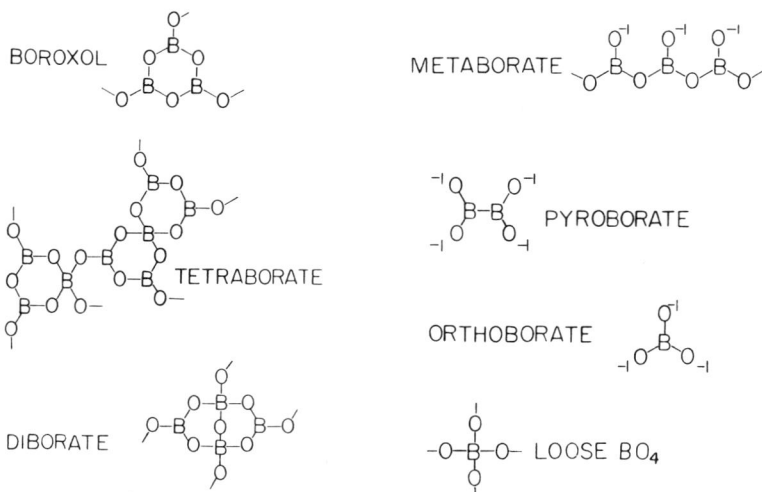

FIG. 15.   Structural groupings found for modified borate glasses.

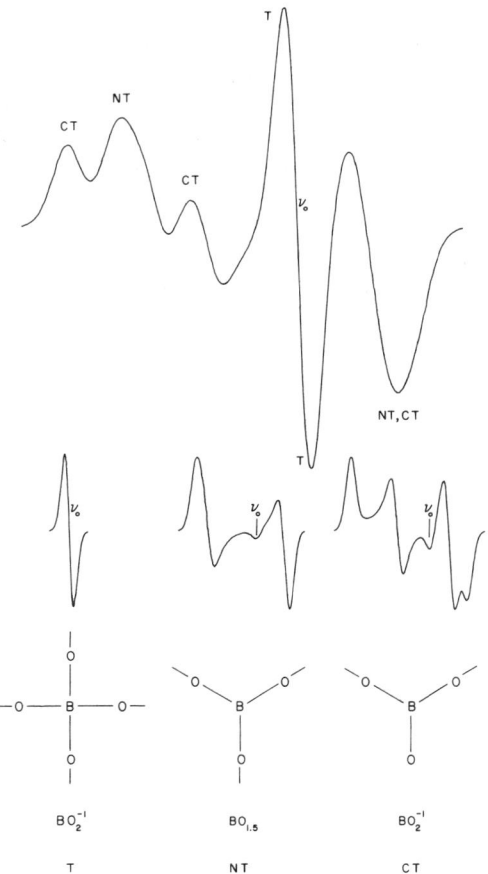

FIG. 16. The derivative spectrum of a dipolar broadened and second-order quadrupolar broadened ^{11}B NMR response. The spectrum is comprised of a weighted sum of spectra due to boron atoms in distinct environments. The contribution due to 4-coordinated boron atoms in $BO_{4/2}^-$ units is represented as T, that due to neutral 3-coordinated boron atoms in $BO_{3/2}$ units as NT (a similar contribution would result for boron atoms in charged BO_3^{-3} units), and that due to charged asymmetric 3-coordinated boron atoms in BO_2^- units as CT. (A similar contribution would result for boron atoms in charged asymmetric 3-coordinated boron atoms in $BO_{5/2}^{-2}$ units.)

^{11}B responses, since the asymmetry in their nuclear environments is comparable. Typical quadrupolar parameters for these units are $Q_{cc} \simeq 2.4$ to 3.0 MHz and $\eta \simeq 0.4$ to 0.6. Since all boron atoms are either bound to three or four oxygen atom neighbors, the fraction of boron atoms present in the glass as $BO_{4/2}^-$ (N_4) can be measured by taking the ratio of the area under the narrow peak of the ^{11}B absorption spectrum, A_4, to the total area under the spectrum, $A_3 + A_4$. (A_3 is the area associated with the all trigonally coordinated boron atoms. Refer to Fig. 17.) Through computer simulation of the lineshape, A_3 can be decomposed into the area arising from uncharged borate or orthoborate units and the area arising from charged asymmetric metaborate and/or pyroborate units. Ratios of these individual areas to the total spectral area yield the fractions of $BO_{3/2}$ or BO_3^{-3} units (N_{3S}) and BO_2^- or $BO_{5/2}^{-2}$ units ($N_{3A} = 1 - N_{3S} - N_4$). Complete structural analysis involves the determination of N_4, N_{3S}, and N_{3A} as functions of the molar compositional parameter R; $R = [X_2O]/[B_2O_3]$, where $[X_2O]$ and $[B_2O_3]$ are the molar contents of alkali oxide and boron oxide, respectively.

1. Structural and Motional Studies of Li_2O-B_2O_3 Glasses

Figure 18 indicates that the behavior of N_4 with R can be understood by dividing the plot into regions (Feller et al., 1982; Yun and Bray, 1981).

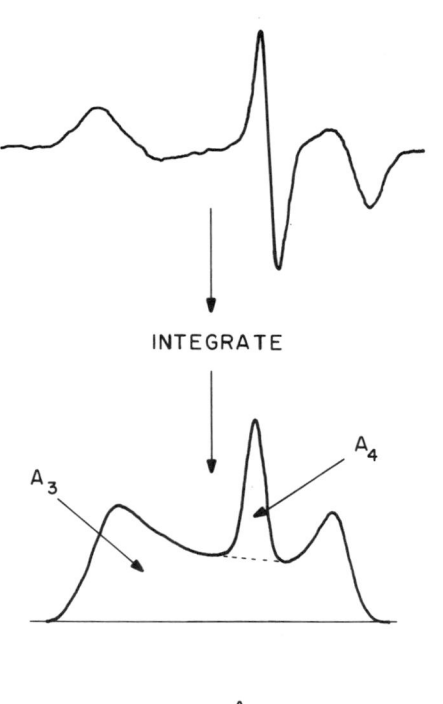

INTEGRATE

A_3

A_4

$$N_4 = \frac{A_4}{A_3 + A_4}$$

FIG. 17.   Determination of N_4 by the area method for a typical ^{11}B NMR spectrum.

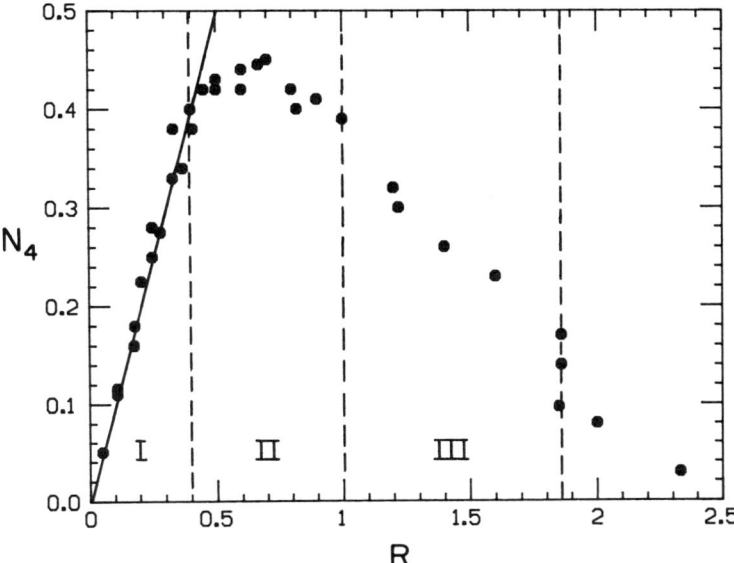

FIG. 18. N_4 versus R for RLi_2O-B_2O_3 glasses. The solid line gives the $N_4 = R$ behavior. The dashed lines designate the three regions discussed in the text.

Region 1 $(0 \leqslant R \leqslant 0.4)$: Glasses in this region are composed of boroxol, tetraborate, and diborate units. As R increases, the number of tetraborate and diborate groups increase at the expense of boroxol groups. (Zhong, 1988, show in Fig. 19 that for alkali borate glasses, the $N_4 = R$ rule in general does not hold for $R > 0.4$.)

Region 2 $(0.4 \leqslant R \leqslant 1.0)$: This region is characterized by the advent and increase of metaborate and loose $BO_{4/2}^-$ units with increasing lithium content. Tetraborate and diborate concentrations gradually decrease to zero at $R = 1$.

Region 3 $(1.0 \leqslant R \leqslant 1.86)$: This region reveals that decreasing metaborate and loose $BO_{4/2}^-$ contents give way to the advent and increase of pyroborate and orthoborate groups. (Other relevant references concerning the structure of lithium borate glasses are Bray and O'Keefe, 1963; Bray and Hendrickson, 1974; and Jellison et al., 1978.)

^{7}Li NMR studies of lithium borate and silicate glasses have been performed by Hendrickson and Bray (1974). Lithium ionic motion in these glasses has been interpreted based on the HB motional-narrowing theory (Eq. (28)). Results indicate the presence of short-range ion motion; however, upon increasing temperature, long-range ion motion sets in. The long-range

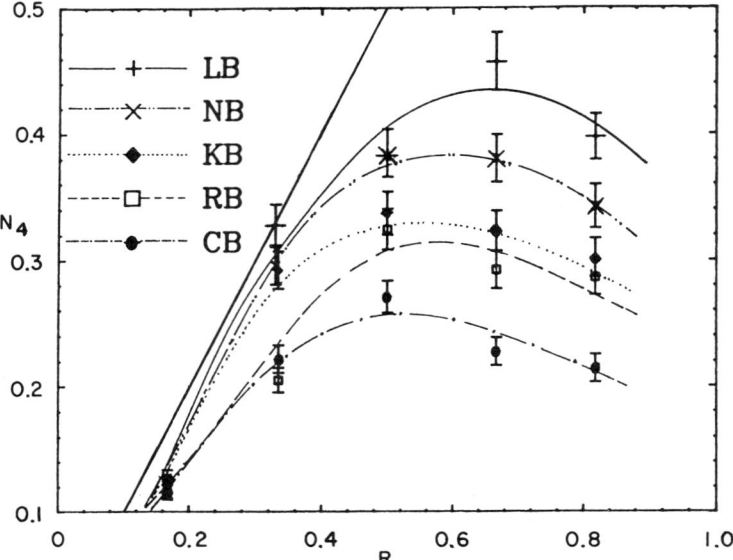

FIG. 19. N_4 versus R for various alkali borate glasses. The linear N_4 behavior deviates more so for increased–atomic-weight alkali ions. (LB, NB, KB, RB, and CB designate lithium, sodium, potassium, rubidium, and cesium borate glasses, respectively.) (Zhong and Bray, Mixed-Alkali-Borate Glasses".)

motion, proceeding through an interstitial ion "hopping" mechanism, is characterized by large activation energies. The dynamic behavior observed in these glasses varies with temperature, composition, and the degree of homogeneity of the glass.

Göbel et al. (1979) give T_1 results for different lithium borate glasses. Their findings indicate two relaxation mechanisms (confirming previous observations) responsible for Li^+ cation motion. These authors were able to describe much of the observed behavior of lithium motion through analysis by a Cole-Davidson distribution of correlation times. (See Eqs. (21) and (22).)

Tabbey and Hendrickson (1980) evaluated quadrupolar effects and the second-moment of the central line for ^7Li in borate glasses. Their results suggest lithium clustering (possibly pairing) since the experimentally derived second moments are larger than those calculated when assuming homogeneously distributed lithium atoms.

2. Structural Studies of Na_2O-B_2O_3 Glasses

The structural model based upon the results of Greenblatt and Bray (1967b) and Yun and Bray (1978a) proposes that N_4 behavior with R be divided into two regimes. NMR results gathered for region 1 ($R \leqslant 0.5$)

support Krogh-Moe's postulate of $N_4 = R$. When $N_4 = N_{4max} \simeq 0.5$, region 2 begins and, as R increases further, N_4 values fall off linearly as $N_4 = N_{4max} - 0.25(R - R_{max})$. N_{3S} values increase as $N_{3S} = 1.25(R - R_{max})$. (Further studies include Bray and O'Keefe, 1963; and Jellison et al., 1978.)

Krämer et al. (1973) observed [23]Na resonances in two borate glasses of large and small Na_2O contents. [23]Na halfwidths of 47 kHz were measured; unfortunately, the spectra were not sensitive to variations in composition and temperature. Analysis was presumably complicated by second-order quadrupolar effects, indicating in these glasses a rather broad distribution in sodium ion sites.

3. Structural Studies of K_2O-B_2O_3 and Rb_2O-B_2O_3 Glasses

Compared to lithium and sodium borate glasses, potassium and rubidium borate glasses are increasingly hygroscopic and have smaller glass-forming regions. Nevertheless, structural studies have been carried out indicating increasing N_4 values for $R < 0.5$. The potassium borate system (Svanson et al., 1962) gives $N_4 = R$ where $0 \leqslant R \leqslant 0.43$. N_4 values for $R > 0.43$ are lower than postulated by the $N_4 = R$ rule, indicating the presence of metaborate and/or pyroborate structural units. The rubidium borate glass system (Bray and O'Keefe, 1963) displays an even greater digression from the $N_4 = R$ rule. Here it seems N_4 increases with R up to about $R = 0.11$. At this point N_4 values increase at a smaller rate than for the potassium, sodium and lithium borate glasses indicating a smaller fraction of tetrahedrally coordinated borons and a greater fraction of asymmetric trigonal boron units with R. (Refer to Fig. 19.)

4. Structural and Motional Studies of Cs_2O-B_2O_3 Glasses

^{11}B NMR results indicate that N_4 values increase with R at a rate lower than for the rubidium borate glass system. Within experimental error it is found that $N_4 \simeq R$ for $R < 0.30$. (Results by Zhong and Bray, 1988), question the exact cutoff and suggest $N_4 = R$ up to $R = 0.2$ for cesium and rubidium borate glasses.) Near $R = 0.66$ the presence of asymmetric BO_2^- and/or $BO_{5/2}^{-2}$ units is clearly seen in the [11]B spectrum (Rhee and Bray, 1971b).

[133]Cs spectra reveal broadening due to chemical-shift anisotropy as well as quadrupolar effects. As pointed out by Rhee and Bray (1971b) the cesium environment is usually understood as being ionic in borate glasses; however, due to its chemical-shift anisotropy (developing due to covalent bonding that destroys the spherical symmetry about the nuclear site), some covalent character appears to be present. The degree of cesium covalency increases with cesium content.

[133]Cs motional studies (Rhee and Bray, 1971a) reveal substantial line narrowing for temperatures greater than 25°C and constant linewidths for

temperatures less than or equal to 25°C. Typical cesium-activation energies in these glasses are comparable to activation energies of lithium atoms in Li_2O-B_2O_3 glasses (0.10 to 0.13 eV).

5. Structural Studies of MgO-B_2O_3 Glasses

The extremely narrow glass-forming region for the MgO-B_2O_3 system allows the incorporation of only 43 to 45 mol% MgO in the glasses. Park et al. (1979) observed that the ^{11}B spectrum for a 45MgO-55B_2O_3 glass ($R \simeq 0.82$) gives responses from both uncharged trigonal $BO_{3/2}$ and tetrahedral $BO_{4/2}^-$ units with $N_4 = 0.20$ and $N_{3S} = 0.80$. ^{11}B quadrupole parameters were found to be $Q_{cc} = 2.61$ MHz and $\eta = 0.18$ for the $BO_{3/2}$ response. The NMR results do not show the broad lines that are associated with asymmetric units (hence no NBOs) and furthermore, since $N_4 < R$, it is concluded that some of the MgO must be incorporated into the glass-forming network covalently. The Mg^{++} ions are employed in converting $BO_{3/2}$ units to $BO_{4/2}^-$ units. Reexamination of the metaborate crystalline compound (MgO-B_2O_3) by Dell and Bray (1982) led these authors to conclude that NBOs are not present in any appreciable amount in this system (confirming previous results) and that the 45 mol% glass is composed of diborate and boroxol structural groups. Similar ^{11}B NMR studies performed on Na_2O-MgO-B_2O_3 glasses (Kim and Bray, 1974b) indicate that magnesium resides ionically below 15 mol% MgO; in higher amounts, it gradually becomes incorporated covalently, possibly in the form of MgO_4 tetrahedra.

6. Structural Studies of CaO-B_2O_3 Glasses

Bishop and Bray (1966) studied six glasses of varying CaO content with R ranging from 0.33 to 1.0. Their findings show that $N_4 \simeq R$ for 30 mol% or less CaO. Above 30 mol%, the N_4 values depart from $N_4 = R$, achieving a maximum value of 0.536 at $R = 0.67$. Unlike the MgO-B_2O_3 glass, CaO-B_2O_3 glasses (with increasing R) reveal the broadened NMR response due to charged trigonal metaborate and/or pyroborate units. No computer simulations of the boron lineshape were performed by these authors; therefore, accurate values for N_{3S} and N_{3A} were not reported.

7. Structural Studies of SrO-B_2O_3 Glasses

Park and Bray (1972) found N_3 and N_4 for strontium borate glasses ($0.37 \leqslant R \leqslant 0.75$). N_4 is observed to increase but fall short of the $N_4 = R$ prediction for $0.37 \leqslant R \leqslant 0.50$. At $R = 0.50$, the slope of the N_4 curve increases such that within experimental error $N_4 = R$ for $0.60 < R < 0.7$. The glass composition 2SrO-B_2O_3 ($R = 0.67$) is found to give $N_4 = 0.67$ and $N_{3A} = 0.09$. Obviously the one-to-two correspondence of strontium ions to singly charged structural units and the one-to-one correspondence of strontium ions to pyroborate structural units is not maintained. This is an interesting result because it implies the presence of trigonally coordinated oxygen atoms (in order to conserve charge). The crystal structure of SrO-$2B_2O_3$ employs oxygen atoms in both two- and threefold coordination with boron; however, the glass of equal composition, upon evaluation of N_4 and N_{3A}, shows no evidence of three-coordinated oxygen. It seems that for SrO-B_2O_3 glasses, threefold coordinated oxygen is favored only for higher SrO content.

8. Structural Studies of BaO-B_2O_3 Glasses

Earlier studies of this glass system (Greenblatt and Bray, 1967a) revealed N_4 behavior similar to that found in the CaO-B_2O_3 system. There were no anomalous results, as with the SrO-B_2O_3 glass system, and N_4 appeared to increase according to the $N_4 = R$ rule up to $R \simeq 0.4$. Stallworth and Bray (1988) have reinvestigated barium borate glasses using both lock-in detection and NASP techniques. It is found that the previous N_4 values are consistently high. The more recent measurements, interprpeted through computer simulation of the ^{11}B response, indicate throughout the entire glass-forming region ($0.30 \leqslant R \leqslant 0.70$) the presence of metaborate and pyroborate structural units. Therefore, the N_4 values fall short of the $N_4 = R$ prediction. The glass of composition $R = 0.70$ yields the highest N_4 and N_{3A} values of 0.33 and 0.25, respectively, and yields the smallest value among the glasses for N_{3S} of 0.42.

9. Structural and Motional Studies of Ag_2O-B_2O_3 and Ag_2O-AgI-B_2O_3 Glasses

Kim and Bray (1974a) reported $N_4 = R$ for Ag_2O-B_2O_3 glasses where $0.05 < R < 0.5$ (i.e., all silver ions are involved in converting an equal number of three-coordinated borons to four-coordinated borons). Therefore, these glasses are structurally similar to the alkali borates. A comparison of the ^{11}B NMR spectra for the glasses and crystalline compounds of the system confirms the proposal (Krogh-Moe, 1965) that the glasses are made of the same structural units found in the crystalline compounds. Later ^{11}B studies of silver borates by Chiodelli et al. (1982) support the previous findings and suggest that the Ag^+ ions are associated with tetrahedral borons of diborate groups such that there is a Ag-Ag pair distance of about 3.5Å.

Upon addition of AgI to Ag_2O-B_2O_3 glasses, conductivities can be increased to the order of $10^{-2}(\Omega cm)^{-1}$. However, it seems that the short-range order of the borate network is not affected by this addition (Chiodelli et al., 1982). Initial ^{109}Ag NMR linewidth studies by Martin et al. (1986) have revealed for Ag_2O-AgI-B_2O_3 glasses that at low temperatures (less than

200K), there are two sites for the Ag^+ ions. At room temperature, all Ag^+ ions are mobile and can be described by a single correlation time.

[109]Ag isotropic chemical-shift measurements for these glasses have also been made by Villa et al. (1986a). However, their interpretation does not employ a two-site description due to a lack of evidence distinguishing Ag^+ ions in purely iodide environments from Ag^+ in borate (oxide) environments. In light of the results of Martin et al. (1986), note is made of the fact that the measurements of Villa et al. were obtained at room temperature or higher. In conclusion, Villa et al. interpret the observed [109]Ag chemical shifts and lineshapes as resulting from Ag^+ sites characterized by a distribution of activation energies.

10. Structural Studies of PbO-B_2O_3–Based Glasses

^{11}B NMR results indicate the presence of only uncharged trigonal $BO_{3/2}$ and tetrahedral $BO_{4/2}^-$ units over the entire glass-forming range ($0.25 \leqslant R \leqslant 3.0$) of the PbO-B_2O_3 system (Kim et al., 1976); the amounts of BO_2^- and $BO_{5/2}^{-2}$ units are negligible. N_4 is approximately equal to R only when $R \leqslant 0.25$. Higher R values yield lower N_4 values than predicted. It is concluded that lead is incorporated into the glass not only as an ionic modifier but as a covalent glass network former. [209]Pb NMR chemical shifts indicate increasing amounts of covalent versus ionic lead with increasing PbO content.

xFe_2O_3-$yPbO$-zB_2O_3 glasses where ($0 \leqslant x \leqslant 15.3\,mol\%$, $z/y = 3$) were studied by Bucholtz and Bray (1983). They determined the effect of paramagnetic ions (Fe^{+++}) on the ^{11}B and [209]Pb NMR responses. The combined effects of paramagnetism and sample magnetization inhomogeneity yield [11]B lineshapes influenced by both Lorentzian and Gaussian broadening interactions. Increasing Lorentzian broadening with larger field strengths, due to the paramagnetic ions, result in distortion of the $BO_{4/2}^-$ NMR response such that N_4 values gathered at higher fields are lower (and less accurate) than values gathered at lower fields. [209]Pb chemical shifts indicated increased Pb^{++} ionic character with Fe_2O_3 content up to $4\,mol\%$. For higher Fe_2O_3 content, chemical shifts were found to correspond to an increased covalent character for lead.

The anionic conducting glass PbF_2-PbO-B_2O_3 has been studied from both structural and dynamical perspectives by Vopilov et al. (1985). ^{11}B NMR suggests that the PbF_2 content does not influence the borate glass structure appreciably. Studies of the ^{19}F NMR lineshape versus temperature reveal broader resonances at low temperatures and a single narrower resonance at high temperatures (450K). The simultaneous presence of both broad and narrow lines at intermediate temperatures is indicative of the presence of both highly mobile (narrow line) and less mobile (broad line) fluorine ions.

11. Studies of Alkali-Halo Borate Glasses

Structural and dynamic studies for Li_2O-LiF-B_2O_3 and Li_2O-$LiCl$-B_2O_3 glasses employing ^7Li, ^{11}B and ^{19}F NMR have been performed by Geissberger et al. (1982). ^{11}B spectra were analyzed for various compositions, yielding the important conclusion that the borate network was not modified by the lithium halide additive; that is, the lithium halide did not cause the formation of $BO_{4/2}^-$, BO_2^-, or $BO_{5/2}^{-2}$ structural units. Magnetic tagging experiments (Bray et al., 1983b) were also performed indicating: (1) that the narrowing of the ^7Li NMR response resulted from the generation of mobile lithium ions as the temperature was increased and (2) that the narrowing of the ^{19}F response resulted not from fluorine motion but from the modulation of the local magnetic field at the fluorine nuclear site due to the moving lithium nuclei. ^7Li T_1 studies of Li_2O-$LiCl$-B_2O_3 glasses by Villeneuve et al. (1979) indicate activation energies near 0.2 eV for lithium ion sites. These results indicate similarities with the Ag_2O-AgI-B_2O_3 glasses and thereby reflect the dual interpretation of lithium ion motional behavior as a two-site (bound and unbound) model and/or as a collection of sites characterized by a distribution of activation energies.

In contrast to the preceding glass system, structural studies performed on Na_2O-NaF-B_2O_3 glasses indicate modification of the borate network due to NaF. Kline and Bray (1966) evaluated ^{11}B and ^{19}F NMR responses in designating BO_3F and BO_2F_2 tetrahedral structural groups in these glasses. These authors observed broadened responses of the $1/2 \leftrightarrow -1/2$ transition for four-coordinated boron units due to B-F heteronuclear dipolar interactions. ^{11}B measurements by Jäger and Haubenreißer (1985) in a reexamination of this system indicate that only the BO_3F is present in the glass. Their interpretation is based on a more thorough study of the dipolar interaction influencing the boron nuclear environment. Expectedly, ^{11}B NMR spectra gathered from corresponding glasses incorporating NaCl instead of NaF do not exhibit the strong heteronuclear dipolar effects since the magnetic moment of chlorine isotopes ^{35}Cl and ^{37}Cl are much smaller than that of fluorine. N_4 measurements performed on glasses in this system (Hintenlang, 1985) indicate that BO_3Cl modification does not occur, and thus the system is believed to behave similarly with the Li_2O-$LiCl$-B_2O_3 glass system discussed previously.

KF-B_2O_3 glasses (10 to 50 mol% KF) studied through NMR by Müller-Warmuth et al. (1970) display N_4 behavior characteristic of K_2O-B_2O_3 glasses (i.e., $N_4 = R$). Surprisingly, borate-network modification due to fluorinated tetrahedral borate units (i.e., BO_3F) is not supported through their NMR results, since B-F dipolar interactions are not reflected in the unexpectedly narrow ^{11}B responses. Hintenlang (1985), however, did observe the broadened ^{11}B line indicative of BO_3F units in spectra taken for a 48 mol% KF borate glass.

12. ^{11}B NMR Studies of Alkali Copper Borate Glasses

Simon and Nicula (1981) used ^{11}B NMR to investigate the influence of glass liquidus temperatures upon the structure of Na_2O-CuO-B_2O_3 glasses for $0.05 \leqslant R \leqslant 0.5$ with CuO content of 0.5 and 1.0 mol%. They found that for glasses fabricated in alumina crucibles at liquidus temperatures above 1000°C, substantial reductions in N_4 values occur as compared to glasses made at lower temperatures. Reinvestigation of this system by Lui (Bray et al., 1983b) using platinum rather than alumina crucibles, failed to confirm the original results. It is probable that the large reduction in N_4 at high temperatures occurred due to the incorporation of Al_2O_3 into the glasses from the alumina crucibles.

$xCuO - (1-x)(K_2O$-$2B_2O_3)$ glasses (where $0 \leqslant x \leqslant 20$ mol%) have been studied by Ardelean et al. (1982) through ^{11}B NMR. Their results show that Q_{cc} values for three-coordinated borons increase from 2.62 to 2.70 MHz as x is increased from 0 to 15 mol%. The increase in Q_{cc} was interpreted as resulting from borate-network distortions as more CuO is incorporated into the glass.

13. Studies of ZnO-B_2O_3 and Tl_2O-B_2O_3 Glasses

Wide-line ^{11}B NMR was employed in determining the structure of ZnO-B_2O_3 glasses (Harris and Bray, 1984). N_4 values range continuously from about 0.25 for $R = 0.8$ to $N_4 = 0.1$ for $R = 2.0$. The N_4 results are generally larger for bulk glasses cooled slowly than for corresponding samples quenched through splat-cooling. These glasses are found not to adhere to the Krogh-Moe theory (1965) used in the structural interpretation of alkali borate glasses.

^{11}B, ^{203}Tl, and ^{205}Tl NMR have been used in structural studies of Tl_2O-B_2O_3 glasses (Baugher and Bray, 1969). For $0 \leqslant R \leqslant 0.25$, Tl^+ ions modify $BO_{3/2}$ units to $BO_{4/2}^-$ units but not as predicted by the $N_4 = R$ rule. Instead, it is found that experimental N_4 values are larger than their corresponding R values. This can be explained by considering the glass to contain threefold coordinated oxygen atoms (in analogy to SrO-B_2O_3 glasses) in association with $BO_{4/2}^-$ units. For $0.25 \leqslant R \leqslant 0.67$, thallium ions convert $BO_{3/2}$ units to $BO_{4/2}^-$ units in one-to-one fashion. Above $R = 0.67$, NBOs are formed at the expense of $BO_{4/2}^-$ units. $^{203,205}Tl$ chemical-shift studies indicate thallium to be distributed among ionic and covalent sites. Ionic thallium is favored at lower Tl_2O content while covalent thallium is more prevalent at higher Tl_2O content. A later [205]Tl study (Panek and Bray, 1977) suggests pairing or clustering of thallium atoms as the thallium content is increased.

C. STUDIES OF MODIFIED SILICATE GLASSES

Investigations of alkali silicate glasses using high fields, MASS, and pulsed NMR techniques for observation of the [29]Si nucleus have indicated that alkali ions are incorporated ionically (interstitially) resulting in the stepwise building of NBOs on silica tetrahedra. The various silica structural groups are designated as Q^4 for the uncharged $SiO_{4/2}$ unit having no NBOs, Q^3 for the singly charged $SiO_{5/2}^-$ unit containing one NBO, Q^2 for the doubly charged SiO_3^{-2} unit containing two NBOs, Q^1 for the triply charged $SiO_{7/2}^{-3}$ unit with three NBOs and Q^0 for the quadruple charged SiO_4^{-4} unit with all NBOs (Fig. 20). ^{29}Si NMR spectra are influenced by dipolar and chemical-shift interactions. MASS techniques allow one to eliminate both dipolar and anisotropic chemical-shift effects thereby leaving all diagnostics in the evaluation of the isotropic chemical shift. Pioneering studies by Lippmaa et al. (1980, 1982) and Grimmer et al. (1984) plus further explorations by Dupree et al. (1984), Smith et al. (1983), and Mägi et al. (1984) have shown that in both silicate crystals and glasses the [29]Si resonances of the various Q^n structural units yield characteristic isotropic chemical shifts. Relative to tetramethyl-silane (TMS), designated chemical shifts range systematically (about 10 ppm for each conversion of a bridging oxygen to a NBO; Table I) from about -110 ppm for the Q^4 unit to about -65 ppm for the Q^0 unit (Fig. 21). The peak position of a particular Q^n site in general varies slightly with the content and type of alkali modifier. Quantitative analysis can be achieved through deconvolution of the [29]Si lineshapes in terms of distributions in Si-O-Si bond angles and isotropic chemical shifts of weighted Q^n sites.

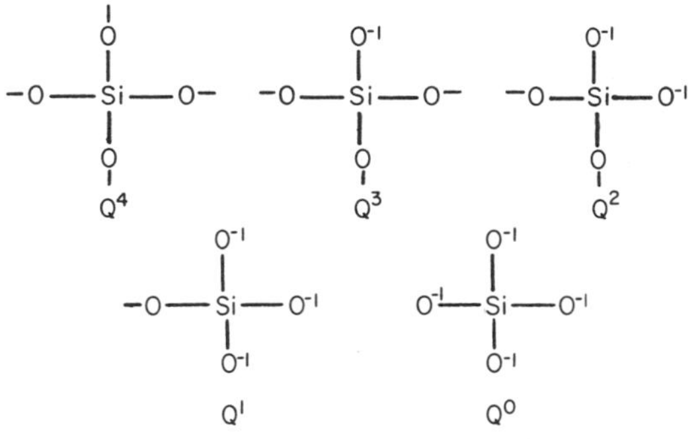

FIG. 20. Q^n silicate structural units. The superscripts indicate the number of associated bridging oxygens.

TABLE I

^{29}Si Isotropic Chemical Shifts Relative to TMS for Alkali-Alkaline Earth-Silicate Crystalline Compounds

Q^n	Compound	Chemical shift (ppm)	Reference
Q^4	SiO_2	-108	Smith et al. (1983)
Q^3	$Li_2Si_2O_5$	-92.5	Murdoch et al. (1985)
Q^3	$Na_2Si_2O_5$	-94.5	Murdoch et al. (1985)
Q^3	$K_2Si_2O_5$	$-91.5, -93, -94.5$	Murdoch et al. (1985)
Q^3	$BaSi_2O_5$	-93.5	Murdoch et al. (1985)
Q^2	Li_2SiO_3	-74.5	Mägi et al. (1985)
Q^2	Na_2SiO_3	-76.8	Mägi et al. (1985)
Q^2	$MgSiO_3$	-81	Smith et al. (1983)
Q^2	$MgCaSi_2O_6$	-84	Mägi et al. (1985)
Q^2	α-$Ca_3(Si_3O_9)$	-83.5	Mägi et al. (1985)
Q^2	$CaSiO_3$	-88.5	Murdoch et al. (1985)
Q^2	$SrSiO_3$	-85	Smith et al. (1983)
Q^2	$BaSiO_3$	-80	Smith et al. (1983)
Q^1	$Li_6Si_2O_7$	-72.4	Grimmer et al. (1984)
Q^1	$Na_6Si_2O_7$	-68.4	Mägi et al. (1985)
Q^1	$Ca_3Si_2O_7$	$-74.5, -76.0$	Mägi et al. (1985)
Q^1	$MgCa_2Si_2O_7$	-73	Smith et al. (1983)
Q^0	Li_4SiO_4	-64.9	Mägi et al. (1985)
Q^0	Be_2SiO_4	-74.2	Mägi et al. (1985)
Q^0	Mg_2SiO_4	-61.9	Mägi et al. (1985)
Q^0	α-Ca_2SiO_4	-70.3	Mägi et al. (1985)
Q^0	$MgCaSiO_4$	-66	Smith et al. (1983)
Q^0	Ba_2SiO_4	-70.3	Mägi et al. (1985)

Much of the observed structural behavior of alkali silicate glasses can be understood by considering the "binary" or "constrained" Q^n distribution model. The model indicates that Q^3 units will continually be formed from Q^4 units as alkali ions become added to the silicate glass matrix. This process continues until all Q^4 units are converted, the glass then being composed of Q^3 units (x = mole fraction alkali oxide = 1/3). Upon further addition of alkali oxide, Q^3 units will be converted to Q^2 units until all Q^3 units become exhausted ($x = 1/2$). This sequential action continues through the advent and saturation of Q^1 ($x = 3/5$) and Q^0 units (ending finally at $x = 2/3$ with only Q^0 units).

1. Structural Studies of Li_2O-SiO_2 Glasses

Earlier studies were concerned, for the most part, with lithium diffusion and relaxation phenomena (Svanson and Johansson, 1970; Hendrickson and Bray, 1974; Göbel et al., 1979) yet it was determined from ^{7}Li lineshape

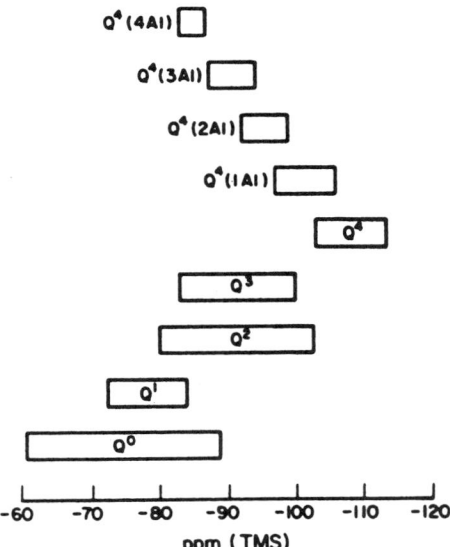

FIG. 21. Ranges of isotropic chemical shifts for tetrahedral silicon sites with aluminum second-nearest neighbors ($Q^4(nAl)$) and silicon second-nearest neighbors. ppm scale relative to TMS. (Smith et al., 1983.)

studies that the lithium sites in silicate glasses are distributed much like those sites in borate glasses (Krämer et al., 1973). High-resolution ^{29}Si studies of various representative calcium silicate crystals composed of only one structural unit Q^n allowed Grimmer et al. (1984) to assign peaks and relative intensities of MASS spectra gathered from Li_2O-SiO_2 glasses. Hence, they were able to monitor silicon atoms involved in Q^4, Q^3, and Q^2 sites. For the four glasses studied spanning the region ($15 \leqslant Li_2O$ mol% $\leqslant 40$), it was found that only two structural units were present for a given composition, except at 33 mol% Li_2O where the glass was comprised entirely of Q^3 units.

Schramm et al. (1984) conducted a thorough investigation into the weightings of the Q^n sites using MASS NMR. It is indicated that for $15 \leqslant Li_2O$ mol% $\leqslant 40$ there is a statistical distribution (Lacy, 1967) of Q^n units present for a given composition. Q^4 units occur for Li_2O contents even as high as 35 mol%. This does not adhere to the preceding (binary) model, which implies that at most two Q^n species are present at any given composition. As indicated by the authors, their lithium silicate glasses were phase-separated into lithium-rich and silica-rich regions (which is especially symptomatic of glasses with lower Li_2O contents). However, with higher lithium contents greater homogeneity is achieved. There is better agreement between experimental and calculated (binary model) Q^n distributions for glasses of higher lithium content (Fig. 22).

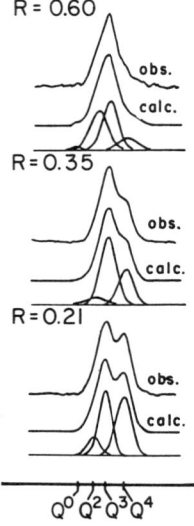

R = 0.60

obs.

calc.

R = 0.35

obs.

calc.

R = 0.21

obs.

calc.

Q^0 Q^2 Q^3 Q^4

FIG. 22. ^{29}Si MAS NMR spectra of three lithium silicate glasses of composition $RLi_2O\text{-}SiO_2$. Also included are the calculated spectra and Gaussians used to construct calculated spectra. ppm scale relative to TMS. (Schramm et al., 1984.)

The width and position of the MASS ^{29}Si peak in glasses not only depend upon the distribution of Q^n units, but can vary subject to changes in the Si-O-Si bond angle. Selvaray et al. (1985) studied distributions in Si-O-Si bond angles in glasses at 15, 33.3, and 40 mol% Li_2O. They assumed Gaussian distributions in isotropic chemical shifts and applied a relation given by Smith and Blackwell (1983) (expressing the shift value as a linear function of the secant of the bond angle, α) to arrive at the most probable values of α for the Q^4, Q^3, and Q^2 units.

2. Structural Studies of $Na_2O\text{-}SiO_2$ Glasses

Unlike the lithium silicate glasses, sodium silicate glasses display homogeneity for the Na^+ ions. However, Mosel et al. (1974) give evidence of nonhomogeneous distribution or cation pairing in sodium silicate glasses (8 to 33 mol% Na_2O). Both Grimmer and Dupree et al. (1984a) have shown through ^{29}Si MASS NMR that the structure of $Na_2O\text{-}SiO_2$ glasses (0 to 50 mol% Na_2O) can be interpreted according to the binary model previously given (homogeneous distribution of sodium ions). As shown in Figs. 23 and 24, the distribution, considering the presence of at most two Q^n units at a time, can be expressed as a function of x (mole fraction alkali oxide):

$$Q^4 = 1 - 2\left(\frac{x}{1-x}\right), \qquad Q^3 = 1 - Q^4, \qquad \text{for} \quad 0 \leqslant x \leqslant \frac{1}{3} \qquad (30a)$$

$$Q^3 = 2 - 2\left(\frac{x}{1-x}\right), \qquad Q^2 = 1 - Q^3, \qquad \text{for} \quad \frac{1}{3} \leqslant x \leqslant \frac{1}{2} \qquad (30b)$$

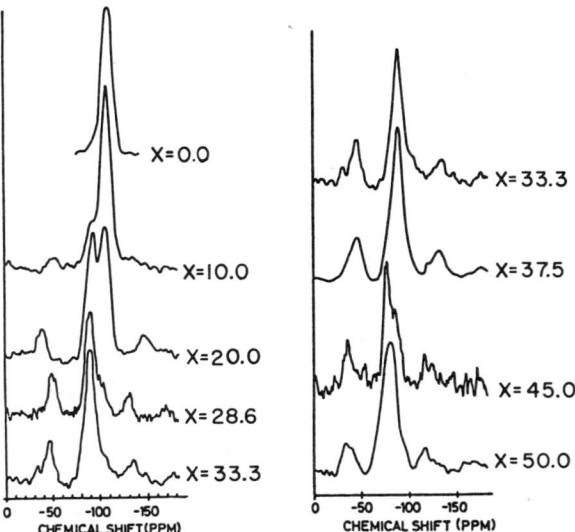

FIG. 23. ^{29}Si spectra obtained by MAS NMR for a series of sodium silicate glasses where X is the mole % Na_2O (ref. TMS). (Dupree et al., 1984.)

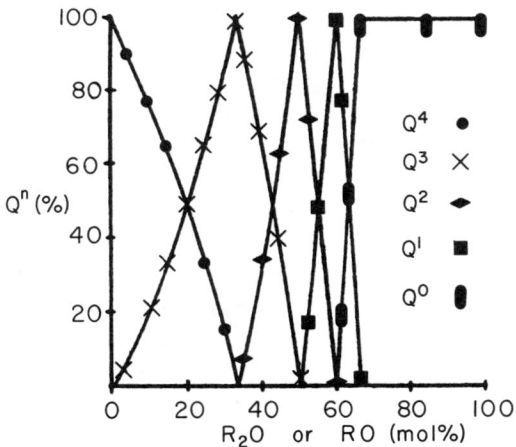

FIG. 24   Binary distribution of Q^n species as a function of modifier oxide mole %. (Dupree et al., 1987.)

$$Q^2 = 3 - 2\left(\frac{x}{1-x}\right), \qquad Q^1 = 1 - Q^2, \qquad \text{for} \quad \frac{1}{2} \leqslant x \leqslant \frac{3}{5} \qquad (30c)$$

$$Q^1 = 4 - 2\left(\frac{x}{1-x}\right), \qquad Q^0 = 1 - Q^1, \qquad \text{for} \quad \frac{3}{5} \leqslant x \leqslant \frac{2}{3} \qquad (30d)$$

The preceding simple expressions are appealing since they have been used with success. However, as shown by Murdoch et al. (1985) through their deconvolution of the ^{29}Si MASS line into Q^4, Q^3, and Q^2 units (weighting of 8:84:8) for a 33.3 mol% Na_2O glass, small deviations from the binary prediction can occur.

Dupree et al. (1984, 1985b) have gathered ^{23}Na NMR responses for Na_2O-SiO_2 glasses. Even though quadrupolar interactions dominate the lineshape, peak positions can be monitored with changing sodium content. Increasing the sodium content from zero to about 30 mol% results in a shift of the position of the sodium peak (from about -2 to -5 ppm). It is found that there is a large discontinuity (from about -5 to 0 ppm) accompanied by a change in slope near 30 mol% Na_2O. Increasing the sodium content further results in larger shifts of up to about $+3$ ppm.

3. Structural Studies of K_2O, Rb_2O, and Cs_2O-SiO_2 Glasses

^{29}Si CW NMR was employed in the Harris and Bray (1980) study of potassium silicate crystals and glasses (0 to 54.5 mol% K_2O). Crystalline lineshapes were gathered and interpreted in terms of the asymmetry of the silicon nuclear site (i.e., Q^4, Q^3, and Q^2 structural units). Comparisons between crystalline and vitreous samples of the same composition reveal similar NMR responses indicating similar structural units comprising the samples. There is a drastic difference between spectra of high potassium content (33, 43.4, and 50 mol% K_2O) and one spectrum gathered from a glass containing 21.5 mol% K_2O. The interpretation (which is consistent with more recent MASS results of other alkali silicate glasses) is that this glass contains both Q^4 and Q^3 structural units and that the NMR response is a superposition of the two contributions.

Dupree et al. (1986) show the ^{29}Si MASS NMR response for a 20K_2O-80SiO_2 glass. Although the central peak appears symmetric, the positions of the sidebands are not symmetrically distributed about the central peak. This suggests that there are at least two Q^n sites. From compositional considerations, Q^4 units are most likely present; however, spinning sidebands are observed in the spectrum which can only be associated with less symmetric Q^3, Q^2, or Q^1 units. In substantiation of these results, Grimmer and Müller (1986) have investigated glasses for 0 to 50 mol% K_2O using MASS NMR. Their results support a binary Q^n distribution and hence suggest that structurally the potassium silicate glass system parallels the sodium silicate glass system.

Rubidium and cesium silicate glasses, for alkali oxide content of 45 mol % or less, appear to be structurally analogous with the sodium silicate glasses (Dupree et al., 1986) and can be described by the preceding binary model. Beyond 45 mol % alkali oxide, ^{29}Si spectra can only be analyzed by simulating with at least three Q^n sites (Q^3, Q^2, and Q^1) thus indicating a departure from the description of stepwise formation of NBOs in the binary model (Fig. 25). No phase separation or devitrification was reported for these samples; however, the authors suggest that disproportions among the Q^n sites occur through compensation by the silicate matrix for steric crowding and Coulombic repulsions between the large Rb^+ and Cs^+ ions.

^{133}Cs spectra gathered from Cs_2O-SiO_2 glasses show broad lineshapes that can be deconvoluted into two superimposed components (Dupree et al., 1986). The peak shifts approximately $+200$ ppm as the Cs_2O content is increased from 11.3 to 51.1 mol %. Since the ^{133}Cs nucleus has a small electric quadrupole moment, asymmetric chemical-shift and dipolar interactions

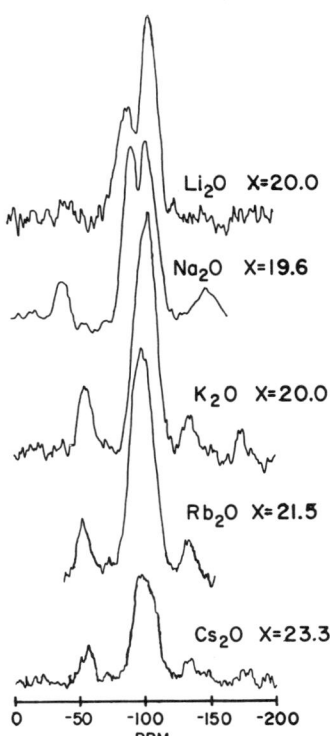

FIG. 25. ^{29}Si MAS NMR spectra for alkali silicate glasses, where X is the mole % alkali oxide (ref. TMS). (Dupree et al., 1986.)

probably give rise to the asymmetry and large breadths of the observed linewidths, especially at high fields. This is substantiated by Milberg et al. (1966) and Otto and Milberg (1967), who observed symmetric nonquadrupolar broadened ^{133}Cs resonances from silicate glasses near 10 kGauss. Furthermore, Milberg's low-field study indicates a discontinuity in cesium chemical shifts near 30 mol% Cs_2O, suggesting a possible similarity with the ^{23}Na response in silicate glasses. As indicated previously, anisotropic chemical-shift broadening in the cesium response of cesium borate glasses may indicate covalent character in the cesium site (Rhee and Bray, 1971b); high-field NMR data may suggest this for silicate glasses as well.

4. Studies of Alkaline Earth Silicate Glasses

Comparisons have been made of ^{29}Si MASS lineshapes and chemical shifts for alkali and alkaline earth disilicate and metasilicate glasses (33 and 50 mol% metal oxide, respectively) (Murdoch et al., 1985). These results show that for disilicates the major structural component is the Q^3 unit while for metasilicates it is the Q^2 unit. This is consistent with previous results from alkali silicate glasses indicating that the crystal and stoichiometric glass contain the same dominant structural unit. As mentioned previously, the positions of the peak maxima are characteristic of the dominant Q^n structural units. Single peaked spectra are found for all glasses (disregarding sidebands; however, spectra of Li_2O and Na_2O-$2SiO_2$ glasses have shoulders about the peak). For the most part, larger linewidths are observed for alkaline earth glasses perhaps implying greater distributions in Q^n species and nonadherence to the binary model. (Nevertheless Engelhardt et al., 1985, interpreted NMR data gathered for CaO-SiO_2 glasses according to a binary distribution.) A possible explanation of these effects, given by Murdoch et al. (1985), is the greater polarizing ability of the alkaline earth ions perturbs the silicate structure causing increased localization of negative charge (influencing the number of NBOs) on silica tetrahedra. This may result in an increased diversity in the Q^n distribution. Since the various Q^n units are characterized by different isotropic chemical shift values, the resultant NMR response will exhibit broadening due to the superposition of responses. Even though there may be broadening rendering ^{29}Si spectra irresolvable, Kirkpatrick et al. (1986a) have alternatively observed the ^{17}O MASS NMR response in MgO-CaO-SiO_2 glasses and crystals. These authors have assigned spectral features as arising from NBO (and associated cation) and bridging oxygen sites and indicate that oxygen structural environments in general differ markedly for the glass as compared to the corresponding crystals.

5. Studies of PbO-SiO_2 and Tl_2O-SiO_2 Glasses

Initial ^{209}Pb chemical-shift studies for PbO-SiO_2 glasses gave constant isotropic values when the lead oxide content is less than 50 mol%. This behavior reflects the similarity in environments of the ionic lead site. For lead content above 50 mol %, there is a continuous change in chemical shifts from ionic to more covalent values (Leventhal and Bray, 1965; Kim et al., 1976) (Fig. 26). Later studies involving the use of high fields and MASS in observation of the structurally sensitive ^{29}Si resonance by Lippmaa et al. (1982) reveal, upon gradual devitrification of 50 mol % PbO glass, that the silicon resonance evolves continuously (with increasing temperature and heat-treatment time) from a single broad peak to a spectrum consisting of three narrow peaks. The three narrow peaks were found to be indicative of the three silicon environments known to occur for the crystal alamosite. Also apparent in the spectra are the advent and gradual disappearance of a broader response characteristic of silica clustering during heat treatment. Employing FT (Fourier transform) NMR techniques, Fujiu and Ogino (1984) gathered low-field (21 kGauss) ^{29}Si resonances from a variety of lead silicate glasses. Their results indicate that the resonances can be quantitatively deconvoluted into a narrow line (-107 ppm) arising from Q^4 units and a broader line (-79 ppm) assigned as the superposition response of Q^3, Q^2, Q^1, and Q^0 type units that have both NBOs and Si-O-Pb bonds. Dupree et al. (1987) have employed ^{29}Si MASS NMR in determining the distribution of Q^n species in lead silicate glasses. Experimental peak positions do not agree with those of Fujiu and Ogino; however, there is agreement in the observation of large spectral widths proposed to arise from distributions of varied Q^n species. Evidence indicates that for glasses containing 30 mol % PbO or less (ionic lead sites), the binary model can account for the observed lineshapes, implying that the silicate network is comprised of Q^4 and Q^3 units. However, for glasses containing larger PbO contents, the broad lineshapes cannot be described through a binary Q^n distribution, but a statistical model (as used by Schramm et al., 1984) can successfully account for the larger distributions in Q^n sites.

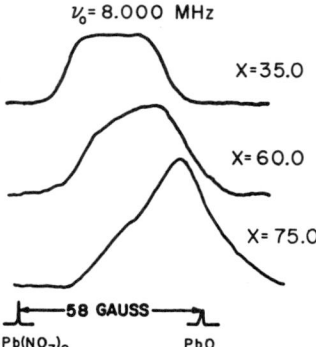

FIG. 26. ^{209}Pb NMR spectra (obtained with CW methods) from binary lead silicate glasses. (Kim et al., 1976.)

^{205}Tl resonances were observed at 24.6 MHz from Tl$_2$O-SiO$_2$ glasses containing about 10 to 40 mol% Tl$_2$O by Otto and Milberg (1967). Thallium lineshapes were observed to be quite broad at lower Tl$_2$O content, decreased linearly by 20% near 22 mol% Tl$_2$O, and maintained a constant breadth of about 29 Gauss for glasses of higher thallium content. These workers suggested that the large widths, due to chemical-shift interactions, reflect large distributions in thallium sites (assuming homogeneously spaced thallium atoms). Panek and Bray (1977) obtained ^{205}Tl resonances for these glasses, confirming previous measurements. These authors proved by studying ^{205}Tl-enriched samples (thereby reducing the ^{203}Tl/^{205}Tl isotopic ratio) that broadening of the line is not solely attributable to chemical-shift effects but that significant broadening due to heteronuclear dipolar exchange interactions between the two thallium isotopes is present as well. These results are consistent with X-ray measurements indicating that thallium ions appear in clusters and become distributed more homogeneously as the Tl$_2$O content increases.

6. Studies of Phosphorus and Fluorine Doped Silicate Glasses

^{31}P high-field FT NMR spectra acquired from two silica samples (prepared by modified chemical vapor deposition, MCVD) containing 1.12 and 0.06 wt% phosphorus were analyzed by Douglass et al. (1985) yielding chemical-shift, relaxational, and dipolar information for the phosphorus environments. Saturation-recovery data yield $T_1 = 38$ minutes for phosphorus. Upon comparison of ^{31}P isotropic chemical shifts gathered from these glasses to known values, it is found that the dominant phosphorus site can be assigned as a pentavalent POO$_{3/2}$ unit with one double-bonded oxygen and three bridging oxygens; resonances of pentavalent phosphorus with five bridging oxygens were not observed in the NMR spectrum. The rate of decay of the spin-echo intensities was modeled with good results yielding a random distribution of phosphorus atoms.

A similar study by the same group (Duncan et al., 1986) was performed in obtaining ^{19}F spectra from MCVD fluorine-doped silicate glasses (1.03 wt% F). Saturation-recovery measurements indicate a fluorine T_1 of 31 minutes. By considering the distribution in F–F homonuclear dipolar couplings through evaluation of the spin-echo decay rate, these authors were able to show that fluorine occurs primarily in the tetrahedral monofluoride SiO$_{3/2}$F species, and, if present, to a lesser degree as the tetrahedral difluoride SiO$_{2/2}$F$_2$ species. As was found with the phosphorus distribution, fluorine is observed to be distributed randomly with an F–F nearest-neighbor distance of 2.56Å.

D. STUDIES OF MODIFIED BOROSILICATE GLASSES

Many ternary glass systems can be formed by inclusion of a metal oxide in boron oxide/silicon oxide melts. The borosilicate matrix accommodates the metal oxide in varying degrees either ionically (e.g., Li^+ and Na^+) or covalently (e.g., PbO at higher lead oxide contents). This can be viewed in terms of a structural evolution as the continuum of glass compositions are realized from the pure modified B_2O_3 glass to the pure modified SiO_2 glass. As previously demonstrated for the binary borate and silicate glasses, NMR can serve as a useful tool in evaluating local structure. The bulk of work done on these systems has been in observation and evaluation of the ^{11}B response using CW or NASP methods (as with the binary borate glasses) in order to find the fraction of boron atoms involved in tetrahedral and various trigonal environments. Glass compositions are represented as R[metal oxide]-$[B_2O_3]$-$K[SiO_2]$; where R and K are given as the molar ratios: $R =$ [metal oxide]/$[B_2O_3]$ and $K = [SiO_2]/[B_2O_3]$. The data are presented as functions of R and K.

1. *Studies of the* $(R)Li_2O$-B_2O_3-$(K)SiO_2$ *Glass System*

The effect upon the glass structure, by the addition of SiO_2 to Li_2O-B_2O_3, is to: (1) "dilute" the borate matrix, (2) stabilize the $BO_{4/2}^-$ unit by allowing the glass to accommodate larger amounts of tetrahedral boron atoms, and (3) alter the fractions of tetrahedral and trigonal boron units for a given value of R (since the silicate network has a different affinity for the accommodation of the metal cation). In general, it is found that the metal cation will preferentially associate itself with the borate network over the silicate network, as is evidenced at low R where $N_4 \simeq R$ independent of K. Hence, as is shown in Fig. 27, N_4 values converge for all K families as R decreases. Beyond $R \simeq 0.5$, N_4 branches off, differentiating the trajectories of the various K families. For constant R values beyond 0.5, the stabilization of the $BO_{4/2}^-$ unit leads to higher N_4 values for increased K. Zhong and Bray (1988), using ^{11}B NMR, model this glass system from the borate perspective. Their model gives N_{3S}, N_4, N_{3A}, and the "break" points designating the advent of new structural units (Fig. 27).

2. *Studies of the* $(R)Na_2O$-B_2O_3-$(K)SiO_2$ *Glass System*

This system has been studied more than other borosilicates (Scheerer et al., 1973; Brungs and McCartney, 1975; Zhdanov and Shmidel, 1975; Yun and Bray, 1978a, 1979; Dell et al., 1983; Xiao and Meng, 1986) perhaps due to the technological importance of glasses based on the sodium borosilicate (e.g., Pyrex and Vycor glass). ^{11}B NMR studies, performed by Bray and coworkers

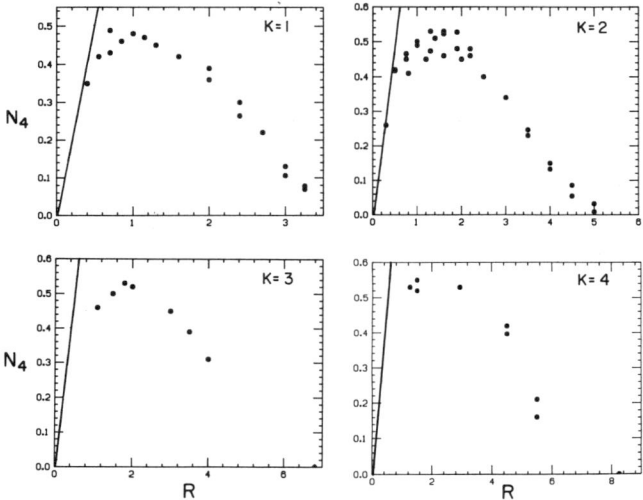

FIG. 27. N_4 versus R for lithium borosilicate glasses ($K = 1$, 2, 3, and 4). The solid straight line gives $N_4 = R$.

(Yun and Bray, 1978a, 1979; Dell et al., 1983), indicate how the borate structure accommodates Na_2O by observing the behavior of N_{3S}, N_4, and N_{3A} with R. As with most modified borosilicate glasses, the characteristic increase in N_4 for $R < 0.5$ is observed. The silicate matrix acts merely as a dilutant and for the most part all Na^+ ions are associated with $BO_{4/2}^-$ units. For R beyond 0.5, the behavior of N_4 reflects the occurrence of N_{3A} and charged Q^n units (Fig. 29). For constant R, N_4 increases with K, indicating the stabilization effect SiO_2 has on the $BO_{4/2}^-$ unit (Fig. 28). Proposed structural models (Dell et al., 1983; Xiao and Meng, 1986) for these glasses are based on the structural groupings (e.g., boroxol, diborate, metaborate) found in alkali borate compounds plus the $BO_{4/2}^-$ unit connected to four $SiO_{4/2}$ tetrahedra. Values of N_4 and N_{3A} (where $N_{3S} = 1 - N_4 - N_{3A}$) are derived from the model as functions of R and K (dashed lines in Figs. 28 and 29).

3. *Studies of the (R) K_2O-B_2O_3-(K)SiO_2 Glass System*

The structure of glasses in this system depends on composition in a manner analogous to that of the sodium borosilicate glasses. However, according to the results of Szu and Bray (1988) the $N_4 = R$ rule holds for R less than about 0.2 for all K. The lower $N_4 = R$ cut-off values are characteristic of alkali borate glasses made with the larger molecular weight alkali cations (Zhong and Bray, to be publ.). For R greater than 0.2, NBOs appear in the borate and silicate networks. Since $N_4 = R$ only when all of the alkali

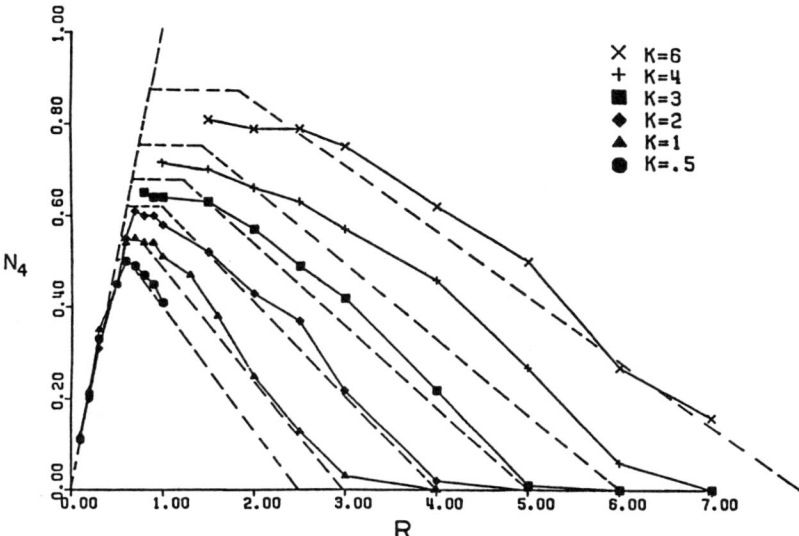

FIG. 28. N_4 versus R for sodium borosilicate glasses shown for different K families. The dashed lines are the predictions based on the model of Dell et al. (1983).

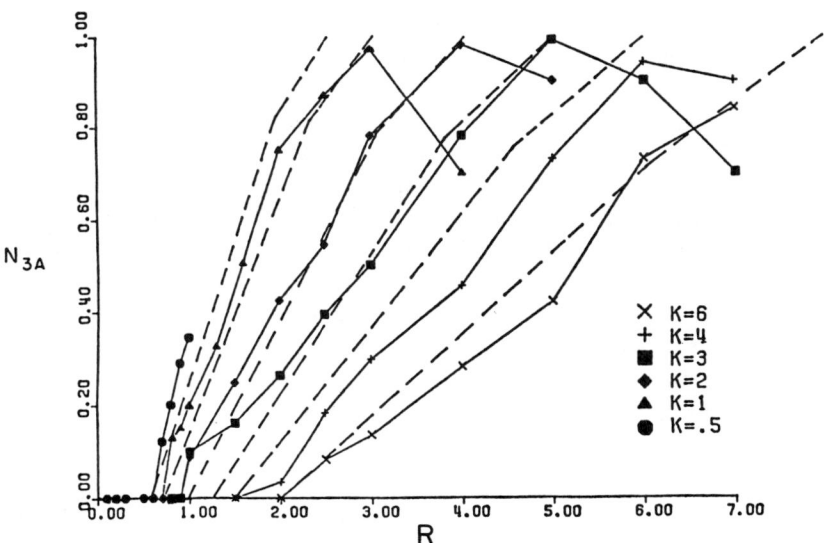

FIG. 29. N_{3A} versus R for sodium borosilicate glasses shown for different K families. The dashed lines are the predictions based on the model of Dell et al. (1983).

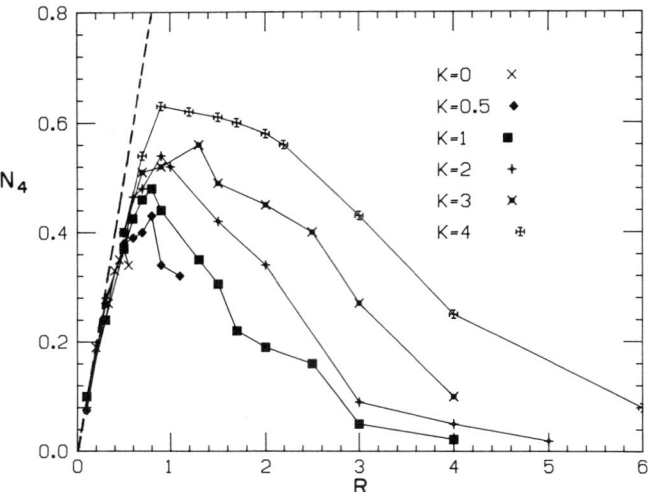

FIG. 30. N_4 versus R for potassium borosilicate glasses shown for different K families. The dashed line gives $N_4 = R$. (Szu and Bray, unpublished.)

oxide in the borate network produces $BO_{4/2}^-$ units, N_4 falls below R when some of the alkali forms NBOs. (Fig. 30).

4. Studies of the $(R)BaO\text{-}B_2O_3\text{-}(K)SiO_2$ Glass System

Stallworth and Bray (1988) have used ^{11}B NMR in studying these glasses for $0 \leqslant K \leqslant 5$. N_4 values fall short of the $N_4 = R$ prediction at low values of R, indicating an early appearance of NBOs associated with boron and silicon as well (Fig. 32). The data indicate a greater probability of Ba^{++} ions being associated with the borate network over the silicate network. As observed in Fig. 31, the usual branching of N_4 versus R according to the value of K is observed, as in the alkali borosilicate systems (Figs. 27–30). The maximum fraction of boron units in tetrahedral coordination ($N_4 \simeq 0.57$) found in the study is for a glass of composition $R \simeq 2.5$ and $K \simeq 5.0$, but larger values may occur for higher values of K. There is no evidence of covalent Ba-O bonds, as might be expected upon consideration of other oxide glasses containing large atomic-weight cations (e.g., Cd^{++}, Tl^+, and Pb^{++}).

5. Studies of the $(R)CdO\text{-}B_2O_3\text{-}(K)SiO_2$ Glass System

Mulkern et al. (1986) measured N_4 for this glass system (Fig. 33) and showed its similarity to N_4 in the $CdO\text{-}B_2O_3\text{-}GeO_2$ glass system. N_4 values between the two systems are identical within experimental error for $K < 1.0$. Silicate glasses of larger K yield smaller N_4 values than the corresponding

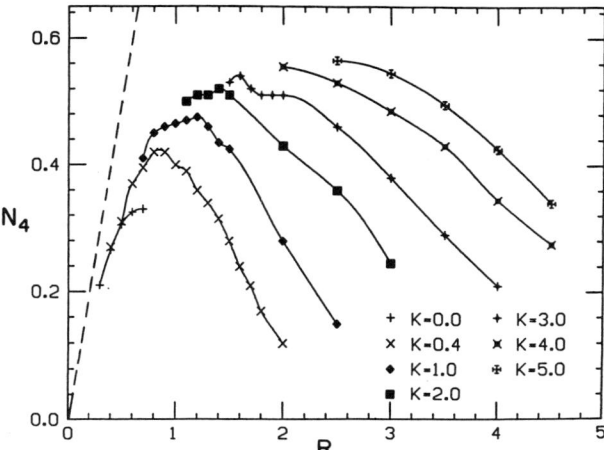

FIG. 31. N_4 versus R for barium borosilicate glasses shown for different K families. The dashed line gives $N_4 = R$. (Stallworth and Bray, unpublished.)

germanates indicating a smaller fraction of boron atoms involved in tetrahedral units and a greater accommodation of cadmium in the silicate matrix than in the germanate matrix. ^{113}Cd chemicals shifts for the silicate glasses show continuous evolution of the resonance peak from high to low field values indicating increased deshielding and increased covalency of the cadmium environment with increased R. The data are consistent with a

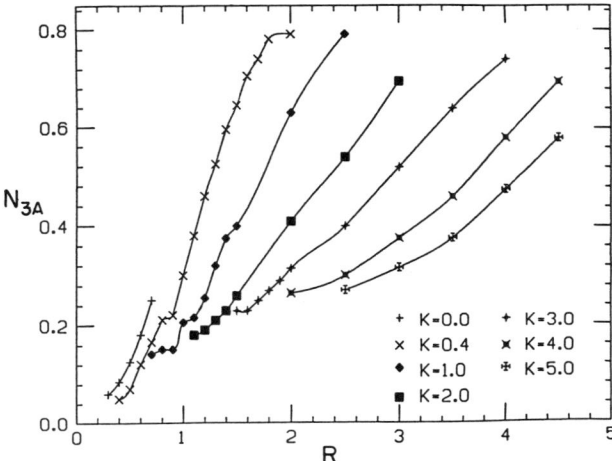

FIG. 32. N_{3A} versus R for barium borosilicate glasses shown for different K families. (Stallworth and Bray, unpublished.)

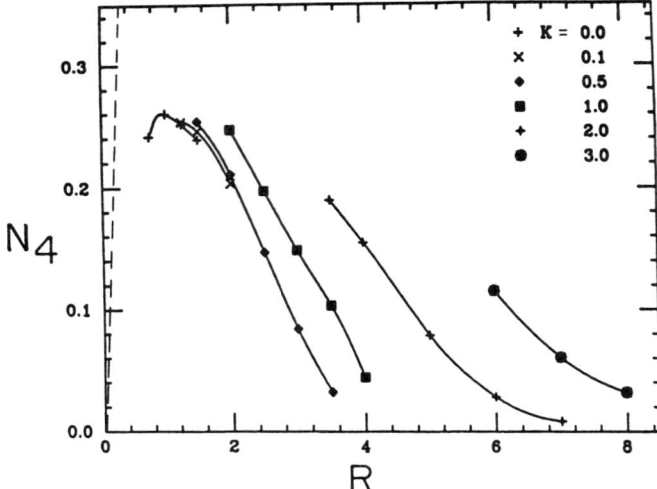

FIG. 33. N_4 versus R for cadmium borosilicate glasses shown for different K families. The dashed line gives $N_4 = R$. (Mulkern et al., 1986.)

model in which the cadmium can be present in either ionic or covalent bonding. The measured chemical shift for the observed broad line is then the weighted average of the values for the unresolved responses from the two sites. At low R, the ionic site is dominant, but covalent cadmium is dominant at high R. It is of interest to note that the chemical shifts indicate cadmium sites of larger ionicity for the binary CdO-B_2O_3 glasses than for the ternary cadmium borogermanates indicating the tendency for cadmium to become more covalent with increased GeO_2 content. (A similar effect may hold for SiO_2 glasses as well.)

6. Studies of the $(R)PbO$-B_2O_3-$(K)SiO_2$ Glass System

^{11}B NMR studies of lead borosilicate glasses by Kim et al. (1976) show (Fig. 34) that N_4 values increase with R independent of K up to $R \simeq 1.0$ ($N_4 \simeq 0.5$). This indicates that lead is predominantly incorporated into the glass as Pb^{++}. Higher R yields branching of N_4 into contours for the individual K families. For a given R, N_4 increases with K in analogy with the other modified borosilicate glasses. The [11]B response shows little ($<10\%$) contribution if any due to BO_2^- or $BO_{5/2}^{-2}$ units; hence, decreasing N_4 with increasing R indicates an increased association of lead with the formation of NBOs on silica tetrahedra and/or PbO being incorporated into the glass-forming network (i.e., increased covalent character of the lead site). [209]Pb chemical shifts follow the trend of increased covalency of the lead site with

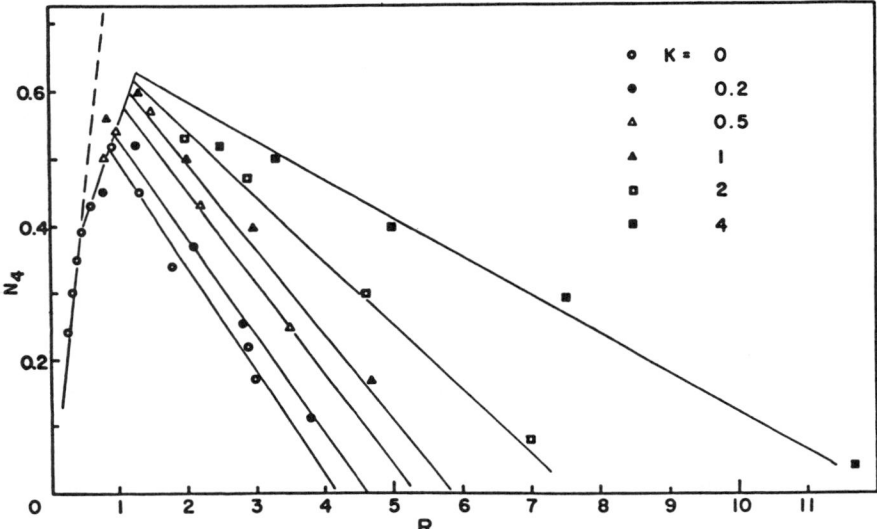

FIG. 34. N_4 versus R for lead borosilicate glasses shown for different K families. The dashed line gives $N_4 = R$. (Kim et al., 1976.)

increased R, as was found in similar studies of the binary lead borate and silicate glasses (Leventhal and Bray, 1965; Kim et al., 1976).

7. High-Field ^{11}B NMR Studies of Commercial Borosilicate Glasses

Corning Pyrex 7740 was one of the first borosilicate glasses to be studied through ^{11}B NMR (Silver and Bray, 1958). More recent studies have involved high-field, MASS, and VASS techniques in observation and characterization of the narrow and broad components of the boron line. As a result of the combined influence of increased chemical-shift effects and decreased second-order quadrupolar broadening, larger field strengths (about 100 kGauss or 10 Tesla) accompanied with MASS allow separation of the three- and four-coordinated boron responses. Consequently, differentiation between the symmetric and asymmetric three-coordinated boron sites is difficult at high fields due to the decreased sensitivity of subtle second-order quadrupole effects (a valuable diagnostic for ^{11}B spectra gathered at 20 kGauss or lower). Fyfe et al. (1982) give the 128.4 MHz MASS ^{11}B response for Corning 7070 glass (composition in wt %: $SiO_2 = 70$; $B_2O_3 = 28$; $K_2O = 1$; $Li_2O = 1$ as given by Scholes (1975)). An excellent demonstration of the VASS technique has been performed by Schramm and Oldfield (1982). They include 115.6 and 48.2 MHz boron spectra of Corning Pyrex 7740 glass (composition in wt % from Scholes: $SiO_2 = 81$; $B_2O_3 = 12.5$; $Al_2O_3 = 2$; $Na_2O = 4.5$) rapidly spun at 36, 54.7 (magic angle), and 75° relative to the external field (Fig. 35). MASS investigations of Pyrex glass have also been performed by Turner et al. (1986). These measurements have resulted in computer simulation of the 160.4 MHz ^{11}B lineshape yielding quadrupolar parameters, chemical shifts, and intensities of three- and four-coordinated boron sites.

E. Studies of Glasses Containing Al_2O_3

The tendency of NBO formation in modified borosilicate glasses can be reduced in varying degrees upon the incorporation of Al_2O_3. NBO suppression with Al_2O_3 content is accompanied by a change in coordination of aluminum such that charged $AlO_{4/2}^-$ tetrahedra (all briding oxygens) are formed. $AlO_{4/2}^-$ units are stabilized by the SiO_2 matrix resulting in higher viscosity melts and increased durability of the glass. For these reasons, Al_2O_3 may be the third most important network-forming oxide for glasses. Equilibria become established between network-forming aluminum in tetrahedral coordination, network-modifying aluminum of higher coordination (Fig. 36), and other glass constituents (e.g., Q^n, $BO_{4/2}^-$, $BO_{3/2}$ units). The fraction of aluminum in a particular environment as a function of glass composition is a desired quantity, for this is needed in determining the glass structure (as in the case of boron and silicon environments in borate and silicate glasses). Even though the aluminum isotope (^{27}Al) is 100% abundant, observation and quantitative evaluation of the resonance in glasses is difficult and often impossible due to large quadrupolar interactions arising from a

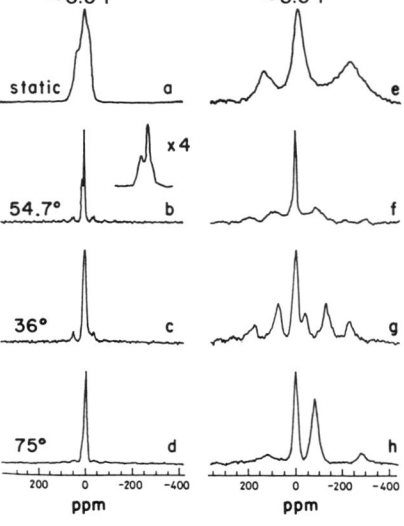

Fig. 35. ^{11}B NMR spectra of Pyrex 7740 glass at 8.45 and 3.52 Tesla under various conditions of sample rotation. (a) Static sample, 8.45 T. (b) MASS at 5.0 kHz, 8.45 T. (c) VASS ($\theta = 36°$) at 5.0 kHz, 8.45 T. (d) VASS ($\theta = 75°$) at 5.0 kHz, 8.45 T. (e) Static sample 3.52 T. (f) MASS at 5.0 kHz, 3.52 T. (g) VASS ($\theta = 36°$) at 5.0 kHz, 3.52 T. (h) VASS ($\theta = 75°$) at 5.0 kHz, 3.52 T. ppm scale relative to BF_3-Et_2O. (Schramm and Oldfield, 1982.)

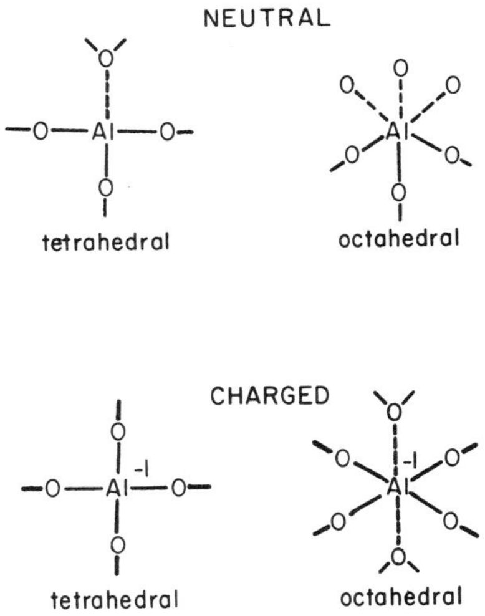

NEUTRAL

tetrahedral   octahedral

FIG. 36. Various aluminate structural units. For most modified aluminosilicate and aluminoborate glasses, charged tetrahedral and to a lesser extent neutral octahedral structural species (compositionally dependent) are most prevalent, the other structural units being present when accompanied by oxygen triclusters.

CHARGED

tetrahedral   octahedral

lack of perfect tetrahedral or octahedral symmetry in the aluminum environment. However, quadrupolar effects can be reduced through the use of higher field strengths and by applying MASS. As a result, ^{27}Al chemical shifts in many cases can be assigned as due to four-, five-, or sixfold coordination sites (Table II). Since the magnitude of the chemical shift is proportional to the external field strength, it is advantageous to work at as high a field as possible in order to separate the NMR responses arising from aluminum in its various environments.

1. Studies of Amorphous Al_2O_3 Films

Anodically formed amorphous Al_2O_3 films have been investigated using ^{27}Al MASS NMR by Dupree et al. (1985a). Observed resonances show three peaks associated with highly symmetric Al sites superimposed upon a much broader line. Presumably the broad line arises due to large quadrupolar interactions characteristic of less symmetric Al sites. Based on the assignments of aluminum resonance peaks gathered from crystals of known structure (0 ± 20 ppm for octahedrally coordinated, $\sim +35$ ppm for pentagonally coordinated Al as in andalusite, and $+50$ to $+80$ ppm for tetrahedrally coordinated Al (all relative to aqueous $Al(H_2O)_6^{+3}$), these authors designate the three narrow peaks as arising from four-, five-, and sixfold aluminum coordination. (The fivefold cationic aluminum is postulated to be associated with 10 to 10 wt% sulphate or ~ 3 wt% carbonate ions.) It was found through integration of the NMR response that larger fractions of aluminum reside in octahedral and pentagonal sites.

TABLE II

^{27}Al Isotropic Chemical Shifts Relative to $Al(H_2O)_6^{+3}$ for Aluminate Crystalline Compounds

Al-O coordination	Compound	Chemical shift (ppm)	Reference
4	β-NaAlO$_2$	$+77$	Müller et al. (1981)
4	KAlO$_2$	$+73$	Müller et al. (1981)
4	CsAlO$_2$	$+77$	Müller et al. (1981)
4	CaAl$_2$O$_4$	$+71$	Müller et al. (1981)
4	BaAl$_2$O$_4$	$+69$	Müller et al. (1981)
4	5BaO-Al$_2$O$_3$	$+81$	Müller et al. (1981)
4	NaAlSiO$_4$	$+58$	de Jong et al. (1984)
4	Mg$_2$Al$_4$Si$_4$O$_{18}$	$+55$	de Jong et al. (1984)
4,6	β-Al$_2$O$_3$	$+64, +9$	Müller et al. (1981)
4,6	γ-Al$_2$O$_3$	$+66, +2$	de Jong et al. (1984)
4,6	Al$_{3.8}$SiO$_{7.7}$	$+60$ and $+44, -3$	de Jong et al. (1984)
4,6	LiAlSi$_2$O$_6$	$+51, +2$	de Jong et al. (1984)
4,6	BaO-6Al$_2$O$_3$	$+74, +5$	Müller et al. (1981)
5	Al$_2$SiO$_5$	$\sim +35$	Dupree et al. (1985a)
6	α-Al$_2$O$_3$	$+5$	Müller et al. (1981)
6	MgAl$_2$O$_4$	0	de Jong et al. (1984)

2. Studies of Aluminoborate Glasses

Low-field ^{11}B CW NMR studies have been performed for various Na$_2$O-B$_2$O$_3$-Al$_2$O$_3$ glasses (Gresch et al., 1976; Zhong and Bray, 1986), K$_2$O-B$_2$O$_3$-Al$_2$O$_3$ glasses (Beekenkamp, 1968), and CaO-B$_2$O$_3$-Al$_2$O$_3$ glasses (Bishop and Bray, 1966). (Compositions are given by the molar ratios: R = [modifier oxide]/[B$_2$O$_3$] and K = [Al$_2$O$_3$]/[B$_2$O$_3$].) In these materials the usual borate structural units are present along with tetrahedral $AlO_{4/2}^-$ units (Fig. 36). Sixfold coordinated aluminum was not taken into account in models of alkali aluminoborate glasses (Beekenkamp, 1968; Gresch et al., 1976); however, its existence in calcium aluminoborate glasses could not be ruled out by Bishop and Bray (1966). If aluminum occurs purely in 4-coordination, then for Al$_2$O$_3$-rich glasses there must be different 4-coordinated environments. One possible description is that aluminum may be distributed among sites of charged $AlO_{4/2}^-$ and neutral tricluster $AlO_{3/2}(O_{1/3})$ units. (The tricluster unit consists of the combination of any three $BO_{4/2}^-$, $AlO_{4/2}^-$, and/or $AlO_{3/2}(O_{1/3})$ units bound to a common oxygen.)

Low-field results also show (for constant R) that as the Al$_2$O$_3$ content increases, the fraction of 4-coordinated boron atoms decreases. Hence, $AlO_{4/2}^-$ tetrahedra are formed at the expense of 4-coordinated borons. Further insight into the various aluminate structural groups was attempted through low-field ^{27}Al NMR measurements. Unfortunately, the aluminum lineshapes appeared broad and featureless indicating very little sensitivity to variations in glass composition (Bishop and Bray, 1966; Gresch et al., 1976).

^{27}Al MASS NMR studies of heat-treated MgO-B_2O_3-Al_2O_3 glasses have been performed by Dupree et al. (1985c). Their spectra show three peaks which vary in intensity and position with composition and heat treatment. Peak assignments indicate the presence of both tetrahedral $AlO_{4/2}^-$ units and octahedral $AlO_{3/2}$ units. It is found that the aluminum tetrahedral unit maintains either all aluminum $Al(OAl)_4$ ($\sim +58$ ppm) or boron $Al(OB)_4$ ($\sim +29$ ppm) next-nearest neighbors, i.e., no intermediate species occur; and the octahedral unit is characterized as having all aluminum next-nearest neighbors ($\sim +1$ ppm). These glasses show no signs of phase separation. NMR results for the devitrified samples relative to the glasses indicate subtle deshielding (positive) shifts for the $Al(OB)_4$ peak and octahedral peak along with decreasing $Al(OAl)_4$ peak intensity and increasing octahedral peak intensity. (The octahedral peaks that arise upon heat treatment are due to Al atoms with at least one boron next-nearest neighbor.)

3. Studies of Aluminosilicate Glasses

Binary roller-quenched Al_2O_3-SiO_2 glasses have been studied by Risbud et al. (1987) using ^{27}Al and ^{29}Si MASS NMR. These glasses, made from 15 to 50 wt% Al_2O_3, are phase separated and yield ^{27}Al spectra with three peaks complicated by spinning sidebands. There are clear indications of both tetrahedral and octahedral aluminum sites, yielding resonance peaks at $\sim +60$ and ~ 0 ppm respectively. A third peak near $+30$ ppm, which is flanked by the 4- and 6-coordinated aluminum responses, is assigned by these authors as arising from Al in fivefold coordinated environments (Fig. 37). These assignments are based upon NMR studies of crystals where aluminum is known to exist in tetrahedrally, pentagonally, and octahedrally symmetric sites. ^{29}Si responses, observed near -109 and -90 ppm, reflect the phase separation of these samples; they are indicative of silicon in SiO_2-rich and Al_2O_3-rich phases respectively and illustrate the sensitivity of the response to second-nearest–neighbor effects.

Initial NMR investigations of sodium, potassium, and calcium aluminosilicates were carried out using wide-line techniques (Müller-Warmuth et al., 1966; Schulz et al., 1968; Dutz and Schultz, 1969). These low-field results show that aluminum linewidths of the calcium aluminosilicate glasses are in general larger than those of the alkali aluminosilicates and do not vary as much with composition. ^{27}Al halfwidths ($v_o \simeq 11$ MHz) of alkali aluminosilicate glasses, unlike alkali aluminoborate glasses, dramatically decrease from over 35 kHz to about 14 kHz for $R' = $ [modifier oxide]/[Al_2O_3] < 1; extension of this ratio beyond 1 yields halfwidths that maintain a constant

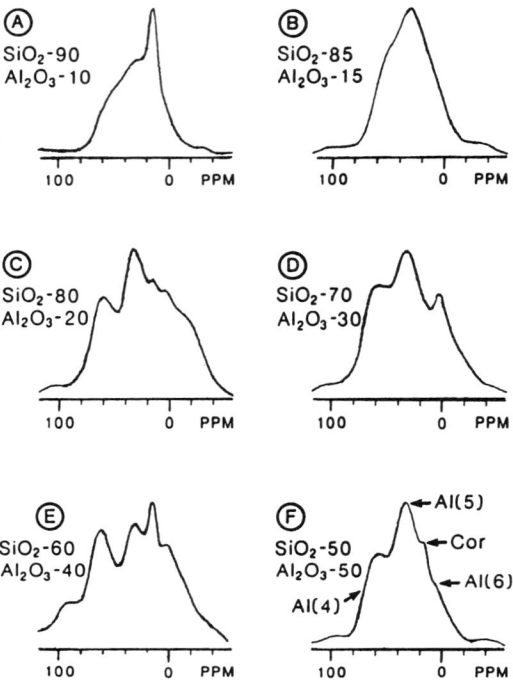

Fig. 37. ^{27}Al MASS NMR spectra of roller-quenched SiO_2-Al_2O_3 glasses (labelled in wt %). All spectra were obtained at 11.7 Tesla with MASS spinning rates greater than 7 kHz. The designations Al(4), Al(5), and Al(6) refer to the responses due to fourfold, fivefold, and sixfold aluminum, respectively. The designation Cor indicates the aluminum response due to undissolved Al_2O_3 in the glass. The ppm scale is relative to the octahedral ^{27}Al response in $AlCl_3$ solution. (Risbud et al., 1987.)

value of about 14 kHz (Fig. 38). As to the status of the octahedral unit these NMR data suggest that for $R' < 1$, there is a larger distribution of Al sites with perhaps both four- and sixfold coordination. A decrease of the Al_2O_3 content is accompanied by a decrease in the distribution of sites such that the tetrahedral configuration is the dominant site.

There have been numerous high-field MASS studies of modified aluminosilicate glasses (Hallas et al., 1983; Thomas et al., 1983; de Jong et al., 1984; Dupree et al., 1985b; Engelhardt et al., 1985; Murdoch et al., 1985; Ohtani et al., 1985; Kirkpatrick et al., 1986b; Oestrike et al., 1987). In general, these do somewhat substantiate the low-field results of Schultz, Müller-Warmuth and coworkers. For some aluminosilicate glasses ($R' < 1$) relatively larger linewidths and/or distinct octahedral peaks have been reported in ^{27}Al MASS spectra (Engelhardt et al., 1985; Hallas et al., 1983). Ohtani et al. (1985) have

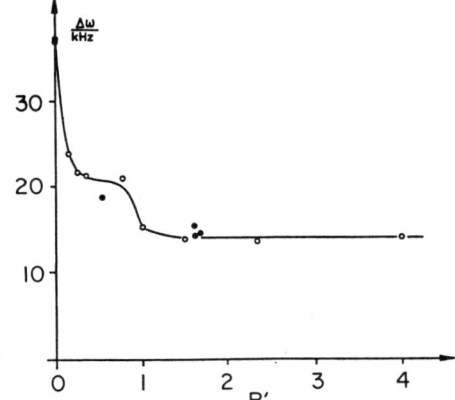

FIG. 38. Half-widths of the quadrupolar broadened asymmetric ^{27}Al NMR
spectra of glasses in the system K_2O-Al_2O_3-
SiO_2 plotter against the ratio $R' = $ mol%
K_2O/mol% Al_2O_3. The open circles belong
to a constant SiO_2 content of 70 mol%.
(Müller-Warmuth and Eckert, 1982.)

indicated that for Na_2O-Al_2O_3-$(6)SiO_2$ glasses (albite glass) fabricated at
pressures of at least 60 kbar, an aluminum peak characteristic of the
octahedral coordination state becomes evident. Even though quantitative
measurements for aluminum sites are at best difficult, it appears that for
$R' > 1$ the tetrahedral site may be the most preferred configuration for
aluminum in alkali and alkaline earth aluminosilicate glasses. Indeed, if
octahedral groups are present they may go undetected, as noted by Hallas et
al. (1983). Their ^{27}Al study of Na_2O-Al_2O_3-SiO_2 glasses showed that
tetrahedrally coordinated Al can be identified but cannot be quantified with
great accuracy due to the presence of inseparable broad features that may
arise from both tetrahedral and octahedral sites. These authors observed that
a significant number of aluminum atoms were unaccounted for in the spectra.
Hence, due to this "missing" contribution, the presence of the octahedral unit
could not be ruled out. (Its response could be strongly broadened through
quadrupolar effects rendering it inobservable.) A similar interpretation is
given for Li_2O, Na_2O, K_2O, MgO, and CaO-Al_2O_3-SiO_2 glasses studied by
de Jong et al. (1983).

Oestrike et al. (1987) have accounted for all the aluminum in their
resonances. Their study involving a variety of aluminosilicate glasses ($R' = 1$)
containing Li^+, Na^+, K^+, Rb^+, Cs^+, Ca^{++}, Sr^{++}, Ba^{++}, and Pb^{++} cations
show that ^{27}Al and ^{29}Si peak maxima shift positively (deshielding) with a
decreasing [Si]/[Si + Al] ratio virtually independent of the cation type. It
appears for these glasses that aluminum is present strictly in tetrahedral
coordination. (However, it is possible that other coordination sites may be
present in insufficient numbers to yield resolved peaks.)

Complementary ^{29}Si NMR results indicate, as shown by Lippmaa et al.
(1981) and Mägi et al. (1984), ^{29}Si chemical shifts are affected by the number

of second-nearest–neighbor aluminum atoms, such that for each Al-O-Si (as opposed to Si-O-Si) bond link there is a deshielding (less negative) shift of about 5 to 6 ppm. Additionally, the effect upon the ^{29}Si response in aluminosilicate glasses by different cations results in increased broadening with cation polarizing ability (i.e., $K^+ < Na^+ < Ca^{++}$) (Murdoch et al., 1985). This is interpreted as reflecting a larger distribution in silicon sites (similar effects were observed for binary silicate glasses). Hence, interpretation of ^{29}Si NMR in modified aluminosilicates can be difficult due to the presence of many types of structural units influencing simultaneously both first- and second-coordination spheres of silicon. Typically, silicon resonances are influenced by NBOs, tricluster type units, $AlO_{4/2}^-$, and/or AlO_6 units. Engelhardt et al. (1985) have performed NMR experiments for CaO-Al_2O_3-SiO_2 glasses. Analysis, based upon their ^{27}Al and ^{29}Si data, has resulted in sorting of the various silica-alumina units in terms of a structural model. Seemingly rare for aluminosilicate glasses, a positive identification of octahedral aluminum (near 0 ppm) in certain samples by ^{27}Al NMR is included in this work (Fig. 39).

Several mixed Na-K aluminosilicate glasses have been studied by ^{23}Na NMR (Oestrike et al., 1987). Interestingly, the sodium peak shifts positively (deshielding) with decreasing $[Na]/[Na + K]$ and with decreasing $[Si]/[Si + Al]$ molar ratios where $[Na + K]$ = constant. In contrast to these results, Dupree et al. (1985b) indicate for $33.3[xNa_2O\text{-}(1 - x)Al_2O_3]$-$66.7SiO_2$ glasses that ^{23}Na chemical shifts become increasingly negative

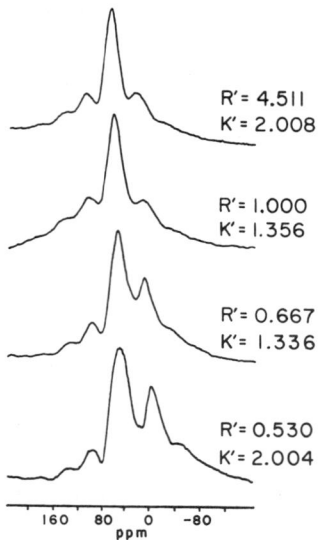

R' = 4.511
K' = 2.008

R' = 1.000
K' = 1.356

R' = 0.667
K' = 1.336

R' = 0.530
K' = 2.004

160   80   0   -80
ppm

FIG. 39. ^{27}Al MASS NMR spectra of selected glasses of composition $R'CaO\text{-}Al_2O_3\text{-}K'SiO_2$. Peak intensities near 60 ppm are indicative of fourfold coordinated aluminum while those near 0 ppm (when $R < 1$) indicate the presence of sixfold coordinated aluminum in the glass (ref. aqueous $AlCl_3$). (Engelhardt et al., 1985.)

(shielded) for decreasing x, indicating the gradual association of Na^+ ions with $AlO_{4/2}^-$ units in preference to NBOs associated with Q^n units.

4. Studies of Aluminoborosilicate Glasses

Kobayashi and Okuma (1976, 1979) have performed structural studies on glasses in the system $K_2O/Na_2O-Al_2O_3-B_2O_3-SiO_2$. Their results show that these glasses have characteristics of both aluminoborate and aluminosilicate glasses in that:

(1) $BO_{4/2}^-$ and $AlO_{4/2}^-$ units compete for Na^+ and/or K^+ cations with more alkali being associated with the alumina network. (N_4 increases to 0.5 as R' decreases to 1 for constant Al_2O_3 content.)

(2) ^{27}Al linewidths increase as R' decreases. Low R' glasses may contain a substantial number of tricluster units and, like aluminosilicate glasses, $AlO_{4/2}^-$ units appear to become prominent in glasses of high R'.

The incorporation of Al_2O_3 into $K_2O-B_2O_3-SiO_2$ glass has been monitored through ^{27}Al MASS NMR by Schiller et al. (1982). This interesting experiment shows the change in aluminum coordination through the gradual demise of the octahedral peak near $+9$ ppm (associated with Al_2O_3-rich regions—prominent for glasses fused at lower temperatures) and the subsequent emergence and eventual dominance of the aluminum tetrahedral peak near $+55$ ppm (associated with more homogeneously distributed aluminum atoms—prominent for glasses fused at higher temperatures).

^{11}B N_4 measurements have been performed on heat-treated $CaO-Al_2O_3-B_2O_3-SiO_2$ glass fibers (Gupta et al., 1985). N_4 is smaller for the rapidly quenched fiber than for the more slowly cooled bulk glass, but the two values become equal when the fiber is annealed. It is clear that N_4 depends upon the fictive temperature of the glass. These authors give an expression for the equilibrium value of N_4 in terms of the transition enthalpy and annealing temperature.

F. STUDIES OF GLASSES CONTAINING P_2O_5

Pure P_2O_5 glass is structurally conceived as a branching network of $POO_{3/2}$ tetrahedra. These neutrally charged units contain one double-bonded oxygen and three bridging oxygen atoms (Fig. 40). Upon addition of metal oxide, the P_2O_5 matrix accommodates metal cations through the formation of $PO_2O_{2/2}^-$ from $POO_{3/2}$ units. The cation charge is balanced by the negative charge which may be delocalized between two (perhaps indistinguishable) terminal oxygen atoms formed from the original double-bonded oxygen and the newly formed "NBO." For instance, the ideal model for alkali metaphosphate glasses ($X_2O-P_2O_5$) constructs the glass from chains of $PO_2O_{2/2}^-$ tetrahedra containing two terminal oxygen atoms.

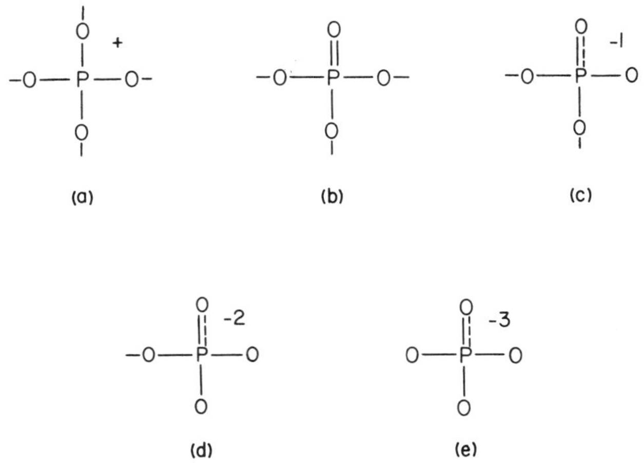

FIG. 40.   Phosphate structural units. (a) $PO_{4/2}^+$ unit as found in borophosphate glasses. (b) Phosphate unit $POO_{3/2}$ as in pure P_2O_5 glass. (c) Metaphosphate unit $PO_2O_{2/2}^-$. (d) Pyrophosphate unit $PO_3O_{1/2}^{-2}$. (e) Orthophosphate unit PO_4^{-3}.

Increased metal cation content (if the glass can be fabricated) results in the formation of phosphate units with higher charge such as $PO_3O_{1/2}^{-2}$ containing one bridging and three terminal oxygens and PO_4^{-3} with four terminal oxygen atoms.

NMR studies of binary phosphate glasses have culminated in the evaluation of various relaxational and structural properties. ^7Li linewidth investigations of Li_2O-P_2O_5 glasses by Göbel et al. (1979) and Müller-Warmuth and Eckert (1982) indicate distinct differences in the lithium cation distribution between these glasses and their borate and silicate analogues. Alkali phosphate glasses, studied by Olyschläger (1977) (Müller-Warmuth and Eckert, 1982), reveal through ^{31}P isotropic chemical shifts the production of $PO_2O_{2/2}^-$ from $POO_{3/2}$ units with increasing alkali oxide content. Jellison (1979) studied ^{31}P spin-lattice relaxation times in CaO-P_2O_5 glass, finding a disorder mode relaxation mechanism. Tl_2O-P_2O_5–based glasses have been studied by observation of ^{205}Tl NMR chemical shifts (Kolditz and Wahner, 1973; Panek et al., 1977b). The NMR data support increased covalency of the Tl-O bond with increasing Tl_2O content. Similar behavior is seen in some borate and silicate glasses where the metal is incorporated into the glass in varying degrees as a modifier (ionically) and network former (covalently).

Phosphorus oxynitride glasses (Na_2O-P_2O_5–based glasses modified by the addition of ammonia) have been studied through ^{31}P, ^{15}N, and ^{23}Na MASS NMR by Bunker et al. (1987). They conclude that the phosphorus network ($PO_2O_{2/2}^-$ chains) is modified primarily through the replacement of varying amounts of both single- and double-bonded oxygen by nitrogen. PO_4-, PO_3N-, and PO_2N_2-type tetrahedral units are present; and there are two nitrogen sites: (1) nitrogen bound twofold by one double and one single bond to phosphorus tetrahedra and (2) nitrogen bound threefold by single bonds to phosphorus tetrahedra.

Villa et al. (1986b, 1987) have studied various lithium and silver phosphate and borophosphate glasses using ^{31}P MASS NMR. Their investigations of binary glasses show clearly separated peaks arising from the different phosphorus environments (Fig. 41). Chemical-shift assignments (relative to aqueous H_3PO_4) for the phosphorus line are -42 to -35 ppm for the $POO_{3/2}$ unit; -28 to -15 ppm for the metaphosphate $PO_2O_{2/2}^-$ unit; -12 to $+3$ ppm for the pyrophosphate $PO_3O_{1/2}^{-2}$ unit; and $+25$ ppm for the orthophosphate PO_4^{-3} unit in crystalline Ag_3PO_4. (Duncan and Douglass (1984) show that the orthophosphate ^{31}P chemical shift can range from -2 to $+21$ ppm.) The effect of substituting boron oxide for phosphorus pentoxide in these glasses indicates increased shielding (more negative shifts) at the phosphorus site. The presence of BPO_4-type units ($BO_{4/2}^-$ and $PO_{4/2}^+$ tetrahedra linked together and occurring in ratios of 1:1) are proposed to account for these effects. The formation of BPO_4 units in glass, proposed by Kreidl and Weyl (1941), is central in the interpretation of borophosphate glass structure. Structural models of Na_2O-B_2O_3-P_2O_5 glass by Beekenkamp and Hardeman (1966) and K_2O-B_2O_3-P_2O_5 glass by Yun and Bray (1978b), based on ^{11}B NMR N_4 and N_{3S} data, employ $POO_{3/2}$, $PO_2O_{2/2}^-$, BPO_4, and $BO_{3/2}$ structural units and $BO_{4/2}^-$ structural units (tetrahedral borate units not involved in BPO_4 units).

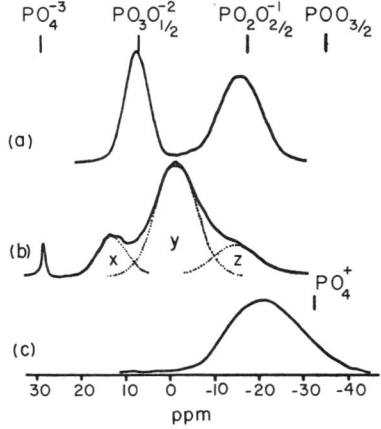

FIG. 41. ^{31}P MASS NMR spectra (ref. H_3PO_4) of selected phosphate glasses. (a) Glass with the nominal composition Ag_2O-$75P_2O_5$: the marks give the average isotropic shifts of the units determined in a number of silver phosphates. (b) Ag_2O-$1.6B_2O_3$-$0.4P_2O_5$ glass: the feature x arises due to $PO_3O_{1/2}^{-2}$ units while features y and z both arise due to different $PO_2O_{1/2}^-$ units. (c) Li_2O-$0.8B_2O_3$-$1.2P_2O_5$ glass. (Villa et al., 1987.)

^{31}P and ^{29}Si MASS NMR studies were performed by Yang et al. (1986) on various alkaline-earth silicate glasses containing up to $10\,wt\%$ P_2O_5. Phosphorus chemical shifts range from 0.4 to $+3.9$ ppm and are designated as due to phosphorus in orthophosphate PO_4^{-3} sites. The association of the alkaline-earth cation with the phosphate unit is further substantiated through interpretation of the silicon NMR chemical shifts. As mentioned previously for modified silicate glasses, ^{29}Si chemical shifts become increasingly positive (deshielding) with modifier oxide content through the number of associated NBOs. ^{29}Si isotropic chemical shifts for these phosphorous silicate glasses become increasingly negative with increasing P_2O_5 content, indicating a preference of the alkaline-earth cations for the phosphate network and an increased reluctance of NBO formation on silicate tetrahedra.

Müller et al. (1983) employ MASS NMR in observation of ^{27}Al in CaO-Al_2O_3-P_2O_5 glasses. Unlike spectra gathered from aluminosilicate glasses, octahedral AlO_6 structural units are clearly evident in ^{27}Al spectra for aluminophosphate glasses. Due to the high sensitivity of aluminum to second-coordination sphere effects, Al-O-Al and Al-O-P bond links can be differentiated. Phosphorus second-nearest neighbors to aluminum shield the aluminum site and give rise to more negative ^{27}Al shifts. The MASS spectra gathered from these glasses reveal three peaks residing near -21, $+4$, and $+36$ ppm in reference to $Al(H_2O)_6^{+3}$. The proposed assignment suggests resonances arising from aluminum in octahedral $Al(OP)_6$, octahedral $Al(OP)_n(OAl)_{6-n}$, and tetrahedral $Al(OP)_4$ sites, respectively.

NMR investigations of various fluorophosphate glasses have been performed yielding interesting results. $Ba(PO_3)_2$-BaF_2/ZnF_2-AlF_3 and $Zn(PO_3)_2$-ZnF_2-AlF_3 glasses have been studied using ^{19}F and ^{31}P CW NMR by Dubiel and Ehrt (1987). Comparisons between crystal and glass second-moment values derived from phosphorus spectra give evidence of both fluorine and proton dipolar couplings to phosphorus. (There is a small but measurable ^{1}H content, occurring perhaps in the form of OH^- groups associated with chain-ending PO_4 units.) Fluorine and phosphorus linewidths suggest P-F dipolar coupling; the presence of monofluoride $[PO_3F]^{-2}$ tetrahedra in these glasses is proposed to account for the observed effects. Vopilov et al. (1986) have studied ^{19}F NMR lineshapes gathered from BaO-P_2O_5-PbF_2-AlF_3 glasses and have drawn some conclusions concerning fluorine ion motion.

V_2O_5-P_2O_5 semiconducting glasses have been investigated via ^{51}V and ^{31}P CW NMR. The dc conductivity of these glasses ($\sim 10^{-6} - 10^{-5}$ $(\Omega cm)^{-1}$ at 300 K) is described through the electron transfer between transition metal ions of different valence. Both diamagnetic V^{+5} and paramagnetic V^{+4}

species are present; hence, by elucidating the glass structure (NMR) and by quantifying the fractions of vanadium in $+4$ and $+5$ valence states (ESR and NMR) the semiconducting properties may be further understood. The ^{51}V NMR response can be analyzed considering both quadrupolar and axially symmetric chemical-shift interactions; furthermore, due to the presence of the V^{+4} species, vanadium spectra may be complicated by paramagnetic shift interactions and/or Knight shift interactions (Alloul, 1977). Lineshape analyses have been performed on glasses of different V_2O_5 content by various workers, and conclusions have been somewhat varied. As indicated by Landsberger and Bray (1970) and France and Hooper (1970), the number ratio of paramagnetic to diamagnetic species $[V^{+4}]/[V^{+5}]$, decreases with V_2O_5 content; however, there is disagreement as to the exact value of this ratio as a function of composition. Lynch et al. (1971) show that this ratio is dependent upon conditions under which samples are prepared (i.e., the atmospheric O_2 concentration in which melts are maintained and quenched), therefore accounting for some discrepancies in the results. Structural interpretations include the presence of VO_5 pyramid groups (Landsberger and Bray, 1970) and, at higher V_2O_5 content, VO_6 octahedral groups (France and Hooper, 1970) interspersed via V-O-P bond links throughout the PO_4 tetrahedral network. Lynch and Sayer (1973) have studied the effects of paramagnetic dopants in WO_3-P_2O_5 glasses through ^{31}P T_1 measurements. Their results indicate for glasses containing W^{+5}, V^{+4}, Mo^{+5}, and Cu^{+2} paramagnetic species that the phosphorus response is characterized by small linewidths and short relaxation times. T_1 decreases with paramagnetic dopant content and appears to follow the spin-diffusion model of de Gennes (1958).

G. STUDIES OF GLASSES CONTAINING Ga_2O_3 AND GeO_2

High-field (7.1 Tesla) ^{69}Ga and ^{71}Ga MASS NMR spectra have been gathered by Zhong and Bray (1987) for xCs_2O-$(1-x)Ga_2O_3$ glasses; $0.3 \leqslant x \leqslant 0.7$. ^{71}Ga responses reveal two peaks which are proposed to arise from different isotropic chemical shifts of fourfold and sixfold coordinated (with oxygen) gallium sites. Upon increasing the Ga_2O_3/Cs_2O molar ratios, the ^{71}Ga spectra display increased intensity of the assigned GaO_6 octahedral site (Fig. 42). This feature appears at lower frequency and implies that the associated gallium site is more shielded. At lower Ga_2O_3/Cs_2O ratios, the response is dominated by a less shielded feature which is assigned as that due to GaO_4 tetrahedra. Support for the identification of the gallium sites was obtained by comparison with the spectra for materials having known gallium coordinations, and by analogy to the chemical shifts for 4 and 6-coordinated aluminum sites in glasses. It is hoped, therefore, that gallium NMR studies

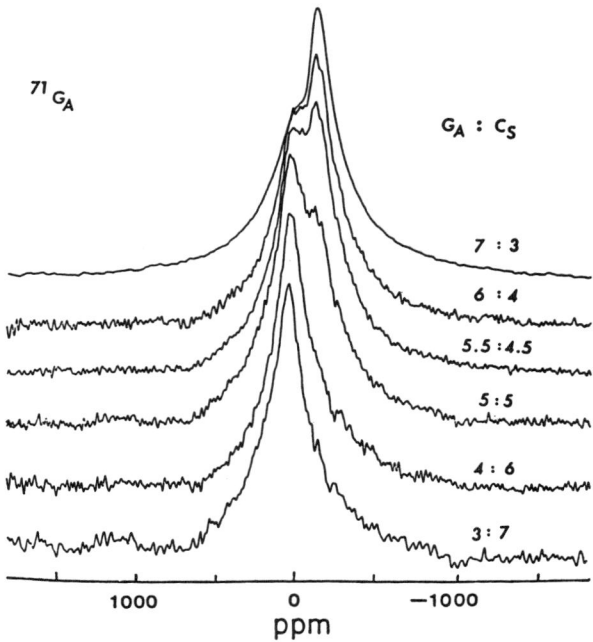

FIG. 42. ^{71}Ga static NMR spectra gathered at 91.53 MHz for Cs_2O-Ga_2O_3 glasses with varying Ga/Cs molar ratios. (Zhong and Bray, 1987.)

will lead to a greater understanding of the roles played by both gallium and aluminum in glass, and may prove invaluable in understanding the structure of aluminate glasses.

^{11}B CW NMR responses have been gathered from binary GeO_2-B_2O_3 glasses by Baugher and Bray (1972) and Göring et al. (1982). The spectra, being second-order quadrupole broadened, are characteristic of trigonal $BO_{3/2}$ boron environments (no responses from $BO_{4/2}^-$, BO_2^-, and $BO_{5/2}^{-2}$ units are seen) and therefore support the contention that this glass network is comprised of intermixed $GeO_{4/2}$ tetrahedra and $BO_{3/2}$ trigonal groups. The ^{11}B coupling constant, e^2qQ/h, is found to increase monotonically from 2.59 to 2.76 MHz as the B_2O_3 content increases from 15 to 100 mol%.

The binary glass system Tl_2O-GeO_2 has been studied using ^{205}Tl NMR (Panek and Bray, 1977). The germanate network is structurally understood as being composed of GeO_4 tetrahedra and GeO_6 octahedra; the latter increases with increasing Tl_2O content. ^{203}Tl chemical shifts have large distributions in these glasses indicating a variety of thallium sites. Low–thallium-content glasses are characterized by narrower linewidths indicating the random incorporation of Tl^+ ions in the glass. Above 15 mol% Tl_2O, the

NMR linewidth is found to increase with Tl_2O content; both ionic and covalent thallium sites are present resulting in larger distributions in chemical-shift values. Higher Tl_2O content glasses, $> 30 \, \text{mol} \%$, are characterized by the combined effect of thallium-atom clustering or pairing and smaller distributions in thallium chemical shifts (indicating the electronic similarity in covalent thallium sites).

Ternary glasses in the system $(0.25)BaO\text{-}B_2O_3\text{-}(K)GeO_2$, for K values up to 2.92, have been studied using ^{11}B NMR by Göring et al. (1982). Typical boron spectra were obtained revealing the presence of $BO_{3/2}$ and $BO_{4/2}^-$ units; no response could be attributed to asymmetric trigonal units. These authors give N_4 values ranging between 0.25 and 0.3 for GeO_2 contents less than about 33 mol %. Glasses studied with higher GeO_2 content reside within a miscibility gap, and associated N_4 values are markedly lower. Coupling constant values decrease with increasing GeO_2 and fall abruptly within the miscibility gap. Another interesting ternary germanate glass system that has been studied via NMR is $CdO\text{-}B_2O_3\text{-}GeO_2$ (Mulkern et al., 1986). As indicated previously, N_4 values parallel those of $CdO\text{-}B_2O_3\text{-}SiO_2$ glasses for $K < 1.0$ and are larger than for the corresponding silicate glasses for $K > 1.0$. ^{113}Cd chemical shifts suggest a slight tendency of cadmium to assume a more covalent role in these glasses with increasing GeO_2 content as well as a more pronounced tendency toward covalency with increasing CdO content.

H. STUDIES OF HALIDE AND SULFIDE GLASSES

BeF_2 glass has been investigated by Aleksandrova and Batsanova (1972) and Dell et al. (1984) employing ^9Be and ^{19}F CW NMR. This glass, being analogous in structure to SiO_2 glass, is formed from BeF_4 tetrahedra with bridging fluorine atoms. The ^9Be response is shown to arise from a combination of two distinct beryllium sites. Be-F heteronuclear dipolar broadening results in the observation of an unsurprising wide line ($\sim 8.5 \, \text{kHz}$); however, superimposed on this wide response exists a narrow line ($\sim 2.2 \, \text{kHz}$) proposed to be due to a fluorine-deficient beryllium site (clustered Be atoms). This anomalous response (associated with about 8% of all Be sites) can be proven not to arise from first- or second-order quadrupole effects or motional narrowing; hence the designation as a distinct beryllium site. ^9Be NMR work (Dell et al., 1984) performed on $NaF\text{-}2BeF_2$ glass displays no anomalous narrow line. The quadrupole coupling constant gathered for ^9Be in the binary glass is much larger than for BeF_2 glass, perhaps indicating a fivefold coordinated beryllium site.

As evidenced by narrowing of the fluorine response at temperatures near 130°C or higher, fluorine ion motion commences (originally thought to occur via fluorine diffusion throughout the glass matrix). However, ^{19}F studies of both BeF_2 glass and the binary glass by Gravina (Bray et al., 1988), reveal that the fluorine linewidth narrows slightly only near the glass transition temperature ($T_g \sim 130°C$) and that below T_g the spin-lattice relaxation time is temperature-independent. The slight narrowing of the NMR response near T_g is indicative of sample melting; hence, it appears that significant fluorine diffusion in the glass is unlikely.

Bray et al. (1983a) have reported fluorine motional narrowing of ^{19}F NMR lineshapes for several ZrF_4-BaF_2-LaF_3–based glasses. As with halo-borate glasses, CW fluorine spectra for these glasses reveal broad and narrow components that are indicative of bound and mobile fluorine ions. Pulsed NMR was used in gathering T_1 values, which in turn were used in finding activation energies of the fluorine sites (ranging between 0.3 and 0.7 eV for appropriate temperatures).

B_2S_3 glass has been studied through ^{11}B CW NMR by Hendrickson and Bishop (1975) and through CW and T_1 measurements by Rubinstein (1976). A temperature-dependent lineshape was observed: at room temperature second-order quadrupolar broadened spectra yielded a coupling constant of 2.51 MHz (which is indicative of a trigonal boron environment); spectra gathered at higher temperatures show a very narrow feature that increases in intensity with temperature. Initial explanations of this behavior claimed that there is a rapid temperature-dependent conversion of BS_3 trigonal units to BS_4 tetrahedra (Hendrickson and Bishop, 1975) and that boron sulfide molecular groups become increasingly mobile with temperature (Rubinstein, 1976). Hintenlang and Bray (1985a), in repeating these measurements on "outgassed" samples, did not observe the narrow feature. They concluded that the narrow line appeared due to the motion of significant quantities of H_2S in the samples.

^{11}B studies of $(R)Li_2S$-B_2S_3 glasses (Hintenlang and Bray, 1985) show N_4 (fraction of boron atoms involved in BS_4 tetrahedra) behavior analogous to binary borate glasses; however, the maximum N_4 value for the sulfide glasses is higher than that for the borate glasses. Asymmetric ^{11}B lineshapes are observed at increased R values indicating the presence of three-coordinated boron units with nonbridging sulfur atoms. N_4 measurements performed on Tl_2S-B_2S_3 glasses (Müller-Warmuth and Eckert, 1982), show analogous behavior to modified borate glasses as well. Interestingly, at low R, the ratio of N_4 to R is greater than unity, indicating the presence of 3-coordinated sulfur atoms (analogous to 3-coordinated oxygen found in certain strontium borate glasses). ^{205}Tl chemical-shift measurements indicate typical thallium behavior of increased covalency with increased Tl (Tl_2S) content.

Ionic lithium motion has been monitored through ^7Li CW and pulsed NMR in Li_2S-GeS_2 glasses by Senegas and Olivier-Fourcade (1983). Second-moment calculations indicate a random Li^+ distribution. T_1 measurements show that there are lithium atoms that are more firmly bound and lithium ions that are free to move about. This fraction of free Li^+ increases with temperature and gives rise (via long-range motion) to the conductivity of the glass.

Temperature-dependent ^7Li linewidths of $LiI-Li_2S-P_2S_5$ and $LiI-Li_2S-As_2S_3$ glasses have been obtained by Visco et al. (1985a, b). These glasses can exhibit large ionic conductivities of about 10^{-3} $(\Omega cm)^{-1}$ at room temperature. Evaluation of the data via the Hendrickson-Bray motional narrowing equation (Eq. (28)) yields activation energies of lithium cation motion. Results indicate the presence of two types of ionic motion: long-range motion (characterized by high activation energies which decrease upon increasing LiI content) and short-range motion (characterized by lower activation energies which appear to be independent of LiI content). Reasonable agreement is attained between activation energies found through NMR and those derived from conductivity measurements.

References

Abragam, A. (1961). "The Principles of Nuclear Magnetism." Oxford University Press.

Aleksandrova, I. P., and Batsanova, L. R. (1972). *Zh. Strukt. Khim.* **13**, 232.

Alloul, H. (1977). "NMR in Metals and Alloys," in "Nuclear Magnetic Resonance in Solids." Plenum Press, New York.

Andrew, E. R. (1955). "Nuclear Magnetic Resonance." Cambridge University Press, Glasgow.

Andrew, E. R., and Eades, R. G. (1953). *Proc. R. Soc. London, A* **216**, 398.

Andrew, E. R., Bradbury, A., and Eades, R. G. (1958). *Arch Sci.* **11**, 223.

Ardelean, I., Coldea, M., and Cozar, O. (1982). *Nucl. Instrum. Methods* **199**, 189.

Bassompierre, A. (1953). *Compt. Rend.* **237**, 39.

Baugher, J. F., and Bray, P. J. (1969). *Phys. Chem. Glasses* **10**, 77.

Baugher, J. F., and Bray, P. J. (1972). *Phys. Chem. Glasses* **13**, 63.

Beekenkamp, P. (1968). *Phys. Chem. Glasses* **9**, 14.

Beekenkamp, P., and Hardeman, G. E. G. (1966). *Verres Réfract.* **20**, 419.

Bishop, S. G., and Bray, P. J. (1966). *Phys. Chem. Glasses* **7**, 73.

Bloembergen, N. (1961). "Nuclear Magnetic Relaxation." W. A. Benjamin, Inc., New York.

Bloembergen, N., Purcell, E. M., and Pound, R. V. (1948). *Phys. Rev.* **73**, 679.

Bray, P. J., and Hendrickson, J. R. (1974). *J. Non-Cryst. Solids* **14**, 300.

Bray, P. J., and O'Keefe, J. G. (1963). *Phys. Chem. Glasses* **4**, 37.

Bray, P. J., Hintenlang, D. E., Mulkern, R. V., Greenbaum, S. G., Tran, D. C., and Drexhage, M. (1983a). *J. Non-Cryst. Solids* **56**, 27.

Bray, P. J., Lui, M. L., and Hintenlang, D. E. (1983b). *Wiss. Ztschr. Fridrich-Schiller Univ., Jena, Math-Naturwiss.* **32**, 409.

Bray, P. J., Gravina, S. J., Hintenlang, D. H., and Mulkern, R. V. (1988). *Magnetic Resonance Rev.* **13**, 263.

Brungs, M. P., and McCartney, E. R. (1975). *Phys. Chem. Glasses* **16**, 48.

Bucholtz, F. (1982). Ph.D. thesis, Brown University, Providence, R.I.

Bucholtz, F., and Bray, P. J. (1983). *J. Non-Cryst. Solids* **54**, 43.

Bunker, B. C., Tallant, D. R., Balfe, C. A. Kirkpatrick, R. J., Turner, G. L., and Reidmeyer, M. R. (1987). *J. Am. Ceram. Soc.* **70**, 675.

Chiodelli, G., Magistris, A., Villa, M., and Bjorkstam, J. L. (1982). *J. Non-Cryst. Solids* **51**, 143.

Cohen, M. H., and Reif, F. (1957). "Quadrupole Effects In Nuclear Magnetic Resonance Studies of Solids," in "Solid State Physics," Vol. 5. Academic Press, New York.

de Gennes, P. G. (1958). *J. Phys. Chem. Solids* **7**, 345.

de Jong, B. H. W. S., Schramm, C. M., and Parziale, V. E. (1983). *Geochim. Cosmochim. Acta* **47**, 1223.

de Jong, B. H. W. S., Schramm, C. M., and Parziale, V. E. (1984). *Geochim. Cosmochim. Acta* **48**, 2619.

Dell, W. J., and Bray, P. J. (1982). *Phys. Chem. Glasses* **23**, 98.

Dell, W., Bray, P. J., and Xiao, S. Z. (1983). *J. Non-Cryst. Solids* **58**, 1.

Dell, W. J., Mulkern, R. V., Bray, P. J., Weber, M. J., and Brawer, S. A. (1984). *Phys. Rev. B* **31**, 2624.

Douglass, D. C., Duncan, T. M., Walker, K. L., and Csencsits, R. (1985). *J. Appl. Phys.* **58**, 197.

Drain, L. E. (1967). *Metall. Rev.* **11**, 195.

Dubiel, M., and Ehrt, D. (1987). *Phys. Status Solidi A* **100**, 415.

Duncan, T. M., and Douglass, D. C. (1984). *Chem. Phys.* **87**, 339.

Duncan, T. M., Douglass, D. C., Csencsits, R., and Walker, K. L. (1986). *J. Appl. Phys.* **60**, 130.

Dupree, R., and Pettifer, R. F. (1984). *Nature* **308**, 523.

Dupree, R., Holland, D., McMillan, P. W., and Pettifer, R. F. (1984). *J. Non-Cryst. Solids* **68**, 399.

Dupree, R., Farnan, I., Forty, A. J., El-Mashri, S., and Bottyan, L. (1985a). *J. Phys.* **46**, C8–C113.

Dupree, R., Holland, D., and Williams, D. S. (1985b). *J. Phys.* **46**, C8–C119.

Dupree, R., Holland, D., and Williams, D. S. (1985c). *Phys. Chem. Glasses* **26**, 50.

Dupree, R., Holland, D., and Williams, D. S. (1986). *J. Non-Cryst. Solids* **81**, 185.

Dupree, R., Ford, N., and Holland, D. (1987). *Phys. Chem. Glasses* **28**, 78.

Dutz, H., and Schultz, G. W. (1969). *Glastech. Ber.* **42**, 89.

Engelhardt, G., Nofz, M., Forkel, K., Wihsmann, F. G., Mägi, M., Samoson, A., and Lippmaa, E. (1985). *Phys. Chem. Glasses* **26**, 157.

Feller, S. A., Dell, W. J., and Bray, P. J. (1982). *J. Non-Cryst. Solids* **51**, 21.

France, P. W., and Hooper, H. O. (1970). *J. Phys. Chem. Solids* **31**, 1307.

France, P. W., and Wadsworth, M. (1982). *J. Magn. Reson.* **49**, 48.

Fujiu, T., and Ogino, M. (1984). *J. Non-Cryst. Solids* **64**, 287.

Fukushima, E., and Roeder, S. B. W. (1981). "Experimental Pulse NMR." Addison-Wesley, Reading, Mass.

Fyfe, C. A. (1983). "Solid State NMR For Chemists." C.F.C. Press, Ontario, Canada.

Fyfe, C. A., Gobbi, G. C., Hartman, J. S., Lenkinski, R. E., O'Brien, J. H., Beange, E. R., and Smith, M. A. R. (1982). *J. Magn. Reson.* **47**, 168.

Ganapathy, S., Schramm, S., and Oldfield, E. (1982). *J. Chem. Phys.* **77**, 4360.

Geissberger, A. E., and Bray, P. J. (1983). *J. Non-Cryst. Solids* **54**, 121.

Geissberger, A. E., Bucholtz, F., and Bray, P. J. (1982). *J. Non-Cryst. Solids* **49**, 117.

Göbel, E., Müller-Warmuth, W., and Olyschläger, H. (1979). *J. Magn. Reson.* **36**, 371.

Göring, R., Kneipp, K., and Nass, H. (1982). *Phys. Status Solidi A* **72**, 623.

Gravina, S. J. (1988). Ph.D. thesis, Brown University, Providence, R.I.

Greenblatt, S., and Bray, P. J. (1967a). *Phys. Chem. Glasses* **8**, 190.

Greenblatt, S., and Bray, P. J. (1967b). *Phys. Chem. Glasses* **8**, 213.

Gresch, R., Müller-Warmuth, W., and Dutz, H. (1976). *J. Non-Cryst. Solids* **21**, 31.

Grimmer, A.-R., and Müller, W. (1986). *Monatsh. Chem.* **117**, 799.

Grimmer, A.-R., Mägi, M., Hähnert, M., Stade, H., Samoson, A., Wieker, W., and Lippmaa, E. (1984). *Phys. Chem. Glasses* **25**, 105.

Gupta, P. K., Lui, M. L., and Bray, P. J. (1985). *J. Am. Ceram. Soc.* **68**, C-82.

Gutowsky, H. S., and McGarvey, B. R. (1952). *J. Chem. Phys.* **20**, 1472.

Gutowsky, H. S., and Pake, G. E. (1950). *J. Chem. Phys.* **18**, 162.

Hallas, E., Haubenreißer, U., Hähnert, M., and Müller, D. (1983). *Glastech. Ber.* **56**, 63.

Harris, I. A. Jr., and Bray, P. J. (1980). *Phys. Chem. Glasses* **21**, 156.

Harris, I. A., and Bray, P. J. (1984). *Phys. Chem. Glasses* **25**, 69.

Hendrickson, J. R., and Bishop, S. G. (1975). *Solid State Commun.* **17**, 301.

Hendrickson, J. R., and Bray, P. J. (1973). *J. Magn. Reson.* **9**, 341.

Hendrickson, J. R., and Bray, P. J. (1974). *J. Chem. Phys.* **61**, 2754.

Hintenlang, D. H. (1985). Ph.D. thesis, Brown University, Providence, R.I.

Hintenlang, D. E., and Bray, P. J. (1985). *J. Non-Cryst. Solids* **69**, 243.

Holzman, G. R., Anderson, J. H., and Koth, W. (1955). *Phys. Rev.* **98**, 542.

Holzman, G. R., Lauterbur, P. C., Anderson, J. H., and Koth, W. (1956). *J. Chem. Phys.* **25**, 172.

Jäger, C., and Haubenreißer, U. (1985). *Phys. Chem. Glasses* **26**, 152.

Jellison, G. E. Jr. (1979). *Solid State Commun.* **30**, 481.

Jellison, G. E. Jr., and Bray, P. J. (1976). *Solid State Commun.* **19**, 517.

Jellison, G. E. Jr., and Bray, P. J. (1978). *J. Non-Cryst. Solids* **29**, 187.

Jellison, G. E. Jr., Panek, L. W., Bray, P. J., and Rouse, G. B. Jr. (1977). *J. Chem. Phys.* **66**, 802.

Jellison, G. E. Jr., Feller, S. A., and Bray, P. J. (1978). *Phys. Chem. Glasses* **19**, 52.

Kim, K. S., and Bray, P. J. (1974a). *J. Nonmetals* **2**, 95.

Kim, K. S., and Bray, P. J. (1974b). *Phys. Chem. Glasses* **15**, 47.

Kim, K. S., Bray, P. J., and Merrin, S. (1976). *J. Chem. Phys.* **64**, 4459.

Kirkpatrick, R. J., Dunn, T., Schramm, S., Smith, K. A., Oestrike, R., and Turner, G. (1986a). "Structure and Bonding in Noncrystalline Solids," G. E. Walrafen and A. G. Revesz, eds. Plenum Press, New York, p. 303.

Kirkpatrick, R. J., Oestrike, R., Weiss, C. A. Jr., Smith, K. A., and Oldfield, E. (1986b). *Am. Mineral.* **71**, 705.

Kline, D., and Bray, P. J. (1966). *Phys. Chem. Glasses* **7**, 41.

Kline, D., Bray, P. J., and Kriz, H. M. (1968). *J. Chem. Phys.* **48**, 5277.

Kobayashi, K. (1979). *J. Am. Ceram. Soc.* **62**, 440.

Kobayashi, K., and Okuma, H. (1976). *J. Am. Ceram. Soc.* **59**, 354.

Kolditz, L., and Wahner, E. (1973). *Z. Anorg. Allg. Chem.* **400**, 161.

Krämer, F., Müller-Warmuth, W., Scheerer, J., and Dutz, H. (1973). *Z. Naturforsch.* **28a**, 1338.

Kreidl, N. K., and Weyl, W. A. (1941). *J. Am. Ceram. Soc.* **24**, 372.

Kriz, H. M., and Bray, P. J. (1971). *J. Non-Cryst. Solids* **6**, 27.

Krogh-Moe, J. (1965). *Phys. Chem. Glasses* **6**, 46.

Lacy, E. D. (1967). *Phys. Chem. Glasses* **8**, 238.

Landsberger, F. R., and Bray, P. J. (1970). *J. Chem. Phys.* **53**, 2757.

Leventhal, M., and Bray, P. J. (1965). *Phys. Chem. Glasses* **6**, 113.

Lippmaa, E., Mägi, M., Samoson, A., Engelhardt, G., and Grimmer, A.-R. (1980). *J. Am. Chem. Soc.* **102**, 4889.

Lippmaa, E., Mägi, M., Samoson, A., Tarmak, M., and Engelhardt, G. (1981). *J. Am. Chem. Soc.* **103**, 4992.

Lippmaa, E., Samoson, A., Mägi, M., Teeäär, R., Schraml, J., and Götz, J. (1982). *J. Non-Cryst. Solids* **50**, 215.

Lynch, G. F., and Sayer, M. (1973). *J. Phys. C: Solid State Phys.* **6**, 3661.

Lynch, G. F., Sayer, M., Segel, S. L., and Rosenblatt, G. (1971). *J. Appl. Phys.* **42**, 2587.

Mägi, M., Lippmaa, E., Samoson, A., Engelhardt, G., and Grimmer, A.-R. (1984). *J. Phys. Chem.* **88**, 1518.

Martin, S., Bischof, H. J., Mali, M., Roos, J., and Brinkmann, D. (1986). *Solid State Ionics* **18–19**, 421.

Milberg, M. E., Otto, K., and Kushida, T. (1966). *Phys. Chem. Glasses* **7**, 14.

Mosel, B. D., Müller-Warmuth, W., and Dutz, H. (1974) *Phys. Chem. Glasses* **15**, 154.

Mulkern, R. V., Chung, S. J., Bray, P. J., Chryssikos, G. D., Turcotte, D. E., Risen, W. M. (1986). *J. Non-Cryst. Solids* **85**, 69.

Müller, D., Gessner, W., Behrens, H.-J., and Scheler, G. (1981). *Chem. Phys. Lett.* **79**, 59.

Müller, D., Berger, G., Grunze, I., Ladwig, G., Hallas, E., and Haubenreißer, U. (1983). *Phys. Chem. Glasses* **24**, 37.

Müller-Warmuth, W., and Eckert, H. (1982). "Physics Reports," **88**.

Müller-Warmuth, W., Poch, W., and Schulz, G. W. (1966). *Glastech. Ber.* **39**, 415.

Müller-Warmuth, W., Poch, W., and Sielaff, G. (1970). *Glastech. Ber.* **43**, 5.

Murdoch, J. B., Stebbins, J. F., and Carmichael, I. S. E. (1985). *Am. Mineral.* **70**, 332.

Noack, F. (1971). "NMR-Basic Principles and Progress," Vol. 3. Springer, Berlin.

Oestrike, R., Yang, W.-H., Kirkpatrick, R. J., Hervig, R. L., Navrotsky, A., and Montez, B. (1987). *Geochim. Cosmochim. Acta* **51**, 2199.

Ohtani, E., Taulelle, F., and Angell, C. A. (1985). *Nature* **314**, 78.

Olyschläger, H. (1977). Thesis, Universität Münster, F.R. Germany.

Otto, K., and Milberg, M. E. (1967). *J. Am. Ceram. Soc.* **50**, 513.

Panek, L. W., and Bray, P. J. (1977). *J. Chem. Phys.* **66**, 3822.

Panek, L. W., Exarhos, G. J., Bray, P. J., and Risen, W. M. (1977). *J. Non-Cryst. Solids* **24**, 51.

Park, M. J., and Bray, P. J. (1972). *Phys. Chem. Glasses* **13**, 50.

Park, M. J., Kim, K. S., and Bray, P. J. (1979). *Phys. Chem. Glasses* **20**, 31.

Rhee, C., and Bray, P. J. (1971a). *Phys. Chem. Glasses* **12**, 156.

Rhee, C., and Bray, P. J. (1971b). *Phys. Chem. Glasses* **12**, 165.

Risbud, S. H., Kirkpatrick, R. J., Taglialavore, A. P., and Montez, B. (1987). *J. Am. Ceram. Soc.* **70**, C-10.

Rubinstein, M. (1976). *Phys. Rev. B* **14**, 2778.

Rubinstein, M., and Resing, H. A. (1976). *Phys. Rev. B* **13**, 959.

Rubinstein, M., Resing, H. A., Reinecke, T. L., and Ngai, K. L. (1975). *Phys. Rev. Lett.* **34**, 1444.

Scheerer, J., Müller-Warmuth, W., and Dutz, H. (1973). *Glastech. Ber.* **46**, 109.

Schiller, W., Müller, D., and Scheler, G. (1982). *Z. Chem.* **22**, 44.

Scholes, S. R. (1975). "Modern Glass Practice," revised by C. H. Greene. Cahners, Boston, Massachusetts.

Schramm, C. M., de Jong, B. H. W. S., and Parziale, V. E. (1984). *J. Am. Chem. Soc.* **106**, 4396.

Schramm, S., and Oldfield, E. (1982). *J. Chem. Soc., Chem. Commun.*, 980.

Schulz, G. W., Müller-Warmuth, W., Poch, W., and Scheerer, J. (1968). *Glastech. Ber.* **41**, 435.

Segel, S. L., Creel, R. B., and Torgeson, D. R. (1983). *J. Mol. Struct.* **111**, 79.

Selvaray, U., Rao, K. J., Rao, C. N. R., Klinowski, J., and Thomas, J. M. (1985). *Chem. Phys. Lett.* **114**, 24.

Senegas, J., and Olivier-Fourcade, J. (1983). *J. Phys. Chem. Solids* **44**, 1033.

Silver, A. H., and Bray, P. J. (1958). *J. Chem. Phys.* **29**, 984.

Simon, S., and Nicula, A. (1981). *Solid State Commun.* **39**, 1251.

Slichter, C. P. (1978). "Principles of Magnetic Resonance," Springer Series in Solid State Sciences 1. Springer, Berlin.

Smith, J. V., and Blackwell, C. S. (1983). *Nature* **303**, 223.

Smith, K. A., Kirkpatrick, R. J., Oldfield, E., and Henderson, D. M. (1983). *Am. Mineral.* **68**, 1206.

Stallworth, P. E., and Bray, P. J. "Barium Borosilicate Glasses," to be published.

Svanson, S. E., and Johansson, R. (1970). *Acta. Chem. Scand.* **24**, 755.

Svanson, S. E., Forslind, E., and Krogh-Moe, J. (1962). *J. Phys. Chem.* **66**, 174.

Szeftel, J., and Alloul, H. (1975). *Phys. Rev. Lett.* **34**, 657.

Szu, S. P., and Bray, P. J. "Potassium Borosilicate Glasses," to be published.

Tabbey, M. P., and Hendrickson, J. R. (1980). *J. Non-Cryst. Solids* **38–39**, 51.

Thomas, J. M., Klinowski, J., Wright, P. A., and Roy, R. (1983). *Angew. Chem.*, Int. Ed., **22**, 614.

Townes, C. H., and Dailey, B. P. (1948). *J. Chem. Phys.* **17**, 782.

Turner, G. L., Smith, K. A., Kirkpatrick, R. J., and Oldfield, E. (1986). *J. Magn. Reson.* **67**, 544.

Van Kronendonk, J. (1954). *Physica* (Utrecht) **20**, 781.

Villa, M., Chiodelli, G., Magistris, A., and Licheri, G. (1986a). *J. Chem. Phys.* **85**, 2392.

Villa, M., Chiodelli, G., and Scagliotti, M. (1986b). *Solid State Ionics* **18–19**, 382.

Villa, M., Carduner, K. R., and Chiodelli, G. (1987). *Phys. Chem. Glasses* **28**, 131.

Villeneuve, G., Echegut, P., Reau, J. M., Levasseur, A., and Brethous, J. C. (1979). *J. Solid State Chem.* **30**, 275.

Visco, S. J., Spellane, P., and Kennedy, J. H. (1985a). *J. Electrochem. Soc.* **132**, 751.

Visco, S. J., Spellane, P., and Kennedy, J. H. (1985b). *J. Electrochem. Soc.* **132**, 1776.

Vopilov, V. A., Buznik, V. M., Matsulev, A. N., Bogdanov, V. L., Karapetyan, A.K., and Khalilev, V. D. (1985). *Fiz. Khim. Stekla* **11**, 162.

Vopilov, V. A., Vopilov, E. A., Buznik, V. M., et al. (1986). *Fiz. Khim. Stekla* **12**, 242.

Waugh, J. S., Huber, L. M., and Haeberlen, U. (1968). *Phys. Rev. Lett.* **20**, 180.

Xiao, S., and Meng, Q. (1986). *J. Non-Cryst. Solids* **80**, 195.

Yang, W.-H., Kirkpatrick, R. J., and Turner, G. (1986). *J. Am. Ceram. Soc.* **69**, C-222.

Yun, Y. A., and Bray, P. J. (1978a). *J. Non-Cryst. Solids* **27**, 363.

Yun, Y. H., and Bray, P. J. (1978b). *J. Non-Cryst. Solids* **30**, 45.

Yun, Y. A., and Bray, P. J. (1981). *J. Non-Cryst. Solids* **44**, 227.

Yun, Y. H., Feller, S. A., and Bray, P. J. (1979). *J. Non-Cryst. Solids* **33**, 273.

Zachariasen, W. H. (1932). *J. Am. Chem. Soc.* **54**, 3841.

Zhong, J., and Bray, P. J. (1986). *J. Non-Cryst. Solids* **84**, 17.

Zhong, J., and Bray, P. J. (1987). *J. Non-Cryst. Solids* **94**, 122.

Zhong, J., and Bray, P. J. (1988) "Mixed Alkali Borate Glasses," to be published.

Zhong, J., Wu, X., Liu, M. U. and Bray, P. J. (1988). *J. Non-Cryst. Solids* **107**, 81.

Zhdanov, S. P., and Shmidel, G. (1975). *Fiz. Khim. Stekla* **1**, 452.

CHAPTER 3

Electron Spin Resonance

D. L Griscom

OPTICAL SCIENCES DIVISION
NAVAL RESEARCH LABORATORY
WASHINGTON, D.C.

I. Introduction

Electron spin resonance (ESR) is in many ways analogous to nuclear magnetic resonance (NMR). In both cases, spin-degenerate energy levels are split by the application of an external magnetic field (the Zeeman effect) and direct transitions between Zeeman levels are observed. In general, these transitions are stimulated by bathing the sample in radio-frequency radiation in the NMR case or microwave radiation in the ESR case, although the precise resonance frequency is dictated in either case by the magnitude of the

151

ISBN 0-12-706707-8

local magnetic field, $\mathbf{H}_{loc}$. The Zeeman energy levels are expressed by the scalar product of the magnetic moment with $\mathbf{H}_{loc}$.

$$E_{Zeeman} = -\boldsymbol{\mu} \cdot \mathbf{H}_{loc}. \tag{1}$$

Neglecting for the present any orbital contributions in the electronic case, the magnetic moments of electrons and nuclei are respectively given by

$$\boldsymbol{\mu}_e = -g\beta\mathbf{S}, \tag{2a}$$

and

$$\boldsymbol{\mu}_N = g_N\beta_N\mathbf{I}. \tag{2b}$$

In Eqs. (2a) and (2b), respectively, g and g_N are the electron and nuclear g factors (by convention g is positive for an electron but g_N assumes the algebraic sign of the magnetogyric ratio of the nucleus in question), $\mathbf{S}$ and $\mathbf{I}$ are the electronic and nuclear spin angular-momentum operators, and β and β_N are the Bohr and nuclear magnetons:

$$\beta = \frac{|e|h}{4\pi mc} \tag{3a}$$

and

$$\beta_N = \frac{|e|h}{4\pi m_p c}. \tag{3b}$$

In the preceding relations, h is Planck's constant, c is the speed of light, e and m are the charge and rest mass of the electron, and m_p is the mass of the proton. Quantum mechanics dictate that the spin vectors $\mathbf{S}$ and $\mathbf{I}$ can have only certain discrete projections on the direction of the local magnetic field: $M_S = S, S - 1, \ldots, -S$ for the electrons and $M_I = I, I - 1, \ldots, -I$ for nuclei. Here, the scalar quantities S and I are termed the "spin" of the electron and the nucleus, respectively. Clearly, the spin can be an integer, half integer, or zero. Selection rules $\Delta M_S = \pm 1$ and $\Delta M_I = \pm 1$ apply in the cases of ESR and NMR, respectively. Thus, assuming $\mathbf{H}_{loc}$ to equal the laboratory applied field, and given a fixed spectrometer frequency v, transitions are stimulated between adjacent states of Eq. (1) when $|\mathbf{H}_{loc}|$ is equal to $|\mathbf{H}_{res}|$ as given by

$$hv = \Delta E_{Zeeman}(\text{ESR}) = g\beta|\mathbf{H}_{res}| \tag{4a}$$

$$hv = \Delta E_{Zeeman}(\text{NMR}) = g_N\beta_N|\mathbf{H}_{res}|. \tag{4b}$$

Since $m_p/m = \beta/\beta_N = 1838$, it is clear from Eqs. (4) why ESR frequencies are generally ~ 2000 times greater than typical NMR frequencies, given that laboratory applied fields of the order of a few hundred milli-Teslas $(1\,\text{mT} = 10\,\text{Gauss})$ tend to be preferred for each type of experiment. This same factor of 2000 also favors ESR *sensitivity* over that of NMR—a particularly fortunate outcome, since the populations of *unpaired* electrons available in condensed materials are generally several orders of magnitude fewer than the numbers of NMR-active nuclei that can be studied. Other parallels and contrasts between the ESR and NMR experiments will be pointed out later.

A single electron has a spin $S = 1/2$. The Zeeman energies of Eq. (1) for this case are diagrammed in Fig. 1a as a function of the magnitude of the applied magnetic field, denoted simply as H. Resonant absorption occurs when $|\mathbf{H}| = |\mathbf{H}_{res}|$, as given by Eq. (4a).

Electron spin resonance as a discipline dates from the early radio-frequency measurements of Zavoisky (1945), but it owes its continuing vitality to the post–World War II availability of microwave sources. Zavoisky (1945) and Cummerow and Halliday (1946) conducted the first measurements of the absorption of microwave energy by systems of unpaired electrons as a function of the magnitude of an externally applied magnetic field. The application of the method to single crystals (where the spectra could be studied also as a function of the angle between the applied field and the crystallographic axes) developed so quickly as to be the subject of lengthy review articles by the early 1950s (e.g., Bleaney and Stevens, 1953; Bowers and Owen, 1955). The possibility of performing ESR in glasses (by definition, an

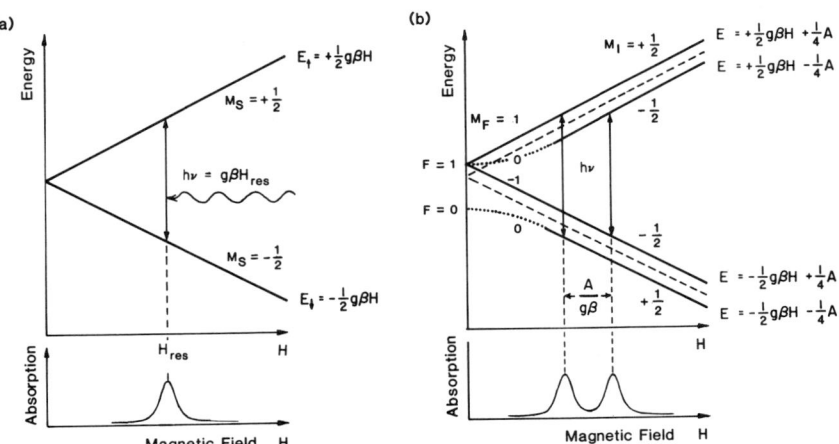

FIG. 1.   Energy levels and ESR resonance absorption schemes for (a) the Zeeman interaction of a single electron ($S = 1/2$) and (b) the Zeeman and hyperfine interactions for $S = 1/2$ and $I = 1/2$. The dashed lines in (b) recapitulate the Zeeman interaction alone; the parallel, unbroken straight lines represent the hyperfine interaction in the high-field limit. The dotted curves are schematic representations of the solution of Breit and Rabi (1931) for the case $|A| \geqslant h\nu$.

isotropic medium where all crystallographic information is absent) was not seriously considered before the work of Sands (1955), who introduced the concept of the powder pattern.

Computer simulation of the powder pattern lineshapes soon opened up the possibility of reliable ESR spectral analysis in glasses. (The earliest example seems to be given by Hochstrasser, 1966.) Still, as will be elaborated in later sections, glasses differ from simple crystalline powders in a fundamental way: The intrinsic randomness of the vitreous state gives rise to statistical distributions of crystal fields that are reflected to various degrees in many spectroscopic properties. Yafaev and Yablokov (1962) may have been the first to recognize frequency-dependent ESR linewidths as probable manifestations of statistical distributions in g values. The use of computer lineshape-simulation methods to deconvolve such statistical information from the glassy-state ESR spectra appears to date from Griscom et al. (1968) and Mackey et al. (1969). From that point onward, the field has blossomed considerably, with many hundreds of papers appearing on the subjects of radiation-induced defect centers, dissolved paramagnetic ions, and even ferromagnetic precipitates in glasses.

The sheer volume and diversity of the literature of ESR in glasses has compelled a high degree of selectivity in many of the longer review articles that have appeared (e.g., Taylor and Bray, 1970, 1972; Griscom, 1973, 1976, 1978a, 1980a; Wong and Angell, 1976; Friebele and Griscom, 1979; Kawazoe, 1985). However, Taylor et al. (1975) and Kliava (1988) provide rather comprehensive mathematical developments together with extensive bibliographies. The approach adopted in the present chapter is to focus strongly on the simplest elements of mathematical theory most commonly required for the analysis of ESR lineshapes in amorphous materials (Section II) but to confine illustrations of the practice of ESR in glasses to selected references drawn only from the literature of dilute transition-group ions (Section III) and defect centers (Section IV) in the traditional oxide glasses. Omitted will be important bodies of work dealing with fluoride glasses, chalcogenide glasses, amorphous elemental semiconductors, aqueous glasses, and polymer glasses, as well as the related subfield of ferromagnetic resonance in glasses. However, the ESR theory to be summarized in the following section applies equally to all of the forementioned systems except those containing ferromagnetic precipitates. (For reviews of the latter, see Griscom, 1980b, 1984a.)

Properly applied, ESR spectroscopy can be used to infer local atomic arrangements and electronic structure at the sites of paramagnetic centers in glasses. Clearly, the ability to extract this type of information provides a powerful window on glass structure. The present chapter critically discusses many of the underlying spectroscopic issues that have arisen and been resolved on the road to realizing the full potential of the technique.

II. ESR Theory

A. INFLUENCE OF ORBITAL ANGULAR MOMENTUM

For convenience of illustration, Eq. (2a) gave the magnetic moment of the electron solely in terms of the spin angular momentum operator S (multiplied by the scalar quantities g and β). But in the general case, the orbital angular momentum L contributes to the electronic magnetic moment operator:

$$\mathbf{\mu} = -\beta(\mathbf{L} + g_e\mathbf{S}), \tag{5}$$

where g_e is the g value of the free electron ($\simeq 2.0023$).

As was presaged in Eqs. (2), it is standard practice to define a g factor to relate the magnetic moment of a particle to its overall angular momentum. Thus, the magnetic moment of an open-shell atom in free space is given by:

$$\mathbf{\mu} = -g_J\beta\mathbf{J}, \tag{6}$$

where $\mathbf{J} = \mathbf{L} + \mathbf{S}$ is the total angular momentum and g_J is the Lande g factor. (See, e.g., Abragam and Bleaney, 1970.) But for most defect centers and many paramagnetic ions embedded in solids, the local "crystal" electric fields act to split orbital degeneracies and leave an orbital singlet ground state. Under these conditions, the orbital angular momentum is said to be "quenched," and the small residual contributions of the orbital angular momentum to the electron paramagnetism can be treated by quantum mechanical perturbation theory. The unperturbed Hamiltonian, $\mathcal{H}_0$, consists of all kinetic energy and Coulomb interaction terms, including the crystal field interactions, while the perturbing Hamiltonian, $\mathcal{H}_1$, is taken to be the sum of the Zeeman interaction, $-\mathbf{\mu} \cdot \mathbf{H}$, and the spin-orbit interaction, $\lambda \mathbf{L} \cdot \mathbf{S}$,:

$$\mathcal{H}_1 = \lambda\mathbf{L} \cdot \mathbf{S} + \beta(\mathbf{L} + g_e\mathbf{S}) \cdot \mathbf{H}, \tag{7}$$

where λ is the spin-orbit coupling constant and H is the local magnetic field (generally understood to be the laboratory applied field).

The result of such a perturbation treatment yields the effective spin Hamiltonian

$$\mathcal{H}_s = \beta\mathbf{H} \cdot \mathbf{g} \cdot \mathbf{S} + \mathbf{S} \cdot \mathbf{D} \cdot \mathbf{S}. \tag{8}$$

In this representation the spin vector S can be viewed as being quantized along the direction of an "effective" local field given by $\mathbf{H} \cdot \mathbf{g}/g_e$, or, alternatively, an "effective" spin $\mathbf{g} \cdot \mathbf{S}/g_e$ can be regarded as being quantized along H. The elements of the g "tensor" (actually not a true tensor since it forms invariants with axial vectors rather than polar vectors) implicitly contain the influences of the spin-orbit and orbit-Zeeman interactions, which admix a small amount of orbital angular momentum back into the orbitally "quenched" ground state.

The **g** tensor takes the form (e.g., Pake and Estle, 1973; Wertz and Bolton, 1972):

$$\mathbf{g} = g_e\mathbf{1} + 2\lambda\mathbf{\Lambda}, \tag{9}$$

where **1** is the unit tensor. For unpaired spins localized on a single atom (or a cluster of like atoms), the tensor $\mathbf{\Lambda}$ is given by

$$\Lambda_{ij} = -\sum_{n=0} \frac{\langle 0|L_i|n\rangle\langle n|L_j|0\rangle}{E_n - E_0}, \tag{10}$$

where $|0\rangle$ is the spacial wavefunction corresponding to the ground state of the unpaired spin and $|n\rangle$ are excited state wavefunctions of energies E_0 and E_n, respectively.

Equations (9) and (10) are exceedingly useful for testing models for defect centers in many inorganic glasses where the molecular orbitals of the unpaired spin frequently comprise linear combinations of s and p atomic orbitals localized on individual atoms. The right-hand side of Eq. (10) can be calculated by recalling that all matrix elements of L_i among s and p states are zero except

$$\langle p_x|L_y|p_z\rangle = \langle p_y|L_z|p_x\rangle = \langle p_z|L_x|p_y\rangle = i. \tag{11}$$

For the case of an unpaired spin delocalized over several unlike atoms (each characterized by a different value of λ), the correct expression for the tensor $\mathbf{\Lambda}$ is more involved (see, e.g., Atkins and Symons, 1967).

The matrix **D** which appears on the right-hand side of Eq. (8) is given by

$$\mathbf{D} = \lambda^2\mathbf{\Lambda}. \tag{12}$$

The $\mathbf{S}\cdot\mathbf{D}\cdot\mathbf{S}$ term is often referred to as the "fine-structure" or "zero-field splitting" term and is most frequently expressed in its alternative form,

$$\mathbf{S}\cdot\mathbf{D}\cdot\mathbf{S} = D[S_z^2 - (1/3)S(S+1)] + E(S_x^2 - S_y^2), \tag{13}$$

where

$$D = D_{zz} - (1/2)(D_{xx} + D_{yy}) \text{ and} \tag{14a}$$

$$E = (1/2)(D_{xx} - D_{yy}). \tag{14b}$$

In analogy with the quadrupole interaction in NMR, the fine-structure splitting appears only in the Hamiltonians of systems for which $S > 1/2$.

B. THE HYPERFINE INTERACTION

When the unpaired electron is found in the vicinity of a magnetic nucleus (an isotope for which $I \neq 0$), the Zeeman energy levels of Eq. (1) each split into $2I + 1$ sublevels due to the interaction that occurs between the electronic and nuclear magnetic moments (given by Eqs. (2a) and (2b), respectively). In the limit that the electronic Zeeman interaction is the dominant energy term (so that $\mathbf{S}$ is quantized along $\mathbf{H}$), it is valid to decompose the local field in Eq. (1) into

$$\mathbf{H}_{loc} = \mathbf{H} + \mathbf{H}_{hf}, \tag{15}$$

where $\mathbf{H}_{hf}$ is an additive field component due to the nuclear moment, which can take on $2I + 1$ discrete values due to the $2I + 1$ equally probable projections of the nuclear spin vector on the axis of quantization. (See Fig. 1b.)

Fermi (1930) has shown that for systems with one unpaired electron of wavefunction $\psi(r)$ with the magnetic nucleus situated at the origin of coordinates, an isotropic hyperfine interaction arises due to the "direct contact" of the electronic and nuclear moments:

$$E_{iso} = -(8\pi/3)|\psi(0)|^2 \boldsymbol{\mu}_e \cdot \boldsymbol{\mu}_N. \tag{16}$$

The operator equation corresponding to Eq. (16) is

$$\mathscr{H}_{iso} = (8\pi/3)g\beta g_N\beta_N|\psi(0)|^2 \mathbf{S} \cdot \mathbf{I} \tag{17a}$$

$$= A_0 \mathbf{S} \cdot \mathbf{I}, \tag{17b}$$

where A_0 is termed the isotropic coupling constant. Clearly A_0 is nonzero only for s state wavefunctions of the magnetic nucleus, as these are the only ones to exhibit finite density at the origin.

In addition to the Fermi contact interaction, electrons and nuclei can interact at a distance via the dipole–dipole interaction, whose classical expression is

$$E_{dipolar} = \frac{\boldsymbol{\mu}_e \cdot \boldsymbol{\mu}_N}{r^3} - \frac{3(\boldsymbol{\mu}_e \cdot \mathbf{r})(\boldsymbol{\mu}_N \cdot \mathbf{r})}{r^5}. \tag{18}$$

Replacing the magnetic moment vectors in Eq. (18) by their correct quantum mechanical operators [see, Eqs. (2)] gives

$$\mathscr{H}_{dipolar} = -g\beta g_N\beta_N \left[\frac{\mathbf{S} \cdot \mathbf{I}}{r^3} - \frac{3(\mathbf{S} \cdot \mathbf{r})(\mathbf{I} \cdot \mathbf{r})}{r^5} \right] \tag{19a}$$

$$= \mathbf{S} \cdot \mathbf{T} \cdot \mathbf{I}, \tag{19b}$$

where $\mathbf{T}$ is the anisotropic hyperfine interaction tensor.

The effective spin Hamiltonian for an $S = 1/2$ electron interaction with a magnetic nucleus is

$$\mathscr{H}_s = \beta \mathbf{S} \cdot \mathbf{g} \cdot \mathbf{H} + \mathbf{S} \cdot \mathbf{A} \cdot \mathbf{I} - g_N\beta_N \mathbf{H} \cdot \mathbf{I}, \tag{20a}$$

where

$$A = A_0\mathbf{1} + \mathbf{T}. \tag{20b}$$

The evaluation of the tensor $\mathbf{T}$ depends greatly on the location of the magnetic nucleus and the nature of the wavefunction of the unpaired spin. Several approximations are customarily employed in treating the hyperfine term of Eqs. (20) (see, e.g., Wertz and Bolton, 1972), namely, (1) that the g factor is effectively isotropic and (2) that the Zeeman term dominates energetically, allowing one to quantize $\mathbf{S}$ along $\mathbf{H}$. In the (not always likely) event that $|H|$ is much larger than the hyperfine field *of the electron at the nucleus*, both $\mathbf{S}$ and $\mathbf{I}$ can be quantized along $\mathbf{H}$. Then in the special case of a p orbital centered on the interacting nucleus, Eq. (19a) reduces to

$$\mathcal{H}_{\text{dipolar}} = g\beta g_N\beta_N \left\langle \frac{3\cos^2\alpha - 1}{2r^3} \right\rangle (3\cos^2\theta - 1)M_SM_I \tag{21a}$$

where θ is the angle between $\mathbf{H}$ and the axis of the p orbital and α is the angle between $\mathbf{r}$ and the same axis. The integral represented by the angular brackets in Eq. (21a) is easily performed, leading to the result

$$\mathcal{H}_{\text{dipolar}} = B(3\cos^2\theta - 1)M_SM_I, \tag{21b}$$

where

$$B = (2/5)g\beta g_N\beta_N\langle r^{-3}\rangle_{np}. \tag{21c}$$

Comparing Eq. (21b) with Eq. (19b), it can be seen that under the present approximations $\mathbf{T}$ is diagonal in a coordinate system aligned with the p orbital of the unpaired spin, with principal axis components $-B, -B, 2B$.

McConnell and Strathdee (1959) evaluated Eq. (19a) for the situation where the magnetic nucleus does not lie at the origin of coordinates of the $2p$ orbital of the unpaired spin, and Atkins (1964) extended their results to the case of a nucleus located at a distance of R from an np orbital. (See Atkins and Symons, 1967.) It must be cautioned, however, that none of the foregoing results are correct for cases where the hyperfine field of the electron at the nucleus is much larger than $|H|$; the correct Hamiltonian in these cases has been worked out by Blinder (1960).

The nuclear Zeeman term on the right-hand side of Eq. (20a) can normally be neglected, as it is usually small and does not affect the energies of the ESR transitions. The electric quadrupole interaction is generally too small to be deconvolved from glassy-state ESR spectra and is therefore omitted from the present discussion.

More detailed accounts of ESR spin Hamiltonian theory can be found in many standard texts, including Abragam and Bleaney (1970), Pake and Estle (1973), and Wertz and Bolton (1972).

C. The Resonance Condition

Whereas the preceding paragraphs have outlined the meaning and derivation of the spin Hamiltonian, the actual evaluations of spin-Hamiltonian parameters from an experimental spectrum require a mathematical form called the resonance condition. The resonance condition is obtained by a process usually involving the following steps: First, the spin Hamiltonian is diagonalized (either exactly or by use of suitable approximations) to obtain the electronic energy levels as a function of the magnitude of the applied field, H. Then, the energies of the allowed electronic transitions ($M_S - 1 \leftrightarrow M_S$, $\Delta M_I = 0$) are written down and equated to the energy of the spectrometer quantum, $h\nu$. Finally, these expressions are generally rearranged to obtain the magnetic field at which resonance occurs, H_{res}, as functions of the spectrometer frequency, ν, the magnetic quantum numbers M_S and M_I, and the angles θ and ϕ relating the direction of the applied field H to the principal axes of the spin Hamiltonian. Taylor et al. (1975) have compiled a rather comprehensive listing of ESR and NMR resonance conditions that have been derived in closed form. A few of the expressions more commonly used in analyzing the ESR spectra of glasses are reproduced here for reference.

For the case $S = 1/2$, $I = 0$, only the Zeeman interaction pertains and the resonance condition is given by

$$H_{res} = h\nu/g\beta, \tag{22}$$

where for axial symmetry

$$g^2 = g_\perp^2 \sin^2 \theta + g_\parallel^2 \cos^2 \theta. \tag{23}$$

In Eq. (23), θ is the angle between the symmetry axis and the direction of H. For orthorhombic symmetry, the expression for g to be used in Eq. (22) is

$$g^2 = g_a^2 \sin^2 \theta + g_3^2 \cos^2 \theta, \tag{24a}$$

where

$$g_a^2 = g_1^2 \sin^2 \phi + g_2^2 \cos^2 \phi, \tag{24b}$$

and g_1, g_2, and g_3 are the principal axis values of the g tensor.

The resonance condition for axially symmetric g and hyperfine tensors with colinear principal axes has been worked out in second-order perturbation theory by Bleaney (1951); for the case $S = 1/2$, this result takes the form

$$H_{res} = \frac{h\nu}{g\beta} - \frac{KM_I}{g\beta} - \frac{A_\perp^2}{4g^2\beta^2 H}\left[\frac{A_\parallel^2 + K^2}{K^2}\right][I(I+1) - M_I^2]$$

$$-(2g^2\beta^2 H)^{-1}\left[\frac{A_\parallel^2 + K^2}{K}\right]^2\left(\frac{g_\parallel g_\perp}{g^2}\right)^2 \sin^2 \theta \cos^2 \theta M_I^2, \tag{25a}$$

where g is given by Eq. (23) and the quantity K is expressed as

$$K^2 = g^{-2}(A_\parallel^2 g_\parallel^2 \cos^2 \theta + A_\perp^2 g_\perp^2 \sin^2 \theta). \tag{25b}$$

In Eq. (25a), the first-order hyperfine term $KM_I/g\beta$ gives rise to a manifold of $2I + 1$ hyperfine lines centered on the mean resonance field of $H_0 = h\nu/g\beta$ (see Fig. 1b); the remaining, second-order terms in this equation account for small shifts in the centers of gravity of each $\pm M_I$ pair of lines with respect to the mean field. This second-order treatment becomes more accurate at higher spectrometer frequencies ν and hence higher fields H. Generally it is a valid approximation to replace the magnetic field appearing in the denominators in Eq. (25a) by the mean resonance field, H_0.

Equations (25), although specialized to the case of axial symmetry, turn out to pertain to a wide range of fundamental defect centers in glass (vide infra), and the extracted values for the parameters $A_\parallel$ and $A_\perp$ provide essential information about the structure of these defects. Specifically, consider the case that the ground-state wavefunction of the unpaired spin is given by

$$\psi_0 = c_s|ns\rangle + c_p|np\rangle + \sum_i c_i|k_i\rangle, \tag{26}$$

where $|ns\rangle$ and $|np\rangle$ are respectively s- and p-state orbitals centered on the magnetic nucleus and $|k_i\rangle$ are ligand atomic orbitals. In that event, the coefficients c_s and c_p can be determined from the experimental hyperfine coupling constants according to

$$c_s^2 = (A_\parallel + 2A_\perp)/3A_0 = A_{\text{iso}}/A_0, \tag{27a}$$

$$c_p^2 = (A_\parallel - A_\perp)/3B = A_{\text{aniso}}/B, \tag{27b}$$

where A_0 and B are the atomic s-state and p-state coupling constants, respectively. Neglecting overlap, the normalization condition gives

$$c_s^2 + c_p^2 + \sum_i c_i^2 = 1. \tag{28}$$

Equations (27) are readily evaluated by using values of A_0 and B calculated from Eqs. (17) and (21), respectively. Tabulations of these "atomic coupling constants" can be found in a number of sources (e.g., Hurd and Coodin, 1967; Atkins and Symons, 1967; Wertz and Bolton, 1972; Morton and Preston, 1978). The accuracy of the calculated atomic coupling constants is generally no better than $\pm 10\%$. However, improved values are sometimes deduced empirically. Table I lists a few representative values employed by authors cited in this chapter. Note that the customary practice of quoting hyperfine coupling constants in units of magnetic field entails a (usually insignificant) approximation when they are used in Eqs. (27). This is because

TABLE I

ELECTRON SPIN RESONANCE PARAMETERS FOR ATOMS

n	Isotope	Abundance (%)	I	$\dfrac{A_0}{g_e\beta}$ (mT)	$\dfrac{B}{g_e\beta}$ (mT)	λ (cm^{-1})
1	^{1}H	99.98	1/2	50.8	—	—
	^{2}H	0.016	1	7.8	—	—
2	^{7}Li	92.57	3/2	10.5	—	0.2
	^{10}B	18.83	3	28	—	—
	^{11}B	81.17	3/2	83.5	1.96	11
	^{14}N	99.64	1	55.0	1.7	76
	^{17}O	0.037	5/2	−166	−4.2	113
	^{19}F	100	1/2	1720	54	270
3	^{23}Na	100	3/2	31.7	—	11
	^{27}Al	100	5/2	98.5	2.1	75
	^{29}Si	4.7	1/2	−171	−3.4	149
	^{31}P	100	1/2	399.2	11.0	299
	^{35}Cl	75.4	3/2	202	6.28	586
	^{37}Cl	24.6	3/2	143.3	4.46	—
4	^{39}K	93.8	3/2	8.3	—	38
	^{73}Ge	7.61	9/2	−84.3	−2.14	940
	^{75}As	100	3/2	314.9	9.1	1550
6	^{207}Pb	21.11	1/2	1068.3	6.55	—

the experimentally measured coupling constants in magnetic field units are $A_\parallel/g_\parallel\beta$ and $A_\perp/g_\perp\beta$, whereas A_0 and B are converted to magnetic-field units by dividing by $g_e\beta$.

In situations where the hyperfine coupling constants are much larger than $h\nu$, the good quantum numbers are $F \equiv S + I$ and M_F, rather than S, M_S, I, and M_I. Then, in order to find the energy levels as a function of magnetic field (Fig. 1b), it becomes necessary to diagonalize the spin Hamiltonian exactly. For the special case of $S = 1/2$ and isotropic g and hyperfine matrices, this solution can be obtained in closed form (Breit and Rabi, 1931). Pake and Estle (1973) discuss the derivation of resonance conditions from the Breit-Rabi formula. A general solution for anisotropic g and hyperfine tensors has been developed by Kawazoe et al. (1982a); some applications of this formalism to specific ESR spectra in glasses were described by Kawazoe et al. (1982b).

A resonance condition for the spin Hamiltonian of Eq. (8) derived in first-order perturbation theory for the special case of isotropic **g** is given by

$$H_{\text{res}} = \frac{h\nu}{g\beta} - \frac{1}{2g\beta}\left[\frac{D}{3}(3\cos^2\theta - 1) - E\sin^2\theta\cos^2 2\phi\right](2M_S - 1). \qquad (29)$$

The fine-structure spin Hamiltonian has the same functional form as the nuclear electric quadrupole Hamiltonian used in NMR. Thus, Eq. (29) could be converted to the familiar first-order relation used in NMR by making the substitutions

$$M_S \leftrightarrow M_I, \tag{30a}$$

$$D \leftrightarrow \frac{3e^2qQ}{4I(2I - 1)}, \tag{30b}$$

$$E \leftrightarrow \frac{\eta}{3} \frac{3e^2qQ}{4I(2I - 1)}, \tag{30c}$$

where the symbols q, Q, and η have their usual meanings in NMR theory. Similarly, the resonance condition for the electronic Zeeman interaction is analogous to the chemical-shift term in the NMR resonance condition. However, within the NMR resonance condition there are no analogies for the hyperfine, quadrupole, or nuclear Zeeman terms appearing in the ESR resonance condition.

D. THE POWDER PATTERN

In single-crystal ESR, it is customary to measure the resonance fields as functions of the angles between the laboratory magnetic field and the crystallographic axes. These data are then fit to resonance conditions such as those listed previously in order to extract the relevant spin Hamiltonian parameters (e.g., Wertz and Bolton, 1972). However, for powdered or glassy samples, this procedure cannot be followed, as the spectrum represents an average over all angles. But the only angular-dependent data that are necessarily lost as a consequence of this powder-sample "smearing" involve the relationship of specific principal-axis directions of the spin Hamiltonian to specific crystallographic directions. The remainder of the angular dependence is not lost; rather it is contained in the shape of the resulting envelope. Thus, in powder or glassy-state ESR (or NMR), the spin Hamiltonian parameters are measured by fitting the experimental spectrum to a specific mathematical shape function, called a powder pattern, which is rigorously derived from the proper resonance condition.

The derivation of the powder pattern is best illustrated by considering the case of an axially symmetric spin Hamiltonian. Following the original treatment by Sands (1955), it is first necessary to recognize that all solid-angle orientations, $d\Omega$, of the symmetry axis are equally probable in a powder sample (i.e., $P(\Omega) = \text{const.}$), whereas the occurrence of a resonance absorption in the field interval $H \rightarrow H + dH$ depends only on the angle, θ, describing the orientation of the symmetry axis with respect to the direction of the applied magnetic field. The fraction of the solid-angle sphere corresponding to the angular interval $\theta \rightarrow \theta + d\theta$ can be expressed. (See Fig. 2.)

$$d\Omega 4\pi = 2\pi|\sin\theta \, d\theta|/4\pi = (1/2)|d\cos\theta|. \tag{31}$$

Thus, the probability of a given crystallite undergoing resonance in the field interval $H_{\text{res}} \rightarrow H_{\text{res}} + dH_{\text{res}}$ can be written

$$P(H_{\text{res}})dH_{\text{res}} = P(\Omega)d\Omega = \text{const.}|d\cos\theta|, \tag{32}$$

where it must be remembered that the resonance field, H_{res}, is a function of θ [given, e.g., by Eqs. (25)] and, in fact, can be cast as a function of $\cos\theta$. Whence, Eq. (32) can be rearranged to yield a simple expression from which powder patterns can be calculated for any axially symmetric spin Hamiltonian:

$$P(H_{\text{res}}) = \text{const.} \left|\frac{dH_{\text{res}}(\cos\theta)}{d\cos\theta}\right|^{-1}. \tag{33}$$

For the simple case of an axially symmetric g tensor and no hyperfine interactions, H_{res} is given by Eqs. (22) and (23). The result of putting these expressions into Eq. (33) and algebraically eliminating $\cos\theta$ is

$$P(H_{\text{res}}) = \frac{\text{const.}}{H_{\text{res}}}\left(\frac{h\nu}{\beta H_{\text{res}}}\right)^2 \left\{(g_\parallel^2 - g_\perp^2)\left[\frac{h\nu^2}{\beta H_{\text{res}}} - g_\perp^2\right]\right\}^{-1/2}, \tag{34}$$

where it now must be understood that H_{res} varies only within the limits set by Eqs. (22) and (23), i.e., between the limits $H_\parallel = h\nu/g_\parallel\beta$ and $H_\perp = h\nu/g_\perp\beta$. Figure 3a displays the ESR powder pattern given by Eq. (34) for the case $g_\parallel > g_\perp$; a mathematical divergence at $H_\perp$ is indicated by the vertical arrows, while a "step shoulder" appears at $H_\parallel$. Figure 3b shows the calculated first derivative of the powder-absorption envelope of Fig. 3a with an experimental first-derivative spectrum in superposition. The experimental spectrum differs from the theoretical prediction due to the existence of a nonzero single-crystal linewidth, thus far neglected in the theoretical treatment.

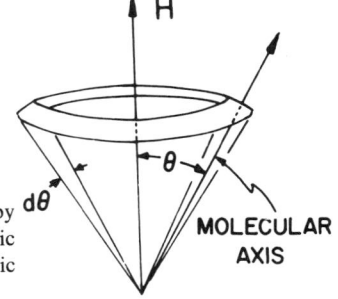

FIG. 2. Differential element of solid angle subtended by the unique axes of an ensemble of identical axially symmetric paramagnetic centers undergoing resonance in the magnetic field interval $H \rightarrow H + dH$.

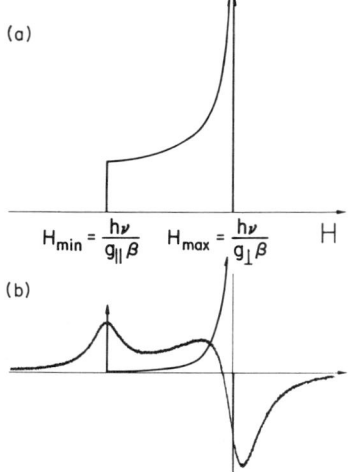

FIG. 3. Powder pattern for a paramagnetic center with $S = 1/2$ and an axially symmetric g matrix $(g_\parallel > g_\perp)$. (a) Absorption curve calculated by means of Eq. (34). (b) First derivative of (a) with superimposed experimental derivative spectrum.

For the case of nonzero nuclear spin, one can obtain $2I + 1$ expressions $P(H_{res}(M_I))$ by substituting Eqs. (25) into Eq. (33). But in contrast to the results of Eq. (34), when the hyperfine interaction is included it is no longer possible to eliminate $\cos \theta$ from the resulting expression. It then becomes necessary to obtain both H_{res} and $P(H_{res})$ as functions of $\cos \theta$. Neiman and Kivelson (1961) carried out such powder-pattern calculations for Cu^{2+} complexes in aqueous glasses. finding that in general mathematical divergences occur at field positions $H_\perp(M_I)$ determined by the condition $\cos \theta = 0$, but noting also the occurrence of so-called extra divergences for certain combinations of spin Hamiltonian parameters. Gersmann and Swalen (1962) have given an approximate relation for the conditions under which such extra divergences might be expected:

$$\cos^2 \theta = [\tau M_I^2(2\sigma - 1)/2 + (\sigma - 1)/(2\sigma - 1)], \tag{35a}$$

where

$$\tau = \frac{(g_\parallel^2 A_\parallel^2 + g_\perp^2 A_\perp^2)(g_\parallel^2 + g_\perp^2)}{(h\nu)^2(g_\parallel^2 - g_\perp^2)^2} \quad \text{and} \tag{35b}$$

$$\sigma = \frac{g_\parallel^2 A_\parallel^2}{g_\parallel^2 A_\parallel^2 + g_\perp^2 A_\perp^2}. \tag{35c}$$

Taylor et al. (1975) provide a more comprehensive tabulation of the various conditions that can lead to the occurrence of "extra" divergences in ESR powder spectra.

Figure 4 displays a set of powder-pattern components calculated point-by-point for Cl_2^- molecular ions in a glass (Griscom and Bray, 1965). Here the effective nuclear spin was $I = 3$ (corresponding to pairs of ^{35}Cl nuclei individually having spin $I = 3/2$). An "extra" divergence is indicated by the arrow labelled "additional absorption."

Closed-form analytical expressions for the powder pattern are generally

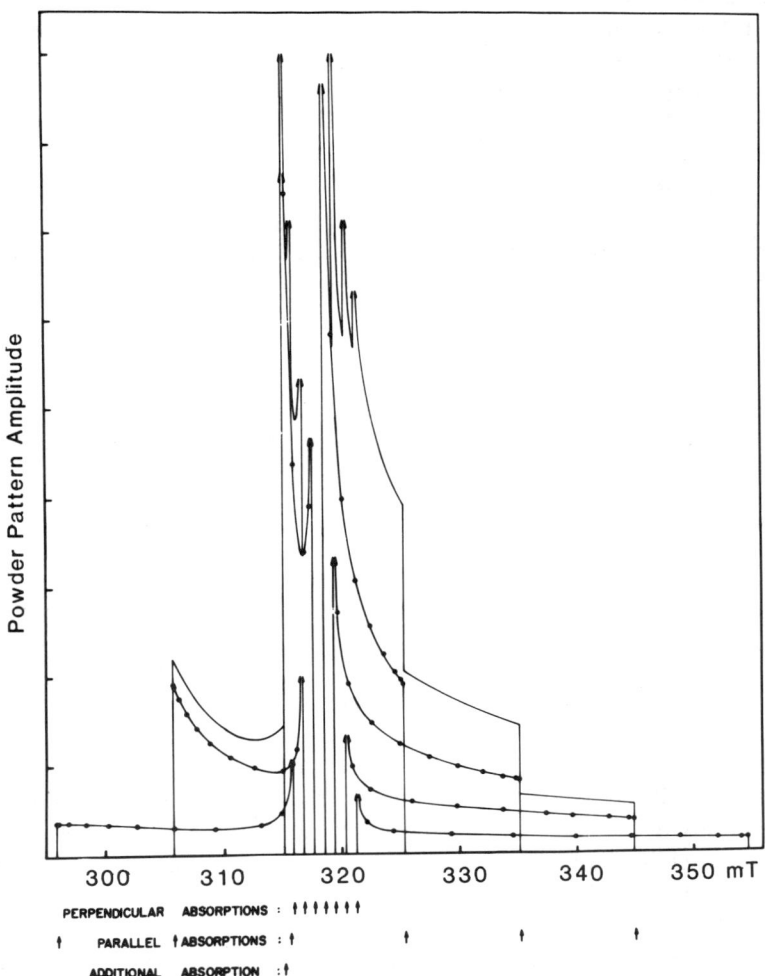

F<small>IG</small>. 4.   Powder pattern for $^{35}Cl^{35}Cl^-$ molecular ions at an X-band frequency. Calculated points are shown for each of seven hyperfine components corresponding to an effective nuclear spin $I = 3$ and magnetic quantum numbers $M_I = +3, +2, \ldots, -3$. $g_\parallel = 2.004$, $g_\perp = 2.035$, $A_\parallel(^{35}Cl)/g\beta = 10.0\,mT$, $A_\perp(^{35}Cl)/g\beta = 0.8\,mT$. (From Griscom and Bray, 1965).

not possible for lower than axial symmetry, in which case the shape function, $P(H)$, must be evaluated by numerical methods (Kneubuhl, 1960). A general expression for this shape function is (e.g., Taylor and Bray, 1970)

$$P(H)dH = \frac{1}{4\pi} \sum_{M_I,M_S} \int_H^{H+dH} W_{M_I, M_S}(\Omega) \, d\Omega(H(M_1, M_S)). \quad (36)$$

Equation (36) may be recognized as a generalization of Eq. (32), valid for any symmetry and any value of the nuclear spin. Here, $W_{M_I, M_S}(\Omega)$ is the transition probability, which was neglected in Eq. (32) and whose weak angular dependence can be safely ignored for most cases of interest (Taylor and Bray, 1970). For convenience, in Eq. (36) the subscript "res" has been dropped from the symbol for the resonance field, which henceforward will be denoted simply by H.

For purposes of locating the field positions of the divergences and step shoulders, it is not always necessary to calculate the complete powder pattern either analytically or numerically. If one visualizes the resonance condition as a surface in $\cos\theta - \phi$ space, the singularities can be shown to occur at critical points where $|\text{grad}_{\theta,\phi}H| = 0$, i.e., where

$$(\partial H/\partial \cos\theta) = (\partial H/\partial\phi) = 0. \quad (37)$$

As discussed, e.g., by Taylor et al. (1975), a saddle point in the surface indicates the presence of a divergence, while a relative extremum indicates a step shoulder.

E. Influences of Amorphous-State Disorder

Equation (37) states that, for powdered-crystalline material, sharp powder-pattern features are to be expected for values of H_{res} that are stationary with respect to small variations in the orientation of the individual crystallites. Clearly, a similar statement can be made for paramagnetic centers in glasses by merely replacing the idea of crystallite orientation by the notion of the orientation of the individual paramagnetic complex, e.g., a paramagnetic ion and its ligands. However, the magnetic resonance spectra of glasses will be influenced by one further factor, namely, statistical distributions of crystal fields arising from the intrinsic randomness in an amorphous solid. Thus, the ESR spectra of paramagnetic centers in glasses having broad distributions of crystal fields will in general exhibit well-resolved features only at orientationally stationary resonance fields that are likewise stationary with respect to variations of the crystal-field–sensitive parameters of the spin Hamiltonian (Griscom and Griscom, 1967). That is, in a glass well-resolved spectral features are encountered when $|\text{grad}_{\theta,\phi,x_1,x_2,...}H| = 0$, where x_1, x_2, ... are the crystal-field–sensitive spin Hamiltonian parameters (Kliava and Purāns, 1980).

The superoxide anion O_2^- serves as an excellent prototype for contrasting the ESR lineshapes encountered in an amorphous material with those manifested in a material that is merely polycrystalline. The experimental spectrum of Fig. 3b is the spectrum of superoxide anions inadvertently present in a commercial polycrystalline sodium peroxide. Känzig and Cohen (1959) have shown that the values of $g_{\parallel}$ and $g_{\perp}$ that characterize O_2^- in a solid depend sensitively on the splitting Δ of the π-antibonding levels resulting from the interaction of the ion with the local crystal fields. (See Fig. 5a.) The expressions for $g_{\parallel}$ and $g_{\perp}$ shown in Fig. 5b are approximations to Känzig and Cohen's formulas; the curves represent plots of these relations as functions of Δ/λ where λ is the spin-orbit coupling constant for the O^- ion. A bold arrow marks the value of Δ/λ that yields the values of $g_{\parallel}$ and $g_{\perp}$ that best characterize the experimental spectrum of Fig. 3b; a computer simulation of this spectrum based on these g values is shown as the dotted curve in Fig. 6b. By contrast, the spectrum of O_2^- ions in an amorphous peroxyborate preparation (Griscom, 1966; Edwards et al., 1969) exhibits an entirely different lineshape (Fig. 6a). This shape was computer-simulated (Griscom, 1978a) using distributions in the values of $g_{\parallel}$ and $g_{\perp}$ generated in Fig. 5b from a posited distribution of Δ values, $P(\Delta)$, by using the theoretical curves as a "transfer function." The resulting simulation is shown as the dotted curve in

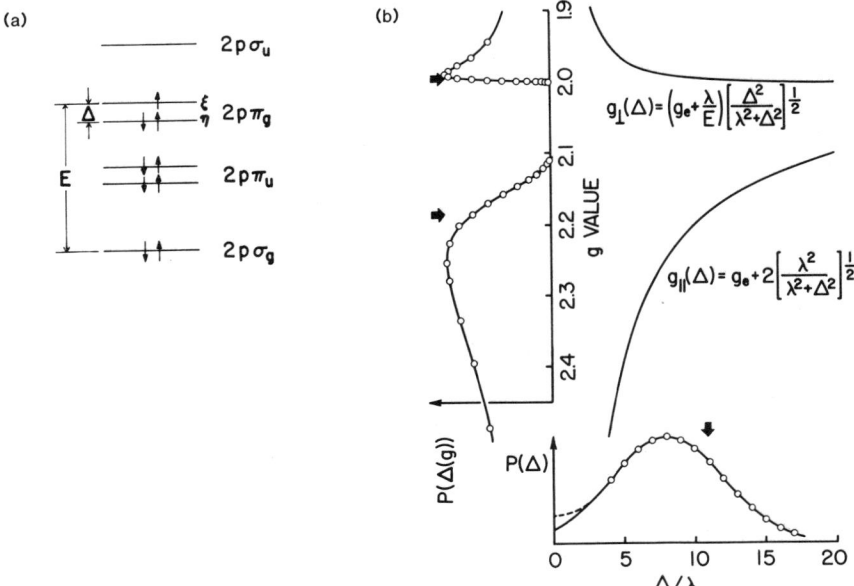

FIG. 5. (a) Energy levels and (b) relationship between g values and the energy-level splitting parameters for the O_2^- molecular ion in solid matrices. (From Griscom, 1978a).

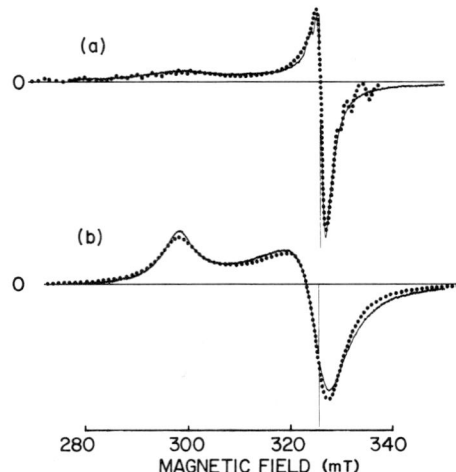

FIG. 6. X-band ESR spectra of O_2^- molecular ions in (a) an amorphous potassium peroxyborate and (b) polycrystalline sodium peroxide. Dotted curves are computer simulations. (From Griscom, 1966, 1978a).

Fig. 6a. It can be seen in Fig. 5b that $g_\perp$ is relatively stationary with respect to wide variations in Δ, whereas $g_\parallel$ is not. Thus, the $H_\parallel$ derivative peak that is so well defined in the polycrystalline sodium peroxide sample (Fig. 6b) is "washed out" almost beyond recognition in the amorphous peroxyborate sample (Fig. 6a). Many other examples of this kind of spectral "smearing" will be given shortly, each of them a consequence of the statistical disorder endemic to the amorphous state.

F. COMPUTER SIMULATION OF ESR SPECTRA IN GLASSES

It is evident from Fig. 4 that the ESR spectra of glasses can be quite complex, even without consideration of the effects of statistical distributions of spin Hamiltonian parameters such as were illustrated in Fig. 6. Thus, while the origin of a glass spectrum sometimes can be inferred from inspection, extraction of highly accurate spin Hamiltonian parameters (not to mention convincing a disinterested reader of their accuracy) generally hinges on performing a computer simulation of the spectrum subject to the rigid constraints of spin-Hamiltonian theory and making use of a reasonable physical model for the effects of vitreous disorder on the relevant parameters.

Simulation of a simple powder spectrum (no vitreous disorder) can be accomplished by first obtaining the powder pattern $P(H)$ of Eq. (36), either in closed form (as in the example of Eq. (34)) or by numerical methods, and convoluting it with a suitable "single-crystal" shape function, $F(H' - H)$, which is usually a Gaussian or a Lorentzian. The convoluted powder pattern, $R(H)$, is given by

$$R(H) = \int P(H')F(H' - H)\,dH'. \tag{38}$$

Since experimental ESR spectra are conventionally obtained as the first derivative of absorption versus the magnitude of the applied magnetic field, the derivative of Eq. (38), $R'(H)$, is normally computed. Figure 7 displays some comparisons of spectra simulated in this way with the experimental first-derivative spectra for Cl_2^- ions in glass (Griscom et al., 1969). Note the influence of frequency.

In Fig. 7, the effects of vitreous-state randomness were approximated by a relatively broad Gaussian convolution function. In general, however, the effects of disorder must be taken into account by forming a weighted sum of the resultant shape functions, $R(x_1, x_2, \ldots, v, H)$, calculated by means of Eq. (38) for an appropriate range of values of some or all spin-Hamiltonian parameters, x_i. Allowing this summation to go over to an integral, the final shape function $K(H)$ can be represented in this case by:

$$K(H) = \Pi_i \int p_i(x_i)dx_i R(x_1, x_2, \ldots, v, H), \tag{39}$$

where the p_i are individual probability density functions describing the distribution of parameters x_i and where Π_i indicates the product over the index i.

Equation (39) takes no explicit account of the correlations that may exist among the various $p_i(x_i)$, e.g., the negative correlation between $p(g_\parallel)$ and $p(g_\perp)$ apparent in Fig. 5b. However, such interrelationships can often be introduced on an ad hoc basis by utilizing a single distribution function to describe the variation of all parameters believed to be linearly interrelated. In any event, a model prescribing specific correlations among random variables (or their statistical independence!) must be assumed in any simulation of a glass spectrum, whether or not it can be deduced a priori. This approach has been amply described by Taylor and Bray (1970) and Taylor et al., (1975).

Peterson, et al., (1974a, b) took a more elegant approach to the computer-simulation problem based on probability theory. Their formalism treated the influences of orientational randomness (the powder pattern) and statistical randomness in spin-Hamiltonian parameters (vitreous state smearing) on an equal footing. Thus, Peterson et al., (1974a, b) employed the notation Ω to denote, not simply the solid-angle sphere with two basis components (θ and ϕ), but an n-dimensional vector space comprising all the parameters that influence the resulting ESR spectrum. For example, in the common special case of axial-site symmetry and no hyperfine or fine-structure interactions,

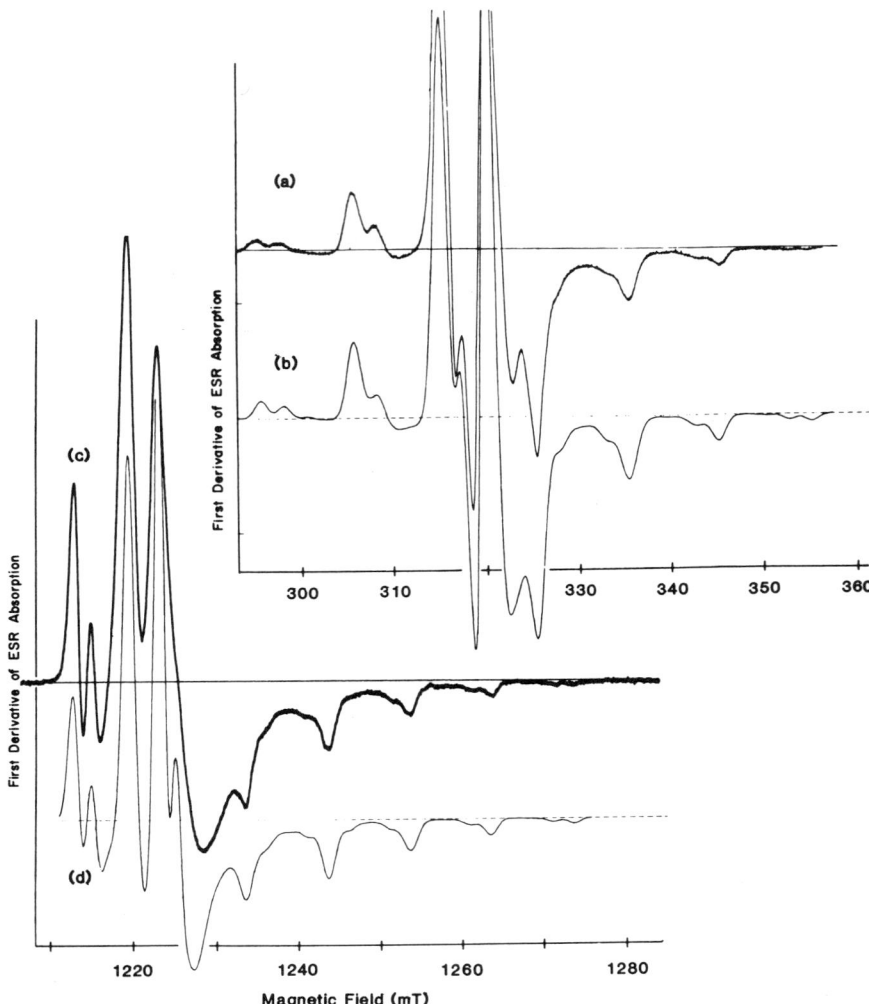

FIG. 7.   ESR spectra of Cl_2^- molecular ions in an irradiated potassium chloride-potassium borate glass. Experimental spectra (a) and (c) were recorded at 9.1 and $\sim 35\,GHz$, respectively. Curves (b) and (d) are computer simulations formed of the properly weighted sum of the spectra of $^{35}Cl\text{-}^{35}Cl^-$, $^{35}Cl\text{-}^{37}Cl^-$, and $^{37}Cl\text{-}^{37}Cl^-$ ions. Spin Hamiltonian parameters were as given in the caption to Fig. 4. (From Griscom et al., 1969).

one can define on Ω a three-dimensional random vector, **x**, whose components are $g_\parallel$, $g_\perp$, and θ. The vector **x** designates a mapping from the sample space, Ω, into a Euclidian three space and assigns a particular point in this space to each sample point ω_i in Ω. The components of **x** can be described by a joint probability density function, which in the example under consideration can be expressed $p_{g_\|,g_\perp,\theta}(\eta_\|,\eta_\perp,\eta_\theta)$, where $\eta_\|$, $\eta_\perp$, and η_θ are dummy variables corresponding to $g_\|$, $g_\perp$, and θ, respectively. Recognizing that θ is statistically independent of $g_\|$ and $g_\perp$ leads to the factoring of the density function

$$p_{g_\|,g_\perp,\theta}(\eta_\|,\eta_\perp,\eta_\theta) = p_{g_\|,g_\perp}(\eta_\|,\eta_\perp)p_\theta(\eta_\theta). \tag{40}$$

Following the same reasoning that led to Eq. (31), it is clear that

$$p_\theta(\eta_\theta) = p_\theta(\theta)|_{\theta=\eta_\theta} = \sin\eta_\theta. \tag{41}$$

In order to obtain the distribution of resonance fields, $K_H(\eta_H)$, Peterson et al. (1974a) first transformed the random vector, $\mathbf{x}$, to a new random vector, $\mathbf{y}$, one component of which was the quantity g given by Eq. (23). Since $\mathbf{x}$ and $\mathbf{y}$ are related by a $1:1$ transformation, the density functions are related by (Wolzencraft and Jacobs, 1977)

$$p_\mathbf{y}(\eta_\mathbf{y}) = p_\mathbf{x}(\eta_\mathbf{x}) \cdot |J_\mathbf{y}(\eta_\mathbf{x})|, \tag{42}$$

where $|J_\mathbf{y}(\eta_\mathbf{x})|$ is the absolute value of the Jacobian associated with the transformation $\mathbf{x} \to \mathbf{y}$. It is to be noted that the Jacobian of Eq. (42) becomes a simple partial derivative when, as in the case under discussion, there is only one independent variable in the probability density function. Thus, for an axial g matrix

$$J_g(\eta_\theta) = \frac{\partial\eta_\theta}{\partial\eta_g} = \left(\frac{\partial g}{\partial\theta}\right)^{-1}_{\theta=\eta_\theta}. \tag{43}$$

The probability function for g is then obtained by substituting Eq. (23) into Eq. (43), followed by substituting Eqs. (41), (42), and (43) into Eq. (40), algebraically eliminating $\cos\theta$, and integrating over the dummy variables corresponding to $g_\|$ and $g_\perp$:

$$p_g(\eta_g) = \iint \frac{\eta_g p_{g_\|g_\perp}(\eta_\|,\eta_\perp)}{[(\eta_\|^2 - \eta_\perp^2)(\eta_g^2 - \eta_\perp^2)]^{1/2}} \, d\eta_\| \, d\eta_\perp. \tag{44}$$

Peterson, et al. (1974a) then performed one further transformation of the random vector $\mathbf{y}$ to a random vector $\mathbf{z}$ having as one of its components the resonance field, H. The relationship of the density function for H to that for g is then given by

$$p_H(\eta_H) = p_g(\eta_g) \cdot \left|\frac{\partial g}{\partial H}\right|_{H=\eta_H} = p_g\left(\frac{h\nu}{\beta\eta_H}\right) \cdot \left|\frac{h\nu}{\beta\eta_H^2}\right|. \tag{45}$$

Combining Eqs. (44) and (45) leads to an integral equation that is formally equivalent to the result of substituting the right-hand side of Eq. (34) for $R(g_\|, g_\perp, \theta, \nu, H)$ in Eq. (39).

As the foregoing example illustrates, the method of Peterson et al., (1974a, b), which treats the computer-simulation problem as a series of trans- formations among vector spaces, is completely equivalent to the original approach of Sands (1955) insofar as the generation of powder patterns is concerned. Although the procedure for introducing "single-crystal" broaden- ing proposed by Peterson et al., (1975) is in principle more correct that the convolution represented by Eq. (38), the practical differences between the two methods are negligible. (See discussion by Jellison, et al., 1976.)

As previously noted, the probability density functions $p_i(x_i)$ to be used in Eq. (39) or the joint probability density functions employed in Eq. (44) must be assumed or deduced before a glassy-state ESR spectrum can be computer- simulated. Indeed, the earliest efforts along these lines generated empirical distribution functions for g values by "cut-and-try" procedures (Griscom, et al., 1968; Mackey, et al., 1969). However, by means of an actual computer- fitted example, Griscom (1972) demonstrated that the experimentally inferred distributions in g values are likely to have their physical origins in statistical distributions in the energy levels E_n which determine the g shifts [c.f., Eq. (10)]. Thus, as already illustrated in Fig. 5b, a Gaussian distribution in energy (the most common statistical form) may generally be expected to lead to skewed distributions in g values.

The computer simulation of the amorphous-state spectrum of Fig. 6a was accomplished by forming a weighted sum of powder patterns calculated according to the nonlinearly spaced "data points" labelled $P(\Delta(g))$ in Fig. 5b. Although physically correct and useful for instructional purposes, this procedure is mathematically unsophisticated and leads to artifactual oscilla- tions in the computed spectrum at both the high- and low-field ends due to the widening spacing of the "data points" corresponding to the lower values of Δ/λ. Peterson et al., (1974a) noted that the true probability density functions in such cases are simply derived from Eq. (42). The example given was that of the Ti^{3+} ion in a tetragonal crystal field (see Section III.A.1), where letting the energy separation of the paramagnetic ground state and the nearest excited state be δ, the value of $g_\perp$ can be expressed

$$g_\perp = 2 - 2\lambda/\delta. \tag{46}$$

If one assumes a Gaussian distribution in δ values, $p_\delta(\eta_\delta)$, then, as illustrated in Fig. 8, the resulting skew-symmetric distribution in g values if given by

$$p_g(\eta) = \left| \frac{2\lambda}{(2 - \eta_\perp)^2} \right| \cdot p_\delta\left(\frac{2\lambda}{2 - \eta_\perp} \right). \tag{47}$$

In deriving Eq. (47), the Jacobian of Eq. (42) could be treated as a simple partial derivative. Friebele et al., (1974) gave an early example of a glassy- state ESR lineshape simulation where it proved necessary to employ a Jacobian of more than one variable.

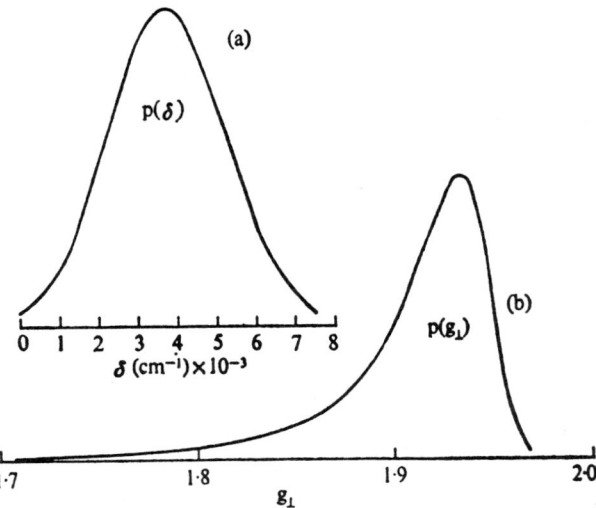

FIG. 8. (a) Gaussian density function representing a spread in crystal-field splittings δ. (b) The skewed density function for $g_\perp$ derived from Eqs. (46) and (47). (From Peterson et al., 1974a).

III. Transition-Group Ions

Transition-group ions, comprising the iron group, rare earths, and actinides, are generally paramagnetic by virtue of their partially filled 3d, 4d, 5d, 4f, or 5f shells. In free space, this paramagnetism has both spin and orbital contributions as expressed by Eq. (5). Given the number of electrons in the incomplete shell, the actual values of the spin and orbital angular momenta for the ground-state term are specified by Hund's rules (e.g., Bleaney and Stevens, 1953; Bowers and Owen, 1955; Abragam and Bleaney, 1970; Pake and Estle, 1973). The first of these rules minimizes the repulsive electron–electron energy, $\mathscr{H}_{\text{Coul}}$, by maximizing S. The second of Hund's rules minimizes the spin-orbit interaction energy, $\mathscr{H}_{\text{SO}}$, by requiring the maximum L consistent with the maximum S. In spectroscopic notation, the values $L = 0, 1, 2, 3, \ldots$, are represented by S, P, D, F, $\ldots$, while the spin multiplicity, $2S + 1$, is indicated by a prefixed superscript. Thus, the iron-group ions, d^1 through d^9, have for their ground-state terms ^{2}D, ^{3}F, ^{4}F, ^{5}D, ^{6}S, ^{5}D, ^{4}F, ^{3}F, and ^{2}D, respectively. These free-space ground-state terms have total degeneracies given by the product $(2S + 1)(2L + 1)$.

With the exception of S-state ions (e.g., d^5 and f^7), when transition-group ions are embedded in a solid their electronic structure may be determined as much by the magnitude and symmetry of the local crystal fields or chemical bonding as by the intrinsic forces which determine the structure of the free ion. In taking into account these solid-state influences, it has proven useful to distinguish three separate situations depending on the magnitude of the "crystal-field" interaction $\mathscr{H}_{\text{xtal}}$ with respect to $\mathscr{H}_{\text{Coul}}$ and $\mathscr{H}_{\text{SO}}$ (Van Vleck, 1935). For iron-group ions, the "medium field" approximation is generally valid, i.e., $\mathscr{H}_{\text{Coul}} > \mathscr{H}_{\text{xtal}} > \mathscr{H}_{\text{SO}}$. Crystal fields of cubic symmetry usually make the dominant contribution to $\mathscr{H}_{\text{xtal}}$, leading immediately to a partial lifting of the orbital degeneracy of the ground term. It is self-evident that d (or f) orbitals which point toward negatively charged ligands are raised in energy with respect to d (or f) orbitals having other orientations. As one consequence, splittings occasioned by tetrahedral ligand coordination are opposite in sign to those due to octahedral coordination. In Fig. 9, an octahedral cubic field is seen to split the fivefold degenerate d-shell manifold into a low-lying orbital triplet and a higher-lying doublet separated by an energy Δ. (This ordering would be reversed in the case of tetrahedral coordination.) The presence of crystal-field components of lower symmetry leads to further removal of orbital degeneracy; for example, tetragonal or trigonal distortions would split the triplet level into a singlet and a doublet (splitting δ in Fig. 9). Whether the singlet lies lower than the doublet or vice versa depends on both the symmetry of the distortion and its magnitude relative to $\mathscr{H}_{\text{SO}}$ (e.g., Bleaney and Stevens, 1953; Bowers and Owen, 1955; Abragam and Bleaney, 1970). Assuming these factors are known, the "zero-order" ground state of a d^n ion embedded in a solid can often be deduced by inserting n electrons into the level scheme of Fig. 9 subject to the constraints of the Pauli principle and Hund's first rule.

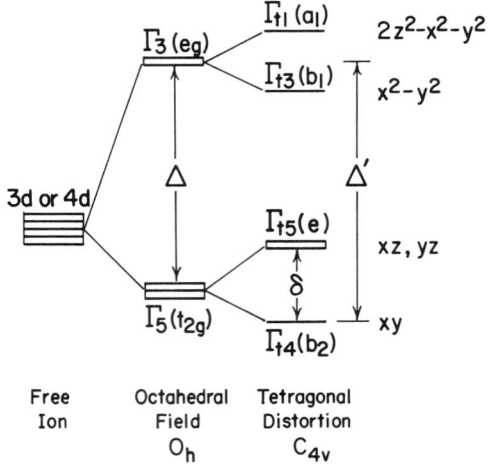

FIG. 9. Influence of crystalline electric fields on the energy levels of D-state ions. The ordering of the levels and the eigenfunction designations at the right assume octahedral coordination with a tetragonal compression along the z axis (From, Griscom, 1980a).

Using the example of an iron-group ion with octahedral coordination and subject to a tetragonal distortion (c.f., Fig. 9), an orbital singlet ground state is predicted in the d^1 case, with $S = 1/2$. The absence of orbital degeneracy implies that the orbital angular momentum has been quenched in zero-order, and hence the state can be assigned an "effective" orbital angular momentum, $L' = 0$. For d^3, Hund's first rule gives $S = 3/2$ and the Pauli principal requires the low-lying triplet "subshell" to be exactly half filled, resulting again in $L' = 0$ (Bowers and Owen, 1955). By the same reasoning, for d^5 all five d orbitals must be singly occupied, giving $L = L' = 0$ and $S = 5/2$. Thus, in the medium field approximation, S-state ions in free space remain S-state ions in a solid. (Instances of the so-called low-spin case occurring for very large values of Δ do not seem to have been reported in glasses.) For d^8, the triplet subshell is completely filled while the doublet is half filled, once again giving $L' = 0$, but with $S = 1$. Similarly, d^9 can be regarded as a hole in an otherwise completely filled d shell and thus equivalent to the d^1 case obtained by inverting the level scheme of Fig. 9 ($L' = 0$, $S = 1/2$).

If the orbital angular momentum is not quenched in zero-order by the local crystal fields of the prevailing symmetry, the Jahn-Teller theorem assures that the local structure will spontaneously distort so as to remove the orbital degeneracy (e.g., Bleaney and Stevens, 1953; Abragam and Bleaney, 1970; Wertz and Bolton, 1972). Nevertheless, this chapter will emphasize cases where the ground states are orbital singlets in the first approximation because these are the cases most easily observed in an ESR experiment. The problem is that minimal quenching of the orbital angular momentum implies the existence of very low lying excited states which in turn lead to exceedingly short spin-lattice relaxation times and large g anisotropies. Both of these effects tend to broaden amorphous-state ESR spectra beyond the limits of observability.

This section emphasizes the pacing issues in ESR spectroscopy of transition-group ions in glasses, using examples drawn from the literature of oxide glasses. The subsections are arranged by zero-order spin and orbital angular momentum quantum numbers. This ion-by-ion approach precludes discussion of glass structural studies employing several different transition ions as probes (e.g., Loveridge and Parke, 1971; Bogomolova et al., 1978a, b; Nicklin et al., 1973).

A. 3d^1 IONS: Ti^{3+}, V^{4+}, Cr^{5+} ($L' = 0$, $S = 1/2$)

1. Ti^{3+}

The ESR spectrum of Ti^{3+} is perhaps more widely reported in glasses than in crystalline hosts. This outcome appears to be due in part to the very short spin-lattice relaxation times encountered in crystals (Abragam and Bleaney,

1970), which are in turn attributable to the relatively small magnitudes of the trigonal-field splittings ($\delta \sim 0.01\,\text{eV}$) of the low-lying orbital triplet which would otherwise be the result of the predominant octahedral field. (See Fig. 9.) By contrast, many glassy hosts turn out to be far more favorable to the ESR observation of Ti^{3+} in consequence of the existence of substantially larger non-cubic "crystal" fields ($\delta \sim 0.25-1.0\,\text{eV}$) (Yafaev and Yablokov, 1962; Arafa, 1972, 1974). In both of the foregoing cases, the estimates of the splitting δ were based on g-value measurements. (See Section III.A.2.) Some evidence supports the notion that the large splittings δ peculiar to Ti^{3+} in glasses may be due to the presence of nonbridging oxygens in the ligand sphere (Kim and Bray, 1970).

Since Ti^{4+} is the preferred valence state of titanium in many oxide glasses melted under "normal" conditions (e.g., Schreiber, 1977), ESR observation of Ti^{3+} is usually facilitated by glass preparation under extremely reducing conditions (Garif'yanov, et al., 1962; Yafaev and Yablokov, 1962) or by irradiation of normally melted glasses (Arafa and Bishay, 1970; Arafa, 1974). On the basis of studies of these types, the ESR "signature" of the Ti^{3+} ion in glasses was determined to be a relatively structureless asymmetric line with $g_\perp > g_\parallel$ and an average g value of $g_{av} \simeq 1.9$. It is to be cautioned, however, that this signature is not unambiguous, as similar resonance lines can arise from other ions, such as Cr^{5+} or Mo^{5+} (Sections III.A.3 and III.E, respectively). Peterson and Kurkjian (1972) provided the final proof that the $g \simeq 1.9$ resonances in the Ti-doped glasses are due in fact to Ti^{3+} by observing the hyperfine structure of ^{47}Ti ($I = 5/2$) in a glass deliberately enriched to 75% in this magnetic isotope, whose natural abundance is only 7.75%. (see Fig. 10.) This accomplishment nevertheless required the employment of very low microwave frequencies to overcome the obscuring effects of the distribution in g values earlier inferred by Yafaev and Yablokov (1962).

It is worth noting that radiation-induced Ti^{3+} centers in glasses not deliberately doped, but containing inadvertent titanium impurities, might potentially masquerade as intrinsic electron or hole traps (Arafa and Assabghy, 1974).

One of the most important applications of Ti^{3+} ESR spectroscopy lies in the establishment of $Ti^{3+}-Ti^{4+}$ equilibria in glasses of both commercial and geochemical interest (Peterson and Kurkjian, 1974a, b; Bell et al., 1976; Schreiber, 1977; Schreiber and Haskin, 1976; Schreiber et al., 1979).

2. V^{4+}

In contrast to the case of the isoelectronic Ti^{3+} ion, the ESR spectrum of V^{4+} in most glasses is rich in hyperfine structure due to the 100% abundant ^{51}V nucleus ($I = 7/2$). It is easily observed in many systems at room temperature (e.g., Hochstrasser, 1966; Hecht and Johnston, 1967; Toyuki and

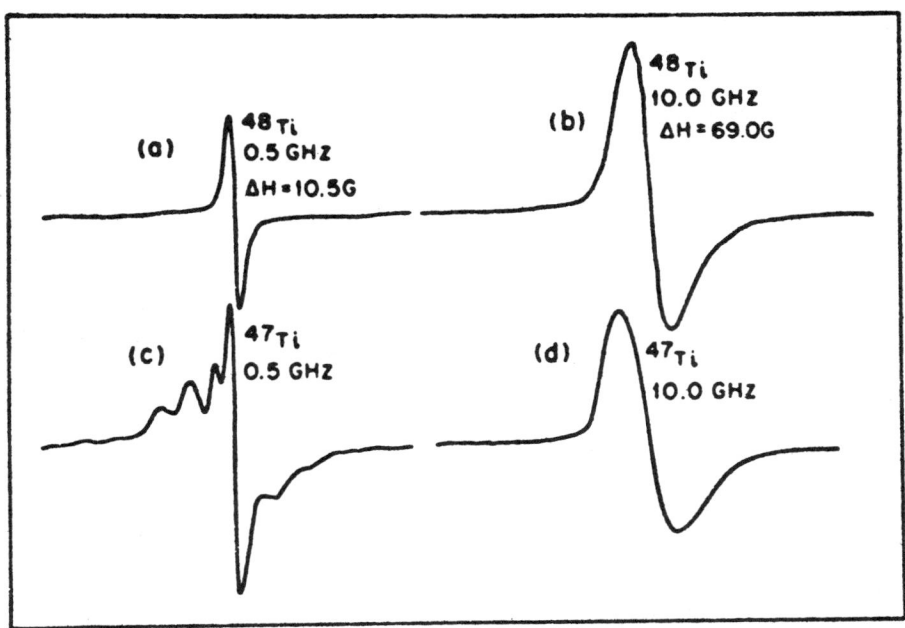

FIG. 10. ESR spectra of $30CaO-69B_2O_3-1TiO_2$ glasses recorded at 0.5 GHz (a and c) and at 10 GHz (b) and d). The samples of (c) and (d) were enriched to 75% ^{47}Ti ($I = 5/2$). (From Peterson and Kurkjian, 1972).

Akagi, 1972; Paul and Assabghy, 1975; Muncaster and Parke, 1977; Bogomolova et al., 1978a, b; Hosono et al., 1979a; Sammet and Bruckner, 1985). Thorough analyses of this ESR spectrum were carried out by Hochstrasser (1966) (see Fig. 11) and Hecht and Johnston (1967). The latter workers also considered theoretically the expected influences of all possible crystal-field environments and ground-state configurations, concluding that V^{4+} in their sodium borate glasses must exist in octahedral coordination with tetragonal compression (i.e., the conditions represented by Fig. 9). The explanation they attached to this outcome was that one of the six V—O bonds may be shorter than the other five (Fig. 12a), whence the ion in the glass might be best regarded as a vanadyl complex, VO^{2+}.

Among the measured-spin Hamiltonian parameters, the principal values of the g tensor are most readily related to the ligand environments of d^1 ions such as Ti^{3+} or V^{4+}. Although Hecht and Johnston (1967) have given more precise expressions for the values of $g_\parallel$ and $g_\perp$, the following more approximate relations are found generally useful:

$$g_\perp = g_e(1 - \lambda\gamma^2/\delta), \tag{48a}$$

$$g_\parallel = g_e(1 - 4\lambda\alpha^2/\Delta'), \tag{48b}$$

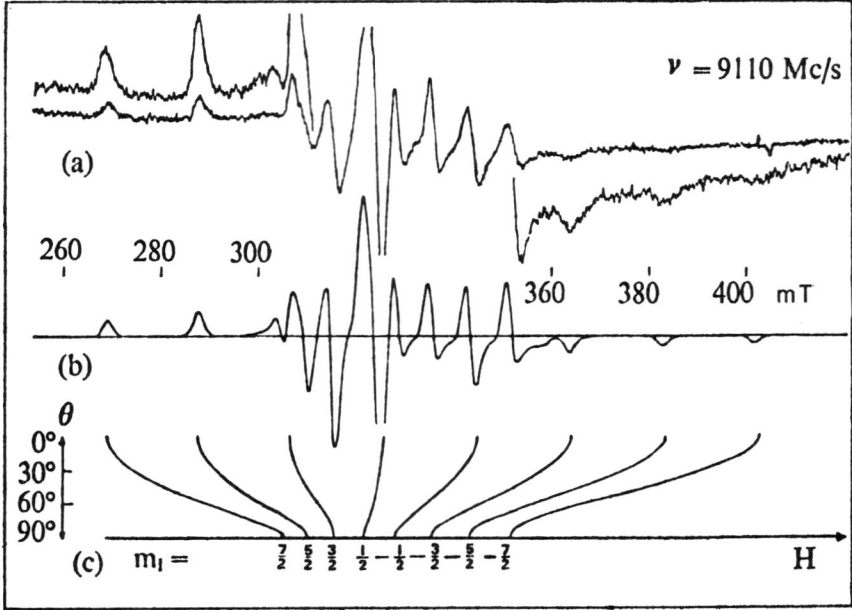

FIG. 11.   ESR of V^{4+} in a borosilicate glass. (a) Experimental first-derivative spectrum. (b) Computer simulation. (c) Single-crystal spectrum computed as a function of the angle between the applied magnetic field and the symmetry axis of the complex. (From Hochstrasser, 1966).

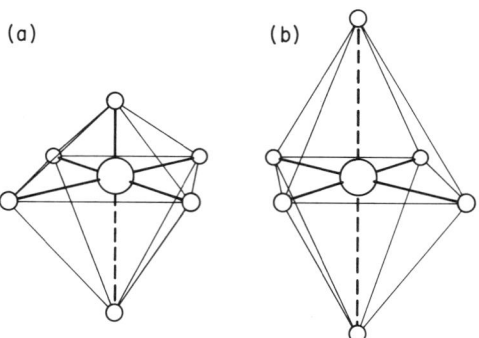

FIG. 12.   Perspective drawing of tetragonally distorted octahedral complexes. (a) The vanadyl complex typical of d^1 ions in oxide glasses. (b) Octahedral complex with tetragonal elongation typical of Cu^{2+} ions in oxide glasses.

where the splittings δ and Δ' are as defined in Fig. 9 and the factors γ^2 and α^2 crudely account for partial covalency. Specifically, $(1 - \alpha^2)$ is a measure of the covalency of the σ bonding between the ion and its equatorial ligands and $(1 - \gamma^2)$ expresses the covalency of the π bonding with the vanadyl oxygen. The energy separations δ and Δ' may be determined from optical studies (e.g., Hecht and Johnston, 1967; Paul and Assabghy, 1975; Sammet and Bruckner, 1985) and with this information Eqs. (48) can be solved for the parameters γ and α. In a number of binary alkali oxide glass systems, increasing alkali oxide contents is found to result in a distinct trend away from strong π bonding with the vanadyl oxygen and toward regular octahedral symmetry (e.g., Toyuki and Akagi, 1972; Paul and Assabghy, 1975).

3. Cr^{5+}

The principal valence states of chromium in solids are 3^+ and 6^+. However, by fusing hexavalent chromium compounds with borax or trivalent compounds with silicates, Garif'yanov (1963) obtained a relatively sharp ESR signal characterized by $g_{\parallel} \simeq 1.94$ and $g_{\perp} \simeq 1.98$ which he interpreted as being due to Cr^{5+}. He also observed a similar signal with $g_{\parallel} \simeq 1.99$ and $g_{\perp} \simeq 1.97$ in a glycerol glass containing chromium ions. In the latter case, apparent hyperfine structure due to 10% abundant ^{53}Cr ($I = 3/2$) was reported. This hyperfine fingerprint has not been observed in the borate and silicate systems, presumably due to the obscuring effects of a distribution of g values such as impeded the observation of hyperfine structure in the Ti^{3+} case (Section III.A.1). Nevertheless, resonance lines characterized by similar g values may be safely regarded as signatures of the Cr^{5+} ion in glasses deliberately doped with chromium and known to contain neither titanium nor molybdenum. As was the case for the isoelectronic Ti^{3+} and V^{4+} ions, the g shifts observed for Cr^{5+} in glasses are interpreted via Eqs. (48) as indicating octahedral coordination with a tetragonal compression, most likely indicating the existence of one very short "vanadyl"-type bond. (See Fig. 12a.)

Brückner et al. (1980) studied the spectra of Cr^{5+} doped into drawn fibers of soda-lime-silicate glasses, noting variations in $g_{\parallel}$ and $g_{\perp}$ as the orientations of the experimental fiber bundles were varied with respect to the direction of the applied magnetic field. It was concluded from these studies that the σ bonds with the four equatorial oxygens are more covalent when their planes lie nearly parallel to the fiber axis and less covalent when perpendicular to it. Similarly, it was deduced that the π bond with the chromadyl oxygen is more covalent when the bond is parallel to the fiber axis and less covalent perpendicular to it. These experiments additionally supported the conclusion that Cr^{5+} is a highly unstable intermediate in the redox reaction $Cr^{3+} \rightleftharpoons Cr^{6+} + 3e^-$.

B. $3d^3$ Ion: Cr^{3+} ($L = 0$, $S = 3/2$)

The ESR spectrum of Cr^{3+} in glasses has been recognized since the work of Garif'yanov and Zaripov (1964). Working in the frequency range 100–600 MHz, these authors observed broad resonances at $g_{eff} \simeq 3.0$ in glasses deliberately doped with chromium. Following the successful approach of Castner et al. (1960) to interpreting the spectrum of Fe^{3+} in glasses (Section III.C), Garif'yanov and Zaripov (1964) accounted for the unusually large g shift, $g_{eff} - g_e$, in terms of the spin Hamiltonian of Eq. (8) by invoking large zero-field splittings. Specifically, they calculated the case $D = 0$, $|E| \gg g_e \beta H$, finding two Kramers doublets (see Fig. 13a), each characterized by an "effective" g tensor with principal axis values $g_1 = 2$, $g_2 = 5.46$, $g_3 = 1.46$. Subsequently, Zakharov and Yudin (1965) obtained spectra at a more conventional X-band frequency (9.3 GHz) which exhibited the now-familiar maximum in the derivative spectrum at $g_{eff} = 5.0$ and zero-crossing at $g_{eff} = 1.78$. To better account for these more distinctive spectral features, the latter workers exactly diagonalized the fine-structure term of Eq. (13), allowing the ratio E/D to take on any value. The results of their approach, also described by Abdrashitova et al. (1972) and Gan Fuxi et al. (1982), are graphically represented in Fig. 14, where the upper-right-hand panel shows as a function of the ratio E/D the calculated resonance fields for the three principal-axis values of the g tensor for the upper and lower Kramers doublets (dashed and unbroken curves, respectively). Note that the influence of the upper Kramers doublet on the experimental spectrum is expected to be minimal because of its greater spectral anisotropy and particularly because of its lower thermal population statistics. Figure 14 was drawn to suggest that

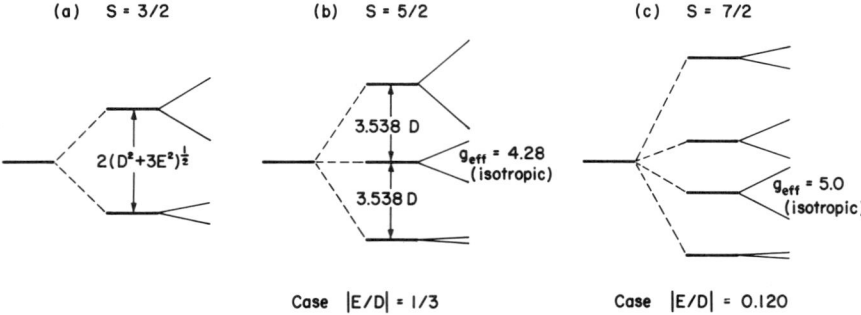

FIG. 13. Schematic fine-structure splittings of spin-degenerate ground states for (a) d^3, (b) d^5, and (c) f^7 ions. Dashed lines represent zero-field splittings due to crystalline electric fields, while diverging unbroken lines represent Zeeman splittings of the remaining Kramers doublets. The specific splittings and isotropic values of g_{eff} indicated in (b) and (c) pertain only to the special cases indicated (From Griscom, 1980a).

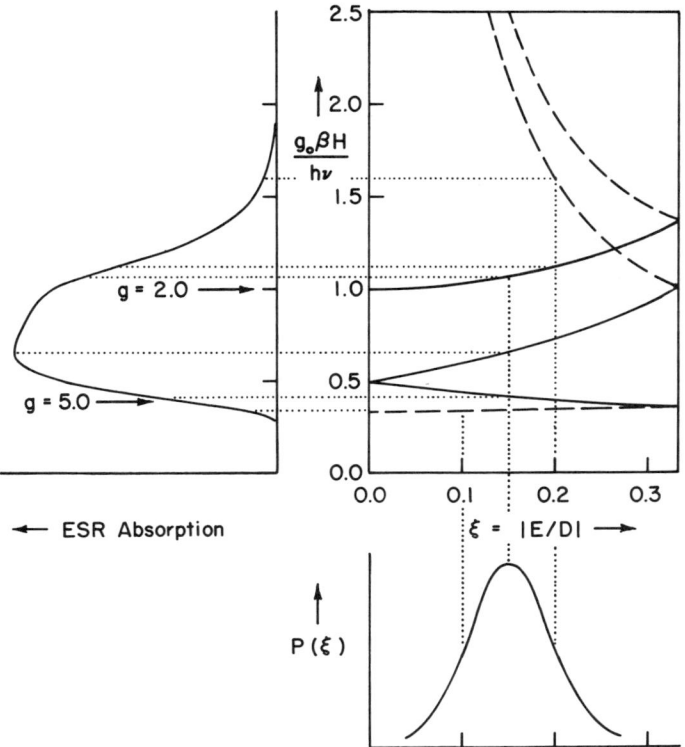

FIG. 14. The X-band ESR absorption curve of Cr^{3+} in an aluminophosphate glass (integrated from the derivative spectrum of Landry, 1968) is interpreted in terms of the g-value theory of Zakharov and Yudin (1965) by assuming a statistical distribution in the parameter $\xi = |E/D|$. (From Griscom, 1980a).

the detailed shapes of the experimental absorption spectra of Cr^{3+} in glasses might be accounted for in terms of a statistical distribution in E/D values (Griscom, 1980a). However, detailed computer simulations to check this hypothesis have apparently not been performed as of early 1989.

The broad asymmetric spectrum illustrated on the left-hand side of Fig. 14 is reasonably typical of oxide glasses with relatively low Cr_2O_3 contents (< 1 wt%). But as noted by Landry et al. (1967), raising the chromium oxide doping level to ~ 9 wt% results in the collapse of this spectrum into a nearly symmetric line centered on $g_{eff} = 1.96$. Landry et al. (1967) argued that the latter manifestation is best explained in terms of exchange-coupled pairs of Cr^{3+} ions. By studying the temperature dependence of the ESR intensity, Fournier et al. (1971) confirmed this view and demonstrated the antiferromagnetic nature of the coupling.

C. 3d^5 Ions: Fe^{3+}, Mn^{2+} ($L = L' = 0$, $S = 5/2$)

1. Fe^{3+}

Sands (1955) reported the observation of several "unknown" resonances in glasses, these signals being characterized by $g_{eff} = 4.2$ or $g_{eff} = 6$. In one of the most significant single contributions to the field of ESR in glasses, Castner et al. (1960) demonstrated a correlation of the intensities of the $g = 4.2$ resonances in several glasses with their analyzed Fe^{3+} contents and then went on to lay the quantum mechanical foundation for the interpretation of this ubiquitous spectral type. Castner et al. (1960) recognized that the observed g shifts, $g_{eff} - g_e$, were unexpectedly large for Fe^{3+} (for which normally $g \simeq g_e$), and this fact led them to conclude that the correct spin Hamiltonian must include fine-structure terms that greatly exceed the Zeeman interaction in absolute magnitude. They considered first the spin Hamiltonian of Eq. (8) in the special case that $D = 0$, $|E| \gg g_e \beta H$, showing that the fine-structure term splits the sixfold spin degeneracy of the $S = 5/2$ ground state into three Kramers doublets with energies $W = 0, \mp 2(7)^{1/2}E$. (See Fig. 13b.) Next they demonstrated by means of degenerate perturbation theory that the central, $W = 0$, doublet is characterized by an isotropic value of $g_{eff} = 30/7 = 4.286$, thus providing a quite general explanation of the spectrally sharp "$g = 4.3$" resonances so often encountered in glasses. The upper and lower doublets of the same manifold were shown to have strongly anisotropic values of g_{eff} with principal values 9.68, 0.86, and 0.61, accounting for the "shoulder" near $g = 10$ and extremely broad "background" normally found in association with the $g = 4.3$ signal. (See Fig. 15.) In a later refinement, Wickman et al. (1965) showed that the condition $D = 0$, $E \neq 0$ is, by the appropriate coordinate transformation, entirely equivalent to the situation $|E/D| = 1/3$. Wickman et al. (1965) calculated the values of g_{eff} to be expected for the entire range of physically distinct possibilities, $0 < |E/D| < 1/3$, when $|D| \gg g_e \beta H$.

Castner et al. (1960) also employed degenerate perturbation theory to investigate the special case $E = 0$, $|D| \gg g_e \beta H$, finding in this event that one of the three resulting Kramers doublets is characterized by effective g values of $g_{\parallel} = 2$ and $g_{\perp} = 6$. Loveridge and Parke (1971) reported the occurrence of resonances fitting this description in amber glasses containing both iron and sulfur and ventured the explanation that the strong axial field term in this case must arise from a single sulfur atom substituting for one of the oxygen ligands.

An alternative explanation of the ubiquitous absorption peak at $g_{eff} = 4.3$ was proposed by Peterson et al. (1974a). Their model was based on an assumed axial local-site symmetry and distributions in both $g_{\parallel}$ and $g_{\perp}$ with a correlation coefficient of $r = -1$. However, Brodbeck (1980) performed

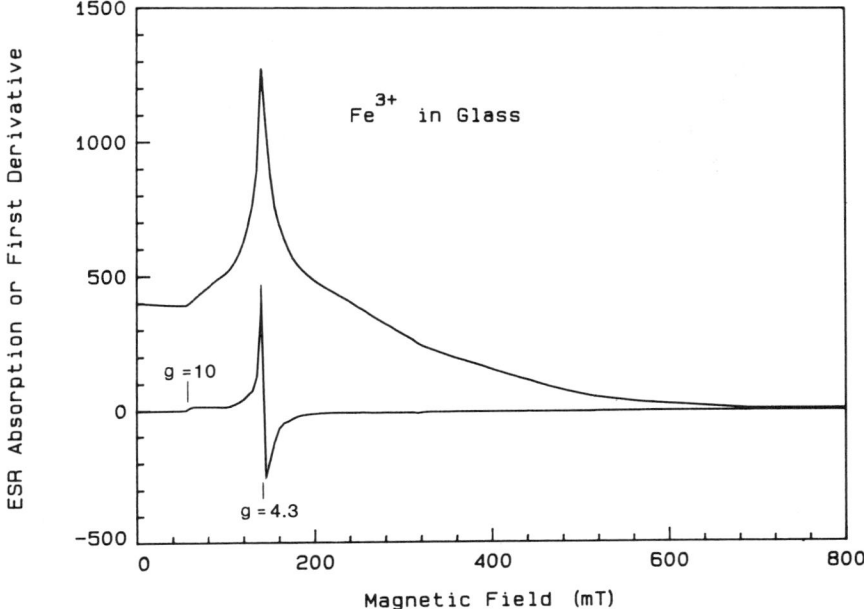

FIG. 15. Spectrum of dilute (0.1%) Fe^{3+} in a complex silicate glass. The experimental derivative spectrum (lower) was numerically integrated to obtain the illustrated absorption curve (upper). Since the integrated absorption of this particular lineshape is extremely sensitive to errors in determining the baseline, the apparent absorption at zero field may be an artifact.

extensive calculations showing that a distribution in the magnitude of D for the case $E = 0$ cannot produce the required joint probability distribution function. Brodbeck was able to conclude that the $g = 4.3$ features commonly observed in glasses could arise from Fe^{3+} subject to rather broad distributions in $|E/D|$, so long as these distributions encompass the value $|E/D| = 1/3$. His plotted isofrequency curves are of assistance in understanding how transitions within the other two Kramers doublets might account for the substantial intensity appearing in Fig. 15 at fields higher than ~ 300 mT.

In analogy to the case for the $g = 5$ resonance due to Cr^{3+} (Section III.B), the $g = 4.3$ signal due to Fe^{3+} in glasses is a characteristic spectral feature only at relatively low concentrations of the paramagnetic ion. Kurkjian and Sigety (1968) were among the first to remark that as the Fe_2O_3 contents of typical oxide glasses are increased from ~ 0.1 wt % toward ~ 5 wt %, the $g = 4.3$ resonance intensity passes through a maximum and is gradually supplanted by a very broad signal (peak-to-peak derivative linewidth $\Delta H_{pp} \sim 150$ mT) centered on $g_{eff} = 2$. Schreiber et al. (1980) have carefully established the proportionalities that exist in certain basalt glass compositions between the intensity of the $g = 4.3$ line and the concentrations of Fe^{3+} below ~ 1 wt%. On the other hand, the broad line at $g_{eff} = 2$ is less well understood, and indeed such manifestations can have a variety of origins depending on the concentrations and valence states of iron and the types of microstructure that may have developed in the glass. (See, e.g., discussions and literature cited in Griscom, 1980a, 1984a.)

2. Mn^{2+}

Although Mn^{2+} is isoelectronic with Fe^{3+}, its ESR spectra are distinctly different due to two main factors. First, the fine-structure splittings for Mn^{2+} are typically an order of magnitude smaller than those of Fe^{3+} in common crystalline hosts (Schneider et al., 1968), and this trend appears to hold in glasses as well. Thus, in contrast to the situation for Fe^{3+} described previously, one typically measures for Mn^{2+} $g_{eff} \simeq 2.0$, as normally expected for an S-state ion. Second, a six-line multiplet spectrum results from the hyperfine interaction of the $^6S_{5/2}$ ground state with the 100%-abundant ^{55}Mn nucleus ($I = 5/2$). The measured hyperfine splittings provide a useful gauge of the relative ionicity of divalent manganese in various hosts (Van Weiringen, 1955) but fall short of indicating coordination numbers (Tsay and Helmholtz, 1969).

The ESR spectra of Mn^{2+} in oxide glasses, though showing some compositional sensitivity, are remarkably similar in their general description from one system to another, consisting at X-band frequencies (~ 9 GHz) of a sharp six-line component centered on $g = 2.0$ flanked by broad "wings," with a weak feature centered on $g_{eff} = 4.3$ and measurable absorption at zero field. A variety of explanations for this spectrum have been offered, beginning with that of Tucker (1962), who proposed the existence of two distinct sites, one with a large value of $|E|$ accounting for the $g = 4.3$ feature and one for which $|D, E| \ll g_e \beta H$ giving rise to the principal features at $g = 2$. On the other hand, De Wijn and van Balderen (1967) advocated a single site with $E = 0$ and $|D| \ll g_e \beta H$ whose allowed transitions accounted for the principal lines at $g = 2$ and whose forbidden transitions were supposed to give the features at $g = 4.3$. The latter view was disputed by Griscom and Griscom (1967) on the basis of the observation that most of the spectral intensity recorded at X-band frequencies lies in the "wings" of the $g = 2$ spectrum and at zero field and, further, that this intensity collapses on $g = 2$ when the measurement frequency is raised to ~ 35 GHz. (See Fig. 16.) Thus, Griscom and Griscom (1967) argued that $|D, E|$ must be comparable to $g_e \beta H$ at X band but smaller than $g_e \beta H$ at Ka band.

Because the inferred intermediate-strength fine-structure splittings lay outside of the ranges of validity of both the second-order perturbation theory result of Eqs. (29) and the degenerate perturbation theory treatment of

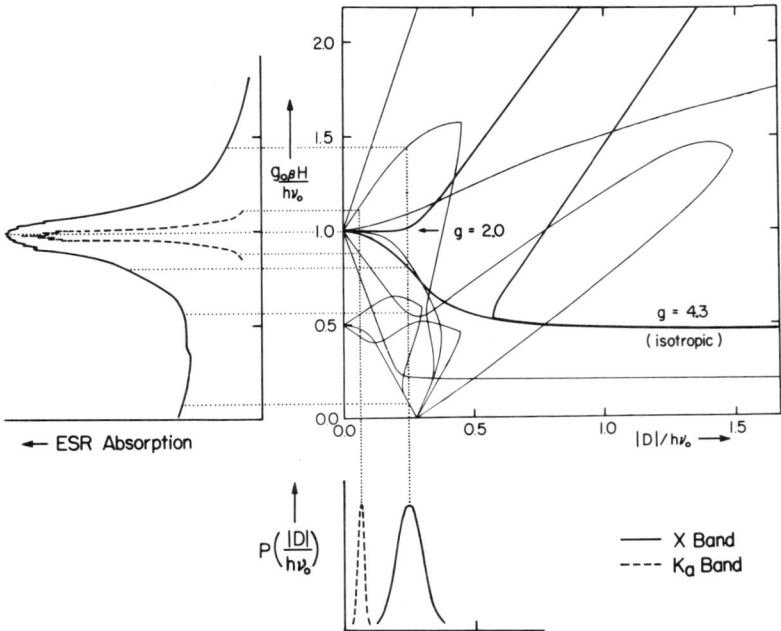

FIG. 16. ESR absorption spectra of Mn^{2+} in an alkali borate glass (left) are interpreted in terms of a distribution in D values for a fixed ratio $|E/D| = 1/3$. Diagram to the upper right shows the normalized resonance fields computed as functions of $|D|/hv$ for the applied magnetic field parallel to each of the three principal axes of the complex. Heavier curves pertain to the cases for which the same result is obtained for H_{app} parallel to two or more principal axes. The heaviest curve segment indicates an isotropic transition that approaches $g_{eff} = 4.3$ for large values of $|D|/hv$. The bell-shaped curves at the bottom represent a single distribution of $|D|$ values plotted as $P(|D|/hv)$ for two different microwave frequencies; such a distribution was proposed to account for the experimentally observed spectra. (From Griscom and Griscom, 1967; Griscom, 1978b).

Castner et al. (1960), Griscom and Griscom (1967) carried out an exact diagonalization of the spin Hamiltonian of Eq. (8) for a range of values of $|D|$ relative to $g_e \beta H$, subject to the constraint $|E/D| = 1/3$. They plotted their calculated values for the critical points in the resonance surface (the expected positions of sharp powder-pattern features) as a function of $|D|$, normalizing the quantities on both axes to the value of the microwave quantum, hv (upper-right-hand panel in Fig. 16). In this way a single diagram suffices to assess the spectra obtained at different microwave frequencies. (More extensive diagrams of this nature for various ratios $|E/D|$ have been published by Dowsing and Gibson, 1969, and Aäsa, 1970.) As illustrated in Fig. 16, Griscom (1978b) argued that the spectra observed for Mn^{2+} in glasses at both X band and Ka band might be explainable in terms of a single-peaked statistical distribution in $|D|$ values with $|D|_{av}/g_e\beta \sim 80$ mT.

Kliava and Purāns (1980) have performed the only thoroughly detailed computer simulations of the spectra of Mn^{2+} in glasses, albeit specialized to the well-resolved hyperfine structure centered on $g = 2$. A resonance condition was employed that treated the fine and hyperfine structure terms to third order in perturbation theory, forbidden transitions ($\Delta M_I \neq 0$) were included, and a reliable expression for the angular-dependent transition probability, $W_{M_I, M_s}(\Omega)$, was used in the calculation of Eq. (36). As apparent in Fig. 17, this sophistication was crucial to obtaining a correct result. A series of spectral simulations was carried out using a joint probability density function for the parameters D and E and various trial values of their mean values D_{av} and E_{av}, their variances ΔD and ΔE, and their correlation coefficient r. On the basis of their results for a K_2O-SiO_2 glass (Fig. 18) and a $ZnO-P_2O_5$ glass,

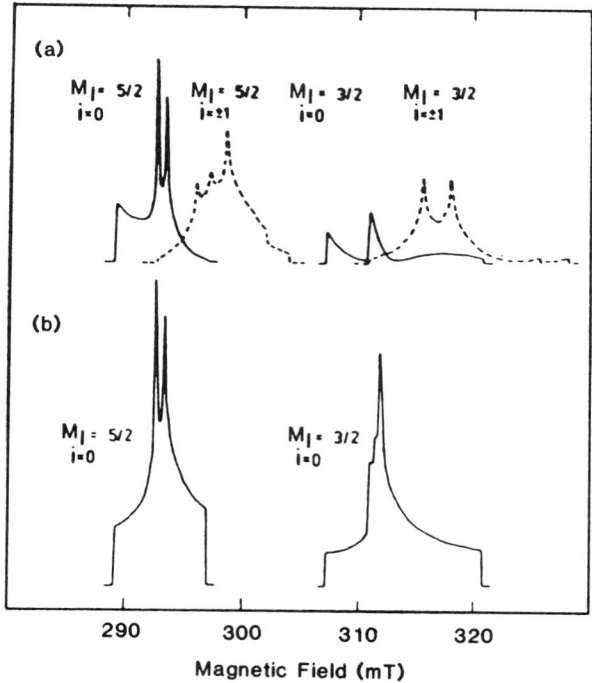

Fig. 17. Mn^{2+} powder patterns calculated to third order for some allowed (unbroken curves) and forbidden (dashed curves) hyperfine transitions belonging to the central, $M_S = -1/2 \leftrightarrow +1/2$ manifold. $g = 2.0$, $A/g\beta = -9.5$ mT, $D/g\beta = 21.0$ mT, $|E/D| = 1/3$, $\nu = 8.9$ GHz. (a) Transition probability included in calculation. (b) Transition probability set equal to one. (From Kliava and Purāns, 1980).

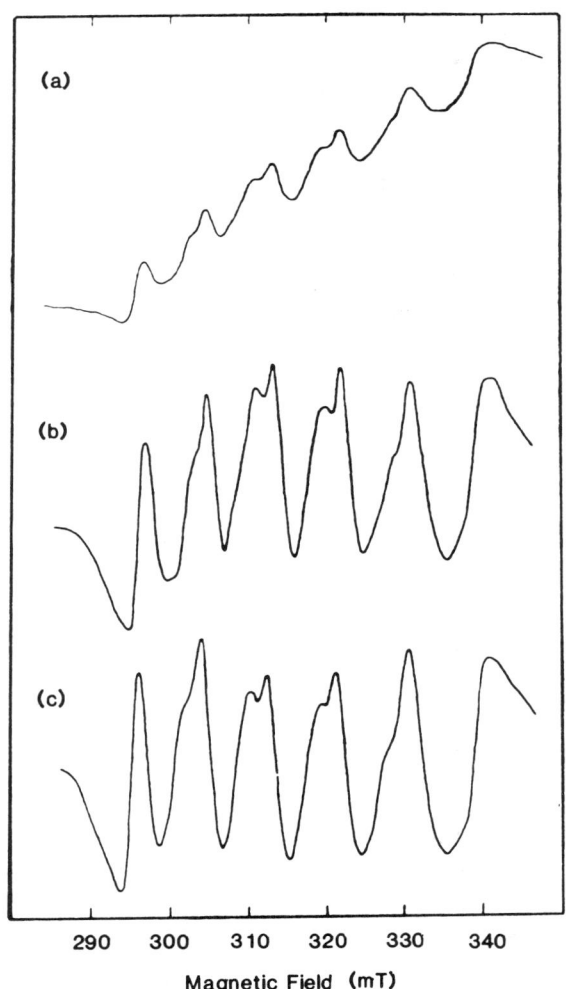

FIG. 18. (a) Experimental spectrum of Mn^{2+} in a K_2O-4SiO_2 glass. (b) Experimental spectrum of (a) from which a computer-synthesized "broad underlying resonance" was subtracted. (c) Computer simulation of (b) employing the following parameters: $g = 2.0$, $A/g\beta = -8.7$ mT, $D/g\beta = 22.0$ mT, $|E/D| = 1/3$, $\Delta D/g\beta = 8.0$ mT, $\Delta E/g\beta = 3.0$ mT, $\nu = 8.9$ GHz. (From Kliava and Purāns, 1980).

Kliava and Purāns (1980) concluded that the sharp-line structure centered on $g = 2$ is characterized by $|D_{av}|/g_e\beta = 22 \pm 2$ mT, $|E/D| \simeq 1/3$, and $r \simeq 0$.

Thus, while agreeing with Griscom and Griscom (1967) on the E/D ratio, Kliava and Purāns (1980) suggested that the simple Gaussian distribution in $|D|$ proposed in Fig. 16 should perhaps be replaced by a two-peaked distribution featuring a sharp spike at a value of $|D| \simeq 22 \, \text{mT}$. However, this suggestion can be fully evaluated only after careful consideration of the implications of having subtracted the "broad underlying resonance" from the spectrum of Fig. 18a to obtain the shape, Fig. 18b, that was actually simulated.

Kliava and Purāns (1980) supported the conclusion of Griscom and Griscom (1967) that the broad underlying resonance, which accounts for approximately 10 times as much intensity as the sharp structure, must be due to sites with $|D|/g_e\beta \gg 22 \, \text{mT}$. Schreurs (1978) has reported ^{55}Mn hyperfine structure associated with the spectral features $g_{\text{eff}} = 4.3$ and $g_{\text{eff}} = 9.4$ in several manganese-doped oxide glasses, confirming that all parts of the X-band spectrum of Fig. 16 are due to Mn^{2+}. Thus, it now appears that most fundamental issues concerning the analysis of the spectrum have been resolved, and progress should soon follow in understanding the dependence of the lineshape on glass composition (as described, e.g., by Griscom and Griscom, 1967).

D. 3d⁹ Ion: Cu^{2+} ($L = 0$, $S = 1/2$)

The study of Cu^{2+} in glasses by ESR spectroscopy dates from the pioneering work of Sands (1955). Its spectrum in oxide glasses is distinctive and easily recognized on the basis of its large axial g anisotropy ($g_\parallel \simeq 2.3$, $g_\perp \simeq 2.06$) and four-line hyperfine structure due to ^{63}Cu and ^{65}Cu ($I = 3/2$ for both isotopes, nuclear moments differing by only 7%, 100% combined abundance).

As discussed by Imagawa (1968), the general nature of the ligand coordination sphere in these glasses can be inferred directly from the fact that $g_\parallel > g_\perp > 2$. Only in an octahedral environment elongated along one of the cube axes does this result obtain. (See Fig. 12b.) In this event, the b_1 orbital singlet lies the lowest (refer to Fig. 9) and the g values are given by (e.g., Sands, 1955)

$$g_\perp = g_e(1 - \lambda\alpha^2/\Delta), \tag{49a}$$

$$g_\parallel = g_e(1 - 4\lambda\alpha^2\beta_1^2/\Delta), \tag{49b}$$

where Δ is the cubic field splitting, λ is the spin-orbit coupling constant (which is negative since the d shell is more than half filled), and α^2 and β_1^2 are parameters introduced (Maki and McGarvey, 1958; Kivelson and Neiman, 1961) to account for covalency. In Eqs. (49), which are approximations to the formulae given by Imagawa (1968), α^2 and β_1^2 represent the contributions of the Cu^{2+} 3d orbitals to the $B_{1g}(\sigma)$ and $B_{2g}(\pi)$ antibonding orbitals, respectively, both involving the equatorial ligands. Again dropping the higher-order terms from Imagawa's formulations, the hyperfine parameters are given theoretically by

$$A_{\perp} = P\left[\alpha^2\left(\frac{2}{7} - \kappa_0\right) + \frac{11}{14}(g_{\perp} - 2)\right] \quad \text{and} \quad (50a)$$

$$A_{\parallel} = P\left[-\alpha^2\left(\frac{4}{7} + \kappa_0\right) + (g_{\parallel} - 2) + \frac{3}{7}(g_{\perp} - 2)\right], \quad (50b)$$

where $P = 2g_N\beta\beta_N\langle r^{-3}\rangle$, β_N is the nuclear g factor, and $\alpha^2\kappa_0$ is the Fermi contact interaction.

Using g values and hyperfine coupling constants scaled from his experimental spectra, Imagawa (1968) was able to solve the more exact forms of Eqs. (49) and (50) for α^2, β_1^2, and κ_0. His results for the sodium and lithium borate systems, illustrated in Fig. 19, were interpreted in terms of the competition between the cupric ion and the network-forming cations for attracting the lone pairs of the intervening oxygen ions. Imagawa (1968) argued that the insensitivity of α^2 to glass composition is likely to be a general consequence, since the actual value of α^2 must in all cases flow from the minimization of the energy of the molecular orbital of the $B_{1g}(\sigma^*)$ ground state. By making this assumption, Imagawa deduced from Eq. (50b) the approximate relation

$$g_{\parallel} - A_{\parallel}/P = \text{const} \quad (51)$$

to hold for any glass system irrespective of the variation of β_1^2 with composition. Kawazoe et al. (1978b) verified the applicability of Eq. (51) to the ESR spectra of Cu^{2+} in potassium borate and sodium phosphate glasses.

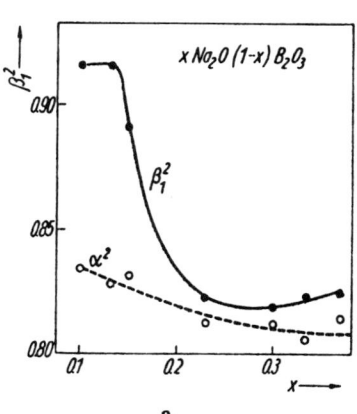

a

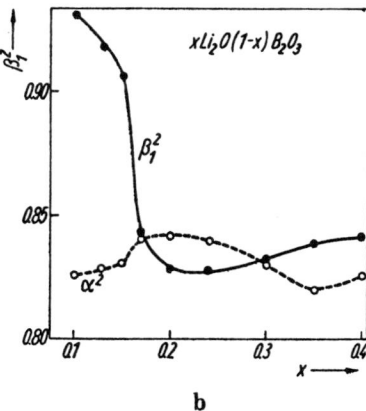

b

FIG. 19. Ionicity of Cu^{2+} bonding in (a) Na_2O-B_2O_3 and (b) Li_2O-B_2O_3 glasses. (From Imagawa, 1968).

Moreover, Kawazoe et al. (1980) confirmed by actual computer lineshape simulations the further conclusion of Imagawa (1968) that, for a single glass of any given composition, statistical distributions in spin Hamiltonian parameters subject to the constraint of Eq. (51) must be the reason for the observed increases in the widths of the $H_\parallel(M_I)$ hyperfine peaks with increasing M_I. (See Fig. 20.)

The physical origins of the compositional variations of β_1^2 have been explored theoretically and experimentally by Kawazoe et al. (1978a, b). In the first of these two papers, the basicities of oxygens in alkali borate glasses were estimated by means of SCF intermediate neglect of differential overlap (INDO) and LCAO-MO calculations. It was found that the basicities could be related to the degree of electronic localization in the highest-occupied-molecular orbitals (HOMOs), which turned out to be out-of-plane π-type orbitals on certain oxygens which were either nonbridging (mole % alkali oxide, $x > 30$) or specific classes of bridging oxygens ($x < 30$). The cal-

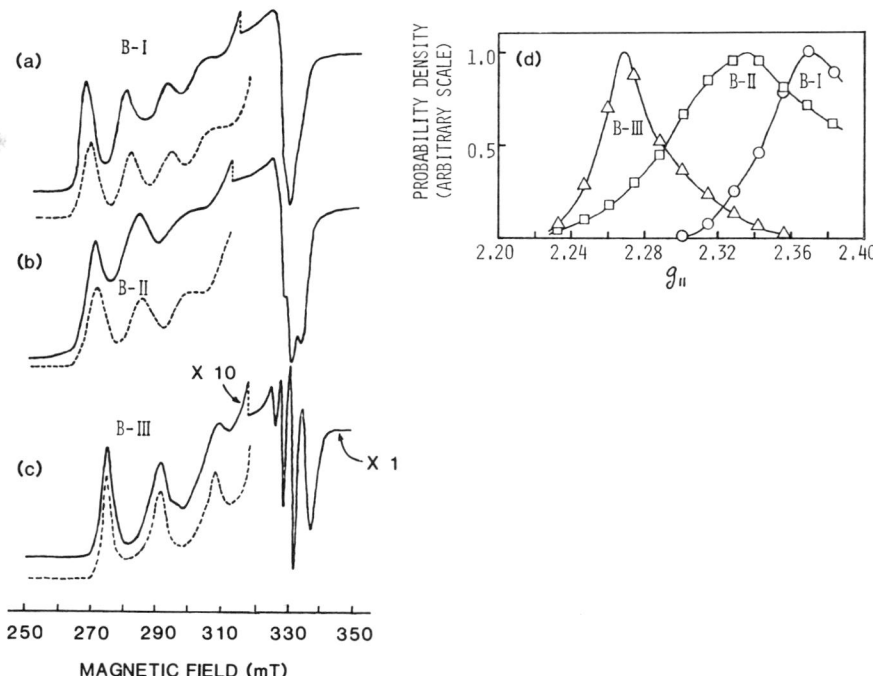

FIG. 20. ESR spectra of Cu^{2+} in (a) $10Na_2O\text{-}90B_2O_3$, (b) $27Na_2O\text{-}73B_2O_3$, and (c) $70Na_2O\text{-}30B_2O_3$ glasses. Unbroken curves are experimental; dashed curves are computer simulations based on the distributions of $g_\parallel$ values given in (d), subject to the constraint of Eq. (51). (From Kawazoe et al., 1980).

culations of Kawazoe et al. (1978a) showed a marked increase in the O 2p electron densities in the out-of-plane π orbitals which were identified as HOMOs in the composition range $x = 15-20$ mol % alkali oxide. Kawazoe et al. (1978b) argued that such an increase would be theoretically expected to result in a decrease in the value of β_1^2 characterizing the dissolved Cu^{2+} probe ions. The sharpness of the change in β_1^2 in a relatively narrow composition range seen for the alkali borates (Fig. 19) is believed to be an artifact of the π-electron delocalization which is a property of borate glasses but not, for example, of phosphates or silicates (Kawazoe et al., 1978b; Hosono et al., 1979b).

E. 4d AND 5d IONS

Partially in consequence of the ions' tendency to incorporate into covalent complexes that are often diamagnetic, ESR observation of 4d and 5d ions is not often reported in crystals (Bowers and Owen, 1955) and is an even more rare occurrence in glasses, where the broadening effects of spectral anisotropies and distributions in local fields further weakens the signal strength.

The $4d^1$ ions Zr^{3+} and Nb^{4+} have been detected by ESR in network oxide glasses by Garif'yanov and Yafaev (1963), Kim et al. (1968), respectively. Mo^{5+} has been studied by Sunch et al. (1971), Baugher and Parke (1977), and Simon and Nicula (1983). From the standpoint of ESR spectral analysis, these ions behaved essentially the same as their $3d^1$ counterparts. In particular, it was found possible to employ Eqs. (48) to analyze the g matrices, leading through the same train of logic as reviewed in Section III.A.2 to the conclusion that all ions are in octahedral coordination with one of the metal-ligand bonds being shorter than the other five. Baugher and Parke (1977) were able to apply Eqs. (50) to the analysis of the hyperfine structure due to the ^{95}Mo and ^{97}Mo isotopes (both $I = 5/2$, nuclear moments differing by only 2%, 25% combined natural abundance). Simon and Nicula (1983) found ESR spectral evidence for three types of Mo^{5+} sites whose relative numbers depended sensitively on the compositions of a series of B_2O_3-Na_2O-MoO_3 glasses. The Nb^{4+} center of Kim et al. (1968) was actually a radiation-induced defect in Nb_2O_5-Na_2O-SiO_2 glasses, and the well-resolved hyperfine spectrum due to ^{93}Nb ($I = 9/2$, 100% abundant) was computer-simulated with considerable precision.

Landry (1968) recorded the ESR spectrum of dilute Mo^{3+} ($4d^3$) in an aluminum-zinc phosphate glass, showing it to be virtually identical to that of Cr^{3+} (Fig. 14) under the same conditions of observation. The remarkable similarity of these two resonances is further support for the conclusion that this distinctive lineshape is a consequence of an $S = 3/2$ ion being subjected to the special condition $|D| \gg g_e\beta H$. (See Section III.B.)

Resonances attributable to W^{5+} ($5d^1$) were reported by Yafaev et al. (1963) in several silicate and phosphate glasses. Although these authors supposed the tungsten ion to be tetrahedrally coordinated, the observed g values (e.g., $g_\parallel = 1.6$, $g_\perp = 1.76$ in the phosphates) are more consistently interpreted in terms of the vanadyl-type model which has successfully explained the spectra of numerous other d^1 ions in glass. (See Section III.A.)

F. 4f IONS: THE RARE EARTHS

The rare-earth ions find their unpaired spins residing in well-screened 4f shells of relatively small radii. Consequently, with the exception of S-state ions, orbital motion tends not to be quenched, and the spin Hamiltonian is treated in the weak-field limit: $\mathcal{H}_{\text{xtal}} < \mathcal{H}_{\text{SO}}$ (e.g., Abragam and Bleaney, 1970). The very small splittings of the orbital degeneracy give rise to extremely short spin-lattice relaxation times, making it imperative to lower the sample temperature below $\sim 20\,\text{K}$ in order to observe the resonance. The use of low temperatures has the added advantage of freezing out all contributions to the spectrum but transitions between the lowest sublevels of the lowest L-S multiplet term (for which $J = L + S$ and J_z are good quantum numbers). On the other hand, large anisotropies are generally encountered in g_{eff}, making observation and unambiguous identification of ESR spectra of rare-earth ions in glasses exceedingly difficult.

1. $4f^1$ Ion: Ce^{3+} ($L = 3$, $S = 1/2$, Ground State $J = 5/2$)

Ce^{3+} may be the only non-S-state rare-earth ion to have been identified by ESR in an oxide glass. This accomplishment (Bishay et al., 1974) was particularly significant in view of the well-known role of cerium in protecting glasses from radiation-induced coloration (e.g., Stroud, 1961; Bishay, 1962). As anticipated, low temperatures ($\sim 5\,\text{K}$) were required to narrow the resonance sufficiently for observation. The extremely broad spectrum that was observed (Fig. 21) was shown by Bishay et al. (1974) to exhibit an intensity proportional to the Ce^{3+} content of the barium aluminoborate base glass as varied by (1) alteration of the total cerium-doping levels of a series of glasses melted under fixed redox conditions, (2) variation of the redox conditions in remelting glasses of fixed total-cerium content, and (3) alteration of the Ce^{3+}/Ce^{4+} ratio by irradiation.

The first step in deriving appropriate spin Hamiltonian parameters involves the recognition that the ground-state term of the free ion is $^2F_{5/2}$ and lies $\sim 2200\,\text{cm}^{-1}$ below the $^2F_{7/2}$ excited state (Abragam and Bleaney, 1970). The principal effect of inserting the ion into a solid matrix is to split the sixfold-degenerate ground-state manifold into three Kramers doublets whose wave functions are $|J_z\rangle = \pm 1/2$, $|J_z\rangle = \pm 3/2$, and $|J_z\rangle = \pm 5/2$. If the latter

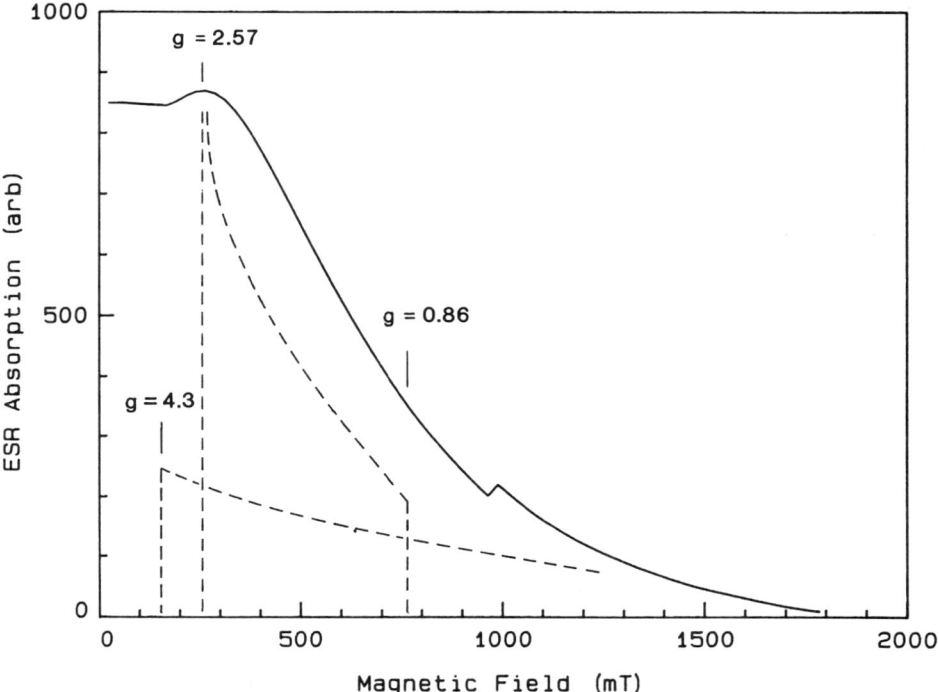

FIG. 21.   ESR of Ce^{3+} in a barium aluminoborate glass. Absorption curve was numerically integrated from an X-band derivative spectrum recorded at 5 K by Bishay et al. (1974). Dashed powder patterns qualitatively illustrate the predictions of Eqs. (53).

splittings are sufficiently large, the Zeeman Hamiltonian

$$\mathscr{H}_Z = g_J \beta (\mathbf{H} \cdot \mathbf{J}) \tag{52}$$

can be treated in degenerate perturbation theory to derive values of g_{eff} for each doublet, assuming each to be characterized by an effective spin $S' = 1/2$. Since the Lande g factor for the ground-state term is $g_J = 6/7$, the effective g matrices are readily shown to be (e.g., Abragam and Bleaney, 1970):

$$|\pm 1/2\rangle: g_{\parallel} = 6/7 = 0.86; \; g_{\perp} = 18/7 = 2.57, \tag{53a}$$

$$|\pm 3/2\rangle: g_{\parallel} = 18/7 = 2.57; \; g_{\perp} = 0, \tag{53b}$$

$$|\pm 5/2\rangle: g_{\parallel} = 30/7 = 4.29; \; g_{\perp} = 0. \tag{53c}$$

Because of the low temperatures involved in the experiment, only the lowest-lying of the three Kramers doublets is expected to contribute to the observed ESR spectrum. In crystals, the lowest doublet is found to be either

$|\pm 1/2\rangle$ or $|\pm 5/2\rangle$, depending on the host (Abragam and Bleaney, 1970). Bishay et al. (1974) interpreted the signal of Ce^{3+} in glass as indicating effective g values $g_{\parallel} \simeq 3$ and $g_{\perp} \simeq 0$. However, as illustrated in Fig. 21, the shape of the integrated absorption curve may be partially accounted for in terms of an axial powder pattern characterized by principal values of g_{eff} approximately the same as those given by Eq. (53a). Such a conclusion is supported by the early computer lineshape simulations of Al'tshuler (1969) and would indicate that at least a fraction of the Ce^{3+} ions occupy $|\pm 1/2\rangle$ ground states.

2. $4f^7$ Ions: Gd^{3+} and Eu^{2+} ($L = 0, S = 7/2$)

When present in glassy hosts of almost any type, both Gd^{3+} and Eu^{2+} tend to exhibit similar ESR spectra, characterized at X-band frequencies by distinct features at effective g values of 2.0, 2.8, and ~ 6.0. Posing a suitable explanation for this ubiquitous "U spectrum" has been a challenge that has occupied many workers over the years (e.g., Nicklin et al., 1973; Chepeleva and Lazukin, 1976; Griscom, 1980a; Cugunov and Kliava, 1982; Koopmans et al., 1983; Brodbeck and Iton, 1985).

In analogy with the approach of Castner et al. (1960) to the $g = 4.3$ resonance of Fe^{3+} in glasses (Section III.C.1), initial activity revolved around the use of degenerate perturbation theory to search out possible explanations of the $g = 6$ feature of the U spectrum. Chepeleva and Lazukin (1976) proposed that the latter feature arises from a large *cubic* fine-structure term in the spin-Hamiltonian, while Nicklin et al. (1973) explored the normal second-order fine-structure Hamiltonian (Eq. (13)) in the limit $|D| \gg g_e \beta H$, noting an isotropic value of $g_{eff} = 5.0$ for the special case of $|E/D| = 0.120$ (Fig. 13c).

Experimentally, Nicklin et al. (1973) recorded the ESR spectra of Gd^{3+} in a sodium-silicate glass at seven separate microwave frequencies. The striking frequency dependence apparent in Fig. 22 was strongly suggestive that the relevant fine-structure terms may be smaller than $h\nu$ at the highest frequencies and comparable to $h\nu$ at the lower frequencies, leading Nicklin et al. (1973) to develop diagrams of normalized resonance field ($g_e \beta H/h\nu$) versus normalized crystal-field parameter (e.g., $D/h\nu$) for $S = 7/2$ ions, in analogy to the one for the $S = 5/2$ case illustrated in Fig. 16. After a careful study of a set of such diagrams calculated for a range of values of $|E/D|$, Nicklin et al. (1973) opted for explaining the features at $g = 6.0$, 2.8, and 2.0 in terms of three distinct sites (multiple-site model). On the other hand, Griscom (1980a) felt that the diagrams of Nicklin et al. (1973) could just as easily support a single-site model incorporating a broad range of distortions (although the detailed explanation of the $g = 6.0$ feature was still not evident on the basis of the calculations then available). Subsequently, Koopmans et al. (1983) achieved a rather good computer lineshape simulation by assuming a broad distribution of $|D|$ values, with $E = 0$.

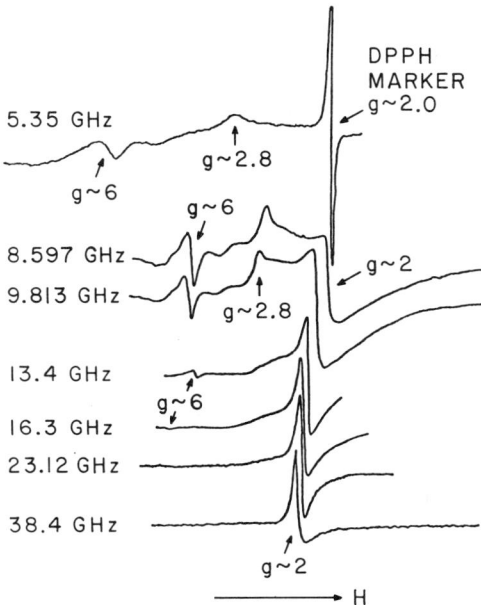

FIG. 22. ESR derivative spectra of Gd^{3+} in a sodium silicate glass as a function of spectrometer frequency. (From Nicklin et al., 1973).

More recently, however, Brodbeck and Iton (1985) critiqued all previous analyses of the spectra of Gd^{3+} and Eu^{2+} ions in glass and went on to perform *ab initio* (from the beginning, without approximations) lineshape simulations under each of the various hypotheses. These simulations, together with a set of validity criteria developed by these authors, enabled Brodbeck and Iton (1985) to conclusively demonstrate that the observed spectra arise from a single-peaked distribution in $|D|/h\nu$ values (centered near 0.19 at X band) with $|E/D|$ ranging broadly (but including values near 1/3). The *ab initio* computed spectra of Fig. 23 (achieved for $|D|_{av} = 0.0560 \, \text{cm}^{-1}$, $\Delta|D| = 0.0192 \, \text{cm}^{-1}$, and a virtually "flat" distribution of $|E/D|$ values) should be compared with the data of Fig. 22. Thus, Brodbeck and Iton (1985) confirmed Griscom's (1980a) belief that many foreign ions, particularly large ones such as Gd^{3+}, tend to dictate their own environments in glass and hence should be described by a single site—albeit a site characterized by a distribution of crystal fields that can be quite broad.

Morris and Haskin (1974), Lauer and Morris (1977), Schreiber (1977), and Schreiber et al. (1979) have employed the ESR technique to determine $(Eu^{2+})/(Eu^{3+})$ ratios in a wide range of silicate melts.

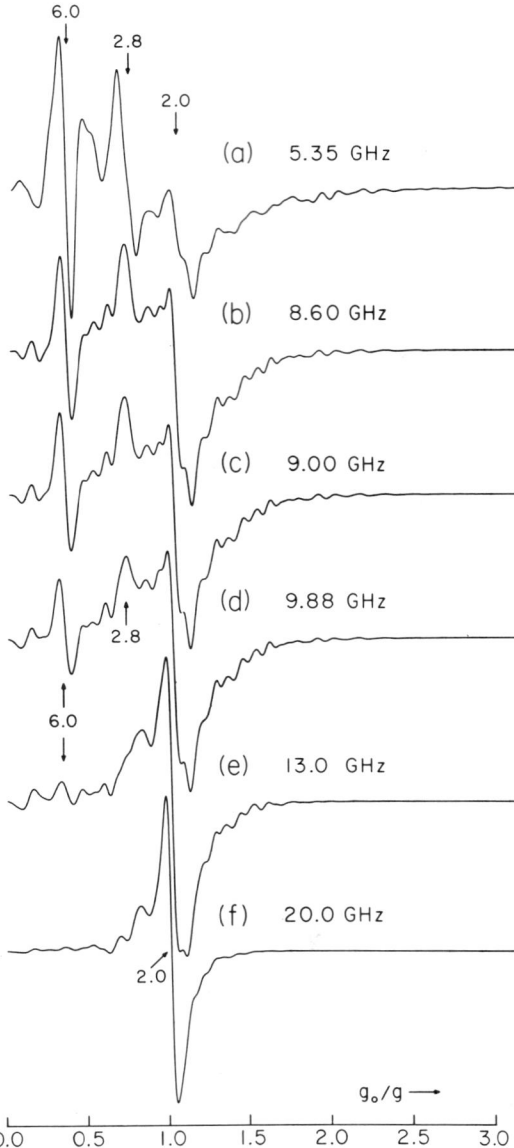

FIG. 23.   Computer-simulated ESR spectra of Gd^{3+} in glass as a function of microwave frequency. All curves were generated assuming a Gaussian distribution of D values with $|D|_{av} = 0.056\,\mathrm{cm}^{-1}$, $\Delta|D| = 0.019\,\mathrm{cm}^{-1}$, and a nearly uniform distribution of $|E/D|$ (From Brodbeck and Iton, 1985). These results should be compared with the experimental spectra of Fig. 22.

IV. Defect Centers in Glasses

As a rule, insulating glasses not containing transition-group ions are diamagnetic and display only weak (and often undetectable) ESR signals in their as-quenched conditions. However, a number of interesting exceptions bear mentioning. Certainly, physical crushing of glasses in vacuo can give rise to paramagnetic surface states detectable by ESR (e.g., Hochstrasser, 1966; Bobyshev and Radtsig, 1981). And it is also becoming apparent that extremely rapid quenching from high temperatures can result in defect formation in glasses that are generally defect-free under normal quenching conditions. Examples include oxygen-vacancy centers in rapidly quenched GeO_2 bulk melts (Kordas et al., 1983), CO_2 pulsed laser-irradiated SiO_2 (Weeks, 1985), GeO_2-doped silica glasses rapidly drawn into fibers (e.g., Hanafusa et al., 1985; Kawazoe et al., 1986b), and interstitial O_2^- molecular ions in splat-quenched $70Na_2O\text{-}30B_2O_3$ glasses (Griscom, 1973). More anomalous are the O_2^- ions observed in certain conventionally quenched calcium aluminate glasses (Hosono and Abe, 1987). Nevertheless, it remains true that high concentrations of quenched-in paramagnetic states are uncommon in glassy dielectrics and that any such quenched-in defects nearly always correspond to generically identical defects that can be created in the same materials by exposure to energetic radiations in the form of nuclear particles, γ-rays, X-rays, or sometimes even UV light (Stathis and Kastner, 1984). For this reason, the focus of the present section will be on radiation-induced defects. In keeping with the theme of this chapter, emphasis will be placed on the use of ESR spectroscopy to correctly identify and characterize the defect species. Some related topics, such as radiation damage processes and thermal bleaching kinetics, have been reviewed elsewhere (e.g., Friebele and Griscom, 1979; Griscom, 1985a).

This section is grouped into subsections treating (1) defects in undoped synthetic silicas of nominally high purities, (2) defects in doped fused silicas developed for fiber optic applications, and (3) defects in selected multicomponent oxide glasses, including oxynitrides. The topic of defects in halide glasses has been reviewed by Griscom and Friebele (1989).

A. UNDOPED SILICAS

Synthetic fused silicas prepared by flame hydrolysis or plasma oxidation of $SiCl_4$ typically contain $\leqslant 1$ ppm total cationic impurities (Brückner, 1970, 1971; Kreidl, 1983). The flame-hydrolyzed (Type III) silicas are characterized by the presence of ~ 1200 ppm water by weight in the form of hydroxyl groups. In contrast, the plasma-deposited (Type IV) synthetic silicas typically contain $\leqslant 5$ ppm water. Virtually all Type-III and Type-IV silicas also contain chloride-ion impurities that may range as high as 1000 ppm (e.g.,

Hetherington, 1967; Rawson, 1967). It will be seen below that the natures of the radiation-induced defects in these materials are strongly influenced by both the OH and Cl contents and to a lesser degree by some occasional organic or nitrogen impurities. However, as opposed to the case of fused natural quartz, where Al impurities are frequently the dominant electron and hole traps (e.g., Schnadt and Räuber, 1971; Brower, 1979), the modern synthetic silicas that are the subject of this section generally exhibit radiation-induced ESR spectra characteristic mainly of oxygen- or silicon-centered defects.

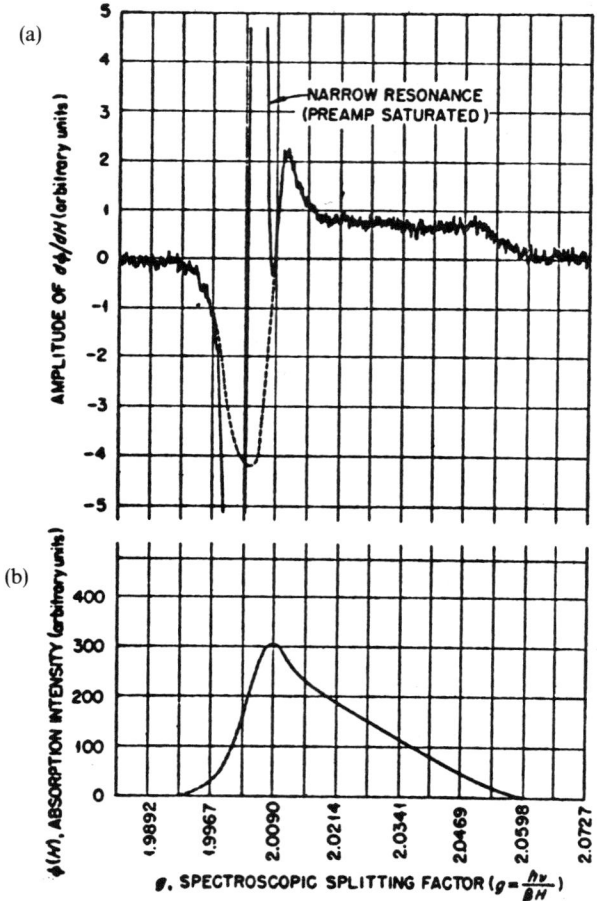

FIG. 24.   ESR spectra of radiation-induced defect centers in fused silica. In the experimental first-derivative spectrum (a), a narrow resonance due to E′ centers is driven off scale. A broader resonance underlying the narrow one has been approximately reconstructed by the dashed curve. The absorption curve of the broader resonance due to oxygen-associated hole centers (b) was obtained by numerical integration. (From Weeks, 1956).

ESR studies of radiation-induced defects in vitreous silica date from the work of Weeks (1956), who reported two distinct resonance lines in neutron-irradiated quartz and fused silica (Fig. 24). A condensed review of the intervening 30 years of progress has been given by Griscom (1986), while some problems and prospects for the coming decades are taken up in Griscom, 1985b).

1. The E' Center

Weeks and Nelson (1960a) adopted the nomenclature "E' center" to designate the defect center responsible for the narrow resonance of Fig. 24a centered near $g = 2.001$. This defect was immediately recognized to be intrinsic to the SiO_2 network, since the large number of centers induced by the heavy neutron irradiation easily outnumbered the known impurity contents of the samples. Conclusive identification of the E' center as an unpaired spin in a dangling sp^3 orbital of a silicon atom awaited the detection and analysis of the ^{29}Si hyperfine structure by Silsbee (1961). (^{29}Si is 4.7% abundant in nature and has a nuclear spin of $I = 1/2$.) Silbee's work was performed in crystalline α quartz—which had the advantage of permitting a determination of the crystallographic orientation of the dangling orbital, which turned out to be parallel to the direction of a normal Si-O bond. (See Weil, 1984, for a review of defects in quartz.)

Weeks and Nelson (1960b) employed powder-pattern analysis of the $g \simeq 2.001$ line to demonstrate a general congruence of the g matrices of the E' centers in quartz and glassy silica. However, it was the observation of the ^{29}Si hyperfine structure of E' centers in the glasses by Griscom et al. (1974) that ultimately confirmed that these defects are essentially identical with their counterparts in crystalline quartz—except for the existence in the glass of a statistical distribution in ^{29}Si hyperfine coupling constants related to vitreous-state disorder (Fig. 25). (An alternative interpretation of the 42-mT hyperfine doublet of Fig. 25a as arising from a separate defect associated with a proton (Shendrick and Yudin, 1978) was disproved by Griscom (1979).)

A crucial factor in achieving the successful computer lineshape simulation of Fig. 25a (dotted curve) was the recognition of the importance of including the second-order hyperfine contributions to the resonance condition of Eq. (25a). As elaborated by Griscom et al. (1974), the high-field hyperfine peak in Fig. 25a is sharper than the low-field peak solely in consequence of the interplay of the first- and second-order hyperfine terms with the statistical distribution of coupling constants represented in Fig. 25c.

Given the generic E'-center model illustrated in the inset to Fig. 25c, the measured ^{29}Si coupling constants can be related to the bond angle ρ at the defect site through the equation (Reinberg, 1964; Griscom et al. 1974)

$$\tan \rho = -\left[2 \left(1 + \frac{A_{aniso}}{B} \cdot \frac{A_0}{A_{iso}} \right) \right]^{1/2}, \qquad (54)$$

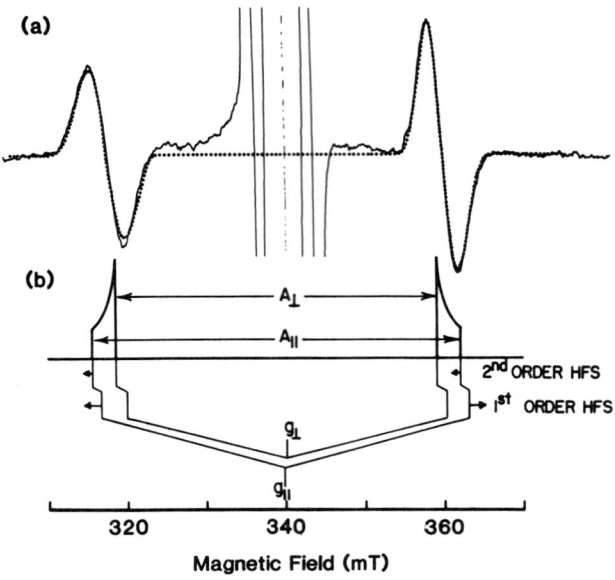

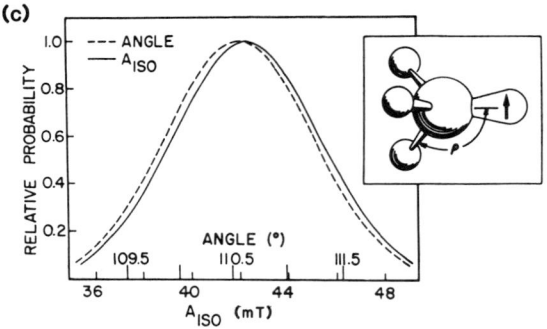

FIG. 25. ^{29}Si hyperfine structure of E' centers in γ-irradiated Corning 7943 fused silica. (a) Experimental spectrum with superimposed computer simulation (dotted curve). (b) single-site powder patterns for ^{29}Si doublet. The simulation of (a) involved a Gaussian distribution in $A_{iso} \equiv (A_{\parallel} + 2A_{\perp})/3$ related to a distribution in the bond angle ρ as illustrated in (c). (From Griscom et al., 1974).

where the relevant quantities are defined in Eqs. (17), (21), and (27). In turn, the probability distribution in A_{iso}, P(A_{iso}), can be converted to a distribution in ρ according to (see Eq. (42))

$$P(\rho) = P(A_{iso}) \left| \frac{\partial \rho}{\partial A_{iso}} \right|^{-1}. \tag{55}$$

Thus, by means of Eq. (55), the distribution in A_{iso} values used in the simulation of Fig. 25a transforms to the slightly skewed distribution in ρ indicated by the dashed curve in Fig. 25c. Devine and Arndt (1987) have used the ^{29}Si hyperfine structure of the E′ center and to determine the bond angle ρ as a function of density in samples densified under high pressure. It is cautioned, however, that Eq. (54) assumes 100% covalency, whereas SiO_2 is known to be $\sim 50\%$ ionic. Indeed, the most recent molecular-orbital calculations (Edwards and Fowler, 1989) suggest that the actual width of the bond-angle distribution may be considerably greater than predicted by Eq. (54) and illustrated in Fig. 25c.

Several distinct species of E′ centers have been delineated in crystalline quartz. (See, e.g., Weil, 1984.) It has been long believed that all of these comprise oxygen vacancies and it is known that several involve associated impurity protons. But the simplest conceivable oxygen-vacancy model appeared to be contradicted by the experimental hyperfine data that showed the unpaired electron to be localized on a single silicon, rather than the expected two. This paradox was resolved with the advant of the asymmetric relaxation model for the E'_1 center (Fig. 26a), as proposed by Feigl et al. (1974) and substantiated by theoretical calculations (Yip and Fowler, 1975; Edwards and Fowler, 1989). An equally promising model for the E'_2 variant (Weeks, 1963) has been advanced by Rudra et al. (1985) (Fig. 26b). Weeks and Sonder (1965) succeeded in correlating the E'_1 center with an optical band centered at 5.85 eV.

Evidence for the existence of more than one E′ center variant in *glassy* SiO_2 was provided by Griscom (1984b), who isolated three distinct g-tensor

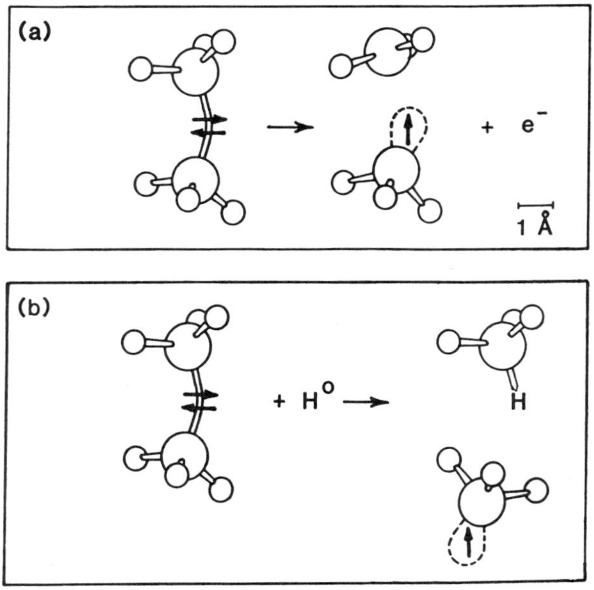

FIG. 26.   Proposed models for E′ centers in α quartz (silica glass). (a) The E'_1 (E'_γ) center. (After Feigl et al., 1974). (b) The E'_2 (E'_β) center. (After Rudra et al., 1985).

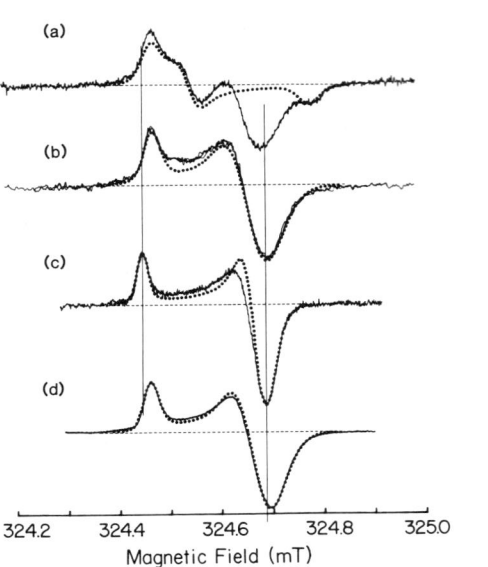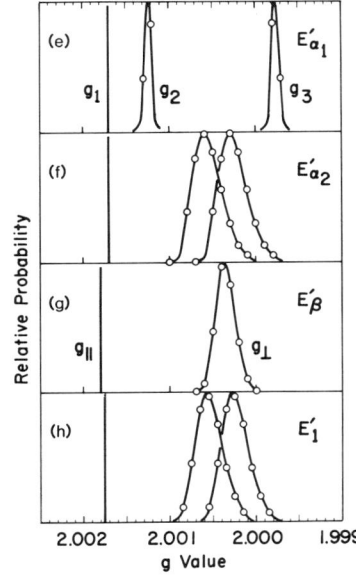

FIG. 27.   E'-center "central-line" spectra and g-value distributions for irradiated high-purity fused silica. Experimental spectra were recorded at 100 K following x-irradiation at 77 K (a) in the dark without warming, (b) exposed to room light without warming, and (c) after 5 min at 210 K. Also, (d) following high-dose γ-irradiation at room temperature. Dotted curves are computer simulations based on the **g**-value distributions of (e), (f), (g), and (h), respectively. (From Griscom, 1984b).

variations that occurred in different relative proportions depending on the oxygen stoichiometry, the OH content of the glass, and the conditions of irradiation. The species denoted E'_α, E'_β, and E'_γ (Fig. 27 and Table II) generally displayed very different thermal bleaching temperatures, but each of these variants was shown to exhibit indistinguishable ^{29}Si 42-mT hyperfine structure, implying that in each case the unpaired spin occupies a single

TABLE II

AVERAGE SPIN HAMILTONIAN PARAMETERS FOR Si E' CENTERS IN IRRADIATED HIGH-PURITY FUSED SILICA

Center	g_1	g_2	g_3	$A_{iso}(^{29}Si)$ (mT)
E'_α	2.0018	2.0013	1.9998	42
E'_β	2.0018	2.0004	2.0004	42
E'_γ	2.0018	2.0006	2.0003	42
E'_δ	2.0018	2.0021	2.0021	10

dangling sp^3 silicon orbital. E'_γ is believed to be essentially identical with the E'_1 center in α quartz (c.f., Fig. 26a). It has been suggested by Griscom (1985b, 1986) that E'_β may be very similar to the E'_2 center, since the anneal kinetics are consistent with the creation process illustrated in Fig. 26b. As yet there has been no substantiation of any of the models proposed for E'_α.

Yet a fourth species of E' center, E'_δ, was reported by Griscom and Friebele (1986) in plasma deposited, low-OH, high-Cl silicas. (See Section IV.A.3.) Available evidence suggests that the ^{29}Si hyperfine splitting associated with E'_δ is about four times smaller than normal (Table II), implying that this latest E' species may comprise an unpaired spin delocalized over several silicon atoms (Griscom and Friebele, 1986).

2. Oxygen-Associated Hole Centers

The broad spectrum of Fig. 24b, characterized by an absorption peak near $g \simeq 2.009$, was eventually recognized as a generic signature of oxygen-associated hole centers (OHCs). But when room-temperature-irradiated samples were investigated at low temperatures (~ 100 K) and their ESR spectra computer simulated, it was demonstrated that the OHC envelope consists of two superimposed components with relative intensities that depend strongly on the OH content of the sample (Griscom, 1978a, Stapelbroek et al., 1979). As will be described shortly, the two basic components illustrated in Fig. 28 were eventually identified as the nonbridging-oxygen hole center, which tends to dominate in high-OH silicas, and the peroxy radical, which is relatively more prevalent in low-OH silicas (or silicas of any water content subjected to sufficiently high-dose irradiations).

Additional OHC-type spectral components are present when either high- or low-OH silicas are irradiated below ~ 100 K and measurements are performed without warming. Two of these components, each of which thermally decays below ~ 200 K, have been computer simulated by Griscom (1989), who attributed them to two varieties of self-trapped holes. Spin Hamiltonian parameters for these "STHs" are listed in Table III.

The peroxy radical (also termed the superoxide radical) can often be recognized by matching the experimentally measured g values with the theoretical predictions of Känzig and Cohen (1959). However, unambiguous identification cannot always be achieved in this way, as the peroxy radical shares very similar g values with the nonbridging-oxygen hole center (Table III). Thus, the final proof of the occurrence of the peroxy radical in fused silica hinged on the observation and analysis of its ^{17}O hyperfine structure by Friebele et al. (1979) in a neutron-irradiation sample enriched to 34% in the isotope ^{17}O. (^{17}O has a natural abundance of only 0.037% and a nuclear spin of $I = 5/2$.) Note that the computer simulation of Fig. 29 demonstrates the

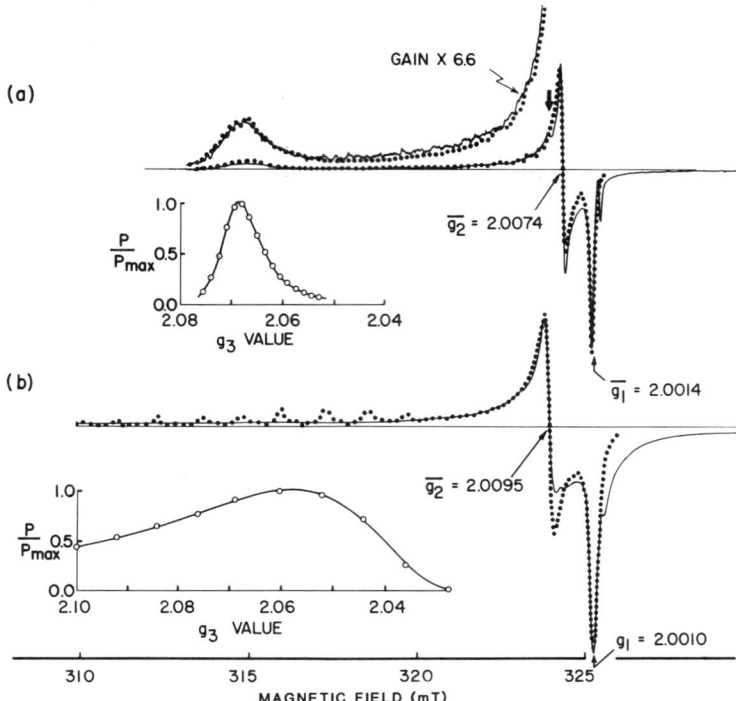

FIG. 28. ESR spectra of oxygen-associated hole centers recorded at 100 K for γ-irradiation high-purity fused silicas. (a) The peroxy radical in a Type-IV (low-OH) silica annealed for ten min at 773 K following a 10^8 rad dose. (b) The nonbridging-oxygen hole center in a Type-III (high-OH) silica irradiated at room temperature. Dotted curves are computer simulations based on the g-value distributions shown in the insets. (From Stapelbroek et al., 1979).

TABLE III

AVERAGE SPIN HAMILTONIAN PARAMETERS FOR OXYGEN-ASSOCIATED HOLE CENTERS IN IRRADIATED HIGH-PURITY SILICAS

Defect	g_1	g_2	g_3	$A_1(^{29}Si)$ (mT)	$A_2(^{29}Si)$ (mT)	$A_1(^{17}O)$ (mT)	$A_{2,3}(^{17}O)$ (mT)
Peroxy[a]	2.0018	2.0078	2.067	0.36	0.42	4.3[c]	1.0[c]
						10.2[c]	0.9[c]
Peroxy[b]	2.0020	2.0085	2.027				
NBOHC[a]	1.9999	2.0095	2.078	1.44	1.44	11.0	1.6
STH$_1$[b]	2.0026	2.0093	2.049				
STH$_2$[b]	2.0054	2.0073	2.013				

[a] γ irradiation at room temperature.
[b] X irradiation at 77 K.
[c] Hyperfine interaction with two *inequivalent* oxygens.

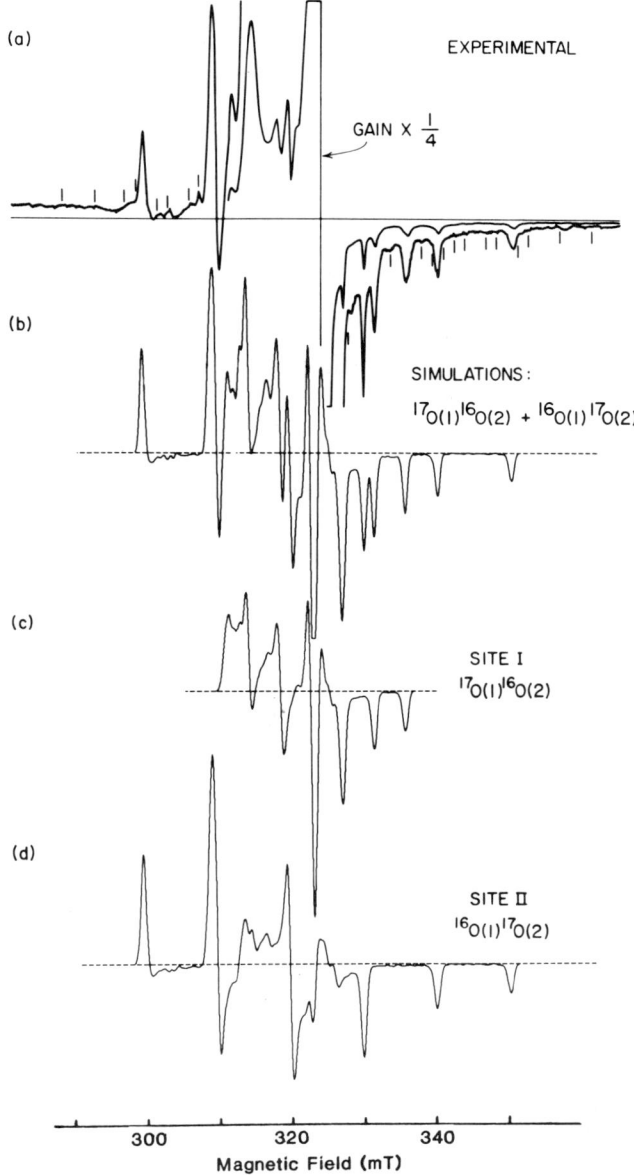

FIG. 29.   (a) ESR spectrum recorded at 110 K for a 36% ^{17}O-enriched silica glass following neutron irradiation and annealing to 773 K. (b) Computer simulation of the highly structured parts of (a) assuming a peroxy-radical model with one ^{17}O and one ^{16}O per defect. The two oxygen sites are seen to be chemically inequivalent, since the best-fit simulation (b) is comprised of an equally weighted sum of component spectra (c) and (d). Tic marks in (a) are the calculated line positions for the case of two ^{17}O's per defect. (From Friebele et al., 1979).

two oxygens that comprise the radical to be chemically inequivalent, i.e., characterized by different ^{17}O hyperfine coupling constants (Table III). The inference that the O_2^- ion should be asymmetrically bonded to a single silicon in the glass network (viz., $\equiv Si-O-O\cdot$) was confirmed by studies of ^{29}Si-enriched samples (Griscom and Friebele, 1981).

The nonbridging-oxygen hole center (NBOHC) may conceivably be created by simple fission of an Si-O bond. However, the predominant NBOHC signals that have been studied to date are believed to result from the radiolysis of OH groups:

$$\equiv Si-OH \rightarrow \equiv Si-O\cdot + H^0. \tag{56}$$

Indeed, if the experiment is performed at temperatures $\leqslant 100\,K$, both the NBOHC and the atomic hydrogen (to be discussed shortly) are observable by ESR (e.g., Griscom, 1984c). Staplebroek et al. (1979) characterized the ^{17}O hyperfine structure of the NBOHC in an isotopically enriched fused silica, finding the principal coupling constants to be in accord with prediction for an R-O· type of radical. However, the analysis was hampered by the presence of an overlapping spectrum due to an aluminum-associated impurity center. Griscom et al. (1987) have obtained an interference-free spectrum for an ^{17}O-enriched sol-gel silica (shown in Fig. 30 together with its computer simulation). The derived spin Hamiltonian parameters are listed in Table III. The NBOHC has been correlated with an optical band centered near 2.0 eV (Nagasawa et al., 1986), although Friebele et al. (1985a) caution that certain 2-eV optical absorptions in silica may have a separate origin. Skuja et al. (1984) discuss additional considerations favoring assignment of a 2.0 eV photoluminescence to NBOHCs produced according to Eq. (56).

3. Impurity-Associated Defects

Atomic hydrogen is by far the most common impurity-related defect encountered in amorphous SiO_2. It is detected as a pair of sharp lines centered on $g = 2.00$ with a splitting of $\sim 50.5\,mT$ due to the hyperfine interaction with ^1H (99.98% abundant in nature, $I = 1/2$.) Silica samples that have been exchanged with deuterium display a corresponding hyperfine tripet with a 15.5-mT overall splitting. (^2H has a nuclear spin of $I = 1$, while $\mu_N(^2H)/\mu_N(^1H) = 0.307$.) H^0 is relatively stable when created by ionizing radiation at temperatures below $\sim 100\,K$, and it has been reported to outnumber even the intrinsic defects induced in Type-III silicas under these conditions (Griscom et al., 1983a; Griscom, 1984c). The production efficiency of H^0 is still higher in crystalline quartz, evidently due to the greater mobility of excitons in a regular defect-free medium (Azuma et al., 1986). Most radiolytic atomic hydrogen anneals out at temperatures $> 130\,K$, mainly as a consequence of dimerization to form H_2 (Griscom, 1984c). Weeks and

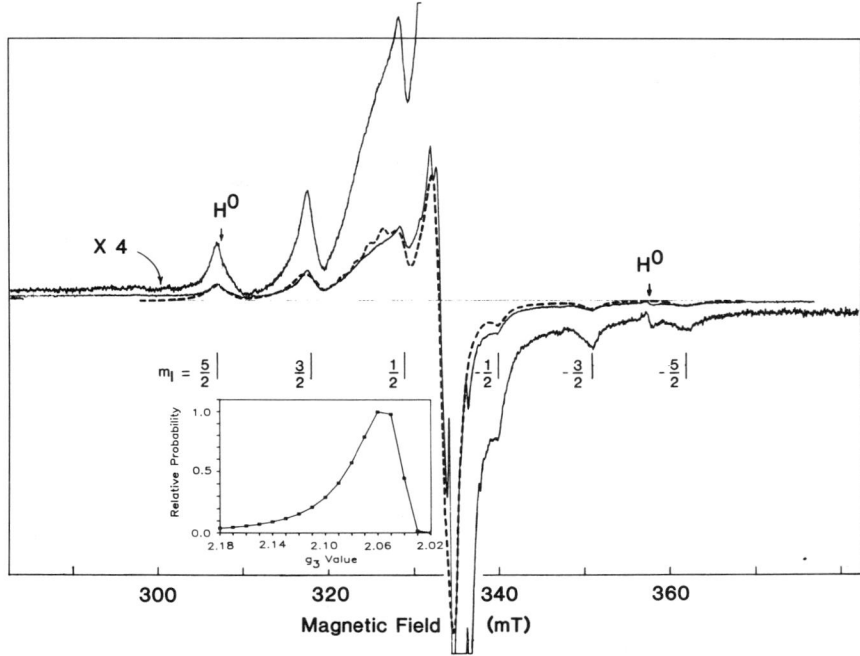

FIG. 30. ESR spectrum recorded at 105 K for the nonbridging-oxygen hole center in an ^{17}O-enriched sol-gel silica x-irradiated at 77 K. Dashed curve is a computer simulation based on the illustrated distribution of g_3 values and the other spin Hamiltonian parameters listed in Table III and assuming 40% isotopic enrichment. (From Griscom et al., 1987).

Abraham (1965) carried out a careful ESR spectroscopic study of H^0 in irradiated α quartz and commented on the slightly different properties of the corresponding center in glassy silica.

Organic radicals have been detected in some (but not all) Type-III silicas following irradiation. The most numerically important of these is the formyl radical, which turns out to be unstable above ~ 300 K, unproducible below ~ 100 K, and unobservable above ~ 250 K (Griscom et al., 1983a). Figure 31 illustrates spectra obtained for both deuterated and normal samples following x-irradiation at 77 K, warming briefly to 300 K, and recooling to 100 K for observation. The computer simulations displayed in Fig. 31 take into account certain motional averaging effects that can be frozen out at lower temperatures (Griscom et al., 1983a). It has been concluded that the formyl radical results from the reaction of radiolytic hydrogen atoms with a carbon monoxide impurity dissolved in the glass in consequence of the use of hydrocarbons in the flame hydrolysis process of manufacture. Postirradiation warming above 100 K is required to mobilize the H^0 in order to initiate

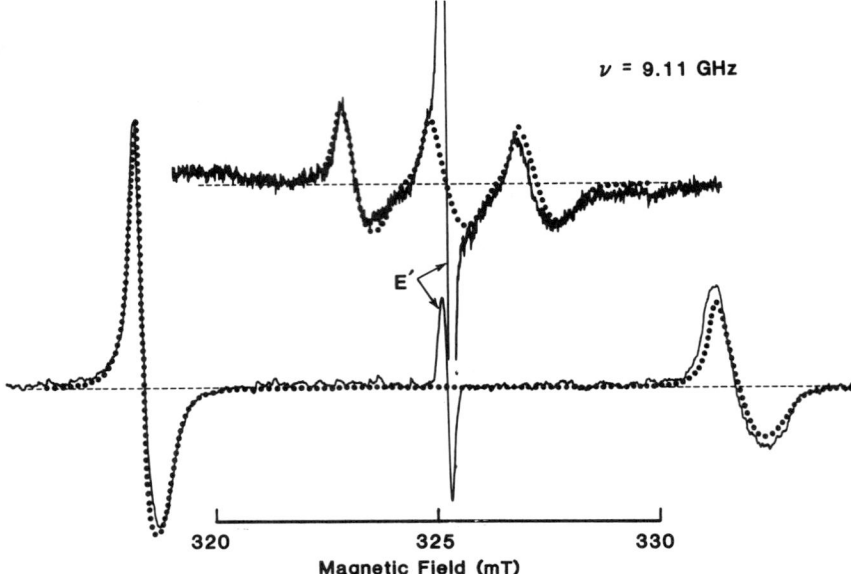

$\nu = 9.11$ GHz

E′

Magnetic Field (mT)

320   325   330

FIG. 31. ESR spectra at 100 K of the formyl radical in x-irradiated high-purity Type-III silicas manufactured by hydrolysis of $SiCl_4$ in a hydrocarbon-oxygen flame. The upper spectrum pertains to a deuterated sample. Dotted curves are computer simulations. (From Griscom et al., 1983a).

the reaction chain. In fact, the formyl radical served serendipitously as a "spin label" demonstrating the motion of radiolytic *molecular* hydrogen above ~ 200 K (Griscom et al., 1983a; Griscom, 1984c). If the same carbon monoxide-containing silicas are γ-irradiated, the resultant formyl radical spectrum diminishes when the accumulated dose exceeds ~ 1 Mrad and is supplanted by a weaker, but much more stable $1:3:3:1$ hyperfine quartet due to the methyl radical, $CH_3\cdot$ (Friebele et al., 1983). Both $CH_3\cdot$ and $CH_2\cdot CH\cdot OH$ radicals have been observed in irradiated sol-gel derived silicas (Wolf et al., 1985; Griscom et al., 1987).

Atomic chlorine has been reported as a radiation-induced defect in certain low-OH plasma-deposited fused silicas (Griscom and Friebele, 1986). This Cl^0 species is thermally bleached above ~ 200 K. The experimental ESR spectrum recorded at 105 K is illustrated in Fig. 32a, where comparison is made with a computer lineshape simulation (Fig. 32b). As indicated by the "stick diagram" in Fig. 32c, the $A_{\parallel}$ hyperfine signatures of ^{35}Cl and ^{37}Cl are clearly manifested as nested pairs of quartets. (^{35}Cl and ^{37}Cl are respectively 74.2 and 24.6% abundant in nature and both have $I = 3/2$; $\mu_N(^{35}Cl)/\mu_N(^{37}Cl) = 1.2$.)

Other defects are observed concomitantly with the atomic chlorine. These include the E'_δ center (which is deconvolved from the experimental spectrum

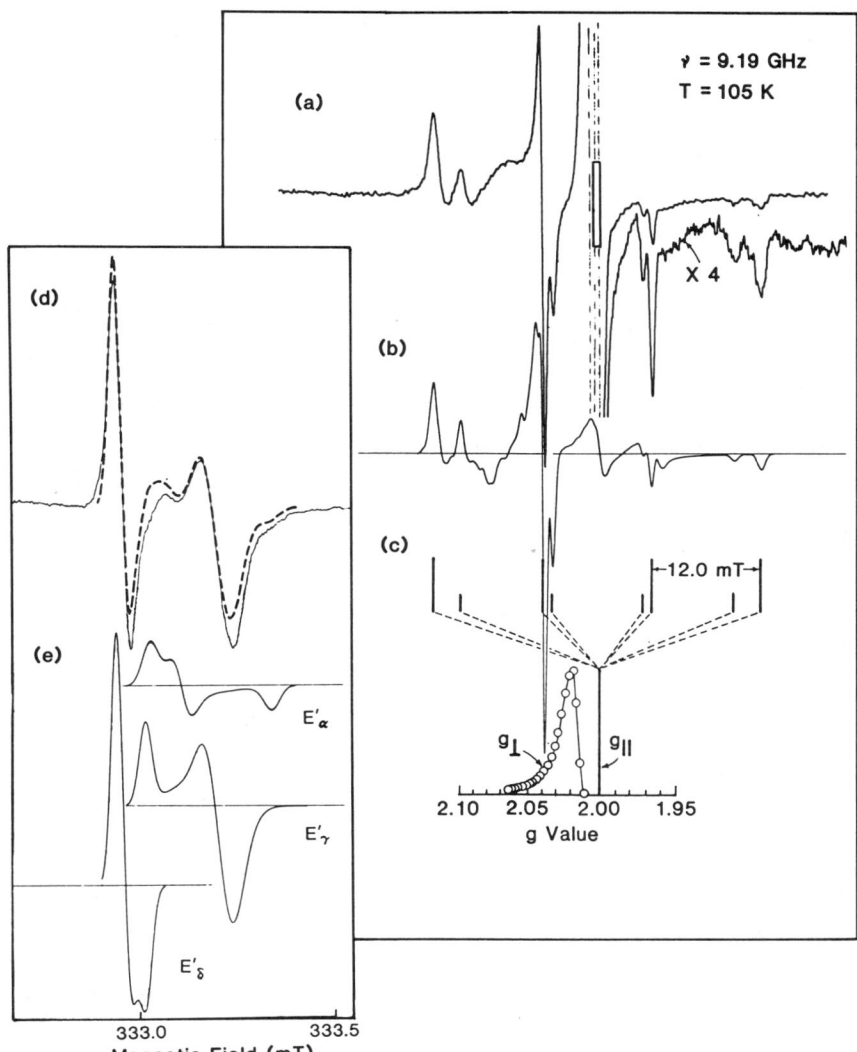

FIG. 32.   ESR spectra of a Cl-containing silica glass x-irradiated at 77 K and measured at 105 K. (a) Spectrum of Cl⁰. (b) Computer simulation. (c) Schematic hyperfine splittings and g values. (d) spectrum of E′ centers from the central region of (a) and its computer simulation (dashed curve). (e) Component spectra for the simulation of (d). (From Griscom and Friebele, 1986).

of Fig. 32d by means of the computer-simulation analysis of Fig. 32e) and a biradical ($S = 1$ state). The $M_S = -1 \leftrightarrow 0$ and $0 \leftrightarrow +1$ transitions of the biradical were computer-simulated (long-dashed curves in Fig. 33) using the resonance condition of Eq. (29). The resulting estimate of the zero-field splitting parameter D (Table IV) was in turn shown to be consistent with

TABLE IV

SPIN HAMILTONIAN PARAMETERS FOR AN $S = 1$ DEFECT IN
IRRADIATED FUSED SILICA CONTAINING CHLORIDE IMPURITIES

g	$|D|/g\beta$ (mT)	$|E|/g\beta$ (mT)
2.002	13.4	0

expectation for a model comprising a pair of E' centers occupying an effective [O–Si–O] trivacancy. The quasi-forbidden $M_S = -1 \leftrightarrow +1$ transition was also observed (Fig. 33c). Griscom and Friebele (1986) concluded that both E'_δ and the biradical are defects formed at specific chloride-decorated precursor sites originally present in the unirradiated materials.

A nitrogen-associated defect center characterized by a ^{14}N hyperfine coupling constant of $A_{iso} = \sim 1.8$ mT was reported by Stathis and Kastner (1984) as a UV-light–induced defect in certain Type-III silicas. (^{14}N is 100% abundant in nature, with $I = 1$.) Tsai et al. (1988a) have succeeded in producing the same center by x-, γ-, and proton-irradiation (Fig. 34) and have modeled the defect as an electron trapped at the site of an N^{3-} ion substitutional for an $[SiO_4]^{4-}$ vacancy. A different nitrogen-related center, the NO_2 molecule, has been detected by ESR only in sputtered SiO_2 films; this trapped molecule has been extensively characterized by Griscom (1978a), Friebele et al. (1985b), and Schwartz et al. (1986).

Other defects with proton hyperfine splittings are abounding. A doublet with a splitting of 11.9 mT is detected only in fused natural quartzes and is evidently due to a proton associated with a germanium impurity (Vitko, 1978; Hibino and Hanafusa, 1984). Doublets with splittings of 7.4 and 1.2 mT commonly observed in high-purity Type-III silicas were also shown by deuterium-substitution experiments to be proton-related. Tsai and Griscom (1987) have detected ^{29}Si hyperfine structure attributable to the 7.4-mT doublet, thus indicating a possible E'-like origin. Triplett et al. (1987) have reported production of the 7.4-mT doublet by soft x-rays in a thermally grown SiO_2 film.

B. DOPED SILICAS

The doping of silica with other network formers has proved difficult when attempted by batch melting techniques in consequence of the common tendency of the dopant to volatilize. However, innovative reactive deposition methods (e.g., Scherer and Shultz, 1983) have opened up the possibility of doping silica with almost any element in the periodic table. The three principal dopants of modern fiber-optic technology are B_2O_3, P_2O_5, and

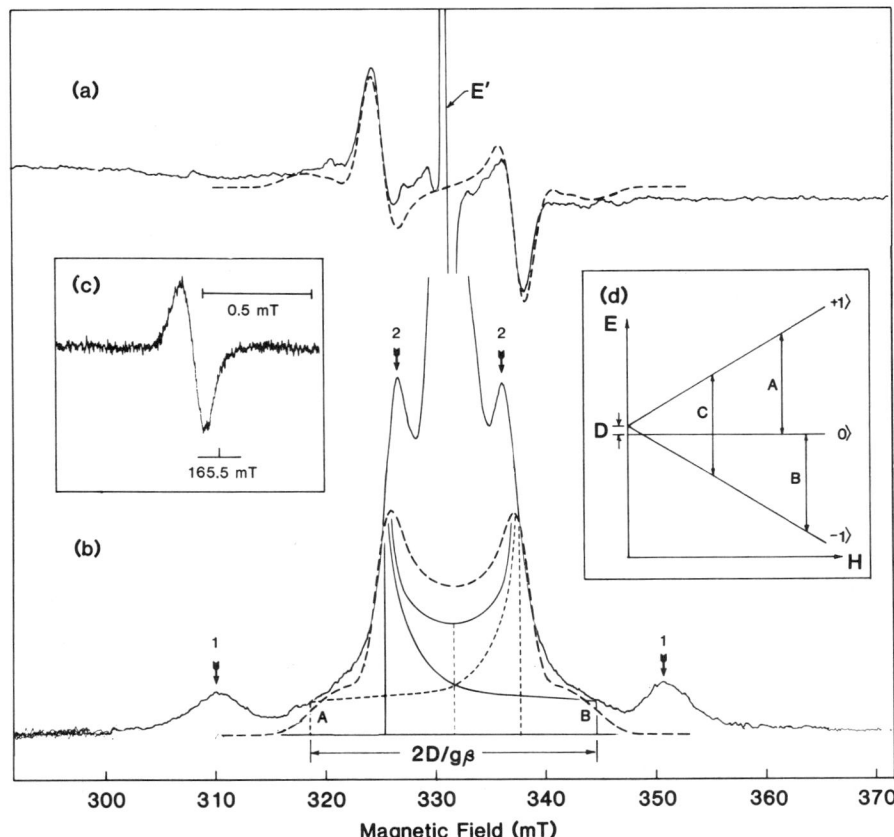

FIG. 33. ESR spectra of a Cl-containing silica glass x-irradiated at 77 K and warmed to room temperature before measurement at 225 K. (a) First-derivative spectrum. (b) High-power second-harmonic mode spectrum and powder pattern for an $S = 1$ state. (c) Very-high-gain first-derivative spectrum at half field. (d) Energy-level scheme for a biradical with zero-field splitting energy D. Dashed curves are computer simulations of the biradical spectrum. Peaks designated 1 and 2 are ^{29}Si hyperfine structure of E'_γ and E'_δ centers, respectively. (From Griscom and Friebele, 1986).

GeO_2. As will be seen shortly, each of these, when doped into SiO_2 to a level of $\sim 1\%$ or more, gives rise to a characteristic manifold of deep electron and hole traps that totally dominate the radiation-damage processes in both the bulk glass and the fibers drawn therefrom.

1. Boron-Doped Silica

Griscom et al. (1976) carried out an ESR investigation of the effects of γ-irradiation on a pure-silica–core borosilicate-clad optical fiber and bulk melt

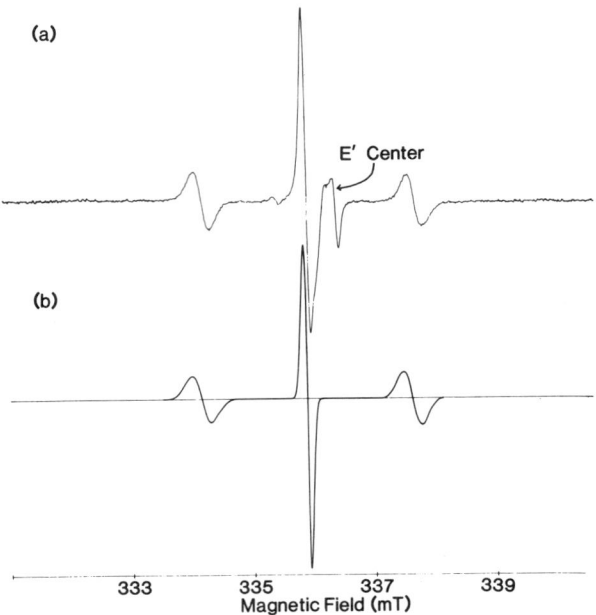

FIG. 34. ESR spectra of a nitrogen-associated center in an x-irradiated fused silica. (a) Experimental spectrum recorded at 310 K. (b) Computer simulation. (From Tsai et al., 1988a).

glasses of the cladding composition $(25B_2O_3\text{-}75SiO_2)$. The observed defect centers were the boron-oxygen hole center (BOHC), the Si E′ center, and a boron analog of the latter, the boron E′ center. The BOHC spectra observed in glasses containing ^{10}B and ^{11}B in their natural abundances (19 and 81%, respectively) and enriched to 95% ^{10}B are illustrated in Fig. 35a and b, respectively. The dotted computer simulations of Fig. 35 include the respective isotopic abundances as constraints, as well as the facts that $I(^{10}B) = 3$, $I(^{11}B) = 3/2$, and $g_N(^{10}B)/g_N(^{11}B) = 0.335$. The derived spin Hamiltonian parameters are listed in Table V. The distributions of g values employed in both simulations is shown in Fig. 35c. The mean values of g_2 and g_3 were used via Eqs. (9) through (11) to calculate the energy spacings between the oxygen $2p_z$ ground state and the near-lying excited states of the model adopted for this defect (Fig. 36). As illustrated in Fig. 36a, the BOHC is believed to comprise a hole trapped on an oxygen bridging between a silicon and a 4-coordinated boron.

Figure 37 depicts the boron E′ center spectra obtained for the same γ-irradiated borosilicate glasses by employing broader field scans and higher spectrometer gains than those pertinent to Fig. 35. Spin Hamiltonian parameters derived for this defect are provided in Table VI. By using values of

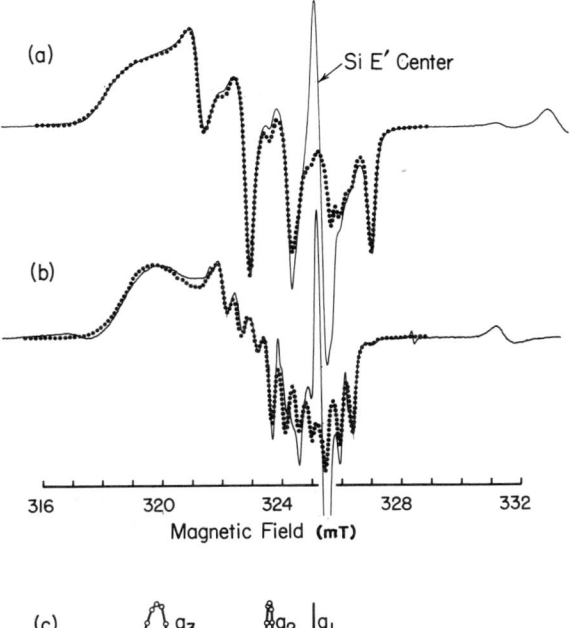

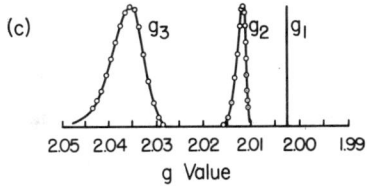

FIG. 35. ESR spectra of irradiated glasses of composition $B_2O_3\text{-}3SiO_2$ with (a) natural boron isotopic abundance (81% ^{11}B, 19% ^{10}B) and (b) 95% enrichment in ^{10}B. Dotted curves are computer simulations of the boron-oxygen hole center component based on the g-value distributions shown in (c). (From Griscom et al., 1976).

TABLE V

AVERAGE SPIN HAMILTONIAN PARAMETERS FOR THE BORON-OXYGEN HOLE CENTER IN IRRADIATED BORON-DOPED SILICA

g_1	g_2	g_3	A_1 (mT)	A_2 (mT)	A_3 (mT)
2.0025	2.0115	2.0355	1.36	1.53	0.87

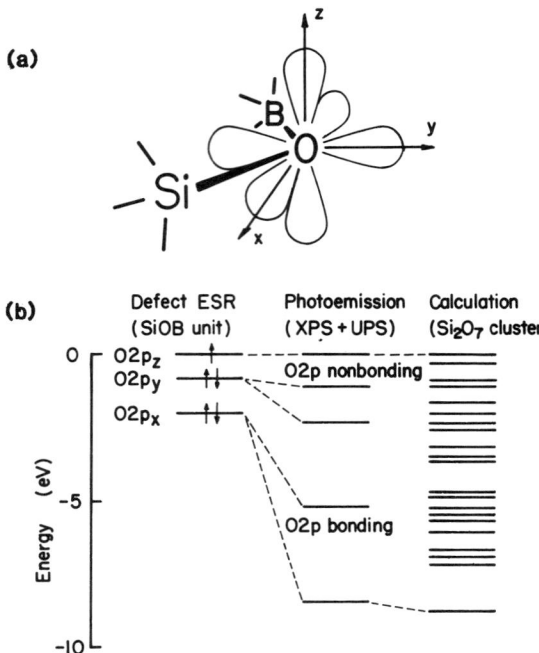

FIG. 36. Model for the boron-oxygen hole center. (a) Steric picture. (b) Energy-level scheme referenced to the top of the valence band. Defect energy levels were calculated from the experimental g shifts via Eq. (10); experimental photoemission data and cluster calculation results were taken from references cited in Griscom et al. (1976).

A_0 and B from Table I in Eqs. (27), it was possible to calculate the orbital composition of the boron E' center, viz., 26% boron 2s and 46% boron 2p (Griscom et al., 1976).

Dose for dose, it was found that the radiation yield of defects in the borosilicate glass was about two orders of magnitude greater than in pure silica. In fact, even the number of *silicon* E' centers induced in the borosilicate glass was over 10 times higher than measured for a Type-IV silica control sample. In possible explanation of this outcome, Griscom et al. (1976) suggested that the tendency of boron to form tetrahedral bonds may result in the presence of a few (negatively charged) 4-coordinated borons in the borosilicate glass network, which for reasons of charge neutrality and overall stoichiometry would require an equal number of (positively charged) 3-coordinated silicons. The 4-coordinated borons would then serve as precursors for the (charge-neutral) BOHC, while the 3-coordinated silicons could trap electrons to form charge-neutral E' center variants.

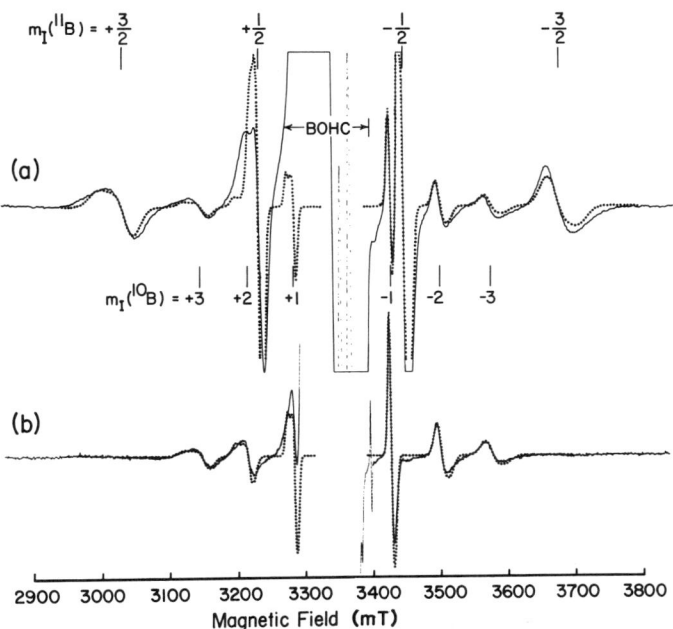

FIG. 37.   ESR spectra of the boron E′ center in irradiated B_2O_3-$3SiO_2$ glasses (a) for boron in its natural abundance and (b) a sample isotopically enriched to 95% ^{10}B. Dotted curves are computer simulations of the respective boron E′ center spectra based on the spin Hamiltonian parameters of Table VI. (From Griscom et al., 1976).

TABLE VI

SPIN HAMILTONIAN PARAMETERS FOR THE BORON ELECTRON CENTER IN IRRADIATED BORON-DOPED SILICA

$g_\parallel$	$g_\perp$	$A_\parallel(^{11}B)$ (mT)	$A(^{11}B)$ (mT)
2.0020	1.9996	23.4	20.7

2. Phosphorus-Doped Silica

Defect centers induced by ionizing radiation in high-purity phosphorus-doped silica glass have been investigated by Griscom et al. (1983b). Samples were of an approximate composition $10P_2O_5$-$90SiO_2$ prepared by a plasma-deposition–of–particle-feed method and subjected to either x-rays at 77 K or γ-rays at room temperature. Four generic-defect species were observed and

TABLE VII

AVERAGE SPIN HAMILTONIAN PARAMETERS FOR SOME PHOSPHORUS-
RELATED DEFECTS IN IRRADIATED PHOSPHORUS-DOPED SILICA

Defect	$g_\parallel$	$g_\perp$	$A_{iso}(^{31}P)$ (mT)	$A_{aniso}(^{31}P)$ (mT)
P_1	2.002	1.999	91	6
P_2	2.001	2.001	120	5
P_4	2.0014	1.9989	8.92	13.3

characterized by ESR on the basis of the observed g values and ^{31}P hyperfine splittings (^{31}P is 100% abundant, $I = 1/2$). These comprised centers analogous to PO_3^{2-} (phosphoryl), PO_4^{4-} (phosphoranyl), PO_2^{2-} (phosphinyl), and PO_4^{2-} radicals. For convenience, and with regard to historical precedence (see Section IV.C.3), these have been respectively designated as the P_1, P_2, P_4, and POHC (phosphorus-oxygen hole center).

Structural identifications were made on the basis of spin-Hamiltonian parameters (Tables VII and VIII) derived from computer lineshape analyses. The P_1 and P_2 centers were determined to be essentially identical with similar species originally designated by Weeks and Bray (1968) in P_2O_5 and alkali phosphate glasses (Section IV.C.3). On the basis of the derived orbital composition (23% P 3s, $\sim 50\%$ P 3p), the P_1 might be properly termed the phosphorus E' center in analogy with the Si and B species discussed already. Presumably it is formed by the trapping of a hole on a 3-coordinated P^{3+} precursor (upper path in Fig. 38c). Griscom et al. (1983b) noted the close agreement between the values of A_{iso} which they measured for P_2 in a phosphosilicate glass and those reported (Uchida et al., 1979) for a silicon-substituted P^{4+} defect in crystalline α quartz. It was therefore concluded that the P_2 defects in the glassy and crystalline media must be essentially identical in nature. The steric model for P_2 illustrated in Fig. 38b is consistent with the relationship of the defect's principal axes to the crystallography of α quartz as determined by Uchida et al. (1979).

Figure 39 displays the P_4 spectrum recorded following an anneal at 673 K to remove overlapping components. The spin Hamiltonian parameters

TABLE VIII

AVERAGE SPIN HAMILTONIAN PARAMETERS FOR THE PHOSPHORUS-OXYGEN HOLE CENTER IN
IRRADIATED PHOSPHORUS-DOPED SILICA

g_1	g_2	g_3	$A_1(^{31}P)$ (mT)	$A_2(^{31}P)$ (mT)	$A_3(^{31}P)$ (mT)
2.0075	2.0097	2.0179	4.8	5.2	5.4

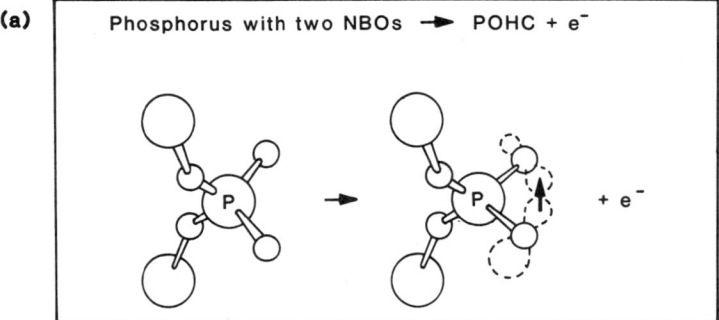

(a)   Phosphorus with two NBOs ➤ POHC + e⁻

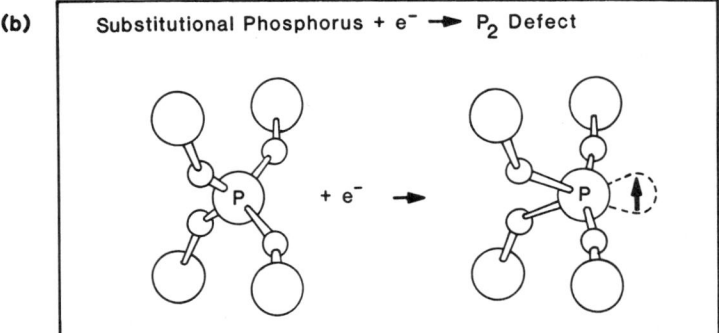

(b)   Substitutional Phosphorus + e⁻ ➤ P_2 Defect

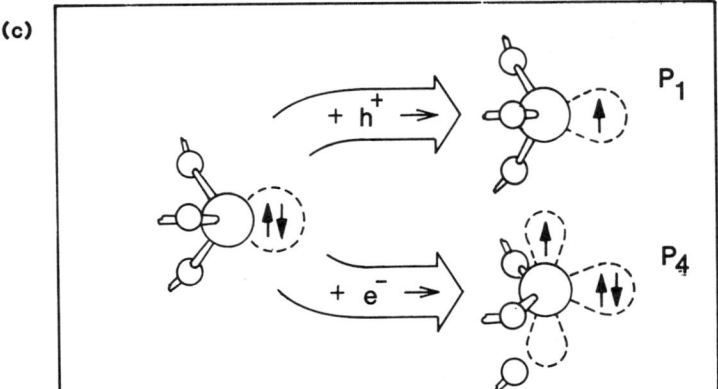

(c)

FIG. 38. Proposed models for some phosphorus-associated defects in irradiated phosphorus-doped silica. (a) The phosphorus-oxygen hole center (POHC). (b) The P_2 center. (c) The P_1 and P_4 centers. The precursor valence states of phosphorus are $5+$ for (a) and (b) and 3^+ for (c). (From Griscom et al., 1983b).

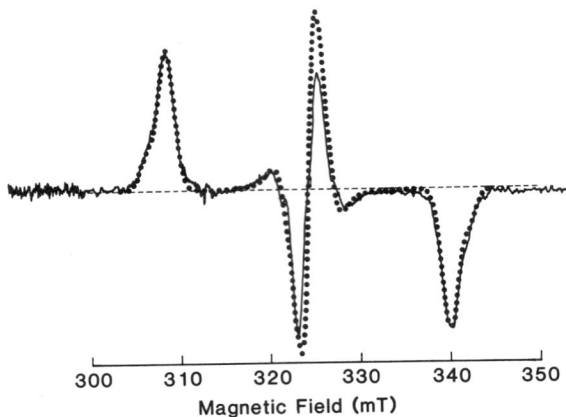

FIG. 39. ESR spectrum of the P_4 center in an irradiated $10\,P_2O_5$-$90\,SiO_2$ glass following heat treatment at 673 K. Dotted curve is a computer simulation based on the parameters of Table VII. (From Griscom et al., 1983b).

(Table VII) derived from the dotted computer simulation proved to comprise the ESR signature of the phosphinyl radical. The hyperfine data indicate the unpaired spin to be $\sim 100\%$ confined to a phosphorus 3p orbital. One possible mechanism for the formation of the P_4 defect in the P_2O_5-SiO_2 glass is represented by the lower path in Fig. 38c.

The POHC is a defect that seems to be ubiquitous to all phosphate glasses. (See also Section IV.C.3.) However, its ESR spectrum is best resolved in P-doped silica (Fig. 40). The derived hyperfine splittings (Table VIII), together with the g-value distributions of Fig. 40b, became the basis for a detailed structural model. This model (Fig. 38a) envisions the trapped hole to be delocalized over two nonbridging oxygens on the same phosphorus atom, in analogy with a model proposed for the so-called HC_2 defect in alkali silicate glasses (Section IV.C.1). Arrows in Fig. 40c denote features attributed to a less stable POHC variant wherein the hole is confined to a single nonbridging oxygen. The latter variant, denoted $POHC_2$, has been studied in more detail by Kawazoe et al. (1986a).

As was previously seen to be the case when boron is doped into silica, the introduction of phosphorus into amorphous SiO_2 is found to result in an aggregate yield of radiation-induced defects approximately two orders of magnitude greater than in a pure silica sample subjected to the same radiation dose. Each of these phosphorus-related defects have been correlated with one or more radiation-induced optical absorption bands (Griscom et al., 1983b).

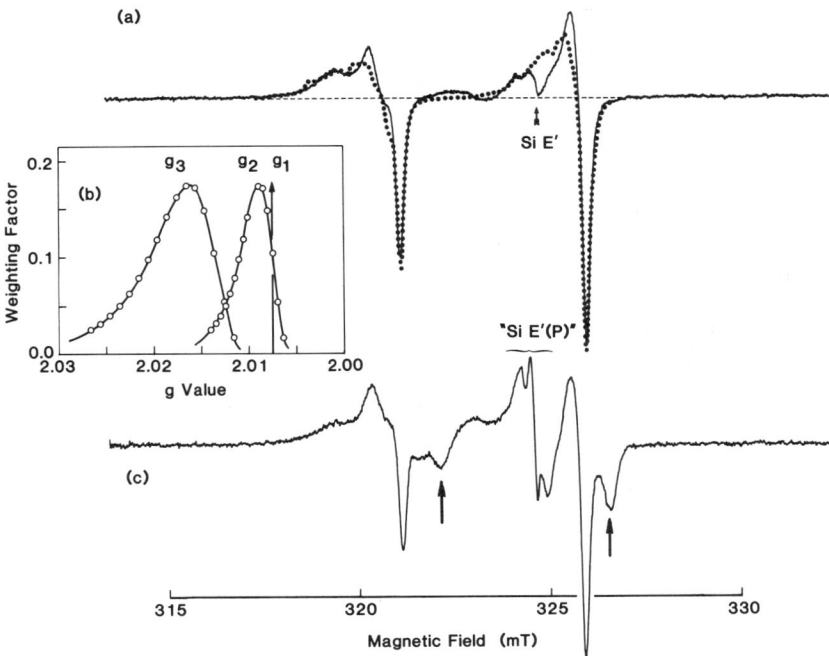

FIG. 40. ESR spectra recorded at 100 K for $10P_2O_5$-$90SiO_2$ glasses following (a) γ-irradiation at room temperature and (c) x-irradiation at 77 K. Dotted curve is a computer simulation of the POHC contribution to the spectrum of (a) using the distribution of g values illustrated in (b) and the hyperfine coupling constants of Table VIII. Additional peaks indicated by arrows are believed to be due to another POHC variant. (From Griscom et al., 1983b).

3. Germanium-Doped Silica and Pure Germania

From most available evidence, it would now appear that the germanium-associated defects in Ge-doped silica are essentially identical to those in pure glassy GeO_2. Thus, dovetailing of the two cases into a single subsection is a convenience that does not seem inappropriate.

Glassy germanium dioxide (together with polycrystalline hexagonal and tetragonal polymorphs of GeO_2) was subjected to electron, reactor, and γ-irradiations and investigated by ESR by Weeks and Purcell (1965), Purcell and Weeks (1969), and Garlick et al. (1971). Two principal resonances were observed in the glasses, one characterized by $g_{\parallel} = 2.0016$ and $g_{\perp} \simeq 1.996$ and a second characterized by $g_1 = 2.002$, $g_2 = 2.08$, and $g_3 = 2.051$. The former was identified as a hole trapped at an oxygen vacancy (what is now termed a Ge E' center) and the latter as a peroxy radical. These identifications have stood the test of time, with the ^{73}Ge hyperfine structure of the radiation-induced Ge E' center finally having been observed by Tsai et al. (1987a). (^{73}Ge is 7.6% abundant in nature and is characterized by a nuclear spin $I = 9/2$.) The ^{73}Ge hyperfine structure of a Ge E' center induced by mechanical grinding was reported by Bobyshev and Radtsig (1981). Although the ^{17}O hyperfine structure of the peroxy radical has not yet been observed, the identification of its spectrum in glassy germania has been greatly reinforced by a detailed study of the analogous center in single-crystal tetragonal GeO$_2$ by Kappers et al. (1978).

Defects in Ge-doped silica were first investigated by Friebele et al. (1974), who isolated several γ-ray–induced defect species—which they denoted in Ge(0), Ge(1), Ge(2), and Ge(3)—in both a bulk glass of a composition 10GeO$_2$-90SiO$_2$ and a Ge-doped–silica-core optical fiber. Of these centers, Ge(0) and Ge(3) displayed g matrices essentially identical to that of the Ge E' center in pure GeO$_2$ glass, and they were tentatively identified on this basis. The thinking (as of 1989) is that Ge(0) and Ge(3) are one and the same center, and it is therefore preferred that these notations be dropped in favor of the nomenclature Ge E'. The spectra of Ge(1) and Ge(2) were disentangled with the aid of annealing experiments, and their g-value distributions were estimated by means of the computer-simulation analyses illustrated in Fig. 41.

Friebele et al. (1974) initially supposed the Ge(1) and Ge(2) centers to be additional Ge E'-center variants that differed only in the number of next-nearest–neighbor germaniums. This view was ultimately disproved by Kawazoe (1985), who made the pivotal first observation of the ^{73}Ge hyperfine structure of these defects. The first-derivative–mode hyperfine spectrum and its computer simulation due to Kawazoe et al. (1986b) is given in Fig. 42a. Friebele and Griscom (1986) took advantage of the superior signal-to-noise ratio afforded by high-power second-harmonic mode to obtain the experimental spectrum of Fig. 42b. The success of the computer lineshape simulation of Fig. 42c was critically dependent on the inclusion of the second-order terms in the resonance condition of Eq. (25a) while simultaneously introducing a statistical distribution in hyperfine coupling constants. (See a discussion of this effect in Section IV.A.1.) As pointed out by Kawazoe et al. (1986b), the dependence of the apparent linewidth on M_I value (which is rigorously reproduced in the simulation of Fig. 42c) can be approximated by separately convoluting each hyperfine component with a Gaussian broadening function whose width is given by

$$\sigma_H^2 = aM_I^2 + bM_I + c, \tag{57}$$

where a, b, and c are suitably chosen constants.

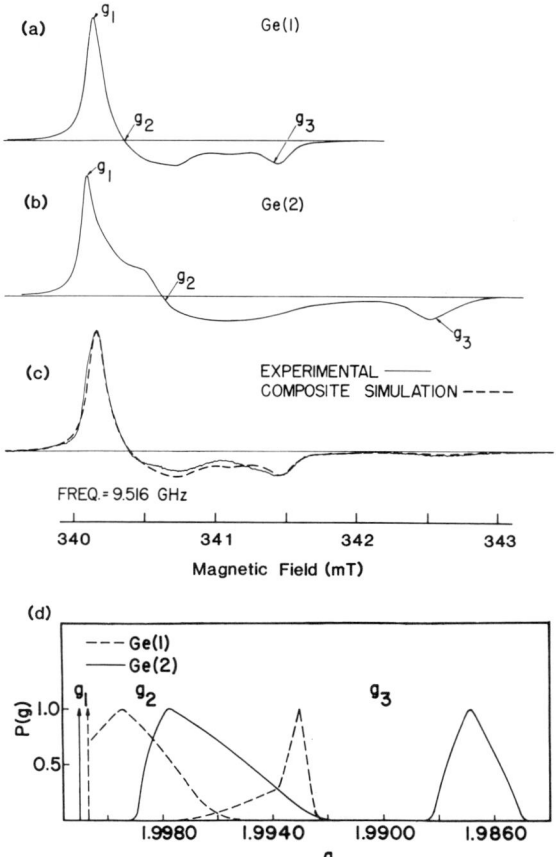

FIG. 41.   ESR spectra of Ge(1,2) centers in irradiated germanium-doped silica. (a) and (b) are computer simulations of the Ge(1) and Ge(2) components, respectively, based on the g-value distributions of (d). The experimental spectrum of (c) (unbroken curve) is computer simulated as a weighted sum of Ge(1) and Ge(2) component spectra (dashed curve). (From Friebele et al., 1974).

On the basis of the foregoing work, it appears that the mean values of A_{iso} for Ge(1) and Ge(2) are indistinguishably, both lying quite close to the value reported for the substitutional-germanium electron center studied in α quartz by Isoya et al. (1978). (See Table IX.) There is now general agreement that the generic structural model for this germanium electron trap is fully analogous to that for the P_2 defect of Fig. 38b. Itoh et al. (1987) and Tsai et al. (1987a) have shown that Ge(1) and Ge(2) are not particular to Ge-doped silica, but

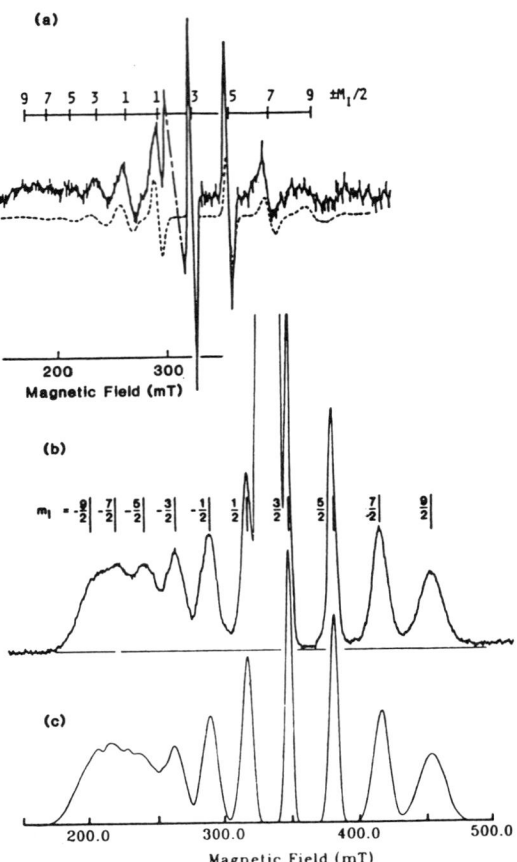

FIG. 42. ^{73}Ge hyperfine structure of Ge(1,2) centers in irradiated GeO$_2$-SiO$_2$ glasses. (a) First-derivative spectrum and its computer simulation (dashed curve.) (From Kawazoe et al., 1986b) (b) High-power second-harmonic mode spectrum. (From Friebele and Griscom, 1986) (c) Computer simulation of (b) assuming the high-power second-harmonic mode spectrum to be indistinguishable from the undifferentiated absorption curve. (From Friebele and Griscom, 1986).

TABLE IX

AVERAGE SPIN HAMILTONIAN PARAMETERS FOR GERMANIUM-ASSOCIATED
DEFECTS IN IRRADIATED GERMANIUM-DOPED SILICA

Defect	g_1	g_2	g_3	$A_{iso}(^{73}Ge)$ (mT)
Ge(1)	2.0007	1.9994	1.9930	28.0
Ge(2)	2.0010	1.9978	1.9868	28.0
Ge E′	2.0011	1.9951	1.9937	23.8

also occur in pure GeO_2 glass. Detailed structural models for Ge(1) and Ge(2) have been proposed by Tsai et al. (1988b).

Kawazoe et al. (1986b) detected the Ge E' center present in unirradiated Ge-doped silica fibers as a consequence of stress-induced bond fission. These authors also succeeded in measuring the ^{73}Ge hyperfine structure of this defect, determining a value of A_{iso} that compares favorably with the values reported by Feigl and Anderson (1970) for the Ge E' centers in Ge-doped α quartz. Spin Hamiltonian parameters for Ge(1), Ge(2), and Ge E' are listed in Table IX. Kawazoe et al. (1986b) contrasted the derived Ge 4s contributions to the orbitals of the two generic defects in germanium-doped silica: $\sim 30\%$ for Ge E' versus $\sim 35\%$ for Ge(1, 2).

C. GLASSES WITH MODIFIER ADDITIONS

Addition of a modifier oxide (typically alkali or alkaline-earth oxide) into a network glass such as SiO_2, GeO_2, or P_2O_5 is well known to result in the formation of nonbridging oxygens that are locally charge-compensated by modifier cations in neighboring interstitial positions. Borate glasses are anomalous in the respect that 4-coordinated borons, rather than nonbridging oxygens, are formed for modifier contents < 25 mole % (e.g., Bray and O'Keefe, 1963). Typically, when a binary glass composed of one network-former oxide and one modifier oxide is subjected to ionizing radiations, it is found to exhibit one to two orders of magnitude more defect centers than the pure network glass equivalently irradiated (e.g., Lee, 1964). Not surprisingly, it will be seen shortly that many of the radiation-induced ESR signals in modifier-containing oxide glasses are traceable to holes trapped on nonbridging oxygens (or on oxygens bonded to 4-coordinated borons). Similarly, studies have demonstrated the involvement of alkalis in virtually all trapped-electron centers that have been well characterized in alkali-oxide–modified glasses.

1. Silicate Glasses

Karapetyan and Yudin (1963) were perhaps the first to recognize that the principal ESR spectra of irradiated alkali silicate glasses comprise super-positions of more than one spectral type. Schreurs (1967) employed a wide range of glass compositions in conjunction with various instrumental techniques to isolate two fundamental components, which he designated HC_1 and HC_2. As illustrated in Fig. 43, HC_1 is predominant in irradiated glasses of very low alkali-oxide contents while HC_1 and HC_2 appear in super-position in glasses containing $\geqslant 25$ mole % modifier oxide. The nomenclature "HC" stands for "hole center" since the trapped-hole natures of these defects were demonstrated by Schreurs and Tucker (1964) on the basis of competitive trapping experiments involving cerium-doped glasses.

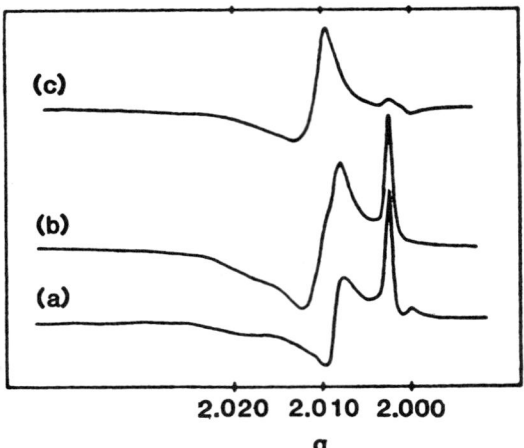

FIG. 43. X-band ESR spectra of irradiated potassium silicate glasses. (a) Mostly HC_1 centers in a K_2O-19SiO_2 glass. (b) HC_1 and HC_2 centers in a K_2O-3SiO_2 glass. (c) Nearly pure HC_2 spectrum obtained as the difference of curve (b) less curve (a). (From Schreurs, 1967).

Figure 44 reproduces the HC_1 spectra of a matched pair of γ-irradiation K_2O-5SiO_2 glasses, one containing ^{29}Si in its natural abundance of 4.7% and one enriched to 95% ^{29}Si (Griscom, 1978a; 1984d). These spectra have been "cleaned" of all HC_2 contributions by use of high microwave power and observation at an elevated temperature. (See Griscom, 1984d.) The dotted computer simulations of Fig. 44 assume a hyperfine interaction with a *single* ^{29}Si nucleus and include the known isotopic abundances as a constraint. The g_3 distribution used in the simulations (Fig. 44c) was modeled by the simple formula

$$g_3 = 2(1 + \lambda/\Delta) \tag{58}$$

by choosing an appropriate Gaussian distribution in the energy denominator Δ. The small value of $A_{iso}(^{29}Si) \sim 1.4\,mT$ used in the simulations (Table X), when compared with the atomic 3s-state value of ~ 170 (Table I), showed the degree of overlap of the unpaired spin with the silicon nucleus to be $\leqslant 1\%$ (Siderov and Tyul'kin, 1967; Cherenda et al., 1975). Similar success has been achieved in recording and computer simulating the HC_1 spectrum in an ^{17}O-enriched glass of the same composition, leading to the determination of ^{17}O hyperfine coupling constants (Table X) characteristic of an unpaired spin in a pure 2p orbital of a single oxygen (Griscom, 1980a; Cases and Griscom, 1984). Together, the ^{29}Si and ^{17}O hyperfine data conclusively demonstrate HC_1 to comprise a hole trapped on a single oxygen bonded to a single silicon in the glass network. In particular, the ^{17}O hyperfine spectra of Cases and Griscom (1984) are devoid of the "extra peaks" necessary to support a model

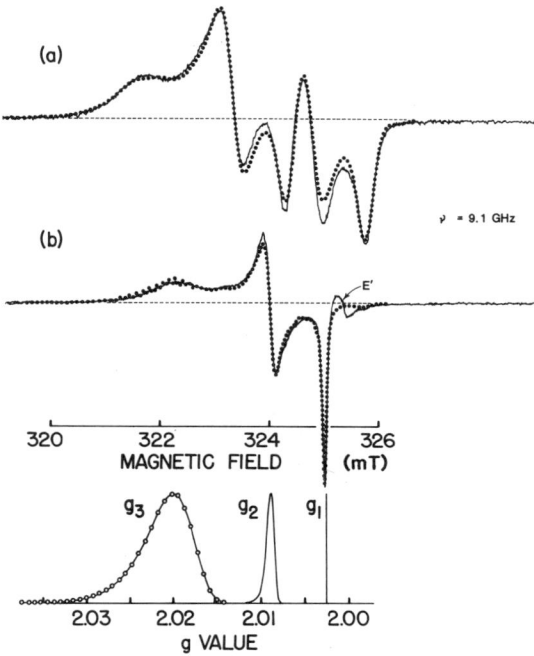

FIG. 44. X-band ESR spectra obtained at 423 K and low microwave power (0.5 mW) for the HC_1 defect in γ-irradiated K_2O-$5SiO_2$ glasses (a) enriched to 95% ^{29}Si and (b) with natural abundance of ^{29}Si (4.7%). Dotted curves are computer simulations optimized by the illustrated distribution in g values and the other parameters listed in Table X. (From Griscom, 1978a).

TABLE X

AVERAGE SPIN HAMILTONIAN PARAMETERS
FOR NONBRIDGING-OXYGEN HOLE CENTERS
IN IRRADIATED ALKALI SILICATE GLASSES

Parameter	HC_1	HC_2
g_1	2.0026	2.0118
g_2	2.0088	2.0127
g_3	2.0213	2.0158
$A_1(^{29}Si)$ (mT)	1.44	1.78
$A_2(^{29}Si)$ (mT)	1.44	1.68
$A_3(^{29}Si)$ (mT)	1.3	2
$A_1(^{17}O)$ (mT)	11.4	—
$A_{2,3}(^{17}O)$ (mT)	1.2	—

for HC_1 proposed by Skuja and Silin (1982) which postulated the formation of an additional bond between the nonbridging oxygen and another oxygen on the same SiO_4 tetrahedron.

Figure 45 illustrates a proposed model for HC_1 (Griscom, 1978a), wherein the hole is trapped in a pure 2p orbital of a nonbridging oxygen. Following Urnes (1961), it is assumed here that all nonbridging oxygens occur in pairs that are initially charge compensated by pairs of alkali ions. In keeping with what is known about the behavior of aluminum-oxygen hole centers in irradiated α quartz (Mackey, 1963), it is further envisioned that one of these compensating ions diffuses away following hole trapping at the site. The energy splitting Δ of the $2p_{x1}$ and $2p_{x2}$ orbitals is suggested to be due to the electrostatic interaction with the remaining alkali ion (indicated as Me^+ in Fig. 45), which together with Si_1 and O_1 defines the x_2-x_3 plane. In support of such a picture, inspection of the data of Schreurs (1967) suggests a possible inverse dependence of $g_3 - g_e \equiv 2\lambda/\Delta$ on the field strength of the modifier ion (Li^+, Na^+, K^+, Rb^+, or Cs^+). The measured value of g_1 (Table X), corresponding to the x_1 direction, is essentially unshifted from the free electron value g_e, as is also predicted by the model. Further theoretical discussions of the g matrix can be found in Zamotrinskaya et al. (1972), Griscom (1978a), and Kordas and Oel (1982).

Two important characteristics of HC_2 are that (1) it becomes relatively more prevalent for glasses of higher-modifier oxide contents and (2) all of the principle values of the g matrix (Table X) differ from the free-electron value g_e by at least $+0.006$. Griscom (1978a, 1984d) reported an additional characteristic of HC_2 in irradiated *potassium* silicate glasses, namely, that (3) it can be reversibly converted to a defect indistinguishable from HC_1 by raising the observation temperature to $\sim 150°C$. (Kordas et al., 1982, reported that no such effect is apparent in *sodium* silicate glasses.) Figure 46 illustrates a model for HC_2 proposed by Griscom (1978a) which accounts for characteristics (1) through (3). In the low-temperature configuration, the hole is envisioned as being delocalized over two nonbridging oxygens on the same silicon, with a single charge-compensating alkali ion being located above or below the plane defined by these three atoms (Fig. 46a). The g-value theory for this type of

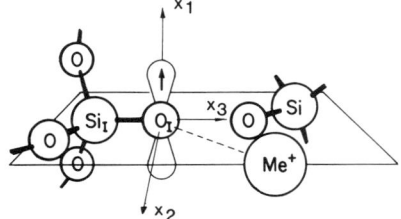

FIG. 45. Structural model for the HC_1 defect after Griscom (1978a). A hole is trapped in the $2p_{x1}$ orbital of nonbridging oxygen O_1. x_1, x_2, and x_3 are principal axes of the g matrix. The x_3 axis is defined by the Si_1-O_1 bond, and the x_2x_3 plane is assumed to be defined by Si_1, O_1, and the position of a nearby alkali ion Me^+.

(a) LOW-TEMP. CONFIGURATION   (b) HIGH-TEMP. CONFIGURATION

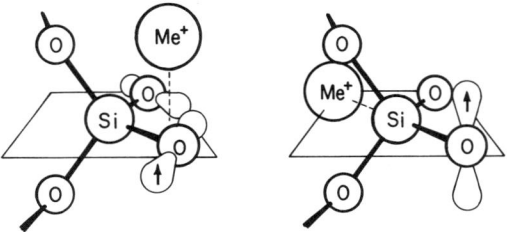

FIG. 46. Structural model for the HC_2 defect after Griscom (1978a). In the low-temperature configuration (a), a hole is trapped on two nonbridging oxygens bonded to the same silicon, and a charge-compensating cation is symmetrically disposed above or below the plane defined by the silicon and its two nonbridging oxygens. At higher temperatures, the cation is envisioned to move to an in-plane position (b), thereby repelling the hole onto the more distant nonbridging oxygen. Note the HC_1-like electronic structure of the high-temperature configuration.

defect has been discussed by Griscom (1978a) and Griscom et al. (1983b). At higher temperatures the model postulates a thermally activated local repositioning of the alkali ion into the aforementioned plane, thereby repelling the hole onto just one of the nonbridging oxygens (Fig. 46b).

Cases and Griscom (1984) have reported ^{17}O hyperfine lines attributable to HC_2 in its low-temperature configuration. Although not thoroughly analyzed, these data are consistent with expectation for an unpaired spin delocalized over two equivalent oxygens, as posited in Fig. 46a. A critical commentary on several proposed models for HC_1 and HC_2 can be found in Griscom (1984d).

A weak singlet signal at $g = 2.000$ was reported by Schreurs (1967) in irradiated potassium silicate glasses of low K_2O contents and tentatively attributed to an "E'-like" defect (Section IV.A.1). An essentially identical feature is denoted E' in Fig. 44b. In the course of their ESR studies of irradiated potassium silicate glasses, Cherenda et al. (1975) reported a 36-mT doublet centered on $g = 2.0$ that did not seem to respond to ^{29}Si isotopic enrichment and was therefore ascribed to a center undergoing a hyperfine interaction with a proton. However, Cases and Griscom (1984), investigating xK_2O-$(1-x)SiO_2$ glasses for which $x \leqslant 0.2$, recorded a radiation-induced 36.8-mT doublet that intensified ~ 10 fold in a sample 95% enriched in ^{29}Si (Fig. 47). This finding led the latter authors to conclude that the 36-mT doublet is the ^{29}Si hyperfine structure associated with Schreurs' E'-like defect. On the basis of the dotted computer simulation of Fig. 47, Cases and Griscom (1984) derived spin Hamiltonian parameters of Table XI. Note that the spin density on silicon for the 36.8-mT doublet is lower than that for the Si E' center ($A_{iso} = 42$ mT, Section IV.A.1) by a factor of $36.8/420 = 0.88$. This

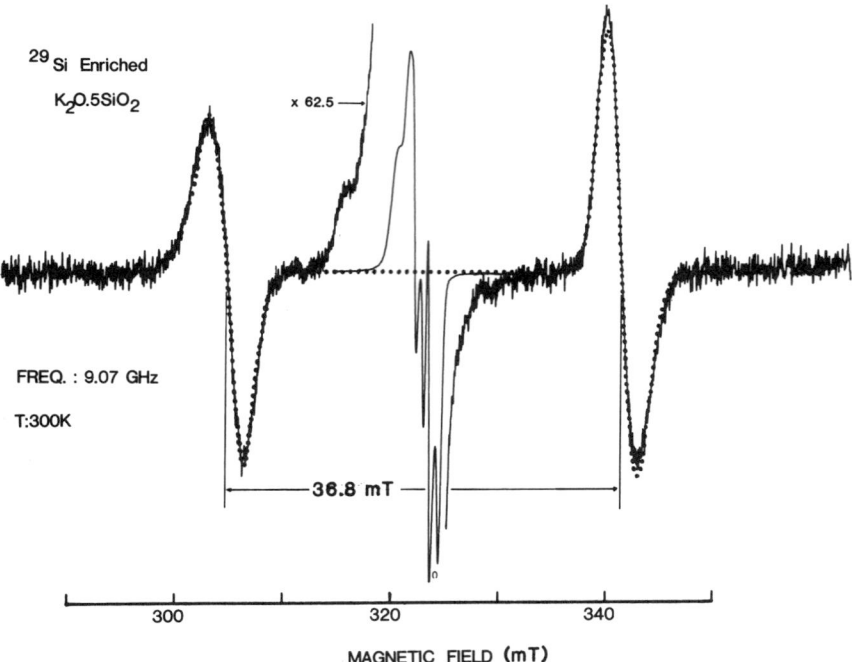

29 Si Enriched

K$_2$O.5SiO$_2$

x 62.5 →

FREQ. : 9.07 GHz

T:300K

— 36.8 mT —

300   320   340

MAGNETIC FIELD (mT)

FIG. 47. Experimental and computer-simulated (dotted curve) ESR spectra of the "E'-alkali" center in a 95%-^{29}Si-enriched K$_2$O-5SiO$_2$ glass γ-irradiated at room temperature. (From Cases and Griscom, 1984).

means that if the E' center were to be considered a pure Si sp^3 dangling orbital defect, then the unpaired spin of the 36.8-mT doublet would be viewed as ~12% delocalized—possibly onto a neighboring alkali ion. On the basis of the Ce^{3+}-doping experiments of Schreurs (1967), this particular "E'-alkali" defect is probably of the trapped-electron type.

Additional defects reported in alkali silicates include peroxy radicals and interstitial O$_2^-$ and O$_3^-$ ions in potassium silicate glasses subjected to high γ-ray doses (Cases and Griscom, 1984) and interstitial OH^0 radicals (stable

TABLE XI

AVERAGE SPIN HAMILTONIAN PARAMETERS FOR THE
E'-ALKALI TRAPPED-ELECTRON CENTER IN
IRRADIATED ALKALI SILICATE GLASSES

$g_{\parallel}$	$g_{\perp}$	$A_{iso}(^{29}Si)$ (mT)
2.0017	2.004	37.0

only below 250 K) in x-irradiated sodium silicate glasses containing high quantities of molecular water (Wolf et al., 1983).

2. Borate Glasses

ESR studies of irradiated glasses containing boron were first reported by Yasaitis and Smaller (1953). They tentatively identified a symmetric four-line spectrum (splitting ~ 1.4 mT) recorded at 373 MHz in a borosilicate glass as arising from a defect undergoing a hyperfine interaction with ^{11}B ($I = 3/2$, 81% abundant). A decade later, Lee and Bray (1963, 1964) carried out systematic studies of the entire alkali borate glass system (xMe$_2$O-(1-x)B$_2$O$_3$, where Me = Li, Na, K, Rb, or Cs) following γ-irradiation at room temperature. The latter authors employed more conventional X-band (~ 9 GHz) and K-band (~ 23 GHz) microwave frequencies and examined for the first time some samples enriched in the ^{10}B isotope. At X band, a "five-line-plus-a-shoulder" spectrum was observed for the normal ^{11}B glasses of low alkali contents, with this lineshape giving way to an asymmetric "four-line" spectrum as the alkali oxide content is increased above $\geqslant 33$ mole % (Fig. 48a, b, c, d). Lee and Bray's (1964) spectra for ^{10}B-enriched samples (e.g., Fig.

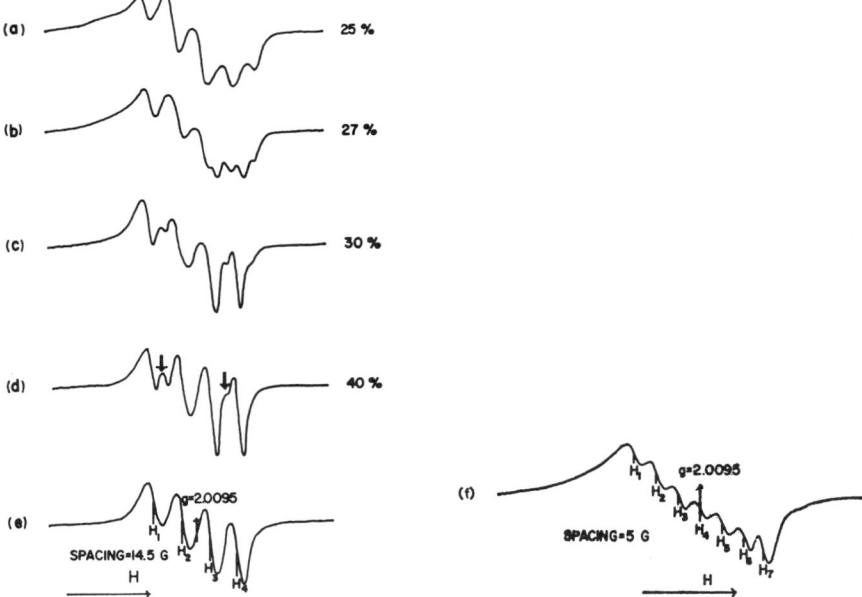

FIG. 48. X-band ESR spectra of a series of potassium borate glasses γ-irradiated at room temperature. (a), (b), (c), and (d) correspond to as-irradiated glasses containing 25, 27, 30, and 40 mole % K$_2$O, respectively. (e) 35 mole % K$_2$O, heat treated at 473 K after irradiation. (f) 35 mole % K$_2$O enriched to 96% ^{10}B. (From Lee and Bray, 1964).

48f) provided compelling proof that both the four- and five-line spectral components represent ^{11}B hyperfine structure. Still, it remained a question as to whether the structureless low-field shoulder arose from the same source as the sharp lines. Indeed, many competing explanations of the overall "five-line-plus-a-shoulder" spectrum were offered (Nakai, 1964; Lee and Bray, 1963, 1964; Karapetyan and Yudin, 1963; Beekenkamp, 1966).

The situation was clarified by Griscom et al. (1968), who studied both glasses and polycrystalline compounds of the lithium borate system following γ-irradiation at 77 K. It was shown that the "five-line-plus-a-shoulder" spectrum also occurs in the $Li_2O \cdot 4B_2O_3$ compound, and it was possible to computer simulate the entire spectra of both the polycrystalline and glassy samples using a single spin Hamiltonian (Table XII). As illustrated in Fig. 49 (Griscom, 1973), the computer simulations showed that the broad low-field shoulder in the glass spectrum exhibits no hyperfine structure because the width W of the distribution in g_3 values (expressed in magnetic field units) is greater than the corresponding ^{11}B hyperfine splitting A_3. (Note that similar results were obtained for corresponding spectrum in a boron-doped silica, Fig. 35.)

For convenience, Taylor and Griscom (1971) proposed that the "five-line-plus-a-shoulder" spectrum be termed the 1 : 3 spectrum, due to its occurrence only in compounds such as $Li_2O \cdot 4B_2O_3$ and $Li_2O \cdot 3B_2O_3$ which contain triborate groups (see, e.g., Griscom, 1978b). Competitive trapping experiments showed the defect to be of the trapped-hole type (Griscom et al., 1968). The best model for the 1 : 3 center is probably that of a hole trapped on an oxygen bridging between a 3- and a 4-coordinated boron in the glass structure. The wavefunction is likely to be a pure oxygen 2p state as portrayed in Fig. 36, rather than either of the delocalized-orbital schemes proposed by Griscom et al. (1968) or Ignat'ev et al. (1972). The hyperfine interaction is with either the 3- or the 4-coordinated boron, but not both. Since only a 4-coordinated boron seems to be responsible for the virtually

TABLE XII

AVERAGE SPIN HAMILTONIAN PARAMETERS FOR BORON-OXYGEN HOLE CENTERS IN IRRADIATED ALKALI BORATE GLASSES

Defect	g_1	g_2	g_3	$A_1(^{11}B)$ (mT)	$A_2(^{11}B)$ (mT)	$A_3(^{11}B)$ (mT)
1 : 3 center	2.0020	2.0103	2.035	1.21	1.42	1.0
1 : 2 center	2.003	2.009	2.020	0.51[a]	0.60[a]	0.29[a]
2 : 1 center	2.0049	2.0092	2.025	1.12	1.29	0.8

[a]Hyperfine interaction with two equivalent borons.

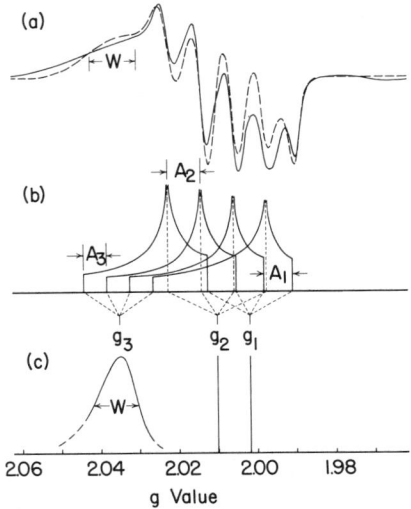

FIG. 49. An analysis of the X-band ESR spectrum of the boron-oxygen hole center in irradiated alkali borate glasses. Unbroken curve in (a) is an experimental spectrum for a 20K$_2$O-80B$_2$O$_3$ glass. Dashed curve in (a) is a computer simulation based on a set of powder ·patterns schematically diagrammed in (b) together with the distributions in g values shown in (c). Because the width W of the g_3 distribution is greater than the hyperfine splitting A_3, no structure is observed on the low-field shoulder in (a). (From Griscom, 1973).

identical hyperfine spectrum in boron-doped silica (Section IV.B.1), it is presumably the 4-coordinated boron that causes the hyperfine splitting in the alkali borate glasses. (There remains a need for theoretical calculations to understand why the 3-coordinated boron at the 1:3-center site should give no measurable splitting.)

Apropos of the preceding discussion, Taylor and Griscom (1971) demonstrated the 1:2 center in irradiated SrO·2B$_2$O$_3$ and Li$_2$O·2B$_2$O$_3$ compounds to exhibit hyperfine interactions with *two equivalent* borons (Table XII). Since all of the borons in the 1:2 strontium compound are 4-coordinated and the 1:2 lithium compound contains diborate structural units wherein a pair of 4-coordinated borons are linked by bridging oxygens, the 1:2 center was modelled as a hole trapped on an oxygen bridging two 4-coordinated borons (Taylor and Griscom, 1971). Taylor and Bray (1972) used computer-simulation methods to deconvolve the 1:3 and 1:2 contributions as a function of glass composition.

The asymmetric "four-line" spectrum occurring in the alkali borate glasses of high alkali contents (e.g., Fig. 48e) was also computer-simulated by Griscom et al. (1968) leading to the spin Hamiltonian parameters listed for it in Table XII. It can be noted that this four-line borate spectrum has crucial features in common with HC_2 in silicates (Section IV.C.1), namely, the facts that (1) none of the principal g values are very close to g_e and (2) the spectrum only occurs in glasses of high alkali contents. Moreover, Griscom and Kriz (1979) found that the only crystalline compound to display a comparable four-line spectrum was calcium pyroborate ($2CaO \cdot B_2O_3$, wherein each boron is bonded to one bridging and two nonbridging oxygens). Thus, the most probable model for the "four-line" center would be a hole trapped on two nonbridging oxygens on the same boron, in analogy to HC_2. In the present compact notation, the four-line center would be specified as a "2:1" center. However, when less specificity is required, the 1:3, 1:2, and 2:1 centers are often referred to simply as boron-oxygen hole centers or "BOHCs."

Two other generic defect types have been identified in irradiated alkali borate glasses. Both of these appear to be of the trapped-electron type and both are stable only below ~100 K. The first is the boron electron center (BEC). Spin Hamiltonian parameters for the BEC listed in Table XIII were extracted by computer-simulation analysis of spectra obtained for ^{10}B and ^{11}B-enriched $18K_2O$-$82B_2O_3$ glasses; variations in A_{iso} with alkali type and glass composition were also reported (Griscom, 1971a). Several models for the BEC were considered by Griscom (1971a), although the preferred one was that of a three-electron bond between a tricoordinated boron and an alkali ion (analogous to the model proposed for the "E'-alkali" defect in alkali silicate glasses in Section IV.C.1). In any event, it is clear that the BEC involves some spin delocalization away from the boron, since the ^{11}B coupling constants for the BEC (Table XIII) are smaller than those of the boron E' center (Table VI) by factors ~0.5. Interestingly, Pontuschka et al. (1987) found that BECs produced in a barium aluminoborate glass by x-irradiation at 77 K can be photo bleached with the concomitant production of atomic hydrogen.

The second trapped electron defect type has been termed the alkali-associated electron center (AEC). The ESR manifestation of the AEC is a broad, relatively featureless resonance with an average breadth that is

TABLE XIII

SPIN HAMILTONIAN PARAMETERS FOR THE BORON-ELECTRON CENTER IN IRRADIATED $18 K_2O$-$82 B_2O_3$ GLASS

g	$A_{iso}(^{11}B)$ (mT)
2.0018	10.8

independent of boron isotope and about equal to the s-state hyperfine coupling constant for Na^0 (K^0) in the case of sodium (potassium) borate glasses (Griscom, 1971b). (The AEC could not be resolved in lithium borate glasses.) Griscom (1971b) demonstrated that the observed AEC spectra could arise from an unpaired electron delocalized over the 3s (4s) orbitals of a cluster of four (six) potassium (sodium) ions. It was argued that still larger clusters may form in the lithium borate glasses, giving rise to a narrower resonance that would be subsumed beneath the envelope of the BOHC and hence become unobservable. Further growth in cluster sizes (independent of alkali type), coupled with the multiple trapping of electrons in paired states, was proposed as the reason for the general inability to observe trapped-electron centers at room temperature. It may be more than coincidental that the AEC spectrum presented by Griscom (1971b) for an irradiated $30Na_2O$-$70B_2O_3$ glass bears a strong resemblance to the photoinduced trapped-electron center reported by Stroud (1961) in a Ce-doped $25Na_2O$-$75SiO_2$ glass.

3. Phosphate Glasses

Karapetyan and Yudin (1962) gave the first report of a now-familiar ESR spectrum in irradiated phosphate glasses characterized by an asymmetric doublet centered on $g \simeq 2.009$ with a splitting ~ 4.0 mT. The relatively small size of the ^{31}P hyperfine coupling constant implies the degree of spin localization on a phosphorus to be $\sim 0.9\%$ (see Eq. (27a)), leading to the hypothesis that the unpaired spin may be confined to neighboring oxygen(s). Schreurs and Tucker (1964) demonstrated the trapped-hole nature of the defect by Ce^{3+} doping experiments. Whence the designation phosphorus-oxygen hole center or POHC. The currently (1989) favored model for the POHC is a hole trapped on two nonbridging oxygens bonded to the same phosphorus (Fig. 38a), as originally proposed by Beekenkamp (1966). Detailed justifications for this model are developed by Griscom et al. (1983b). The POHC spectrum is illustrated in Fig. 40 for the case of phosphorus-doped silica.

In the course of his studies of γ-irradiated $50RO$-$50P_2O_5$ glasses ($R = Li_2$, Na_2, Ca, Ba), Nakai (1964) recorded a doublet structure centered on $g = 2.0$ with mean splittings of ~ 77.5 mT. Weeks and Bray (1968) demonstrated that the field separation and central g value of this doublet in a $LiPO_3$ glass were independent of microwave frequency (Fig. 50), thereby proving the pair of lines to be ^{31}P hyperfine structure of a single defect which they designated P_3. Also investigated by Weeks and Bray (1968) were glassy and polycrystalline P_2O_5 and a series of alkali phosphate glasses with RO/P_2O_5 ratios ranging from 0.3 to 1.6. Upon γ-irradiation, the pure P_2O_5 materials all displayed ESR spectra comprised of the POHC and an additional pair of doublets, P_1

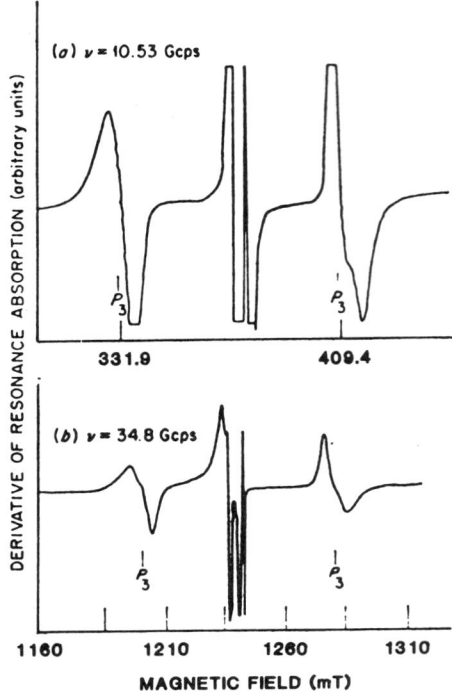

FIG. 50. ESR spectra of irradiated $50Li_2O$-$50P_2O_5$ glasses recorded at microwave frequencies of (a) 10.53 and (b) 34.8 GHz. (From Weeks and Bray, 1968).

and P_2, with mean splittings of ~ 95 and 140 mT, respectively. These splittings are virtually identical with those for the defects respectively designated P_1 and P_2 in irradiated phosphorus-doped silica (Section IV.B.2, Table V), and the appropriate structural models are almost certainly those illustrated in Fig. 38. By analogy with the E'-alkali center in alkali silicate glasses (Section IV.C.1) and the BEC in alkali borate glasses (Section IV.C.2), the P_3 center is probably best regarded as a defect involving a three-electron bond between a network phosphorus and an alkali ion in the glass structure.

Hosono et al. (1985) studied the ESR spectra of $50RO$-$50P_2O_5$ glasses (R = Li_2, Na_2, Be, Mg, Ca, Ba) γ-irradiated at 77 K. ^{31}P hyperfine doublets were reported that clearly correspond to the P_1, P_2, and P_3 doublets described previously (although the authors employed a different notation). Only the P_2 center (in the present notation) was observed in the $BeO \cdot P_2O_5$ glasses immediately following irradiation; this appeared to be converted to P_1 by warming or illumination with UV light. Hosono et al. (1985) also reported observation of a P_4 (phosphinyl-radical-type) defect in calcium metaphosphate glasses that had been melted under reducing conditions. A tri-coordinated P^{3+} ion resulting from the partial reduction of the melt was envisioned to form the defect by dissociative electron capture (lower path in

TABLE XIV

SPIN HAMILTONIAN PARAMETERS FOR SOME PHOSPHORUS-
RELATED DEFECTS IN IRRADIATED P_2O_5 AND ALKALI
PHOSPHATE GLASSES

Glass	Center	g	$A_{iso}(^{31}P)$ (mT)
P_2O_5	P_1	2.01	98.5
	P_2	2.03	150
P_2O_5-0.3Li_2O	P_1	2.03	96.3
	P_2	2.03	130.5
P_2O_5-1.0Li_2O	P_3	2.013	78.4

Fig. 38c). Typical spin Hamiltonian parameters reported for P_1, P_2, P_3, and P_4 in P_2O_5 and phosphate glasses are listed in Table XIV.

4. Oxynitride Glasses

Mackey et al. (1970) carried out ESR and optical studies of the influences of oxidation and reduction on the radiation sensitivity of sodium silicate glasses. Comparisons were made of the x-ray–induced ESR spectra in $25Na_2O$-$75SiO_2$ glasses prepared under various redox conditions, including (1) melting in platinum in air, (2) melting in graphite in argon, and (3) bubbling dry N_2 through the graphite melts. The standard superposition of HC_1 and HC_2 spectra (Section IV.C.1) was observed in both case (1) and case (2), proving that reduction in a totally inert atmosphere has no qualitative influence on the ESR-active radiation-induced defects. But in irradiated glasses prepared by method (3), HC_1 and HC_2 were found to be reduced in intensity and supplanted by two new spectral components, one of which comprised a triplet of equally spaced lines with an average splitting of 3.6 mT. Spectra obtained at three microwave frequencies (Fig. 51) showed the line spacing to be frequency-independent and hence due to a hyperfine interaction with a nucleus of spin $I = 1$. This nucleus was concluded to be ^{14}N because the occurrence of the resonance only in nitrided glasses and because ^{14}N is the only abundant $I = 1$ nuclide in nature. The spin Hamiltonian parameters of Table XV were derived from the successful computer simulations of Fig. 51. The model for this nitrogen-associated defect proposed by Mackey et al. (1970) comprises a hole trapped on a nitrogen bridging two silicons in the glass network. The broad distribution in g_3 values (inset to Fig. 51) which was required to achieve the computer simulations was interpreted as arising from a distribution in Si-N-Si bond angles. It can be noted that the defect center is charge-neutral after hole trapping. Mackey et al. (1970) gave evidence that the precursor to this center in alkali silicate glasses is probably charge-

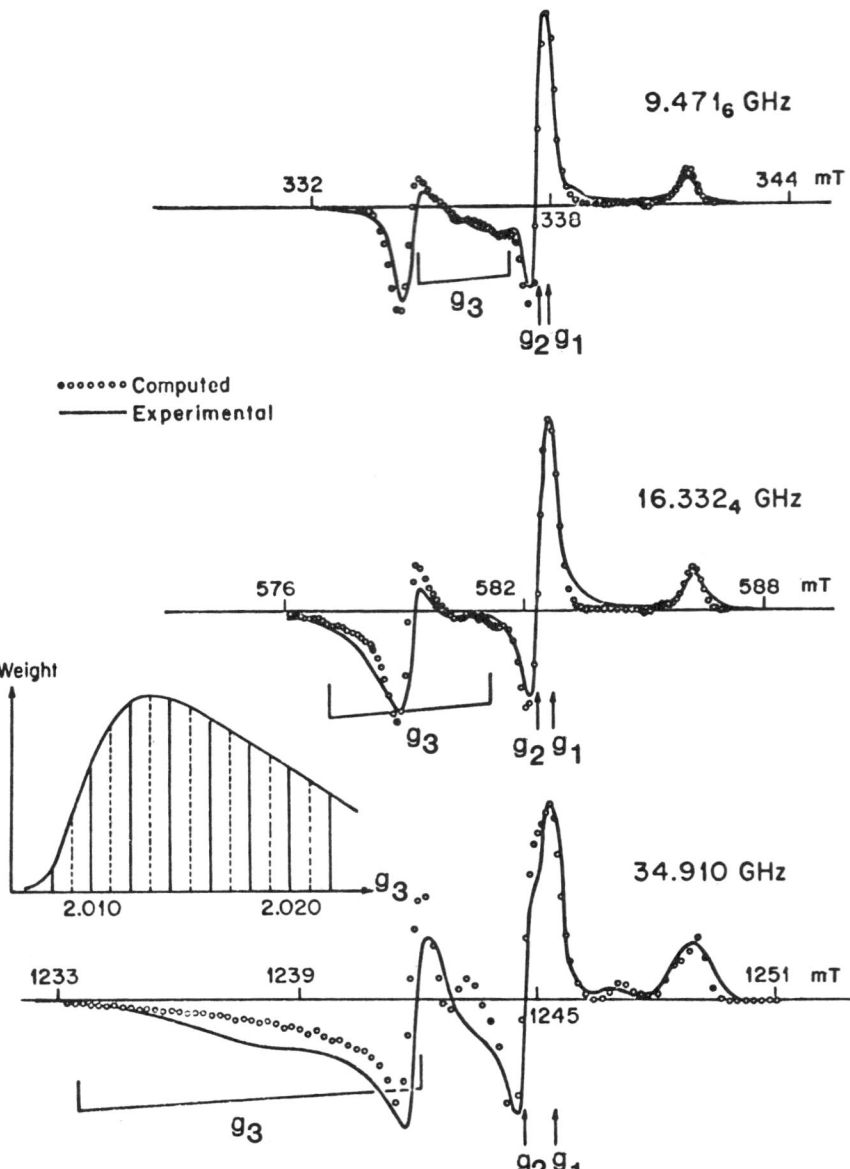

FIG. 51. ESR spectra of nitrogen hole centers in irradiated nitrided alkali-silicate glasses recorded at three microwave frequencies. Fully drawn curves are experimental; circles represent computer simulations based on the distribution of g values illustrated. (From Mackey et al., 1970).

TABLE XV

SPIN HAMILTONIAN PARAMETERS FOR THE BRIDGING NITROGEN HOLE
CENTER IN IRRADIATED NITROGEN-CONTAINING SILICATE GLASSES

g_1	g_2	g_3	$A_1(^{14}N)$ (mT)	$A_{2,3}(^{14}N)$ (mT)
2.0026	2.0039	2.0017	3.60	0.2

compensated by an alkali ion which diffuses away after hole trapping. However, the same nitrogen hole center (NHC) has also been observed in alkali-free silicon oxynitride glasses (Griscom, 1978a; Friebele et al., 1985b), suggesting that some negatively charged 2-coordinated nitrogens might exist as defects in pure Si_3N_4-SiO_2 glasses.

Radiation-induced defects in nitrided phosphate glasses and polycrystal-line compounds were investigated by Watanabe et al. (1985). In addition to the P_1, P_2, P_3, and POHC defects commonly observed in phosphate glasses, a new "five-line" spectrum (average splitting $\sim 3\,mT$) was recorded at X-band frequencies. Competitive trapping experiments involving samples doped with Cd^{2+} or Ag^+ suggested that the new center is of the trapped-hole type. Isotopic substitution experiments (^2H had no effect, ^{15}N changed the number of lines and their spacings) demonstrated the new five-line resonance to involve hyperfine interactions with two ^{31}P and one ^{14}N nuclei. The decomposition of the various hyperfine components and a computer simula-tion of the spectrum in polycrystalline magnesium metaborate are illustrated in Fig. 52; the corresponding spin Hamiltonian parameters are listed in Table XVI. On the basis of these results, the structural model proposed by Watanabe et al. (1985) is a hole trapped on a nitrogen bridging between two phosphorus atoms in the glass network. An analysis employing Eq. (54) indicated the unpaired spin to occupy a pure N 2p orbital essentially perpendicular to the plane of the P-N-P bond.

5. Glasses Containing Arsenic or Antimony

Arsenic and antimony are common additives to silicate glasses to promote the removal of bubbles (e.g., Kreidl, 1983). The action of these polyvalent fining agents is generally to enhance resolution of gaseous oxygen when the glass is quenched, e.g., by oxidation of As_2O_3 to As_3O_5. Vitko and Shelby (1982) noted a decided influence of fining agents on the solarization of a borosilicate glass and reported an unidentified ESR spectrum that occurred only in solarized glasses containing As_2O_3.

Imagawa et al. (1982) investigated the effects of oxidation and reduction on the γ-ray induced ESR spectra of a $16Na_2O$-$11CaO$-$73SiO_2$: $0.1AsO_x$ glass.

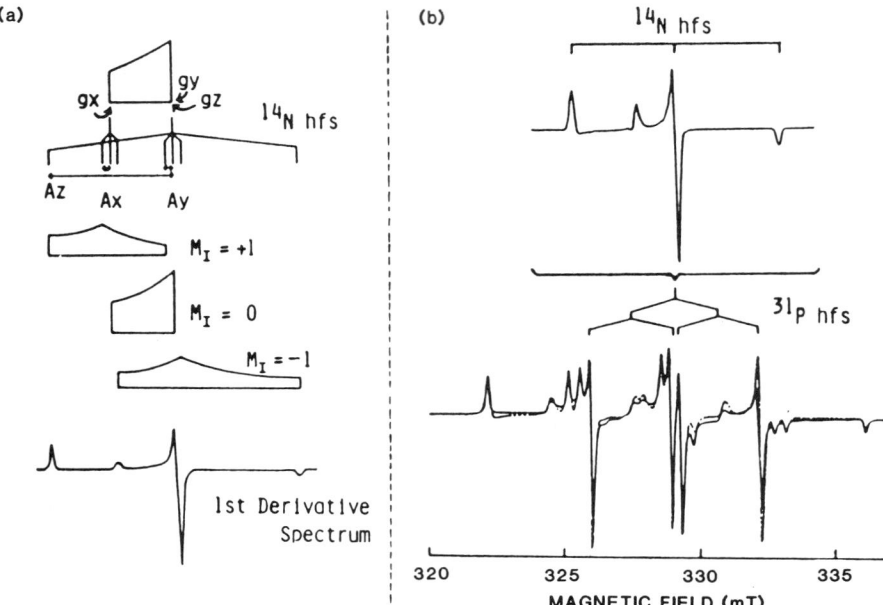

FIG. 52. ESR spectrum of irradiated polycrystalline magnesium metaphosphate containing nitrogen. Powder-pattern reconstruction and computer-simulated lineshape in (a) consider only a hyperfine interaction with a single ^{14}N nucleus ($I = 1$). In (b), the ^{14}N hyperfine components are further split by interactions with two ^{31}P nuclei with $I = 1/2$. The unbroken curve in (b) is experimental; the dashed curve is a computer simulation. (From Watanabe et al., 1985).

TABLE XVI

SPIN HAMILTONIAN PARAMETERS FOR THE
BRIDGING-NITROGEN HOLE CENTER IN
IRRADIATED SODIUM METAPHOSPHATE (SMP)
AND MAGNESIUM METAPHOSPHATE (MMP)

Parameter	SMP	MMP
g_x	2.0076	2.0128
g_y	2.0031	2.0044
g_z	2.0035	2.0042
$A_x(^{31}P)$ (mT)	2.92	2.93
$A_y(^{31}P)$ (mT)	2.78	2.83
$A_z(^{31}P)$ (mT)	2.92	2.83
$A_x(^{31}P)$ (mT)	2.92	3.22
$A_y(^{31}P)$ (mT)	2.78	3.11
$A_z(^{31}P)$ (mT)	2.92	3.22
$A_x(^{14}N)$ (mT)	0.2	0.2
$A_y(^{14}N)$ (mT)	0.2	0.2
$A_z(^{14}N)$ (mT)	3.54	3.66

The spectra of Figs. 53a and 53c were obtained for the oxidized and reduced glasses, respectively. Two different hyperfine quartets due to ^{75}As (I = 3/2, 100% abundant) were deconvolved by means of the dotted computer simulations, leading to the spin Hamiltonian parameters of Table XVII. For present purposes, these two arsenic-related defect centers are here labelled As_1 and As_2 to suggest probable analogies with the P_1 and P_2 centers of Section IV.B.3. Indeed, the models proposed by Imagawa et al. (1982) for As_1 and As_2 are essentially the same as those of the P_1 and P_2 defects as diagrammed in Fig. 38c and b, respectively, if arsenic is substituted for phosphorus.

Hosono et al. (1984) studied the UV solarization of a 16Na$_2$O-11CaO-73SiO$_2$ glass codoped with 0.15AsO$_x$ and 0.015CeO$_x$. Both As_1 and As_2 were created and found to grow during the first three hours of illumination. But when the illumination time was extended beyond three hours, As_2 decreased back to zero while As_1 gained in intensity by a factor of ~ 2 before saturating after 200 hours. In interpreting the results, it was assumed that the photochemistry is driven by the release of photoelectrons from Ce^{3+} (Stroud, 1961). The defect centers that formed are all As^{4+}, since this is the only known paramagnetic state of arsenic. Hosono et al. (1984) interpreted $\acute{A}s_2$ as an electron trapped on a 4-coordinated As^{5+} in the glass network, and they noted that the growth kinetics imply that As_2 must be a chemical intermediate, with As_1 being the final product. Since it can be inferred from the spin Hamiltonian parameters (Table XVII) that As_1 comprises a 3-coordinated As^{4+}, Hosono et al. (1984) concluded that As_2 must convert to As_1 by detachment of a nonbridging oxygen from its coordination sphere. It is noted in passing that essentially the same As_1 and As_2 defects have been

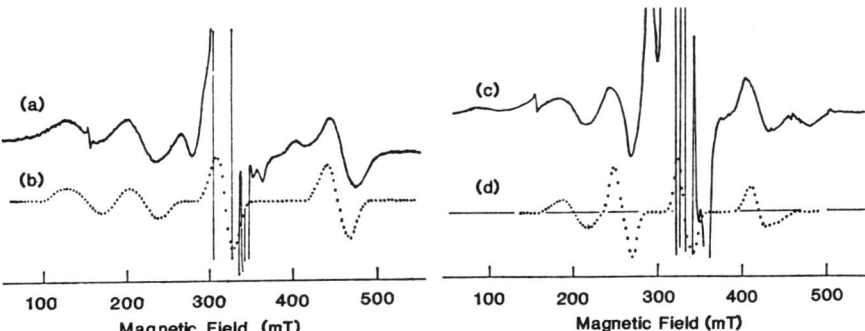

FIG. 53. ESR spectra of irradiated soda-lime silicate glasses containing arsenic. (a) and (c) are experimental spectra for the oxidized and reduced glasses, respectively. (b) and (d) are computer simulations of the As_2 and As_1 resonances, respectively. (From Imagawa et al., 1982).

TABLE XVII
SPIN HAMILTONIAN PARAMETERS FOR SOME ARSENIC-
ASSOCIATED CENTERS IN IRRADIATED ARSENIC-
CONTAINING GLASSES

Center	g	$A_{iso}(^{75}As)$ (mT)
As_1	2.0	74.8
As_2	2.0	99.0

observed in an x-irradiated low-thermal-expansion glass ceramic (Tsai et al., 1987b).

The photochemistry of a *reduced* soda-lime silica glass containing Ce^{3+} and As^{3+} (not As^{5+}) was studied by Hosono et al. (1983), who isolated a defect center that for present purposes will be denoted As_4. In analogy to the P_4 center of Fig. 38c, As_4 is believed to result from dissociative electron capture at the site of a 3-coordinated As^{3+}.

In the course of developing and characterizing a family of polychromic glasses, Schreurs and Davis (1979) recorded a UV-induced ESR spectrum (Fig. 54) that they were able to attribute to Sb^{4+} created by the transfer of an

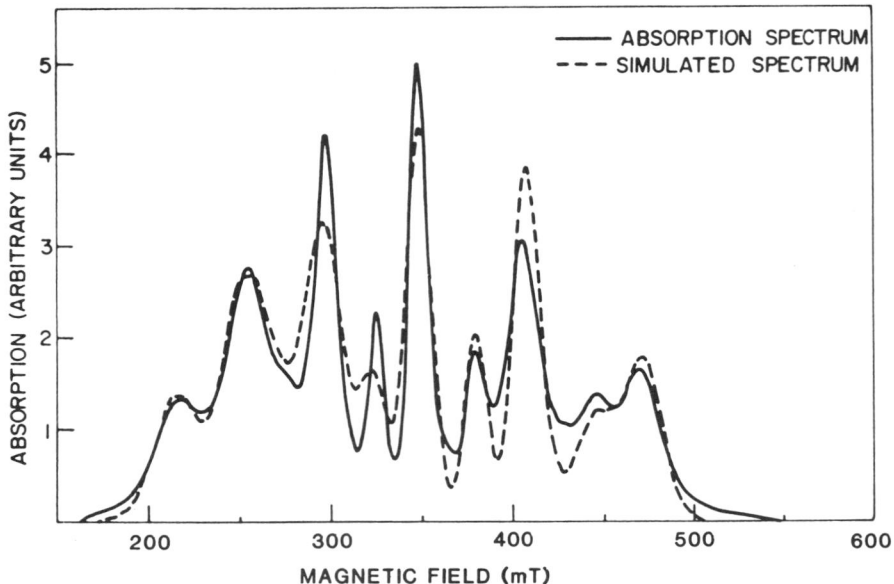

FIG. 54.   ESR spectrum of photoinduced Sb^{4+} in an $18K_2O$-$82SiO_2$: $0.02CeO_2$-$0.044Sb_2O_3$ glass. Dashed curve is a computer simulation. (From Schreurs and Davis, 1979).

electron to an Sb^{5+} from a Ce^{3+}. The dashed computer simulation of Fig. 54 takes into account the two isotopes of antimony, ^{121}Sb and ^{123}Sb ($I = 5/2$, 57% abundant and $I = 7/2$, 43% abundant, respectively) and leads to the determination of the mean value of $|\psi(0)|_{5s}$ to be five to ten times smaller than free-atom value extrapolated from tabulated results for isoelectronic 5s-state atoms. This result was interpreted as implying the unpaired spin to reside in some kind of a hybrid orbital.

6. ns^1 Defect Centers

With the exception of atomic hydrogen, all of the defect centers just described have been seen to comprise unpaired spins trapped either in pure p states or in sp^n ($n \geqslant 2$) hybrid orbitals. This final section will briefly catalog a number of defects associated with common oxide glass components or dopants wherein the unpaired spin occupies a nearly pure s state on the central atom.

$Zn^+(4s^1)$. Radiation-induced ESR signals assignable to monovalent zinc have been reported in zinc-containing alkali borate glasses (Berger, 1973), in a zinc silicate glass (Sigel et al., 1974), and in a multicomponent glass ceramic (Tsai et al., 1987c). The principal g values in the zinc silicate glass were $g_1 \simeq 2.000$, $g_2 \simeq 1.997$, and $g_3 \simeq 1.994$. Hyperfine structure due to the 4%-abundant ^{67}Zn has been reported in a glass ceramic by Tsai et al. (1989).

$Ag^+(5s^1)$. A variety of radiation-induced Ag^0 centers have been reported in glasses of many different compositions (e.g., Shields, 1966; Zhitnikov and Mel'nikov, 1968; Imagawa, 1969a; Assabghy et al., 1977; Hosono et al., 1980). The simplest and best characterized of these is the isolated silver atom (observed, e.g., in nearly pure B_2O_3 glass), which gives rise to a nested pair of sharp-line doublets with splittings ~ 61 and $70\,mT$ due to the hyperfine interaction with the ^{107}Ag and ^{109}Ag nuclides (51 and 49% abundant, respectively; both $I = 1/2$). These values are 97–100% of the free-atom values A_0. Arafa et al. (1980) computer-simulated the spectra for $10K_2O$-$90B_2O_3$ glasses containing 51 and 98% ^{107}Ag, thereby demonstrating that $|A_\perp| > |A_\|$|. This condition is evidence of a slight admixing of 4d silver orbitals into the 5s ground state. (See Laman and Weil, 1977.) In borate or silicate glasses containing $\geqslant 20$ mole% alkali oxide, the sharp Ag^0 spectrum is supplanted by a broader signal with smaller hyperfine splittings that are strongly dependent on alkali type (Imagawa, 1969a).

$Cd^+(5s^1)$. Imagawa (1969b) investigated γ-ray–induced ESR signals in a wide range of borate, silicate, and phosphate glasses doped with cadmium. Figure 55 illustrates the influence of composition on the spectra of glasses of the system $(x - 1)Li_2O$-$(100 - x)B_2O_3$-$1CdO$. The signals centered near

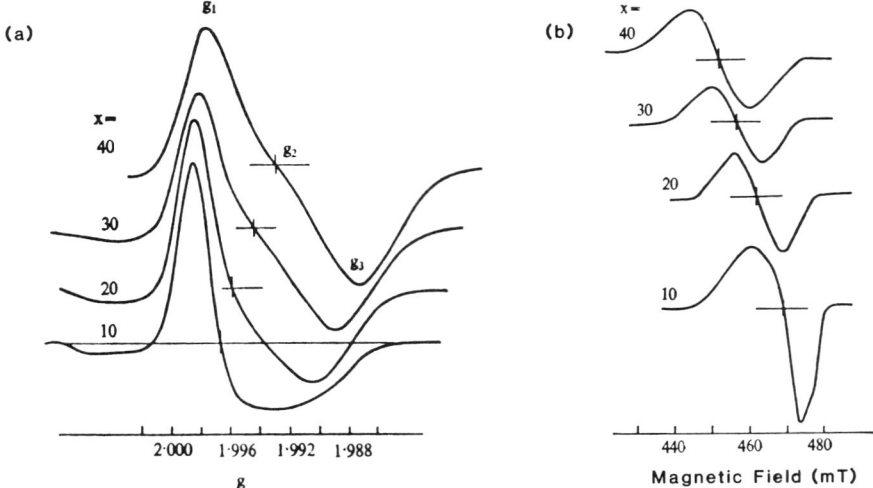

FIG. 55.   X-band ESR spectra of Cd^+ in irradiated $(x-1)Li_2O$-$(100-x)B_2O_3$-$1CdO$ glasses. (a) Spectra due to nonmagnetic isotopes of Cd. (b) High-field spectra due to hyperfine interactions with ^{111}Cd and ^{113}Cd. (See Fig. 56.) (From Imagawa, 1969b).

$g = 2$ (Fig. 55a) arise from those centers associated with the nonmagnetic isotopes of cadmium, while the high-field manifestations can be ascribed to those centers that incorporate ^{111}Cd ($I = 1/2$, $g_N = -0.5922$, 12.9% abundant) and ^{113}Cd ($I = 1/2$, $g_N = -0.6195$, 12.3% abundant). The ratio of their nuclear g factors is so close to 1.0 that for the broad lines typically observed in glasses the effects of the two isotopes are effectively indistinguishable. However, isotope effects are clearly observable for Cd^+ in CaO and β-$Ca(PO_3)_2$ (Hosono et al., 1981). For each glass composition of Fig. 55, the observed high-field signal comprises one of two hyperfine lines predicted by the solutions of Breit and Rabi (1931) for the case $A_{iso} \simeq hv$, the other line being too broad for observation. Figure 56 schematically illustrates for the case of glassy $Ca(PO_3)_2$ how a small distribution in A_{iso} can transform into a pair of transitions of vastly different widths. As the chemical properties and ionic radius of Cd^{2+} are very similar to those of Ca^{2+}, Hosono et al. (1981) emphasized that the extreme sensitivity of the Cd^+ hyperfine absorptions to crystal-field variations renders the cadmium ion a excellent probe of the local disorder around Ca^{2+} sites in solids.

$Au^0(6s^1)$. Imagawa (1969c) reported the UV-induced ESR spectra of the gold atom in a photosensitive soda-lime silicate glass co-doped with Ce and Au. The four-line hyperfine spectrum due to ^{197}Au ($I = 3/2$, 100% abundant) was centered on $g = 2.10$ with a mean splitting of 127 mT. The rather large

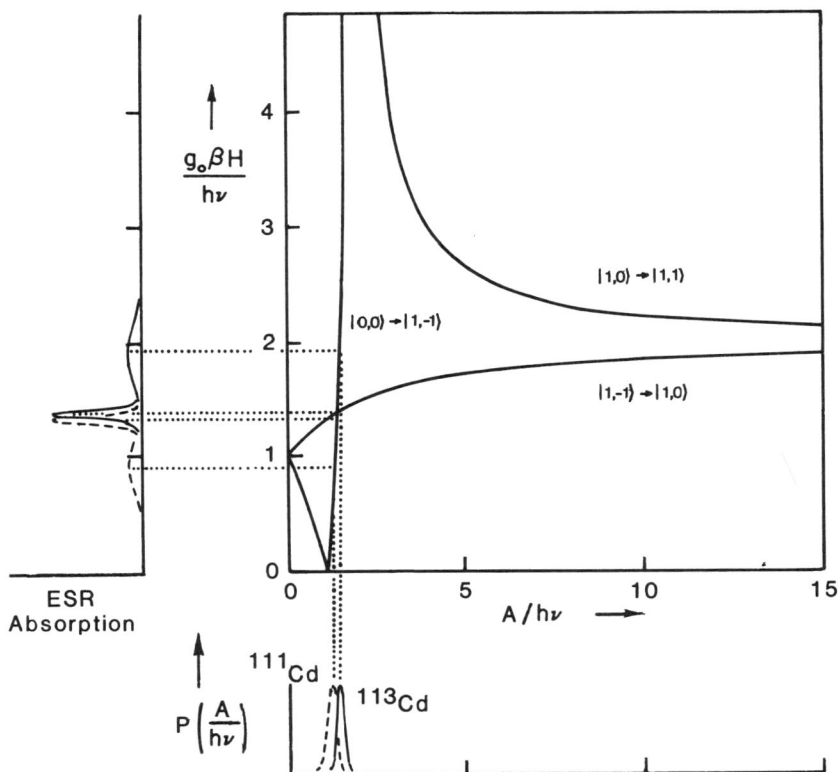

FIG. 56. Breit–Rabi solutions for an $I = 1/2$ nucleus (upper-right-hand panel). The example of $^{111,113}Cd^+$ in $Ca(PO_3)$ glass is used to show schematically how a distribution in A values can give rise to hyperfine lines of disparate widths. (Centers associated with nonmagnetic isotopes undergo resonance at $g_e\beta H/h\nu \simeq 1.0$.) (Adapted from Hosono et al., 1981.)

positive g shift and the fact that A_{iso} is $\sim 42\%$ smaller than for the free atom was explained in terms of hybridization with a 5d hole orbital.

$Tl^{2+}(6s^1)$. Radiation effects in Tl^+-containing alkali silicate glasses were explored by Hosono et al. (1982), who observed two overlapping ESR signals that they designated α and β. At X-band frequencies, both of these signals comprised doublets falling at substantially higher fields than $H_0 \equiv h\nu/g_e\beta$. Each of these doublets could be analyzed assuming hyperfine interactions with ^{203}Tl (29.5% abundant, $g_N = 1.5960$) and ^{205}Tl (70.5% abundant, $g_N = 1.6115$), both of these isotopes having $I = 1/2$. Because the experimental spectra were relatively broad while the nuclear moments of the two isotopes differ by only $\sim 1\%$, no isotope effects were expected or observed. The α and β

pairs of lines therefore correspond to Tl^{2+} at two distinctly different sites. For glasses along the join $xNa_2O\text{-}3T_2O\text{-}(97\text{-}x)SiO_2$ γ-irradiated at 77 K and observed before warming, the intensity ratio $\beta:\alpha$ increased with increasing x. Hosono et al. (1982) noted that upon warming to 300 K, the α signal converted to the β signal. Spin Hamiltonian parameters were derived by computer lineshape simulations that used first-order perturbation theory to introduce anisotropy into the Breit–Rabi analysis. On this basis, the unpaired spin was deduced to reside in orbitals that were 27% Tl 6s and 32% Tl 6p for the α center and 64% Tl 6s for the β center. Models for the two defects were proposed that involved the trapping of holes on Tl^+ ions in axial (α) and spherically symmetric (β) ligand field environments.

$Pb^{3+}(6s^1)$. Kim and Bray (1968) demonstrated that a structureless asymmetric resonance, apparently characterized by $g_\parallel \simeq 1.97$ and $g_\perp \simeq 2.005$, is manifested by possibly all irradiated oxide glasses containing a substantial PbO component. Figure 57 illustrates this characteristic signal in a γ-irradiated $13Na_2O\text{-}18PbO\text{-}69SiO_2$ glass; the dashed curve is a computer simulation (Friebele, 1977) that demonstrates the g-value distribution to be quite different than the simple axial symmetry originally inferred. Friebele (1977) also recorded and analyzed by exact matrix diagonalization one of two predicted hyperfine lines due to ^{207}Pb ($I = 1/2$, 21% abundant), thereby demonstrating the involvement of a lead ion in the defect. The second ^{207}Pb hyperfine line at higher field was reported by Hosono et al. (1980) in a γ-

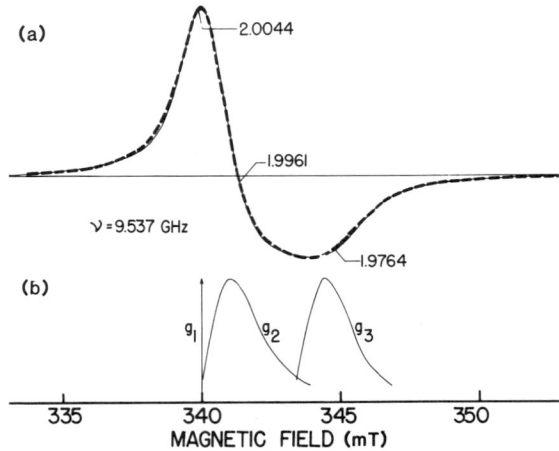

FIG. 57. X-band ESR spectrum of the Pb^{3+} defect center (associated with nonmagnetic Pb isotopes) in γ-irradiated lead silicate glasses. Unbroken curve in (a) is the experimental spectrum; dashed curve is a computer simulation based on the distribution of g values shown in (b). (From Friebele, 1977).

irradiated lead phosphate glass. Both Friebele (1977) and Hosono et al. (1980) analyzed the shapes of the hyperfine lines in terms of statistical distributions in hyperfine coupling constants, although observation of the broader, high-field line proved decisive in extracting the distribution function. The mean values of $A_{iso} = 1068\,\text{mT}$ and $A_{aniso} = 6.5\,\text{mT}$ (Friebele, 1977) imply the orbital of unpaired spin to be 38% Pb 6s and $\sim 5\%$ Pb 6p. For a lead metaphosphate glass, the Pb 6s contribution was found to be 53% (Hosono et al., 1980). The remaining spin density is believed to be delocalized over neighboring oxygen atoms.

References

Aäsa, R. (1970). *J. Chem. Phys.* **52**, 3919.

Abdrashitova, E. I., Khodakovskaya, R. Y., and Khar'kova, N. I.((1972). *Inorgan. Mat. USSR* **8**, 1966.

Abragam, A., and Bleaney, B. (1970). "Electron Paramagnetic Resonance of Transition Ions." Oxford University Press, London.

Al'tshuler, T. S. (1969). *Sov. Phys. JETP* **28**, 962.

Amosov, A. V., Petrovskii, G. T., and Yudin, D. M. (1969). *Dokl. Akad. Nauk, SSSR* **188**, 1020.

Arafa, S. (1972). *J. Am. Ceram. Soc.* **55**, 137.

Arafa, S. (1974). *Phys. Chem. Glasses* **15**, 42.

Arafa, S. and Assabghy, F. (1974). *J. Appl. Phys.* **45**, 5269.

Arafa, S., and Bishay, A. (1970). *Phys. Chem. Glasses* **11**, 75.

Arafa, S., Boulos, E., and Griscom, D. L. (1980). (unpublished).

Assabghy, F., Arafa, S., Boulos, E., Bishay, A., and Kreidl, N. J. (1977). *J. Non-Cryst. Solids* **23**, 81.

Atkins, P. W. (1964). Ph.D. thesis, University of Leicester, England.

Atkins, P. W., and Symons, M. C. R. (1967). "The Structure of Inorganic Radicals." Elsevier, Amsterdam.

Azuma, N., Miazaki, T., Fueki, K., Sakaguchi, I., and Hirano, S.-I. (1986). *J. Am. Ceram. Soc.* **69**, 19.

Baugher, J., and Parke, S. (1977). In "Amorphous Materials," R. W. Douglas and B. Ellis, eds. Wiley-Interscience, London, p. 399.

Beekenkamp, P. (1966). *Philips Res. Rep.* Suppl 4, 1.

Bell, P. M., Mao, H. K., and Weeks, R. A. (1976). Proc. Lunar Sci. Conf., 5th, p. 2543.

Berger, R. (1973). *C. R. Acad. Sci., Ser. B* **276**, 425.

Bishay, A. (1962). *J. Am. Ceram. Soc.* **45**, 389.

Bishay, A., Quadros, C., and Piccini, A. (1974). *Phys. Chem. Glasses* **15**, 109.

Bleaney, B. (1951). *Phil. Mag.* **42**, 441.

Bleaney, B., and Stevens, K. W. H. (1953). *Rep. Prog. Phys.* **16**, 108.

Blinder, S. M. (1960). *J. Chem. Phys.* **33**, 748.

Bobyshev, A. A., and Radtsig, V. A. (1981). *Kinetika i -Kataliz* **22**, 1540.

Bogomolova, L. D., Jachkin, V. A., Lazukin, V. N., Pavlushkina, T. K., and Shmuckler, V. A. (1978a). *J. Non-Cryst. Solids* **28**, 375.

Bogomolova, L. D., Jachkin, V. A., Lazukin, V. N., and Shmuckler, V. A. (1978b). *J. Non-Cryst. Solids* **27**, 427.

Bowers, K. D., and Owen, J. (1955). *Rept. Prog. Phys.* **18**, 304.

Bray, P. J., and O'Keefe, J. G. (1963). *Phys. Chem. Glasses* **4**, 37.

Breit, G., and Rabi, I. I. (1931). *Phys. Rev.* **38**, 2082.

Brodbeck, C. M. (1980). *J. Non-Cryst. Solids* 40, 305.

Brodbeck, C. M., and Iton, L. E. (1985). *J. Chem. Phys.* 83, 4285.

Brower, K. L. (1979). *Phys. Rev. B* 20, 1799.

Brückner, R. (1970). *J. Non-Cryst. Solids* 5, 123.

Brückner, R. (1971). *J. Non-Cryst. Solids* 5, 177.

Brückner, R., Sammet, M., and Stockhorst, H. (1980). *J. Non-Cryst. Solids* 40, 273.

Cases, R., and Griscom, D. L. (1984). *Nuc. Inst. & Methods* **B1**, 503.

Castner, T. Jr., Newell, G. S., Holton, W. C., and Slichter, C. P. (1960). *J. Chem. Phys.* 32, 668.

Chepeleva, I. V., and Lazukin, V. N. (1976). *Dokl. Akad. Nauk SSSR* 226, 311.

Cherenda, N. G., Shendrik, A. V., and Yudin, D. M. (1975). *Phys. Status Solidi B* 69, 687.

Cugunov, L., and Kliava, J. (1982). *J. Phys. C* **15**, L933.

Cummerow, R. L., and Halliday, D. (1946). *Phys. Rev.* 70, 433.

Devine, R. A. B., and Arndt, J. (1987). *Phys. Rev. B* 35, 9376.

de Wijn, H. W., and van Balderen, R. F. (1967). *J. Chem. Phys.* 46, 1381.

Dowsing, R. D., and Gibson, J. F. (1969). *J. Chem. Phys.* 50, 294.

Edwards, J. O., Griscom, D. L., Jones, R. B., Watters, K. L., and Weeks, R. A. (1969). *J. Am. Chem. Soc.* 91, 1095.

Edwards, A. H., and Fowler, W. B. (1989). *Phys. Rev. B* (to be published).

Feigl, F. J., and Anderson, J. H. (1970). *J. Phys. Chem. Solids* 31, 575.

Feigl, F. J., Fowler, W. B., and Yip, K. L. (1974). *Solid State Comm.* 14, 225.

Fermi, E. (1930). *Z. Physik* 60, 320.

Fournier, J. T., Landry, R. J., and Bartram, R. H. (1971). *J. Chem. Phys.* 55, 2522.

Friebele, E. J. (1977). "Proc. Int. Congr. Glass, 11th," Prague 3, 87.

Friebele, E. J., and Griscom, D. L. (1979). In "Treatise on Materials Science and Technology, Vol. 17: Glass II," M. Tomozawa and R. H. Doremus, eds. Academic Press, New York, p. 257.

Friebele, E. J., and Griscom, D. L. (1986). In "Defects in Glasses" (MRS Proceedings, vol. 61), F. L. Galeener, D. L. Griscom, and M. J. Weber, eds. Materials Research Society, Pittsburgh, Penn., p. 319.

Friebele, E. J., Griscom, D. L., and Sigel, G. H. Jr. (1974). *J. Appl. Phys.* 45, 3424.

Friebele, E. J., Griscom, D. L., Stapelbroek, M., and Weeks, R. A. (1979). *Phys. Rev. Lett.* 42, 1346.

Friebele, E. J., Griscom, D. L., and Rau, K. (1983). *J. Non-Cryst. Solids* 57, 176.

Friebele, E. J., Griscom, D. L., and Marrone, M. J. (1985a). *J. Non-Cryst. Solids* 71, 133.

Friebele, E. J., Griscom, D. L., and Hickmott, T. W. (1985b). *J. Non-Cryst. Solids* 71, 351.

Gan Fuxi, Deng He, and Liu Huiming (1982). *J. Non-Cryst. Solids* 52, 135.

Garif'yanov, N. S. (1963). *Sov. Phys. Solid State* 4, 1795.

Garif'yanov, N. S., and Yafaev, N. R. (1963). *Sov. Phys.–JETP* 16, 1392.

Garif'yanov, N. S., and Zaripov, M. M. (1964). *Sov. Phys. Solid State* 6, 1209.

Garif'yanov, N. S., Rubtsov, M. I., and Ryzhmanov, Y. M. (1962). *Steklo i Keramika* 7.

Garlick, G. F. J., Nichols, J. E., and Ozer, A. M. (1971). *J. Phys. C* 4, 2230.

Gersmann, H. R., and Swalen, J. D. (1962). *J. Chem. Phys.* 36, 3221.

Griscom, D. L. (1966). Ph.D. thesis, Brown University, Providence, R.I.

Griscom, D. L. (1971a). *J. Chem. Phys.* 55, 1113.

Griscom, D. L. (1971b). *J. Non-Cryst. Solids* 6, 275.

Griscom, D. L. (1972). *Sol. State Commun.* 11, 899.

Griscom, D. L. (1973/74). *J. Non-Cryst. Solids* 13, 241.

Griscom, D. L. (1976). In "Defects and Their Structure in Nonmetallic Solids," B. Henderson and A. E. Hughes, eds. Plenum, New York, p. 323.

Griscom, D. L. (1978a). *J. Non-Cryst. Solids* 31, 241.

Griscom, D. L. (1978b). In "Borate Glasses: Structure, Properties, Applications," L. D. Pye, V. D. Frechette, and N. J. Kreidl, eds. Plenum, New York, p. 11.

Griscom, D. L. (1978c). In "The Physics of SiO_2 and Its Interfaces," S. T. Pantelides, Pergamon, New York, p. 232.

Griscom, D. L. (1979). *Phys. Rev.* **B 20**, 1823.

Griscom, D. L. (1980a). *J. Non-Cryst. Solids* 40, 211.

Griscom, D. L. (1980b). *J. Non-Cryst. Solids* 42, 287.

Griscom, D. L. (1980c). *Phys. Rev.* **B22**, 4192.

Griscom, D. L. (1984a). *J. Non-Cryst. Solids* 67, 81.

Griscom, D. L. (1984b). *Nucl. Inst. & Methods* **B1**, 481.

Griscom, D. L. (1984c). *J. Non-Cryst. Solids* 68, 301.

Griscom, D. L. (1984d). *J. Non-Cryst. Solids* 64, 229.

Griscom, D. L. (1985a). In "Radiation Effects in Optical Materials" (SPIE vol. 541), P. W. Levy, ed. SPIE, Bellingham, Wash., p. 38.

Griscom, D. L. (1985b). *J. Non-Cryst. Solids* 73, 51.

Griscom, D. L. (1986). In "Defects in Glasses" (MRS Symposium Proceedings vol. 61), F. L. Galeener, D. L. Griscom, and M. J. Weber, eds. Materials Research Society, Pittsburgh, Penn., p. 213.

Griscom, D. L. (1989). *Phys. Rev. B* (in press).

Griscom, D. L., and Bray, P. J. (1965). (unpublished).

Griscom, D. L., and Friebele, E. J. (1981). *Phys. Rev. B* **24**, 4896.

Griscom, D. L., and Friebele, E. J. (1986). *Phys. Rev. B* **34**, 7524.

Griscom, D. L., and Friebele, E. J. (1989). In "Halide Glass Optical Fibers," I. Aggarwál and G. Lu, eds., (to be published).

Griscom, D. L., and Griscom, R. E. (1967). *J. Chem. Phys.* 47, 2711.

Griscom, D. L., and Kriz, H. M. (1979). *Am. Cer. Soc. Bull.* 58, 381.

Griscom, D. L., Taylor, P. C., Ware, D. A., and Bray, P. J. (1968). *J. Chem. Phys.* 48, 5158.

Griscom, D. L., Taylor, P. C., and Bray, P. J. (1969). *J. Chem. Phys.* 50, 977.

Griscom, D. L., Friebele, E. J., and Sigel, G. H. Jr. (1974). *Sol. State Commun.* 15, 479.

Griscom, D. L., Sigel, G. H. Jr., and Ginther, R. J. (1976). *J. Appl. Phys.* 47, 960.

Griscom, D. L., Stapelbroek, M., and Friebele, E. J. (1983a). *J. Chem. Phys.* 78, 1638.

Griscom, D. L., Friebele, E. J., Long, K. J., and Fleming, J. W. (1983b). *J. Appl. Phys.* 54, 3743.

Griscom, D. L., Brinker, C. J., and Ashley, C. S. (1987). *J. Non-Cryst. Solids* 92, 295.

Hanafusa, H., Hibino, Y., and Yamamoto, F. (1985). *J. Appl. Phys.* 58, 1356.

Hecht, H. G., and Johnston, T. S. (1967). *J. Chem. Phys.* 46, 23.

Hetherington, G. (1967). *J. Brit. Ceram. Soc.* 3, 595.

Hibino, Y., and Hanafusa, H. (1984). *Appl. Phys. Lett.* 45, 614.

Hochstrasser, G. (1966). *Phys. Chem. Glasses* 7, 178.

Hochstrasser, G., and Antonini, J. F. (1972). *Surf. Sci.* 32, 644.

Hosono, H., and Abe, Y. (1987). *J. Am. Ceram. Soc.* **70**, C38.

Hosono, H., Kawazoe, H., and Kanazawa, T. (1979a). *J. Non-Cryst. Solids* 33, 125.

Hosono, H., Kawazoe, H., and Kanazawa, T. (1979b). *J. Non-Cryst. Solids* 33, 103.

Hosono, H., Nishii, J., Kawazoe, H., Kanazawa, T., and Ametani, K. (1980). *J. Phys. Chem.* 84, 2316.

Hosono, H., Kawazoe, H., Nishii, J., and Kanazawa, T. (1981). *J. Non-Cryst. Solids* 44, 149.

Hosono, H. Kawazoe, H. Nishii, J., and Kanazawa, T. (1982). *J. Phys. Chem.* 86, 161.

Hosono, H., Abe, Y., Kawazoe, H., and Imagawa, H. (1983). *J. Am. Ceram. Soc.* 66, C192.

Hosono, H., Abe, Y., Kawazoe, H., and Imagawa, H. (1984). *J. Non-Cryst. Solids* 63, 357.

Hosono, H., Abe, Y., and Kawazoe, H. (1985). *J. Non-Cryst. Solids* 71, 261.

Hurd, C. M., and Coodin, P. (1967). *J. Phys. Chem. Solids* 28, 523.

Ignat'ev, E. G., Kasymova, S. S., Petrovskii, G. T., and Yudin, D. M. (1972). *Izv. Akad. Nauk, SSSR, Neorg. Mat.* 8, 552.

Imagawa, H. (1968). *Phys. Stat. Solidi* 30, 469.

Imagawa, H. (1969a). *J. Non-Cryst. Solids* 1, 335.

Imagawa, H. (1969b). *Phys. Chem. Glasses* 10, 187.

Imagawa, H. (1969c). *J. Non-Cryst. Solids* 1, 262.

Imagawa, H., Hosono, H., and Kawazoe, H. (1982). *J. de Physique* C9, C169.

Isoya, J., Weil, J. A., and Claridge, R. F. C. (1978). *J. Chem. Phys.* 69, 4876.

Itoh, H., Shimizu, M., Ohmori, Y., and Nakahara, M. (1987). *J. Lightwave Tech.* LT-5, 134.

Jellison, G. E. Jr., Bray, P. J., and Taylor, P. C. (1976). *Phys. Chem. Glasses* 17, 35.

Känzig, W., and Cohen, M. H. (1959). *Phys. Rev. Lett.* 3, 509.

Kappers, L. A., Gilliam, O. R., and Stapelbroek, M. (1978). *Phys. Rev. B* 17, 4199.

Karapetyan, G. O., and Yudin, D. M. (1962). *Sov. Phys.—Solid State* 3, 2063.

Karapetyan, G. O., and Yudin, D. M. (1963). *Sov. Phys. Solid State* 4, 1943.

Kawazoe, H. (1985). *J. Non-Cryst. Solids* 71, 231.

Kawazoe, H., Hosono, H., and Kanazawa, T. (1978a). *J. Non-Cryst. Solids* 29, 173.

Kawazoe, H., Hosono, H. and Kanazawa, T. (1978b). *J. Non-Cryst. Solids* 29, 159.

Kawazoe, H., Hosono, H., Kokumai, H., Nishii, J., and Kanazawa, T. (1980). *J. Non-Cryst. Solids* 40, 291.

Kawazoe, H., Hosono, H., and Nishii, J. (1982a). *J. Chem. Phys.* 76, 3422.

Kawazoe, H., Hosono, H., Nishii, J., and Kanazawa, T. (1982b). *J. Chem. Phys.* 76, 3429.

Kawazoe, H., Kohketsu, M., Watanabe, Y., Shibuya, K., and Muta, K. (1986a). In "Defects in Glasses" (MRS Symposium Proceedings vol. 61), F. L. Galeener, D. L. Griscom, and M. J. Weber, eds. Materials Research Society, Pittsburgh, Penn., p. 339.

Kawazoe, H., Watanabe, Y., Shibuya, K., and Muta, K. (1986b). In "Defects in Glasses" (MRS Symposium Proceedings vol. 61), F. L. Galeener, D. L. Griscom, and M. J. Weber, eds. Materials Research Society, Pittsburgh, Penn., p. 349.

Kim, Y. M., and Bray, P. J. (1968). *J. Chem. Phys.* 49, 1298.

Kim, Y. M., and Bray, P. J. (1970). *J. Chem. Phys.* 53, 716.

Kim, Y. M., Reardon, D. E., and Bray, P. J. (1968). *J. Chem. Phys.* 48, 3396.

Kivelson, D., and Neiman, R. (1961). *J. Chem. Phys.* 35, 149.

Kliava, J. (1988). "EPR Spectroscopy of Disordered Solids", Zinante Publishing House, Latvian SSR Academy of Sciences.

Kliava, J., and Purāns, J. (1980). *J. Magn. Res.* 40, 33.

Kneubuhl, E. K. (1960). *J. Chem. Phys.* 33, 1074.

Koopmanns, H. J. A., Perik, M. M. A., Nieuwenhuijse, B., and Gellings, P. J. (1983). *Phys. Stat. Solidi* B120, 745.

Kordas, G., and Oel, H. J. (1982). *Phys. Chem. Glasses* 23, 179.

Kordas, G., Camara, B., and Oel, H. J. (1982). *J. Non-Cryst. Solids* 50, 79.

Kordas, G., Weeks, R. A., and Kinser, D. L. (1983). *J. Appl. Phys.* 54, 5394.

Kreidl, N. J., (1983). In "Glass Science and Technology, Vol. 1: Glass-Forming Systems," D. R. Uhlmann and N. J. Kreidl, eds. Academic Press, New York, p. 105.

Kurkjian, C. R., and Sigety, E. A. (1968). *Phys. Chem. Glasses* 9, 73.

Laman, F. C., and Weil, J. A. (1977). *J. Phys. Chem. Solids* 38, 949.

Landry, R. J. (1968). *J. Chem. Phys.* 48, 1422.

Landry, R. J., Fournier, J. T., and Young, C. G. (1967). *J. Chem. Phys.* 46, 1285.

Lauer, H. V. Jr. and Morris, R. V. (1977). *J. Am. Ceram. Soc.* 60, 443.

Lee, S. (1964). Ph.D. thesis, Brown University, Providence, R.I.

Lee, S., and Bray, P. J. (1963). *J. Chem. Phys.* 39, 2863.

Lee, S., and Bray, P. J. (1964). *J. Chem. Phys.* 40, 2982.

Loveridge, D., and Parke, S. (1971). *Phys. Chem. Glasses* 12, 19.

Mackey, J. H. Jr. (1963). *J. Chem. Phys.* 39, 74.

Mackey, J. H., Jr. Kopp, M., Tynan, E. C., and Yen, T. F. (1969). In "Electron Spin Resonance of Metal Complexes." Plenum Press, New York, p. 33.

Mackey, J. H., Jr. Boss, J. W., and Kopp, M. (1970). *Phys. Chem. Glasses* 11, 205.

Maki, A. H., and McGarvey, B. R. (1958). *J. Chem. Phys.* 29, 31.

McConnell, H. M., and Strathdee, J. (1959). *Mol. Phys.* 2, 129.

Morris, R. V., and Haskin, L. A. (1974). *Geochim. Cosmochim. Acta* 38, 1435.

Morton, J. R., and Preston, K. F. (1978). *J. Magn. Res.* 30, 577.

Muncaster, R., and Parke, S. (1977). *J. Non-Cryst. Solids* 24, 399.

Nagasawa, K., Hoshi, Y., Ohki, Y., and Yahagi, K. (1986). *Japn. J. Appl. Phys.* 25, 464.

Nakai, Y. (1964). *Bull. Chem. Soc. Jpn.* 37, 1089.

Neiman, R., and Kivelson, D. (1961). *J. Chem. Phys.* 35, 156.

Nicklin, R. C., Johnstone, J. K., Barnes, R. G., and Wilder, D. R. (1973). *J. Chem. Phys.* 59, 1652.

Pake, G. E., and Estle, T. L. (1973). "The Physical Principles of Electron Paramagnetic Resonance," 2nd ed. W. A. Benjamin, Inc., Reading, Mass.

Paul, A., and Assabghy, F. (1975). *J. Mat. Sci.* 10, 613.

Peterson, G. E., and Kurkjian, C. R. (1972). *Sol. State Commun.* 11, 1105.

Peterson, G. E., Kurkjian, C. R., and Carnavale, A. (1974a). *Phys. Chem. Glasses* 15, 52.

Peterson, G. E., Kurkjian, C. R., and Carnavale, A. (1974b). *Phys. Chem. Glasses* 15, 59.

Peterson, G. E., Kurkjian, C. R., and Carnavale, A. (1975). *Phys. Chem. Glasses* 16, 63.

Pontuschka, W. M., Isotani, S., and Piccini, A. (1987). *J. Am. Ceram. Soc.* 70, 59.

Purcell, T., and Weeks, R. A. (1969). *Phys. Chem. Glasses* 10, 198.

Rawson, H. (1967). "Inorganic Glass-Forming Systems." Academic Press, New York.

Reinberg, A. R. (1964). *J. Chem. Phys.* 41, 850.

Rudra, J. K., Fowler, W. B., and Feigl, F. J. (1985). *Phys. Rev. Lett.* 55, 2614.

Sammet, M., and Brückner, R. (1985). *J. Non-Cryst. Solids* 71, 253.

Sands, R. H. (1955). *Phys. Rev.* 99, 1222.

Scherer, G. W., and Schultz, P. C. (1983). "Glass Science and Technology, Vol. 1: Glass-Forming Systems" D. R. Uhlmann and N. J. Kreidl, eds. Academic Press, New York, p. 49.

Schnadt, R., and Räuber, A. (1971). *Sol. State Commun.* 9, 159.

Schneider, J., Dischler, B., and Räuber, A. (1968). *J. Phys. Chem. Solids* 29, 118.

Schreiber, H. D. (1977). Proc. Lunar Sci. Conf., 8th, pp. 1785. ff.

Schreiber, H. D., and Haskin, L. A. (1976). Proc. Lunar Sci. Conf., 7th, pp. 1221. ff.

Schreiber, H. D., Thanyasiri, T., Lach, J. J., and Legere, R. A. (1978). *Phys. Chem. Glasses* 19, 126.

Schreiber, H. D., Sobota, M. S., and Laur, H. V. Jr. (1979). *Am. Ceram. Soc. Bull.* 58, 877.

Schreiber, H. D., Lauer, H. V. Jr., and Thanyasiri, T. (1980). *Geochim. Cosmochim. Acta* 44, 1599.

Schreurs, J. W. H. (1967). *J. Chem. Phys.* 47, 818.

Schreurs, J. W. H. (1978). *J. Chem. Phys.* 69, 2151.

Schreurs, J. W. H., and Davis, D. H. (1979). *J. Chem. Phys.* **71**, 557.

Schreurs, J. W. H., and Tucker, R. F. (1964). In "Proc. Int'l. Conf. on the Physics of Non-Crystalline Solids, Delft, 1964," J. A. Prins, ed. North Holland, Amsterdam, p. 616.

Schwartz, R. N., Clark, M. D., Chamulitrat, W., and Kevan, L. (1986). In "Defects in Glasses" Symposium Proceedings (MRS vol. 61), F. L. Galeener, D. L. Griscom, and M. J. Weber, eds. Materials Research Society, Pittsburgh, Penn., p. 359.

Shendrik, A. V., and Yudin, D. M. (1978). *Phys. Status Solidi* **B85**, 343.

Shields, L., (1966). *J. Chem. Phys.* 45, 2332.

Siderov, T. A., and Tyul'kin, B. A. (1967). *Dokl. Akad. Nauk, SSSR* 175, 872.

Sigel, G. H. Jr., Evans, B. D., Ginther, R. J., Friebele, E. J., Griscom, D. L., and Babiskin, J. (1974). NRL Memorandum Report 2934, Naval Research Laboratory, Washington, D.C.

Silsbee, R. H. (1961). *J. Appl. Phys.* 32, 1459.

Simon, S., and Nicula, Al. (1983). *J. Non-Cryst. Solids* 57, 23.

Skuja, L. N., and Silin, A. R. (1982). *Phys. Stat. Sol. A* 70, 43.

Skuja, L. N., Silin, A. R., and Boganov, A. G. (1984). *J. Non-Cryst. Solids* 63, 431.

Stapelbroek, M., Griscom, D. L., Friebele, E. J., and Sigel, G. H. Jr. (1979). *J. Non-Cryst. Solids* 32, 313.

Stathis, J. H., and Kastner, M. A. (1984). *Phys. Rev. B* 29, 7079.

Stroud, J., (1961). *J. Chem. Phys.* 35, 844.

Sunch, J. F., Sayer, M., Sigel, S. L., and Rosenblatt, G. (1971). *J. Appl. Phys.* 42, 2587.

Taylor, P. C., and Bray, P. J. (1970). *J. Mag. Res.* 2, 305.

Taylor, P. C., and Bray, P. J. (1972). *J. Am. Ceram. Soc.* 51, 234.

Taylor, P. C., and Griscom, D. L. (1971). *J. Chem. Phys.* 55, 3610.

Taylor, P. C., Baugher, J. F., and Kriz, H. M. (1975). *Chem. Rev.* 75, 203.

Toyuki, H., and Akagi, S. (1972). *Phys. Chem. Glasses* 13, 15.

Triplett, B. B., Takahashi, T., and Sugano, T. (1987). *Appl. Phys. Lett.* 50, 1663.

Tsai, T.-E., and Griscom, D. L. (1987). *J. Non-Cryst. Solids* 91, 170.

Tsai, T.-E., Griscom, D. L., Friebele, E. J., and Fleming, J. W. (1987a). *J. Appl. Phys.* 62, 2264.

Tsai, T.-E., Griscom, D. L., and Friebele, E. J. (1987b). "Diffusion and Defect Data," Vol. 53–54, 469.

Tsai, T.-E., Higby, P. L., Friebele, E. J., and Griscom, D. L. (1987c). *J. Appl. Phys.* 62, 3488.

Tsai, T.-E., Griscom, D. L., and Friebele, E. J. (1988a). *Phys. Rev. B* 38, 2140.

Tsai, T.-E., Griscom, D. L., and Friebele, E. J. (1988b). *Phys. Rev. Lett.* **61**, 444.

Tsai, T.-E., Friebele, E. J., Griscom, D. L., and Pannhorst, W. (1989). *J. Appl. Phys.* **65**, 507.

Tsay, F.-D., and Helmholtz, L. (1969). *J. Chem. Phys.* 50, 2642.

Tucker, R. F. (1962). In "Advances in Glass Technology." Plenum, New York, p. 103.

Uchida, Y., Isoya, J., and Weil, J. A. (1979). *J. Phys. Chem.* 83, 3462.

Urnes, S. (1961). *Trans. Brit. Ceram. Soc.* 60, 85.

Van Vleck, J. H. (1935). "Electric and Magnetic Susceptibilities." Oxford University Press, London.

Vanf Weiringen, J. S. (1955). *Disc. Faraday Soc.* 19, 118.

Vitko, J. Jr. (1978). *J. Appl. Phys.* 49, 5530.

Vitko, J. Jr. and Shelby, J. E. (1982). *J. Am. Ceram. Soc.* **65**, C86.

Watanabe, Y., Kanazawa, T., and Kawazoe, H. (1985). *J. Non-Cryst. Solids* 71, 279.

Weeks, R. A. (1956). *J. Appl. Phys.* 27, 1376.

Weeks, R. A. (1963). *Phys. Rev.* 130, 570.

Weeks, R. A. (1985). *J. Non-Cryst. Solids* 71, 435.

Weeks, R. A., and Abraham, M. M. (1965). *J. Chem. Phys.* 42, 68.

Weeks, R. A., and Bray, P. J. (1968). *J. Chem. Phys.* 48, 5.

Weeks, R. A., and Nelson, C. M. (1960a). *J. Am. Ceram. Soc.* 43, 399.

Weeks, R. A., and Nelson, C. M. (1960b). *J. Appl. Phys.* 31, 1555.

Weeks, R. A., and Purcell, T. (1965). *J. Chem. Phys.* 43, 483.

Weeks, R. A., and Sonder, E. (1963). In "Paramagnetic Resonance," vol. 2, W. Low, ed. Academic Press, New York, p. 869.

Weil, J. A. (1984). *Phys. Chem. Minerals* 10, 149.

Wertz, J. E., and Bolton, J. R. (1972). "Electron Spin Resonance: Elementary Theory and Practical Applications." McGraw-Hill, New York.

Wickman, H. H., Klein, M. P., and Shirley, D. A. (1965). *J. Chem. Phys.* 42, 2113.

Wolf, A. A., Friebele, E. J., Griscom, D. L., Acocella, J., and Tomozawa, M. (1983). *J. Non-Cryst. Solids* 56, 349.

Wolf, A. A., Friebele, E. J., and Tran, D. C. (1985). *J. Non-Cryst. Solids* 71, 345.

Wong, J., and Angel, C. A. (1976). "Glass Structure by Spectroscopy." Marcel Dekker, New York.

Wolzencraft, J. M., and Jacobs, I. M. (1967). "Principles of Communication Engineering." Wiley, New York, p. 163.

Yafaev, N. R., and Yablokov, Yu. V. (1962). *Sov. Phys. Solid State* 4, 1123.

Yafaev, N. R., Garif'yanov, N. S., and Yablokov, Yu. V. (1963). *Sov. Phys. Solid State* 5, 1216.

Yasaitis, E. L., and Smaller, B. (1953). *Phys. Rev.* 92, 1068.

Yip, K. L., and Fowler, W. B. (1975). *Phys. Rev.* **B11**, 2327.

Zakharov, V. K., and Yudin, D. M. (1965). *Sov. Phys. Solid State* 7, 1267.

Zamotrinskaya, E. A., Torgashinova, L. A., and Anufrienko, V. F. (1972). *Izvestiya Akademii Nauk SSSR, Neorg. Mat.* **8**, 1136.

Zavoisky, E. (1945). *J. Phys. USSR* 9, 211.

Zhitnikov, R. A., and Mel'nikov, N. I. (1968). *Sov. Phys.—Solid State* 10, 80.

CHAPTER 4

Electron-Microscope Studies of Glass Structure

J. Zarzycki

LABORATORY OF SCIENCE OF VITREOUS MATERIALS
UNIVERSITY OF MONTPELLIER, FRANCE

I. Introduction: Scope of Electron-Microscope Studies in Glasses

The structure of glasses can be studied at different levels of resolution. The diffusion of X-rays and neutrons (Chapter 5, Volume 4A) and different spectroscopic methods (Chapter 5, Volume 4A) confirm the existence of a short-range order, which, in a great majority of cases, is seen to be similar to that of parent crystalline phases. In the most favorable cases this local order is known up to 5 to 10Å and constant efforts are made to extend this limit to higher values.

On the other hand, in some glassy systems, phase separations may introduce textures the finest of which are in the range of 20 to 50Å and which go up to several thousand angströms. They can be followed by small-angle scattering of X-rays (SAXS) or of neutrons (SANS) (Chapter 5, Volume 4A). They are also currently studied by conventional electron microscopy, which is well adapted to this resolution range.

Between these two limits is situated a region that may be defined as corresponding to a "middle-range" or "intermediate" order in which a very characteristic part of the structural organization of glasses is to be found.

253

Unfortunately very few methods are at our disposal to study this range—one of the most direct is high-resolution electron microscopy (HREM) and considerable efforts have been made to evaluate how much information could be effectively gained from this technique.

Glasses may also crystallize (devitrify) and the process of formation of crystalline phases by nucleation and growth within a glassy matrix was the object of numerous studies using electron-microscopic techniques.

Electron-microscopic (EM) investigation of glasses may thus be divided into two major groups:

1. Low-resolution studies of textures resulting from phase separation and devitrification processes and
2. High-resolution studies of "middle-range" order in amorphous phases.

Observations of surface alteration of glasses under different chemical agressions also belong to the first group, and to the second group belong studies of initial stages of the formation of nuclei or of spinodal decomposition in phase transformation: unmixing or crystallization processes.

As of the late 1980s, electron microscopy represents a whole group of techniques which permit not only the imaging of the structure at high magnification and resolution levels but also the determination of detailed information on the structure (microdiffraction) and *quantitative* chemical analysis of locally observed regions by various associated spectroscopical techniques (X-ray and electron emission, etc.). This constitutes the basis of Analytical Electron Microscopy. It is therefore necessary to start by a brief description and classification of the various methods of observation. For more detailed information, treatises by Williams and Joy (1984) and Hren et al. (1979) should be consulted.

II. Imaging Modes

In the conventional transmission electron microscope (TEM) a more or less parallel beam (typically of 100 to 200 KeV energy) is passed through the specimen and focused by an objective lens to form in successive planes, first, a diffraction pattern and then an image (Fig. 1). Subsequent lenses of the microscope can be used to focus at either of these planes; the instrument may thus be used in imaging or diffraction modes.

The microscope is analogous to a classic (photonic) microscope using visible light; the glass lenses are, however, replaced by magnetic coils and the whole system must be placed in vacuum in order to ensure the propagation of electrons. Only the transmitted and elastically scattered electrons are used to form images.

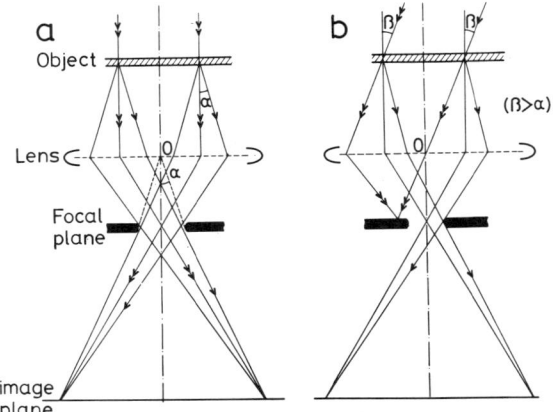

FIG. 1.   Image formation in a transmission electron microscope. (a): bright field. (b): dark field.

The magnified images are usually produced from only a portion of the diffraction pattern selected by an objective aperture. When the directly transmitted beam is included in the aperture, bright field images are produced (Fig. 1a). Dark-field images are obtained when the illumination axis is tilted so that the direct beam is excluded by the objective aperture (Fig. 1b).

In the scanning electron microscope (SEM), the lenses act on the incident beam forming a very fine probe that is scanned over the specimen. The interaction of the beam with the material excites simultaneously a number of physical processes (Fig. 2). The variation of intensity of any of these excitations may be converted by suitable transducers into an electrical signal that is monitored on a display system synchronously with the position of the scanning beam.

In the classical SEM, the specimen is generally thick and only back-scattered and secondary electrons are used as well as characteristic X-rays emitted.

In the scanning transmission electron microscope (STEM) the specimen is thin enough to permit observation of transmitted electrons as well as elastically and inelastically scattered electrons.

The modern analytical electron microscopy (AEM) is based on the STEM mode and all the preceding signals are made available.

The dedicated AEM instruments are less numerous than the classical TEM and SEM, but a number of modern TEM can also operate on the STEM mode.

The resulting image contains information about the spatial variation of the efficiency of a chosen process across the specimen's surface.

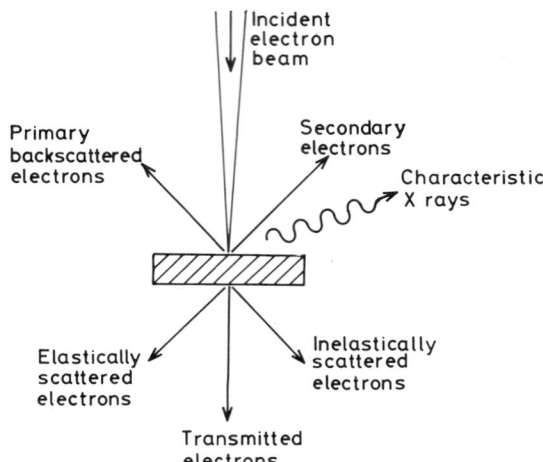

FIG. 2. Schematic representation of the most common signals used for imaging in a scanning electron microscope.

The observed contrast depends on the nature of the specimen, the type of interaction detected, the number of electrons involved, and the characteristics of the transducer. This permits various operational modes:

1. *The emissive mode* is based on the observation of electrons emitted from the specimen: low energy (0–50 eV) secondary emitted electrons or back-scattered electrons of energy close to that of the excitation beam. The back-scattered electrons give information about the elemental constituents of the specimen and its surface topography. The secondary electrons are related to surface topography and the presence of electric and magnetic fields close to the specimen surface. When only back-scattered electrons are collected, the microscope is said to be operated in the reflective mode. Energy analysis of Auger electrons permits identification of the nature of the emitting atom.

2. *The X-ray mode:* If X-ray radiation emitted when the specimen is scanned with the electron beam is collected, the microscope works similarly to the electron-probe X-ray microanalyzer.

If the characteristic X-ray wavelength of a given element is selected, the microscope permits the distribution of this element across the area of the specimen to be studied. When various emitted wavelengths are successfully used, quantitative information about the elemental constitution of the specimen may be obtained. This is the basis of the powerful technique of energy dispersive analysis of X-rays (EDAX), which is used in modern electron microscopes as of the late 1980s.

3. Other modes, less used in glass research, are possible: absorbed current mode, beam-induced conductivity mode (useful for semiconductor specimens), cathodoluminescence mode, and even acoustic mode.

4. The crystallographic orientation contrast may be used to produce electron channelling patterns in a way similar to X-ray Kikuchi-patterns, which give complementary information on orientation and perfection of crystalline regions of the specimen.

The scanning modes of electron microscopes are versatile because information may be digitized, stored, and treated to improve contrast of the images by standard image processing techniques (Rez and Williams, 1982).

Logarithmic amplification, background-intensity subtraction, Fourier treatments to remove effects due to incorrect focusing, and "chromatic" aberration of the system are possible.

These methods are usually more efficient than optical Fourier transform methods applied to image processing in standard (TEM) microscopes.

Finally, time-resolved scanning electron microscopy may be used to observe contrast changes in time-varying phenomena (e.g., fracture processes).

III. Low-Resolution Studies

It can be seen that the coupling of electron microscopes with various spectroscopic methods resulted in a powerful set of techniques of fine analysis of materials. Their application in glass science met with varied success.

The low-resolution methods became an indispensable and standard way of observation of surface or volume modification of specimens during phase separation and devitrification.

Studies of these phenomena invariably involve systematic EM observation supplementing other analytical techniques.

A. DIRECT ELECTRON MICROSCOPIC OBSERVATIONS

Direct observation of the specimens fully utilizes the resolving power of the microscope and presents the advantage that structural information (obtained by diffraction) as well as elemental analysis are possible. In the TEM mode very thin specimens are required (~ 100–300Å). In practice, thin-blown foils, edges of chips or splinters, and fragments obtained by microtomy with diamond knives were used. Special methods of embedding were developed. To avoid surface attack effects, ultrathin fibers were successfully drawn inside the electron microscope and directly transferred to the observation stage (Zarzycki and Mezard, 1962).

Ion thinning by bombarding with Ar^+ ions permits sufficiently thin regions to be obtained to examine multiphasic specimens without any danger of altering the coexisting phases. Techniques for obtaining thin glass sections for TEM were described by Turkalo (1968). Bach (1970) and Barber (1970) used ion sputtering to obtain thin foils of nonmetal glasses.

Chemical etching was used to partly or totally remove one of the phases for contrast enhancement.

Chemical dissolution followed by abrasion and ion beam thinning was used by Hing and McMillan (1973) in their study of glass ceramics.

In-situ ion etching in a SEM microscope was developed by Dharival and Fitch (1977).

A hot-stage TEM for crystallization studies of glasses is described by Kinser and Hench (1970).

The volume distribution of particles can be observed by stereoscopic methods to avoid overlap problems.

Calculations of tilt angles for obtaining stereo TEM micrographs were given by Tambuyser (1984).

The contrast depends on the nature of the glass phases (e.g., presence of heavy or light atoms, the beam voltage, and the thickness of the specimen. The optimum acceleration voltage is generally situated between 50 and 120 kV. Higher voltage permits thinner specimens to be used but at the cost of reduced contrast.

Jennings and Pratt (1980) describe the use of a high-voltage EM and a gas reaction cell for microstructural investigations of wet Portland cement.

Care must be taken to avoid specimen modification by interaction with the electron beam.

Beam-heating effects ending in crystallization were observed in amorphous chalcogenide films by Bagley and Northover (1970).

Bright- and dark-field observations and moiré patterns were used to study the crystallization of certain chalcogenide glasses (Chaudhari and Herd, 1972).

Filament formation in "Ovonic" glasses was studied by high-voltage TEM, SEM, and microprobe analysis to elucidate the mechanism of switching (Sie et al., 1972; Allinson and Barry, 1979).

Risbud (1982) describes STEM studies coupled with electron energy loss spectroscopy (EELS) to evaluate multiphase microstructures in oxide and non-oxide glasses.

Direct observation of "columnar" structures in glassy $GeSe_2$ films by TEM is reported by Chen (1981). Sherman et al. (1981) describe dark-field imaging of semicrystalline polymers by STEM.

Bandyopadhyay (1984) used SEM to study crack propagation in bulk polymeric materials.

Studies of phase separation in glasses currently use TEM techniques and optical transforms of the micrographs may be used to evaluate spinodal-type decomposition textures (Zarzycki and Naudin, 1969).

EM study of nucleation and growth of silver particles in a thin glass film is reported by Bando and Kiriyama (1975). Shulin and Jingwei (1985) made a study of coloration of glass by CdSe. Borg (1973) reports some applications of TEM to study lunar glasses.

B. REPLICA TECHNIQUES

The present development of scanning electron microscopes with improved resolving power tends to render the replica techniques obsolete.

In the absence of scanning microscopy, the introduction of replicas is necessary when the specimen thinning is difficult or when the material's susceptibility to electron beam makes direct observation impracticable.

The resolution of replicas is lower than that of direct observation by TEM, and, moreover, the possibility of analysis of the material itself is lost except in extraction replicas when adherent particles of the material are removed and examined.

Numerous techniques for replication developed for electron-microscopic examination of glasses have been described (Vogel, 1977: Vogel et al., 1982). For a detailed description of preparation techniques, Glavert (1972) should be consulted.

The main problem in the observation of glasses is the initial cleanliness of the surface to be replicated. In some cases, because of an extremely rapid interaction, in particular with atmospheric moisture, fresh surfaces are preferred—they may be obtained by fracturing the specimen in high vacuum. The contrast enhancement of surface topography is generally produced by chemical etching; care must be taken not to introduce unwanted artifacts (e.g., by secondary redeposition of substances generated during etching).

Simple replicas are generally most satisfactorily obtained using carbon-evaporation techniques. Surface topography may be advantageously suggested by standard shadowing techniques. The mixed C-Pt-Ir method produces excellent shadowed replica in one operation. It has very fine grain, and the residual "noise" due to the structure of the replica itself may be estimated using ideally smooth crystals deposited on the specimen to be replicated, e.g., MO_3 crystals (Vogel, 1977).

For systems containing crystals with prominent features that tend to break the replicas, double replicas were used in the past.

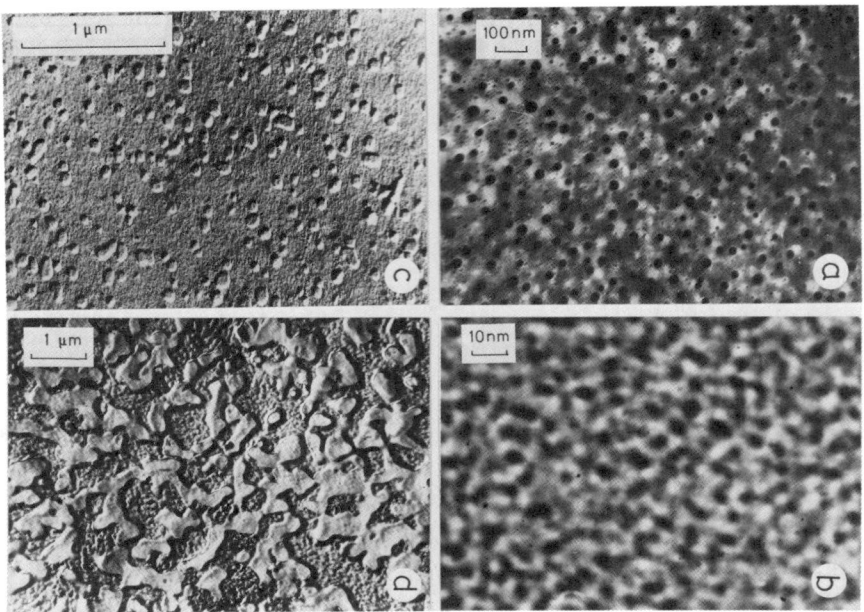

FIG. 3. Examples of phase separation in B_2O_3-PbO-Al_2O_3 glasses observed by direct transmission (a,b) and replica techniques (c,d).

Wu et al. (1973) describe a metallic replica technique for SEM that is indestructible under electron beam and may be used for soft materials.

Figures 3 and 4 show representative examples of microstructures observed in the field of phase separation in glasses by TEM and SEM. Figure 5 illustrates the effect of atmospheric moisture on fine structure observed by TEM.

IV. High-Resolution Studies

Obtaining direct real space images of amorphous solids down to a resolution of $\sim 3\text{Å}$ is now possible with the advent of high-resolution electron microscopy methods. Before I discuss more recent work on imaging and interpretation, a short summary of the theory of image formation will be presented. For more detailed treatment, see, e.g., Cowley (1975) and Spence (1981).

A. DIFFRACTION THEORY OF IMAGE FORMATION

The process of image formation in the electron microscope can be described by an extended form of Abbe's theory. Two successive stages can be distinguished (Fig. 6). In the first stage the electronic beam interacts with the

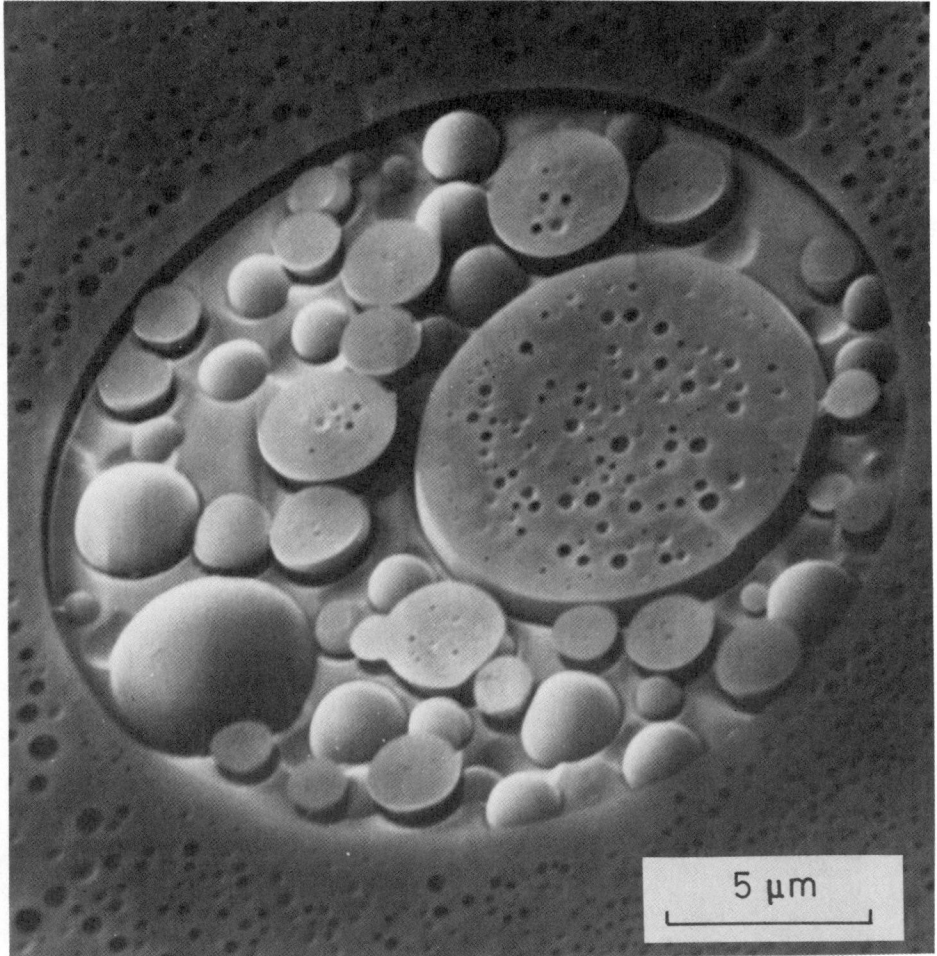

FIG. 4. Multiple-phase separation in a BaO-B_2O_3-SiO_2 glass observed by SEM with secondary electrons. (From Vogel et al., 1982.)

object. The incident plane wave $\psi_0(x, y)$ is modified by the electrostatic potential of the object, which introduces a phase shift

$$-\frac{\pi}{\lambda E} \phi(x, y) = \sigma\phi(x, y)$$

where E is the acceleration potential, λ the wavelength, and $\phi(x, y)$ the projection of the potential distribution on a plane normal to the incident beam. The specimen which is supposed to be thin enough not to modify the amplitude acts as a phase object. The distribution of the amplitude of the

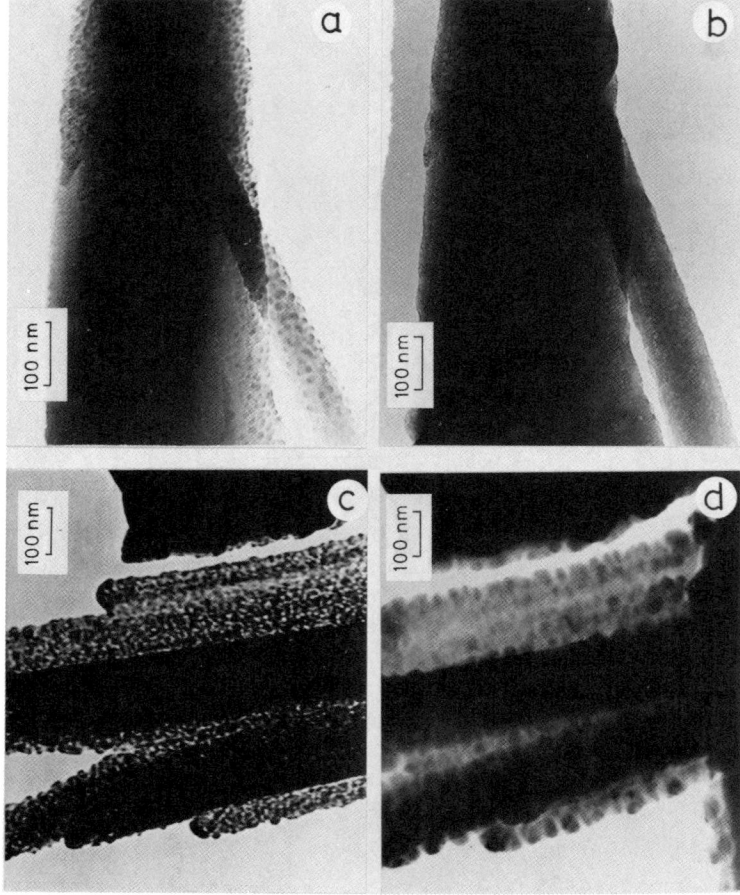

FIG. 5.   Effect of atmospheric moisture on fine structure of glass observed by TEM. a,c: glass fibrils drawn from the melt inside the column of electron microscope. b,d: the same after a brief contact with ambient atmosphere. (From Zarzycki and Mezard, 1962.)

wave function at the exit plane of the specimen is then:

$$\psi(x, y) = \psi_0(x, y)q(x, y)$$

with

$$q(x, y) = \exp[-i\sigma\Phi(x, y)].$$

In a second stage this wave is treated by the objective lens of the microscope (followed by other projection lenses) and transferred on to the observation plane where the amplified image is formed. This second stage is important as the imperfections of the imaging system introduce distortions and modifica-

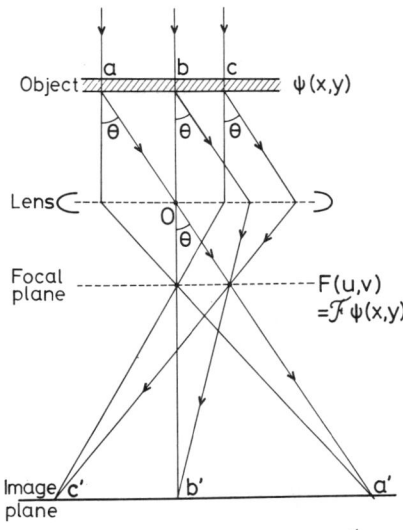

FIG. 6.   Image formation in transmission electron microscope (schematic).

tions of the transmitted information that affect the final image. The discussion can be limited to the action of the objective lens; the subsequent lenses only amplify the image produced by the objective lens. In Fig. 6, the mechanism of image formation is shown schematically; three important planes may be distinguished: object, focal, and image planes. The radiation diffused by the specimen under an angle θ is focalized by the lens (L) in a point of the focal plane depending on θ. In the focal plane an amplitude distribution is formed, the amplitude in each point being the sum of all the amplitudes diffused under a given angle. The corresponding intensity distribution is Frauenhofer's spectrum of the specimen.

Formally this amplitude distribution is the Fourier transform of the amplitude of the wave leaving the specimen:

$$F(u,v) = \mathscr{F}[\psi_0 q(x, y)]$$

$$= \psi_0 \iint q(x, y)\exp[2\pi i(ux + vy)] \, dxdy$$

where u and v are angular variables that define the positions in the focal plane

$$u = x/f\lambda$$

and

$$v = y/f\lambda,$$

f being the focal length.

The lenses transfer the radiation from the focal plane into the image plane, which in optical terms represents a second Frauenhofer diffraction process with interferences between different waves reaching each point of the image plane. Mathematically this corresponds to a second Fourier transform,

$$\psi_0(x, y) = \mathscr{F}\{\mathscr{F}[\psi_0 q(x, y)]\} = \psi_0 q\left[-\frac{x}{M}, -\frac{y}{M}\right],$$

which reconstructs, inverted and amplified with the ratio M, the transmission function q of the specimen. This perfect reconstruction supposes that all the spatial components are present and that correct phase relationships are conserved during the transfer, which is not possible in practice. The highest frequency components are suppressed by the diaphragm, which limits the spatial resolution. The defocalization and spherical aberration of the objective lens introduce supplementary phase shifts.

The defocalization (i.e., imaging of a plane situated Δz from the exit plane of the specimen) introduces a phase change $\pi \Delta z \theta^2 / \lambda$. The third-order spherical aberration (impossible to suppress) introduces a phase change $\pi C_s \theta^4 / 2\lambda$ where C_s is the coefficient of spherical aberration.

These phase changes are additive and so the amplitude of the wave is multiplied by a term $\exp[i\gamma(\theta)]$ where

$$\gamma(\theta) = \frac{\pi \Delta z \theta^2}{\lambda} + \frac{\pi C_s \theta^4}{2\lambda}.$$

The objective lens then produces the amplified image combining the different waves with appropriate $\gamma(\theta)$ phase changes. In the bright-field operation mode all the electrons, whether diffused or not, contribute to the image. If the optical system were perfect, no image contrast would be observed. In a real system, this contrast is due to the function $\gamma(\theta)$. By changing the focusing the contrast will be increased but the different spatial components will be affected to a different extent. This introduces effects that must be taken into account in the interpretation of the image. In the case of axial illumination and thin specimens, a Fourier component of periodicity $d = 2\lambda \sin(\theta/2)$ is transferred with a contrast $\sin \gamma(\theta)$.

In the case of amorphous specimens there exists a continuous spectrum of diffused waves; several of these intensity components may show no contrast or even a reversed contrast depending on the defocus conditions.

This spatial filtering depends on the transfer function $\sin \gamma$ and defocus conditions may be used to visualize a given range of periodicities of the object. Figure 7 shows $\sin \gamma$ calculated for different defocus values Δz as a function of d/d_0 where d is the periodicity and $d_0 = (C_s \lambda^3)^{1/4}$, a constant which for 100 kV electrons is equal to 5.6Å. For a defocus $\Delta z_1 = -1.2\sqrt{C_s \lambda}$

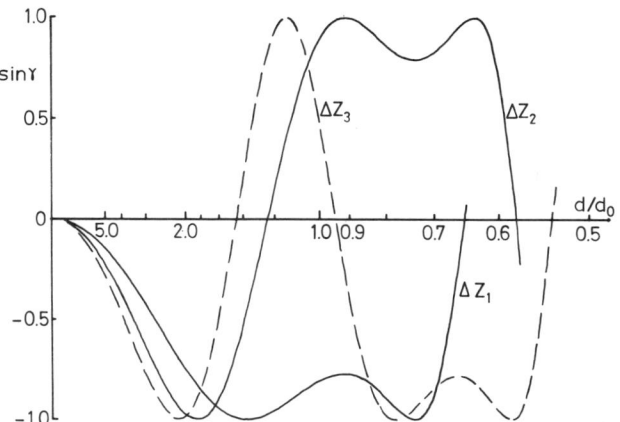

FIG. 7.   Transfer function $\sin \gamma$ calculated for different defocus values ΔZ. (From Howie, 1978.)

(Scherzer's defocus), the microscope will enhance the transmission of period-icities between 4 and 10Å, while for a greater defocus $\Delta z_2 = -1,85\sqrt{C_s \lambda}$, periodicities comprised between 8 and 10Å as well as between 3.3 and 6Å will be enhanced, the latter with contrast reversal. For a still greater defocus Δz_3, three transmitted bands are seen to exist that will strongly affect the aspect of the image.

B. APPLICATION TO AMORPHOUS STRUCTURES

The preceding discussion shows that the electron microscope acts as a strong spatial filter and that image formation at a high resolution depends on the optical characteristics of the system. The importance of "noise" and its relevance on the imaging problem of amorphous structures was not however immediately realized.

The early work was centered on the proof of "continuous random network" versus "crystallite" models of the structure. Computer-generated structures from different models were tested to see if theoretically the presence of "crystallites" in an amorphous matrix could be detected.

In the scattering theory of disordered structures the amplitude A scattered by a collection of N atoms with scattering factors f_i is

$$A = \sum_i^n f_i \exp(i \mathbf{k} \cdot \mathbf{r}_i)$$

where $\mathbf{k}$ is the scattering vector and $\mathbf{r}_i$ the position vector of atom i.

Diffraction experiments give only the scattered intensity

$$I = AA^* = \sum_i \sum_j f_i f_j \exp[i\mathbf{k}(\mathbf{r}_i - \mathbf{r}_j)]$$

and all the information concerning the origin of vectors $\mathbf{r}_i$ and $\mathbf{r}_j$ contained in the phase of the scattered amplitude is lost. This is the classic problem in crystallography. From the Fourier transform of I, a radial distribution function may be obtained which is, however, insufficient and higher-order correlations would be required to fix unambiguously the structure. (See Chapter 1, Volume 4A.)

The electron microscope is precious since the imaging process retains the phase information of the amplitudes and through the second Fourier transform the lens gives the spatial distribution of the atoms.

In the case of scattering from disordered structures, the problem is how to distinguish the useful information from the "noise" present in the image.

This can be reduced to the validity of the central limit theorem (CLT) (see, e.g., Papoulis (1965)), in which case the electron wave at the image plane can be considered as a random signal and the phase information contains no structural information.

The formalism developed for laser "speckle" statistics can be used (Goodman, 1975) and it can be shown that if the CLT holds, the intensity distribution for phase-contrast imaging in the weak-phase object approximation is Gaussian and the phase distribution is uniform.

This is the base for treating high-resolution images by phase randomization: if phases are originally random, image reconstruction after randomization of phases in the Fourier space produces similar image features.

To obtain a situation where CLT no longer holds and where images might contain structural information, thin specimens of about 10Å must be used.

However, it has been shown by Smith et al. (1981) that such thin foils no longer behave as a bulk solid, and that image quality is degraded, probably because of phonon dispersion. Thus specimens of about 100Å should be used in practice. The best conditions correspond to dark-field imaging with a relatively large objective aperture. The deviations from random images can thus be detected for coherently scattering regions of more than about 7Å in size. Larger regions are also detectable in the axial bright field but the contrast is poorer and the noise enhanced (Chevalier, 1982).

Reducing the coherence of illumination for dark-field images can be achieved by an annular dark-field detector (Howie, 1978).

Stereo imaging in the dark field avoids the problem of overlap which reduces the applicability of the CLT. It is, however, an extremely difficult technique for amorphous solids and necessitates specially equipped high-resolution microscopes with side-entry goniometers coupled with computer-controlled lenses to eliminate diffraction or image rotation (Chevalier, 1980).

However, even if the images are optically correct, their interpretation may give rise to doubt since the structure of very thin films may be particular and does not reflect that of bulk solid.

Also high-chemical reactivity of some glasses may produce possible artifacts. These were observed by Fukamashi et al. (1979) for $Pd_{60}Ni_{20}Si_{20}$ and by Chevalier (1982) for $Cu_{60}Zr_{40}$ glasses.

In conclusion, it can be said that the mechanism of image formation is now fairly well understood and that HREM is capable of detecting "ordered" inhomogeneities of about 7Å in size in foils of about 100Å. Failure to detect such inhomogeneities can be ascribed to the presence of continuous random network or dense random packing of hard-spheres types of structure.

For very thin specimens (30–50Å), the CLT may no longer hold and images may have different statistics, but such images have not yet been adequately interpreted in terms of models.

It seems unlikely that electron microscopy can, as of the late 1980s, produce information on a smaller scale (~ 3Å) in amorphous solids—i.e., to detect the position of individual atoms.

C. Progress of HREM Studies in Glasses

Early HREM work mostly concerned amorphous semiconducting solids Ge, Si, chalcogenide glasses, and occasionally SiO_2. The existence of bright specks in dark-field images was interpreted as being due to the presence of microcrystallites (Rudee, 1971) or of ordered regions (Chaudhari et al., 1972a,b,c). Irregular fringe patterns in tilted bright-field images were also thought to support the microcrystallite model (Rudee and Howie, 1972; Howie et al., 1973). Further work showed, however, that the fringe pattern depended on the angle of illumination tilt (Herd and Chaudhari, 1974) and was thus an instrumental effect. In thicker films, inelastically scattered electrons contribute to fringe contrast when tilted illumination is employed and this effect arises from the transfer function of the microscope (Parsons and Hoelke, 1974; McFarlane, 1975; McFarlane and Cochran, 1975; Krakow et al., 1976).

The importance of "noise" was fully realized by Berry and Doyle (1973) and Krivanek (1976) in the interpretation of HREM micrographs.

The scattered electron wave was treated as a random signal. Using this concept, image statistics can be predicted as in the case of laser speckle (Lowenthal and Arsenault, 1970).

After 1970 the main effort was concentrated on the possibilities of imaging computer-modeled structures (Krivanek, 1975; Krivanek and Howie, 1975; Graczyk and Chaudhari, 1975). This work has shown the virtual impossibility of detecting "crystallites" in an amorphous structure. On the other hand, quasi-periodic structures might sometimes arise even from continuous-random network structures. Krivanek et al. (1976) suggested an optical transform to phase-randomize electron micrographs.

An analogous procedure using digitized images and a computer is described by Saxton et al. (1977). The idea is that if phases are initially random, then after randomization in Fourier space, similar image features are obtained.

Axial high-resolution imaging was used by Gaskell and Mistry (1979), but results could not be interpreted in terms of any particular model. Results obtained by Gaskell and Mistry (1979) and Gaskell et al. (1979) were the first in which the extended transfer interval ensured a faithful representation of the electron wave intensity (Figs. 8 and 9).

The work of Gibson (1980) and Gibson et al. (1977) is related to the problem of "noise" in the dark-field mode and to methods of reducing "speckle" in micrographs.

Difficulties arise when images computed from models are compared to experimental micrographs (Howie, 1978). Instrumental aberration terms are often not included in the calculations and calculated images do not display the optimum defocus conditions (Krivanek and Howie, 1975).

Clusters used in model calculations are not sufficiently large, which decreases overlap problems and increases the probability of fringe structures.

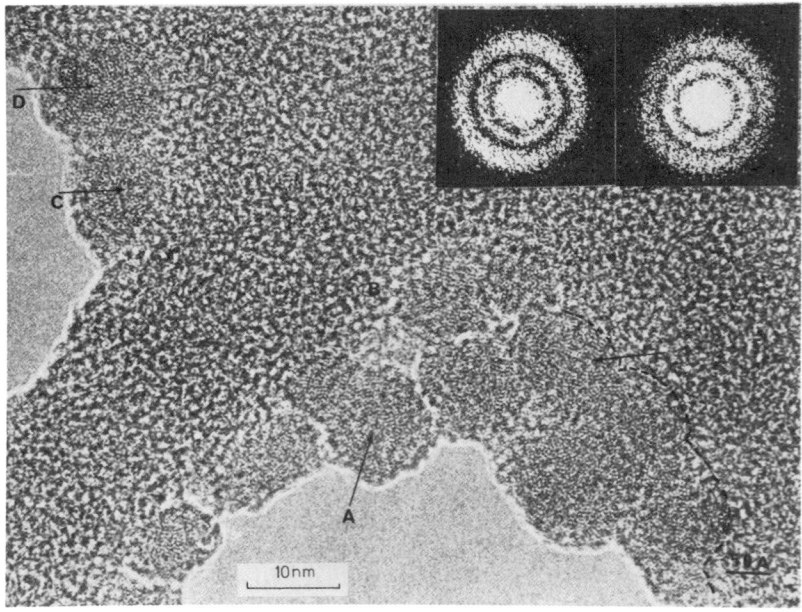

FIG. 8.   High-resolution transmission electron micrograph of solution-precipitated amorphous silica particles (Ludox SM 30). The particles A to D measure between 11 and 16 nm. A boundary separates particles and the amorphous carbon support film. The inserts show optical diffractograms of silica (left) and carbon (right). (From Gaskell and Mistry, 1979.)

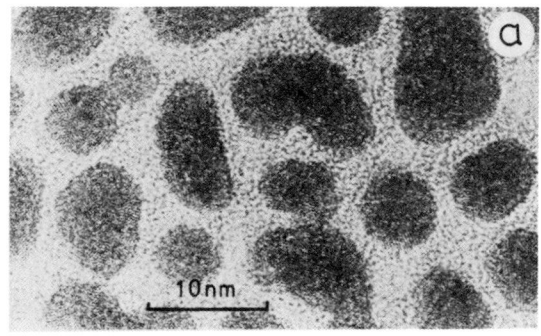

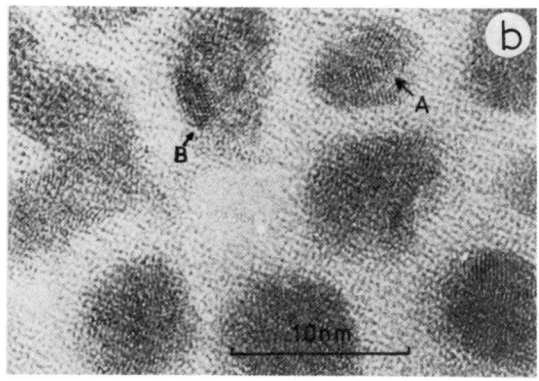

FIG. 9.   High-resolution images of a particulate Pd-Si alloy recorded with axial illumination at 500 KV. (a) Typically amorphous. (b) Showing some signs of crystallinity; particles A and B show low-contrast fringe systems inclined at angles of 140° and 120°. (From Gaskell and Smith, 1980.)

Model calculation should be performed for the actual specimen's thickness and effects of thermal vibration and inelastic scattering should be included.

An optical bench for generating simulated images from random objects with variable spatial filtering was used by Spaepen and Meyer (1974) and Krivanek (1975). It is, however, difficult to ensure that the optical simulation is sufficiently close for detailed comparisons to be possible in practice. Striking low-resolution beaded structures were observed in thin films of Si deposited on NaCl substrates. In silicate glasses fringes with a tendency to run parallel to the fracture edge as well as "birds' nests" structures reminiscent of those found in some carbon blacks have been seen by Freeman et al. (1977).

The problem of transfer functions for low-contrast objects has been examined with an improved formalism by Lannes et al. (1981).

V. Conclusions

Electron microscopy has become a standard tool for studying phase transformations and devitrification phenomena in glasses. High-resolution studies of disordered structures meet, however, with inherent difficulties of interpretation of resulting images. The detection of partial crystalline order seems possible in some cases but, in general, the thickness of the specimens is too great and resulting overlap of individual atomic positions renders the structural interpretation ambiguous. This is in striking contrast with HREM studies of crystalline specimens where the lattice-imaging leads to detailed and unequivocal conclusions.

References

Allinson, D. U., Barry, T. I., Clinton, D. J., Hughes, A. J., Lettington, A. H., and Savage, J. A. (1979). *J. Non-Cryst. Sol.* **31**, 307–331.

Bach, H. (1970). *J. Non-Cryst. Sol.* **3**, 1–32.

Bagley, B. G., and Northover, W. R. (1970). *J. Non-Cryst. Sol.* **2**, 161–169.

Bando, Y., and Kiriyama, R. (1975). *J. Non-Cryst. Sol.* **18**, 119–127.

Bandyopadhyay, S. (1984). *J. Mat. Sc. Lett.* **3**, 39–43.

Barber, D. J. (1970). *J. Mat. Sc.* **5**, 1–8.

Beamish, A., and Hourston, D. J. (1976). *J. Mat. Sci.* **11**, 1581–1582.

Berry, M. V., and Doyle, P. A. (1973). *J. Phys. C: Solid State Phys.* **6**, 46.

Borg, J. (1973). *J. Mat. Sci.* **8**, 484–489.

Chaudhari, P., and Herd, S. R. (1972). *J. Non-Cryst. Sol.* **8–10**, 56–63.

Chaudhari, P., Graczyk, J. F., and Charbnau, H. P. (1972a). *Phys. Rev. Lett.* **29**, 425.

Chaudhari, P., Graczyk, J. F., and Herd, S. R. (1972b). *Phys. Stat. Sol.* **B52**, 801.

Chaudhari, P., Herd, S. R., Ast, D., Brodsky, M. H., and von Gutfeld, R. J. (1972c). *J. Non-Cryst. Sol.* **8–10**, 900–908.

Chen, C. H. (1981). *J. Non-Cryst. Sol.* **44**, 391–395.

Chevalier, J. P. (1980). *J. Microscopy* **119**, 113.

Chevalier, J. P. (1982). "Electron Microscopy and Analysis 1981," M. J. Goringe, ed. Institute of Physics, London and Bristol, England, p. 391.

Clarke, D. R., Breakwell, P. R., and Sims, G. D. (1970). *J. Mat. Sci.* **5**, 873–880.

Cowley, Y. M. (1975). "Diffraction Physics." North Holland, Amsterdam.

Dhariwal, R. S., and Fitch, R. K. (1977). *J. Mat. Sci.* **12**, 1225–1232.

Freeman, L. A., Howie, A., Mistry, A. B., and Gaskel, P. H. (1977). In "The structure of Non-Crystalline Materials," P. H. Gaskell, ed. Taylor and Francis, London, p. 245.

Fukamashi, M., Hoshimoto, K., and Yoshida, H. (1979). *Scripta Met.* **13**, 807.

Gaskell, P. H., and Mistry, A. B. (1979). *Phil. Mag. A* **39**, 245.

Gaskell, P. H., and Smith, D. J. (1980). *J. of Microscopy* **119**, no. 1, 63–72.

Gaskell, P. H., Smith, D. J., Catto, C. J. D., and Cleaver, J. R. A. (1979). *Nature* **281**, 465.

Gibson, J. M. (1980). In "Electron Microscopy and Analysis 1979," T. Mulvey, ed. Institute of Physics, London and Bristol, England, p. 273.

Gibson, J. M., Howie, A., and Stobbs, W. M. (1977). In "Electron Microscopy and Analysis 1977," D. L. Misell, ed. Institute of Physics, London and Bristol, p. 275.

Goodhew, P. Y. (1972). In "Practical Methods in Electron Microscopy," vol. 1, A. M. Glauert, ed. North Holland, Amsterdam and London, pp. 7–180.

Goodman, J. W. (1975). In "Laser Speckle and Related Phenomena," J. C. Dainty, ed. Springer Verlag, Berlin, chap. 2.

Graczyk, J. F., and Chaudhari, P. (1975). *J. Non-Cryst. Sol.* **17**, 299.

Graczyk, J. F., and Chaudhari, P. (1976). *Phys. Stat. Sol.* (B) **75**, 593.

Graczyk, J. F., and Chaudhari, P. (1978). *Phys. Stat. Sol.* (B) **58**, 501.

Herd, S. R., and Chaudhari, P. (1974). *Phys. Stat. Sol.* (A) **26**, 627.

Hing, P., and McMillan, P. W. (1973). *J. Mat. Sci.* **8**, 340–348.

Howie, A. (1978). *J. Non-Cryst. Sol.* **31**, 41.

Howie, A., Krivanek, O. L., and Rudee, M. L. (1973). *Phil. Mag.* **27**, 235.

Hren, J. J., Goldstein, J. I., Joy, D. C. eds. (1979). "Introduction to Analytical Electron Microscopy," Plenum, N.Y.

Jennings, H. M. and Pratt, P. U. (1980). *J. Mat. Sci.* **15**, 250–253.

Kinser, D. L., and Hench, L. L. (1970). *J. Mat. Sci.* **5**, 369–373.

Krakow, W., Ast, D. G., Goldfarb, W., and Siegel, B. M. (1976). *Phil. Mag.* **33**, 985.

Krivanek, O. L. (1975). Ph.D. thesis, University of Cambridge.

Krivanek, O. L. (1976). "Electron Microscopy 1976," D. Brandon, ed. Tal International, Jerusalem, p. 275.

Krivanek, O. L., and Howie, A. (1975). *J. Appl. Cryst.* **8**, 213.

Krivanek, O. L., Gaskell, P. H., and Howie, A. (1976). *Nature* **262**, 454.

Lannes, A., Tanaka, M., and Temple, P. (1981). *Optik* **60**, 1–28.

Lowenthal, S., and Arsenault, H. (1970). *J. Opt. Soc. Am.* **60**, 1478.

McFarlane, S. C. (1975). *J. Phys. C: Solid State Phys.* **8**, 2819.

McFarlane, S. C., and Cochran, W. (1975). *J. Phys. C: Solid State Phys.* **8**, 1311.

Papoulis, A. (1965). "Probability, Random Variables and Stochastic Processes." McGraw-Hill, New York.

Parsons, J. R., and Hoelke, W. (1974). *Phil. Mag.* **30**, 135.

Risbud, S. H. (1982). *J. Non-Cryst. Sol.* **49**, 241–251.

Rudee, M. L. (1971). *Phys. Stat. Sol.* **B46**, K1.

Rudee, M. L., and Howie, A. (1972). *Phil. Mag.* **25**, 1001.

Saxton, W. O., Howie, A., Mistry, A., and Pitt, A. (1977). In "Developments in Electron Microscopy and Analysis 1977," D. L. Misell, ed. Institute of Physics, London and Bristol, p. 119.

Sherman, E. S., Adams, W. W., Thomas, E. L. (1981). *J. Mat. Sci.* **16**, 1–9.

Shulin, W., and Jingwei, F. (1985). *J. Mat. Sci. Lett.* **4**, 568–570.

Sie, C. H., Dugan, M. P., and Moss, S. C. (1972). *J. Non-Cryst. Sol.* **8–10**, 877–884.

Smith, D. J., Stobbs, W. M., and Saxton, W. O. (1981). *Phil. Mag. B* **43**, 907.

Spaepen, F., and Meyer, R. B. (1974). *J. Non-Cryst. Sol.* **13**, 440.

Spence, J. C. H. (1981). "Experimental High Resolution Electron Microscopy." Clarendon Press, Oxford, England.

Stobbs, W. M. (1977). "The Structure of Non-Crystalline Materials." Taylor and Francis, London, p. 253.

Tambuyser, P. (1984). *J. Mat. Sci. Lett.* **3**, 184–186.

Turkalo, A. M. (1968). *J. Am. Chem. Soc.* **51**, 470.

Vogel, W. (1971). "Structure and Crystallisation of Glasses." Pergamon, Oxford, England.

Vogel, W. (1977). *J. Non-Cryst. Sol.* **25**, 172.

Vogel, W., Horn, L., Reiss, H., and Völksch, G. (1982). *J. Non-Cryst. Sol.* **49**, 221–240.

Williams, D. B., and Joy, D. C. eds. (1984). "Analytical Electron Microscopy," San Francisco Press, San Francisco.

Wu, W., Argon, A. S., and Turner, A. P. L. (1973). *J. Mat. Sci.* **8**, 1670–1672.

Zarzycki, J., and Mezard, R. (1962). *Phys. Chem. Glasses* **3**, 163.

Zarzycki, J., and Naudin, F. (1969). *J. Non-Cryst. Sol.* **1**, 215–234.

CHAPTER 5

Mössbauer Effect in Glasses

G. Tomandl

INSTITUT FÜR WERKSTOFFWISSENSCHAFTEN III (GLAS UND KERAMIK)
UNIVERSITÄT ERLANGEN, WEST GERMANY

I. Introduction

Since the discovery of the recoil-free γ-resonance absorption by Rudolph Mössbauer in 1985 (who was awarded the Nobel prize for his work) (Mössbauer, 1958a,b, 1959) an exponential increase of papers dealing with this effect could be observed. The so-called Mössbauer effect (ME) originally was developed as a method for measuring nuclear physical properties. But very soon it was found out that the real value of this effect consisted of the numerous possibilities of applications in such different fields as relativity theory, solid-state physics, metallurgy, chemistry, and biology.

Mössbauer originally discovered his effect using ^{191}Ir. This isotope gives only a small effect and the measurement needs a considerable experimental expense, e.g., low temperature. Fortunately, soon other isotopes such as ^{57}Fe and ^{119}Sn were discovered with a much better resolution and much more

273

comfortable experimental conditions. Because of the extremely high resolution, hyperfine interactions can be readily observed, thus giving valuable information about the closer environment of the Mössbauer nucleus within the structure. The ME is only sensitive to the next neighbors of the atom. Therefore a periodic structure in the solid need not necessarily be present, but also the effect can be observed in amorphous systems, such as glasses.

Pollak et al. (1962) were the first who observed the ME in iron-containing SiO_2 glasses. General descriptions of the ME theoretically and experimentally can be found in the books of Frauenfelder (1962), Wertheim (1965), Wegener (1966), Greenwood and Gibb (1971), Vertes et al. (1979), and Barb (1980).

A complete bibliography for the years 1958–1965 was collected in the Mössbauer Data Index by Muir et al. (1966). The continuation of the bibliography for the years 1966–1976 was performed by Stevens and Stevens (1977). Additionally, the Mössbauer Effect Reference and Data Journal (1967–) can be used to stay up to date with all literature about ME.

II. Theory

Reviews presenting the theoretical aspects of the ME include Tomandl et al. (1967), Tomandl and Oel (1970), Kurkjian (1970), Tomandl (1974), Gonser (1975), and, more recently, Müller-Warmuth and Eckert (1982).

An atomic nucleus that has been excited in any way emits γ-radiation during the transition into the ground state. During the emission of this photon, the free nucleus experiences a recoil according to the laws of conservation of energy and momentum, which reduces the energy of the emitted quants by

$$E_R = \frac{E_o^2}{2Mc^2}.$$ (1)

E_0 is the original energy of the γ-quant, M the mass of the nucleus, c light velocity. The energy of the emitted line is therefore

$$E_{em} = E_0 - E_R.$$ (2)

The reverse process, the absorption of γ-rays with exciting of the nucleus, gives the energy of the absorption line

$$E_{ab} = E_0 + E_R.$$ (3)

In general, each line intensity is Lorentzian:

$$I(E) \sim \frac{1}{1 + \left(\dfrac{E - E_0}{\Gamma/2}\right)^2},$$ (4)

$I(E)$ being the intensity, E the energy, E_0 maximum energy, and Γ the halfwidth of the line. Emission and absorption lines almost do not overlap, because $\Gamma < \delta E$. Therefore, resonant absorption with free nuclei is generally not possible. For example, the 14.4 keV γ-radiation of the excited ^{57}Fe gives $\Gamma = 4.6 \, 10^{-9}$ eV, $E_R = 2 \, 10^{-3}$ eV.

Mössbauer found out that in a solid under certain conditions a recoil-free line the "Mössbauer-line," is emitted. In solids, the energy consumption and emission can happen only in small portions, phonons. If the phonon energy E_P is greater than E_R and E_R is smaller than the energy necessary to push an atom from its lattice site, then a final probability exists that the whole lattice accepts the recoil energy and the γ-radiation is emitted or absorbed without any shift of E_R. Formally, in Eq. (1), instead of the mass of one nucleus M, the mass of all atoms in the specimen has to be inserted (at least 10^{15} atoms).

The probability of recoil-free emission or absorption is called Mössbauer-Lamb factor (sometimes also Debye-Waller factor), f. According to the Debye model, E_P corresponds to lattice vibrations with frequencies between 0 and a maximum frequency ω_D. The theory results in a formula for f as a function of temperature T:

$$f(T) = \exp\left\{ -\frac{3E_R}{2k\theta}\left[1 + 4\left(\frac{T}{\theta}\right)^2 \int_0^{\theta/T} \frac{x \, dx}{e^x - 1} \right] \right\} \qquad (5)$$

or in the limit for low temperatures $(T < \theta)$

$$f(T) \approx \exp\left\{ -\frac{E_R}{k\theta}\left(\frac{3}{2} + \frac{\pi^2 T^2}{\theta^2} \right) \right\}, \qquad (6)$$

θ being the Debye temperature. For high temperatures, the approximation of (5) is:

$$f(T) \approx \exp\left\{ -\frac{6E_R T}{k\theta^2} \right\}. \qquad (7)$$

The recoil-free line, the Mössbauer line, can be used to perform resonance-absorption experiments. For this purpose a Mössbauer source, most often a radioactive source, is mounted on a velocity drive, which moves the source with a velocity against the absorber. The absorber normally is the material being investigated. Because of the Doppler effect, the emitted γ-rays are shifted somewhat within the energy scale of E_D:

$$E_D = \frac{v}{c} E_0. \qquad (8)$$

In the case of ^{57}Fe, 1 mm/s corresponds to $4.8 \, 10^{-8}$ eV. For ^{119}Sn, it corresponds to $7.97 \, 10^{-8}$ eV.

The number of photons transmitted by the absorber is counted and recorded as a function of velocity (v). For high velocities, no resonant absorption can occur; therefore, we get the highest count rate, which is not dependent on v. At lower velocities a resonant absorption reduces the count rate, thus giving negative peaks. Mathematically speaking, the measured spectrum is a convolution of the emission with the absorption spectrum, weighted with the absorption due to the thickness of the absorber. In the special case of Lorentzian lines for both emitter and absorber and negligible thickness of the absorber, again Lorentzian lines result with a halfwidth equal to the sum of the two halfwidths of emitter and absorber, respectively. In the case of nonnegligible absorber thickness as a first approximation also Lorentzian functions can be used for the mathematical description, but with correction factors for the halfwidth, giving larger values than in the simple case of zero thickness. (See, e.g., Frauenfelder, 1962.)

Sometimes the role of source and absorber is changed. For these cases the source contains the material under investigation. The spectra obtained are reflected images of the "normal" absorption spectra. The advantage of this arrangement is the possibility of a smaller minimum concentration of the Mössbauer isotopes. On the other hand, sometimes it is not always clear whether the influence of the parent nucleus or of the Mössbauer isotope is effective in the spectra, e.g., ^{57}Co or ^{57}Fe in the case of iron spectroscopy.

Another possibility is the measurement of the scattered Mössbauer γ-radiation or the conversion electrons. In this way, information can be achieved coming just from a thin layer at the surface of the sample. Positive peaks are recorded. The only problem is the scatter geometry, which either gives a poor resolution (an apparent large linewidth) or makes large scatterers with noncomfortable shapes necessary.

A. Interactions Influencing the Spectra

For the most useful Mössbauer nuclei, the natural linewidths of the recoil lines are so small that even the small effects of the chemical environment of the atoms on the levels of the energies of the nuclei can be detected.

Three main interactions have to be considered: isomer shift (IS), quadrupole splitting (QS) and magnetic hyperfine splitting (HFS). In Fig. 1 the term scheme for a $3/2 \to 1/2$ transition (as, for example, ^{57}Fe or ^{119}Sn) is shown. Each level can be slightly shifted and/or split by the influence of its electronic environment, which again is also determined by the next surrounding, e.g., the next neighbors. A transition from the excited level to the ground level corresponds to the emission γ-line of the Mössbauer source. Vice versa, the transition from the ground to the excited level describes the absorption in the sample.

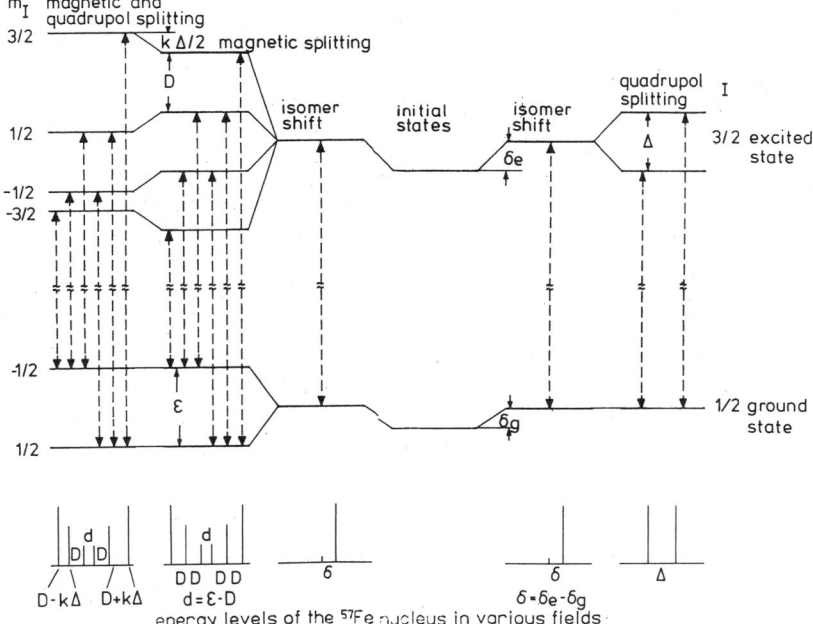

FIG. 1.   Energy levels of the ^{57}Fe nucleus in various environments.

1. Isomer Shift

A shift of the levels arises from the Coulombic interaction between the nucleus and the electrons. Only an influence of the electron density at the nucleus is effective. Therefore only the s-electrons contribute to the isomer shift.

There are different shifts for the ground and excited levels depending on the different nuclear radii. In the Mössbauer experiment, the difference of the shifts in the source and the absorber is always observed. This difference is called chemical isomer shift, which can be expressed in velocity scale by

$$\delta = \frac{2}{3}\pi Z e^2 \{\langle r_e^2 \rangle - \langle r_g^2 \rangle\}\{|\psi_s(0)|^2 A - |\psi_s(0)|^2 s\}\frac{c}{E_0} \qquad (9)$$

where Z is atomic number, e is elementary charge, and $\langle r_e^2 \rangle$ and $\langle r_g^2 \rangle$ are the mean square nuclear radii in the excited and ground states. $|\psi_s(O^2)|$ are the electron densities at the nucleus for the source and the absorber, respectively. This term contains the information about the valency states and chemical bonding.

Isomer shifts cannot be given in an absolute scale but only relative to a specific reference material. Therefore, the source that has been used must always be given in connection with any value for IS.

2. Quadrupole Splitting

The quadrupole splitting can be calculated from the eigenvalues of the Hamiltonian. It results in a splitting of the energy levels. (See Fig. 1.) The number of the splittings depends on the quantum number I (nuclear spin) and the magnetic quantum number m_I ($m_I = -I, -I + 1, \ldots, I - 1, I$). The energy levels are split into $I + 1/2$ doubly degenerated levels because the splitting depends on m_I^2.

The energy levels in the case of the absence of an inner magnetic field and an axially symmetric electric field gradient (EFG) tensor are:

$$E_Q = \frac{eV_{zz}Q}{4I(2I - 1)} [3m_I^2 - I(I + 1)]. \qquad (10)$$

For example, for ^{57}Fe and ^{119}Sn, the nuclear spin is $I = 2/3$. Therefore, the ground state remains unsplit and the excited state becomes twofold split, one level for $m_I = \pm 1/2$, another for $m_I = \pm 3/2$. The energy separation is for this case:

$$\Delta E_Q = eV_{zz}Q/2. \qquad (11)$$

For the transition $I = 3/2 \rightarrow 1/2$ also in the case of a nonsymmetric EFG tensor, a solution for E_Q is possible: instead of $eV_{zz}Q$ $eV_{zz}Q(1 + \eta^2/3)^{1/2}$, with asymmetry parameter $\eta = (V_{xx} - V_{yy})/V_{zz}$ can be written. V_{xx}, V_{yy}, and V_{zz} are diagonal elements of the diagonalized EFG tensor with $|V_{zz}| > |V_{yy}| \geqslant |V_{xx}|$. V_{zz} is the maximum value of the electric field gradient.

The Laplace equation holds: $V_{xx} + V_{yy} + V_{zz} = 0$. Since η can never be greater than unity, the term $(1 + \eta^2/3)^{1/2}$ is 1.16 at maximum and can be disregarded for many cases.

The separation of the two lines Δ in the MB spectrum in the velocity scale is

$$\Delta = \frac{1}{2} eV_{zz}Q(1 + \eta^2/3)^{1/2} \frac{c}{E_0}. \qquad (12)$$

Normally sources are used showing only one unsplit line. Therefore only the QS of the absorber must be considered. The QS is caused by an electric field gradient at the nucleus. Filled electron shells give no contribution to the EFG like Fe^{3+}. Electrons outside of filled shells like in Fe^{2+} contribute to the EFG. In addition to this, the influence of the next neighbors give a nonnegligible contribution to the QS via the polarization of the electrons.

3. Magnetic Interaction

In the presence of the magnetic field at the nucleus, a magnetic hyperfine splitting (HFS) of the energy levels occurs. (See Fig. 1.) The levels are split into $2I + 1$ niveaus. In the absence of an EFG, E_H can be expressed by

$$E_{\mathrm{H}} = \frac{\mu}{I} H m_I. \tag{13}$$

μ is the magnetic moment of the nucleus, different for the ground and excited state.

The distances between the equidistant levels are:

$$\Delta E_{\mathrm{H}} = \mu H / I. \tag{14}$$

The conversion to the velocity scale gives

$$\varepsilon = \frac{c}{E_\mathrm{o}} \frac{\mu_\mathrm{g}}{I_\mathrm{g}} H \quad \text{and} \quad D = \frac{c}{E_\mathrm{o}} \frac{\mu_\mathrm{e}}{I_\mathrm{e}} H \tag{15}$$

for the ground and excited states respectively. ε and D as shown in Fig. 1. For ^{57}Fe a splitting into two levels for the ground state and into four levels for the excited state can be specified. Due to the selection rule of $|\Delta m_I| \leqslant 1$, only six lines (instead of eight) can be observed. The values for the magnetic moments for ^{57}Fe are

$$\mu_\mathrm{g} = 0.09024 \; \mu N$$

$$\mu_\mathrm{e} = -0.12547 \; \mu_N$$

μ_N (nuclear magnetons) $= 3.15149 \; 10^{-12} eV/G$ with $G =$ Gauss

$$\varepsilon = 1.185 \; 10^{-5} \frac{mm/s}{G} H$$

$$D = 0.677 \; 10^{-5} \frac{mm/s}{G} H. \tag{16}$$

In the presence of combined electric and magnetic interaction, each level is shifted as shown in the example of Fig. 1. The exact values for the energy levels cannot be given in closed form. Only certain approximations are possible.

For example, for an axially symmetric EFG tensor with its principal axis having an angle θ with the magnetic direction:

$$E_H = -\mu H / I + (-1)|m_I| + 1/2 \frac{e V_{zz} Q}{8} (3 \cos^2 \theta - 1). \tag{17}$$

Note that for $\cos^2 \ominus = 1/3$, the quadrupole effect is apparently absent. For ^{57}Fe and converted to velocity scale (see Fig. 1, left-most),

$$\Delta = \frac{1}{2} e V_{zz} Q \frac{c}{E_0} \quad \text{and} \quad k = \frac{1}{2}(3 \cos^2 \theta - 1). \tag{18}$$

For a nonsymmetric EFG tensor, but $\ominus = 0$, also a closed solution for $I = 3/2 \rightarrow 1/2$ can be given. (See Greenwood and Gibb, 1971.)

For $I = 3/2 \rightarrow 1/2$ in almost any case (for small QS), the following relations hold:

$$v_3 - v_2 = v_5 - v_4 \tag{19}$$

$$\delta = (2v_1 + v_2 + v_3 + v_4 + v_5 + 2v_6)/8 \tag{20}$$

$$H = 73.86 \frac{kG}{mm/s}(v_6 - v_5 + v_2 - v_1) \quad (\text{for } ^{57}\text{Fe}) \tag{21}$$

$$k\Delta = (2v_1 - v_2 - v_3 - v_4 - v_5 + 2v_6)/4 \quad (\text{only for a symmetric tensor}). \tag{22}$$

v_1 to v_6 are the positions of the six lines in velocity scale. These positions can be determined from measured MB spectra by fitting methods. (See II.B.) In any case, the distances $v_3 - v_2$ and $v_5 - v_4$ should be kept equal.

An inner magnetic field can be found in ferro-, antiferro-, and ferrimagnetic materials. But also paramagnetic materials sometimes show HFS in the case of a long spin-lattice relaxation time compared to the lifetime of the excited state of the Mössbauer niveau. (See III.C.2.) On the other hand, the HFS can degenerate if this relaxation time becomes too small as in the case of superparamagnetic particles. The anisotropy constant for these particles is dependent on the particle volume. The frequency of the fluctuations of the magnetization therefore becomes too high and the inner magnetic field during the Mössbauer measuring time is more or less averaged out. (See III.C.1.)

The intensities of the six lines should be for a polycrystalline or glassy material as 3:2:1:1:2:3. These ratios hold only for a random orientation of the internal magnetic field to the direction of advance of the γ-photons. For polycrystalline materials exhibiting textures or absorbers having a too-large thickness, these ratios can be changed remarkably.

4. Goldanskii-Karyagin Effect

In the absence of an inner or external magnetic field, a pure quadrupole splitting is expected. For a $3/2 \rightarrow 1/2$ transition (e.g., ^{57}Fe) a doublet with equal linewidth and height most often is observed. In very few cases, however, an asymmetric doublet is obtained. One reason might be a texture in a polycrystalline material. But also in glasses, typically the Fe^{2+}-doublet, asymmetric doublets can be observed.

A possible explanation for this was given by Goldanskii et al. (1962) and developed mathematically by Karyagin (1963). The recoilless fractions for the transitions of the two levels of the excited state to the ground state may be different. This anisotropy of the f-factor for the two transitions gives rise to a different height of the two peaks of the doublet, but also to different areas of both peaks. The so-called Goldanski-Karyagin effect should be more pronounced at higher temperatures.

Experimentally it is difficult to decide whether this effect is present or a texture exists. In the case of glasses, a different distribution of IS and QS at the same time could also explain this asymmetry. (See next section.)

5. Line Broadening

Especially in glasses, the measured lines most often are remarkably broadened and sometimes no more Lorentzian. One reason is a variation in the arrangement of the next neighbors, thus giving rise to a distribution of IS and/or QS.

Also a distribution of a magnetic field may result in a broadening. But very often relaxation effects related to inner magnetic fields may be the cause for a line broadening.

B. EVALUATION OF MB-SPECTRA BY FITTING METHODS

Most often the MB lines do overlap and therefore a numerical separation is necessary. In the past very sophisticated programs have been developed, most of them using the method of least squares. A set of functions (in general Lorentzian functions) is assumed. Fitting parameters are line position, height, and width and one for the background. These parameters are varied until the standard deviation has reached a minimum. This is a nonlinear problem that only can be solved by using iterative methods. Therefore, in many cases, convergence fails. One reason may be that the set of the initial guess values is too far away from the best solution. But very often, convergence only can be forced by introducing constraints between the parameters; e.g., all linewidths are kept equal, some peak positions are fixed, or the distance between the lines 2 and 3 are kept equal to 4 and 5 for six line splittings.

For these cases, a very critical valuation is necessary. Sometimes the result is already predefined with the initial parameter set and the constraints. For example, Nolet (1980) shows in an example the ambiguity of the fit of the asymmetric Fe^{2+} doublet, which he also could fit with a set of three doublets of equal height.

A different approach using measured MB spectra instead of synthetic mathematical functions for the peaks is also sometimes applied. These evaluation techniques are generally denoted as stripping methods. Varret and Naudin (1979) separated in this way the Fe^{2+} doublet from the complicated spectrum of Fe^{3+} showing a relaxation effect in silicate glasses. (See III.C.2.) Levitz et al. (1980) and Massiot (1985) proposed similar methods for their evaluations.

C. MÖSSBAUER SOURCES

Figure 2 shows the elements and isotopes that exhibit ME. Unfortunately most of them either show a too small effect (small f factor) or show a too-large or too-small linewidth (because of a not suitable lifetime of the excited state) or else there is not an easy-to-make radioactive isotope with a sufficient halflifetime, which exhibits a populated Mössbauer niveau. In some cases, a particle accelerator must be available. In many cases, the measurement must be performed at low temperatures (e.g., He-temperature). Figure 2 also shows those elements that are the most suitable ones for measurements, especially on glasses. Unfortunately no elements such as Si, B, Al, O and the network modifiers are available. Among the latter, only ^{40}K in principle is possible but only with experimental difficulties and very poor resolution and sensitivity for structural effects.

MÖSSBAUER PERIODIC TABLE

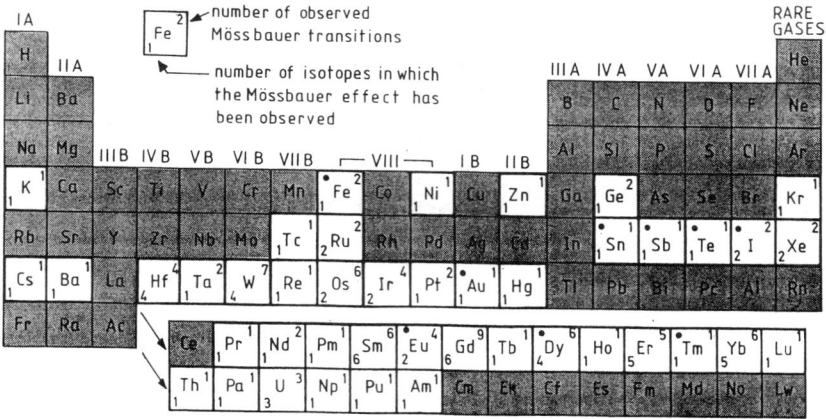

FIG. 2. Mössbauer periodic table. The white boxes denote the Mössbauer nuclides. The most suitable nuclides are marked with a dot. (Revised version from Mössbauer data index (Muir et al., 1966). See also Gonser, 1975.)

Let us consider the remaining useful isotopes:

By far most of all measurements are made with the ^{57}Fe isotope. As a source ^{57}Co diffused into Cu-, Pt-, Pd-, and Cr-foils is used. The decay scheme is shown in Fig. 3. The halflifetime of 270 days is long enough for practical purposes. The minimum linewidth of 0.2 mm/sec for source and absorber together is most often sufficient to resolve QS and HFS. Due to $I_g = 1/2$ and $I_e = 3/2$, the spectra are not too complicated. The spectra can be readily measured at room temperature and above. ^{57}Fe has a natural abundance of 2.19%. This is sufficient to measure the MB down to a concentration of about 2% Fe_{nat}. Using iron enriched with ^{57}Fe, measurements down to 0.1% Fe are possible depending on the type of the spectra (e.g., the presence of relaxation effects).

^{119}Sn is almost as useful as ^{57}Fe. The source is ^{119}Sn, which can be readily made in a nuclear reactor. The halflifetime of 245 days is in the same range as of ^{57}Co. Also, I_g and I_e are the same. Only the lifetime of the Mössbauer level is somewhat shorter, therefore giving a minimum linewidth of 0.63 mm/sec. The natural abundance is 8.6%.

Other MB isotopes possibly useful for studies on glasses are ^{121}Sb, ^{125}Te, ^{127}I, ^{151}Eu, ^{161}Dy, ^{169}Tm, and ^{197}Au. Sb and Te are important for investigations on nonoxide glasses. All these isotopes have certain disadvantages: Te, I, Eu, Tm, and Au have large linewidths. Sb, I, Eu, and Dy show very complicated spectra due to higher numbers of the nuclear spins. I, Te, Dy, Tm, and Au have short lifetimes of the parent nuclides.

D. EXPERIMENTAL TECHNIQUES

Figure 4 presents a scheme of the standard Mössbauer measuring arrangement. The source is mounted on an accelerator which is in most cases

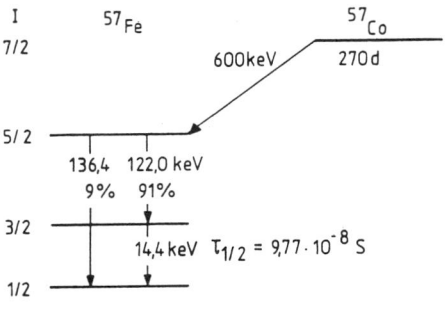

FIG. 3.   Decay scheme of ^{57}Co.

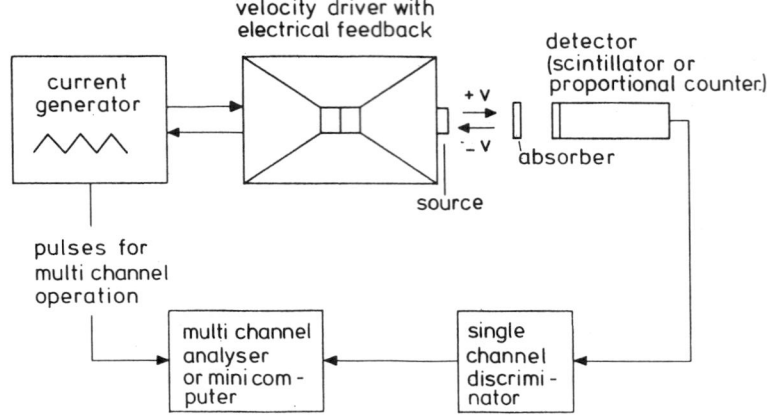

FIG. 4.   Schematic arrangement of a Mössbauer spectrometer.

a loudspeaker system with electronic feedback. The velocity function versus time is most often a symmetric triangle. Sometimes, especially in the case of heavy parts to be moved (e.g., for low temperature measurements), a sine function is also used. To collect the data, either a multichannel-analyzer in multiscale mode or a minicomputer with direct memory access is used. The kind of detector is dependent on the energy of the Mössbauer isotope. Proportional and scintillation counters are used most often. Lithium-drifted germanium or silicon detectors are also sometimes used. They enjoy the advantage of having the best energy resolution, but often the active areas are too small—or detectors with larger areas are too expensive.

E. Other Methods Utilizing the Mössbauer Effect

1. Selective-Excitation Double-Mössbauer Spectroscopy

A single line source of ^{57}Co is attached to a constant velocity drive. The velocity of the γ-source is adjusted to the energy of the desired absorption resonant energy in the sample. (See Balko and Hoy, 1974, 1983.) The radiation is scattered on the sample and analyzed using a constant acceleration drive with a single line absorber. In this way, only one energy niveau is excited; therefore, only those transitions can be observed that start from the only excited niveau.

The MB spectra become simpler. Especially relaxation effects can be studied in more detail because the niveaus, which relax at different temperatures, can be excited separately. (See II.C.) Problems with this method arise from the low count rate, which makes strong MB sources necessary, and from a rather complex theoretical treatment.

2. Mössbauer Scattering

Scattering experiments utilizing the extremely monochromatic Mössbauer radiation can be used to determine the fraction of Rayleigh to Rayleigh and Compton scattering. The temperature dependence of this fraction can be related to the atomic motion in the glassy state (Champeney and Sedgwick, 1972).

Another type of Mössbauer scattering is high-energy resolution X-ray spectroscopy (HERXS). Albanese and Deriu (1979) describe this method as a combination of the ME and conventional X-ray diffraction. For glasses this method should give the opportunity to determine the radial density distribution (RDD) solely for Rayleigh scattered γ-rays. Therefore, the movement of the atoms smearing out the conventionally obtained RDD should no more disturb the RDD obtained by HERXS. Unfortunately, the experimental difficulties—especially with the very low count rate—prevented these authors from achieving good RDDs (Albanese and Ghezzi, 1974).

A third type of Mössbauer scattering was proposed by Champeney and Woodhams (1966). In a conventional MB experiment using a source and a single line absorber, a sheet of any inhomogeneous material is placed between source and absorber and rotated transversely to the γ-ray beam with a velocity of up to 2500 cm/s. A broadening of the MB line results, which can be interpreted as a narrow angle scattering at angles of $\leqslant 10^{-5}$ radians. Inhomogeneities (e.g., phase separation and crystallization) should be detectable this way.

All MB-scattering experiments are useful with materials *not* containing the MB-isotopes. Therefore they also do not show the extreme selectivity as the normal MB experiments do.

III. Applications

Reviews about applications of the ME on glasses can be found by Tomandl et al. (1967), Kurkjian (1970), Taneja et al. (1973), Coey (1974), and Müller-Warmuth and Eckert (1982). A general review about using the ME to determine valency states (in high and low spin iron) and coordination numbers from IS and QS was given by Duncan and Golding (1965).

A. INVESTIGATIONS OF THE GLASS STRUCTURE

1. Valency State

As explained in Section II.A.1, the exact value of IS can be used to determine the valency state of the MB atom. In contrast to EPR, all valency states can be detected with the ME. In EPR only those atoms having an electron in excess (e.g., a paramagnetic atom) can give rise to a peak in the spectrum. Using the ME, distinctions between Fe^{2+} and Fe^{3+}, Sn^{2+} and Sn^{4+}, Sb^{3+} and Sb^{5+} (Paul et al. 1977), Np^{2+} (or Np^{3+}) and Np^{5+} (Jove et al. 1979), and Eu^{2+} and Eu^{3+} (Coey et al. 1981) is possible. Figure 5 shows, as an example, the dependence of the MB spectra on the ratio of Fe^{2+}/Fe_{total}. The alkali silicate glasses were melted at atmospheres varying from strongly-reducing to oxidizing conditions. With help of computer fits, each spectrum can be resolved into two doublets as shown in Fig. 6. Assuming equal f factors for the two valency states, the area ratios between the doublets can be attributed to the ratio of the valency states of iron Fe^{2+}/Fe_{total}. Figure 7 shows a comparison of these ratios determined by MB evaluation and by chemical analysis.

In the same graph another method of evaluating the area ratios was used that gives at least the same accuracy with much less expense in calculations. A formula was derived that correlates the ratios of the peak heights of the peak 3 to peak 1 to the area ratios of the two doublets, assuming constant peak positions and widths. Also, Goldman and Bewley (1985) proposed a similar method. This method is rather accurate especially for glasses containing mainly only one valency state of iron, provided there is no change in the parameters of the doublets within the series under evaluation (e.g., change in coordination, see III.A.2).

Lewis and Drickamer (1968) found a change of Fe^{3+} to Fe^{2+} with an external applied pressure of up to 200 kbar. The ratio of the two valency states followed a power law with pressure. The exponent was correlated to the difference between the volumes of the ferric and ferrous sites. This difference is smaller in glasses than in crystals.

In Fig. 8, the MB-spectra for tin in borate glasses with varying ratio of Sn^{2+}/Sn_{total} is shown. In contrast to the MB spectra for iron, the two valency states of tin are clearly resolved and therefore very easily evaluated with computer fits.

There are several possible sources of errors in the evaluation of the valency ratios. Due to the overlapping, the computer fits always involve some uncertainties, especially if one doublet is very small. The main problem, however, is the fact that it cannot be generally assumed that the f factor is equal for the different valency states. For instance, Lechtenböhmer et al. (1982) showed a very different temperature dependence of the f factors for

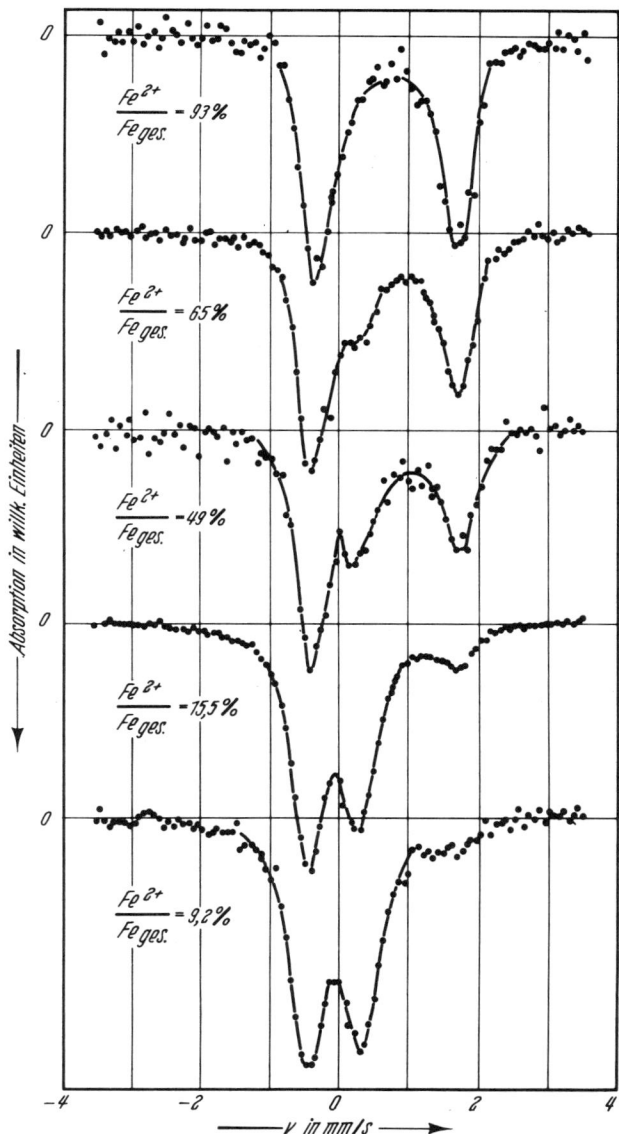

FIG. 5. Series of Mössbauer spectra for silicate glasses containing Fe with various ratios of Fe^{2+}/Fe_{total}. (Tomandl et al., 1967.)

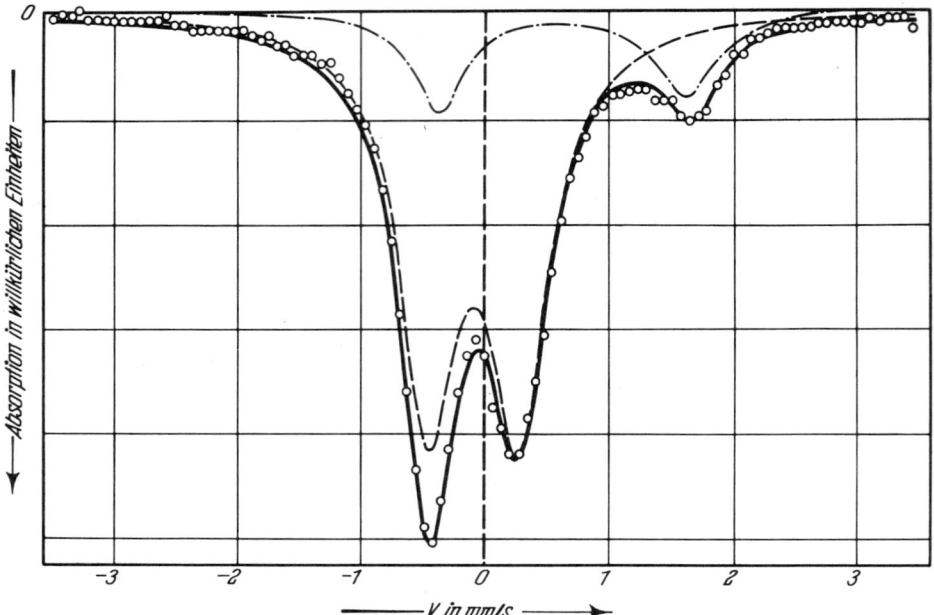

FIG. 6.   Example for a separation of a Mössbauer spectrum into doublets belonging to Fe^{3+} (narrow) and Fe^{2+} (wide). (Fourth spectrum of Fig. 5, $Fe^{2+}/Fe_{total} = 15.5\%$.) (Frischat and Tomandl, 1969.)

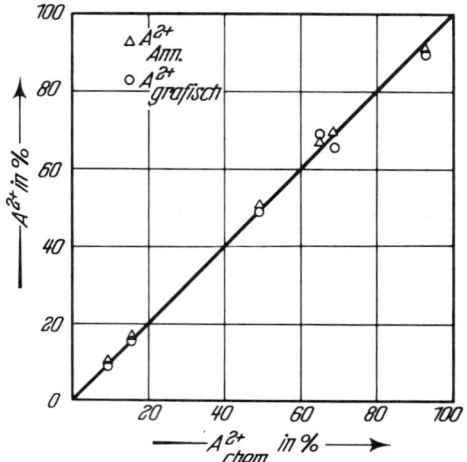

FIG. 7.   Comparison of evaluations of the valency ratios of iron in silicate glasses with Mössbauer and chemical analysis. (Frischat and Tomandl, 1969.)

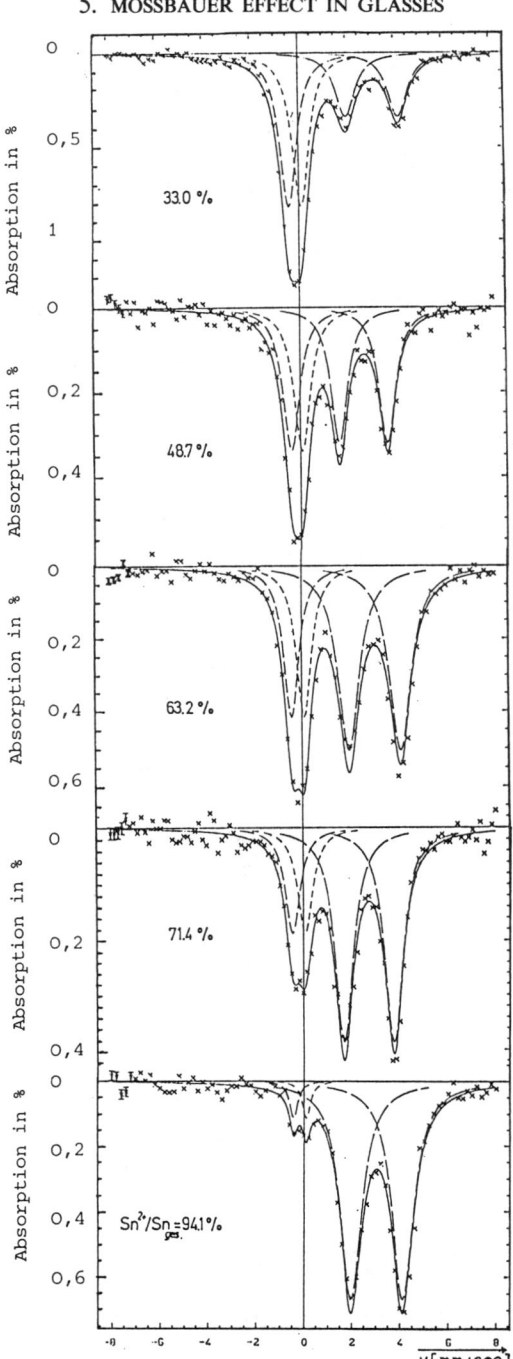

FIG. 8.   Series of Mössbauer spectra for borate glasses containing Sn for different ratios Sn^{2+}/Sn_{total}. (Dannheim and Frey, 1978.)

Sn^{2+} and Sn^{4+} in borate glasses. (See Fig. 9.) On the other hand, the same group (Cremers et al., 1984) found equal f factors for Sn^{2+} and Sn^{4+} in chalcogenide glasses.

Coey et al. (1981) determined the f factors for Eu^{2+} and Eu^{3+} and found large differences. Therefore, the valency ratio also has to be corrected with these f factors.

Another source of errors may be the occurrence of relaxation effects. (See III.C.2.) For example, this might have been the case in Labar and Gielen 1973–1974). At low iron concentration, the MB-spectra for Fe^{3+} are changed remarkably. Furthermore, it must be stated that also the wet chemical analysis of iron and tin in different valency states sometimes can be rather erroneous.

2. Coordination

Walker et al. (1961) calculated the IS as a function of x, the s-electron density for the configurations $3d^5 4s^x (Fe^{3+})$ and $3d^6 4s^x (Fe^{2+})$. The portion of x of the s-electron may be either the result of a contribution from the ligand

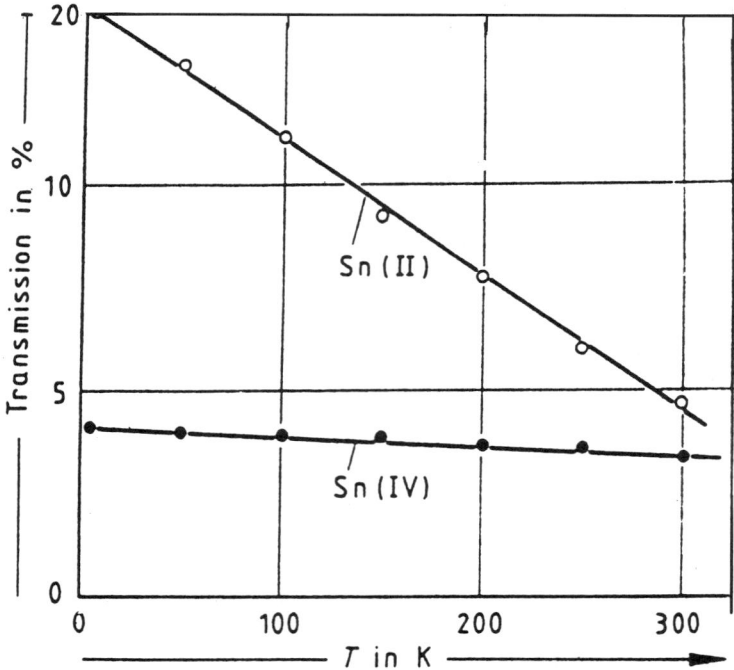

FIG. 9.   Transmission for Sn^{2+} and Sn^{4+} as a function of temperature for borate glasses, showing the different magnitude of the f factors. (Lechtenböhmer et al., 1982.)

ions (e.g., oxygen) or a compression of the 4s wave function due to the smaller Fe-O distance in a tetrahedral Fe configuration compared to an octahedral one. Kurkjian and Sigety (1968) were the first who listed the IS parameters of crystalline Fe^{3+} compounds, having tetrahedral or octahedral Fe coordinations (TC or OC). They found well-defined regions of IS values for both coordinations. The octahedral and tetrahedral shifts are consistent with $x \sim 0$ and $x \sim 0.1$ respectively. Assuming that the same limits also hold for glasses, they concluded that Fe^{3+} is tetrahedrally coordinated in silicate glasses and octahedrally coordinated in phosphate glasses.

The same results were found by Frischat and Tomandl (1969). Figure 10 shows the IS of Fe^{3+} as a function of the ratio of $Fe^{3+}/$ $(Na^+ + 1/2Ca^{2+} + Fe^{3+})$ for a silicate glass. For low Fe^{3+} concentrations, tetrahedral coordination is favored. For higher Fe^{3+} concentrations, the number of network formers Na and Ca is no longer sufficient to hold Fe^{3+} in tetrahedral coordination. Therefore, an increasing amount of Fe^{3+} goes into octahedral coordination. In Fig. 11, an attempt is shown to fit the MB spectrum of a silicate glass with a higher content of Fe (65 PL in Fig. 10) with two doublets for Fe_t^{3+} and Fe_o^{3+} and one doublet for Fe^{2+}. This fit can only be made convergent if constraints between several parameters are intróduced (e.g., equal heights and widths for each doublet and a fixed difference between the ISs of the two Fe^{3+} doublets).

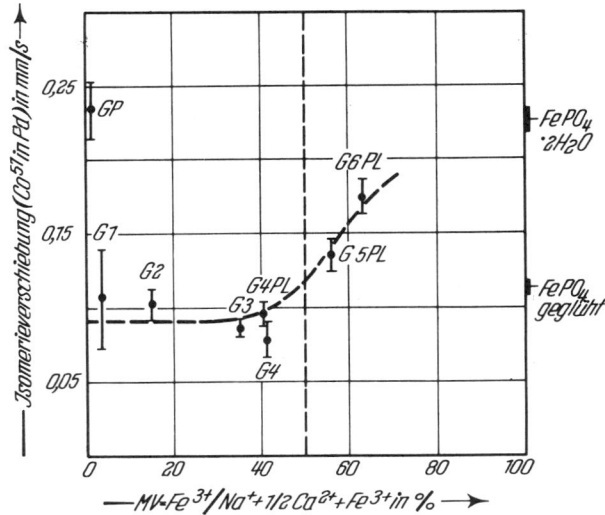

FIG. 10.   Isomer shift as a function of $Fe^{3+}/$ network modifier for silicate glasses (G1 to G6) and for phosphate glass (GP). For comparison Fe^{3+} in octahedral coordination ($FePO_4 \, 2H_2O$) and in tetrahedral coordination ($FePO_4$ annealed). (Frischat and Tomandl, 1969.)

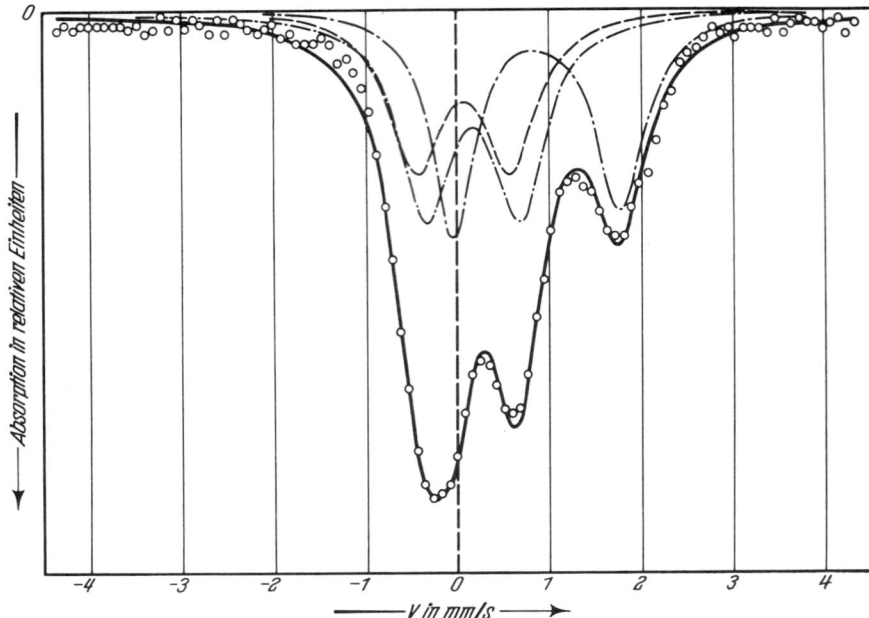

FIG. 11. Attempt for fitting doublets for Fe_t^{3+}, Fe_o^{3+}, and Fe^{2+} (glass G5PL of Fig. 10). (Frischat and Tomandl, 1969.)

A little later, many investigations were started to find out the coordination numbers and ratios for many different glasses. In addition to the changes in IS with coordination number, a dependence of the QS for Fe^{3+} and Fe^{2+} in different coordinations was also found. Generally an octahedrally coordinated Fe ion possesses a less distorted environment and consequently smaller QS than a tetrahedrally coordinated ion. The general rule for a distinction between Fe^{2+} and Fe^{3+} in both coordinations can be given in the following limits for IS and QS:

$$Fe_t^{3+} \quad 0.0 < \delta < 0.2 \quad 0.7 < \Delta < 1.0$$
$$Fe_o^{3+} \quad 0.2 < \delta < 0.4 \quad 0.3 < \Delta < 0.9$$
$$Fe_t^{2+} \quad 0.6 < \delta < 0.8 \quad 2 \;\; < \Delta$$
$$Fe_o^{2+} \quad 0.8 < \delta < 1.2 \quad \Delta \;\; < 2$$

δ is the IS in mm/s for a source $^{54}Co(Pd)$ and Δ is the QS in mm/s. The subscripts t and o refer to tetragonal and octahedral. The Δs cannot be generally given because there is a strong dependence on the composition of the glass, especially the kind of network formers. However, there is another serious difficulty in giving a general valid rule for the MB parameters as a function of the coordination number.

a. Fits with two doublets. Most of the papers perform a fit with just two doublets for Fe^{3+} and Fe^{2+}, respectively. The parameters δ and Δ obtained in this way usually show a monotonical dependence on the concentration of any sum of network modifiers. In this way, an attribution to coordination numbers can be performed.

Kurkjian and Sigety (1968 and earlier) were the first who discussed the coordination number of Fe^{3+} derived from IS in silicate and phosphate glasses. In the following years, a tremendous number of papers dealing with the same subject appeared. Gosselin et al. (1967) discussed the coordination number of Fe^{3+} in silicate glasses, Mattern (1965) in phosphate glasses. Lewis and Drickamer (1968) investigated the coordination of Fe^{2+} in silicate and phosphate glasses. As an additional parameter for their arguments they also used an external applied pressure during the MB measurements. Other papers about coordination numbers in iron-containing glasses are Hirayama et al. (1968) for alkali and earth alkali phosphate glasses; Taragin and Eisenstein (1970) for borate glasses; Hirao et al. (1979) for leucite-type silicate glasses and crystals; Takeda et al. (1979) for obsidian glasses; Kozhukharov et al. (1979) for telluride glasses; Maksimov et al. (1979) for all kinds of oxide glasses (they also speak of trigonal Fe^{3+}-surrounding in borate glasses); Hirao et al. (1980), Bandyopadhyay et al. (1980a), and Levitz et al. (1980) for silicate glasses; Eissa et al. (1981) for aluminoborate glasses containing alkali and earth alkali as network modifiers; and Nassar et al. (1983) for lithium borosilicate glasses.

Vidau et al. (1983) in fluorine-containing glasses, Yan-Fu et al. (1983) in silicate and borosilicate glasses, and Kawamoto et al. (1984a,b) in Zr-Ba-Fe-fluoride glasses investigated the structural behavior of Fe^{2+} and Fe^{3+} by considering δ and Δ and made conclusions on the coordination number in these different glasses. Another paper about amorphous fluorides is Lopez-Herrera et al. (1983).

The most important results are the following. In silicate glasses, Fe^{3+} and Fe^{2+} is in fourfold (tetrahedral) coordination for a high content of network modifiers, otherwise in sixfold (octahedral) coordination. In phosphate glasses, Fe is always in OC (possibly with the exception of very low Fe content). In telluride glasses, Fe^{3+} is octahedrally coordinated.

In fluoride glasses, Fe^{3+} is suggested as a glass former in OC, Fe^{2+} as a modifier in OC. Germanate and borate glasses behave in similar manner.

Dannheim and Frey (1978) investigated the incorporation of tin and iron in sodium borate glasses. Fe^{3+} and Fe^{2+} show changes of the coordination number from 6 to 4 with increasing Na content. The critical value is at about 20 at % Na. No influence of the borate anomaly could be detected. In this paper and also in Dannheim et al. (1976), the changes in coordination number of tin-containing borate and silicate glasses are investigated. The changes in δ are much more pronounced than for Fe: for Sn^{4+} $-0.25 < \delta < -0.02$, and for Sn^{2+} $2.6 < \delta < 3.5$, giving the range of change from fourfold to sixfold coordination. (Reference is a barium stannate source.) In silicate glasses, both Sn^{4+} and Sn^{2+} undergo a change in coordination state from 4 to 6 with increasing ratio of Sn^{4+} or Sn^{2+} to ($1/4$ Sn^{4+} + $1/2$ Sn^{2+} + $Na+$). In borate glasses, no change in coordination for Sn^{4+} could be observed. But, as can be seen from Fig. 12, a slight variation in the isomer shift for Sn^{4+} as a function of the Na^2O content could be detected that runs

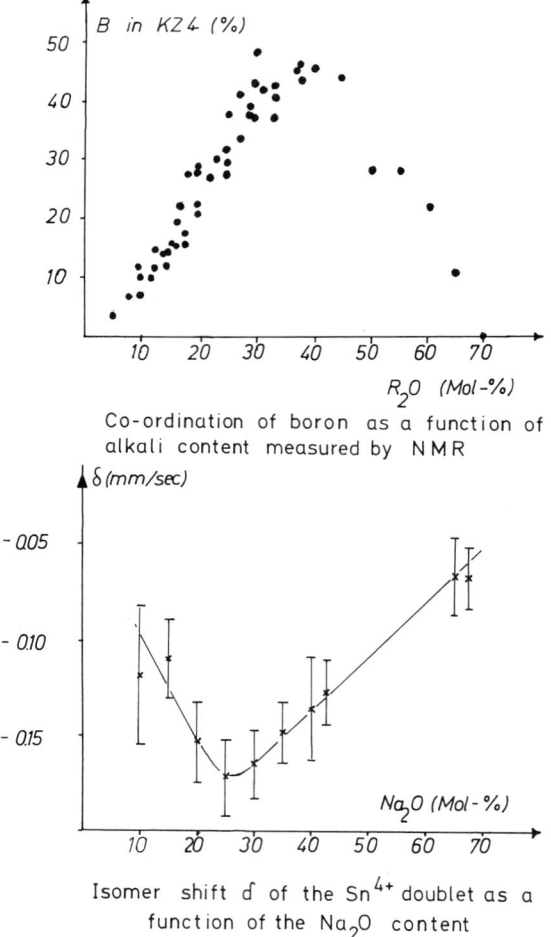

Co-ordination of boron as a function of alkali content measured by NMR

Isomer shift δ of the Sn^{4+} doublet as a function of the Na_2O content

FIG. 12. Isomer shift for Sn^{4+} in a borate glass compared to the amount of fourfold coordinated boron, determined by NMR. (Dannheim and Frey, 1978.)

essentially antiparallel to the percentage of the fourfold coordination of boron, determined by nuclear magnetic resonance (NMR). (See, e.g., Müller-Warmuth and Eckert, 1982.) In general, valency states and coordination numbers can be determined much easier for Sn rather than Fe, because the peaks are much better separated and the variations in the IS are larger. Other investigations on Sn in glasses were performed by Eissa et al. (1974), by Lechtenböhmer et al. (1982), by Archakov and Seregina (1983), Nishida et al. (1984) and in germanate glasses by Kurbanova (1985). Cremers et al. (1984) detected Sn^{4+} in Se-Sn-As glasses.

   b. Fits with up to four doublets. A different approach to determine the coordination numbers is a more detailed fit of the spectra. As an extreme example, fits with up to four doublets in one spectrum are shown in Fig. 13. The spectra are obtained with SiO_2-Fe_2O_3 glasses prepared from gels. Due to the strong overlapping of the very left four peaks of the four doublets giving just one sum peak in Fig. 13c, the fit is no longer unique. Therefore, constraints must be introduced that force the peaks to their desired positions. However, for these cases the ISs and QSs are no longer free variables and therefore not fully definite. In the example of Fig. 13, fortunately also the spectra for pure Fe^{3+} (a) and Fe^{2+} (b) are available, thus giving the exact values for the eight peak positions. But in many cases these standard spectra are not available.

   In principle, there are two ways of setting the two doublets corresponding to the two coordination states for fitting one experimental doublet belonging to one of the two valency states of iron: two interlocked doublets with two different QSs, but approximately the same IS, or two shifted doublets with two different ISs and about the same QS. An example for the first case was already described in Fig. 13. Figure 11, also already mentioned, illustrates the second case. The second case apparently applies (but unfortunately no graphs are shown) in Iwamoto et al. (1978, 1979). For Ca-silicate glasses they fitted two shifted doublets into the Fe^{2+} doublet and one into the Fe^{3+} doublet for the low–iron-concentration regime (~ 0.2 mol %). The coordination of Fe^{3+} could not be determined exactly. The two other doublets were belonging to Fe^{2+} in both coordinations, the ratio depending on the content of the iron concentration. For glasses with higher iron content, two shifted doublets for the Fe^{3+} doublet again (the ratio depending on the iron concentration), and one for the Fe^{2+} doublet were used. The Fe^{2+} proved to be in OC. Also, Kishore et al. (1985) seemed to have fitted their spectra for Ba-borate glasses in a similar way.

   The following authors treated their evaluations according to the first case: Levy et al. (1976) and O'Horo and Levy (1978) for silicate glasses, Burzo and Ardelean (1979a,b) for lead borate glasses, Grave et al. (1980) for complex

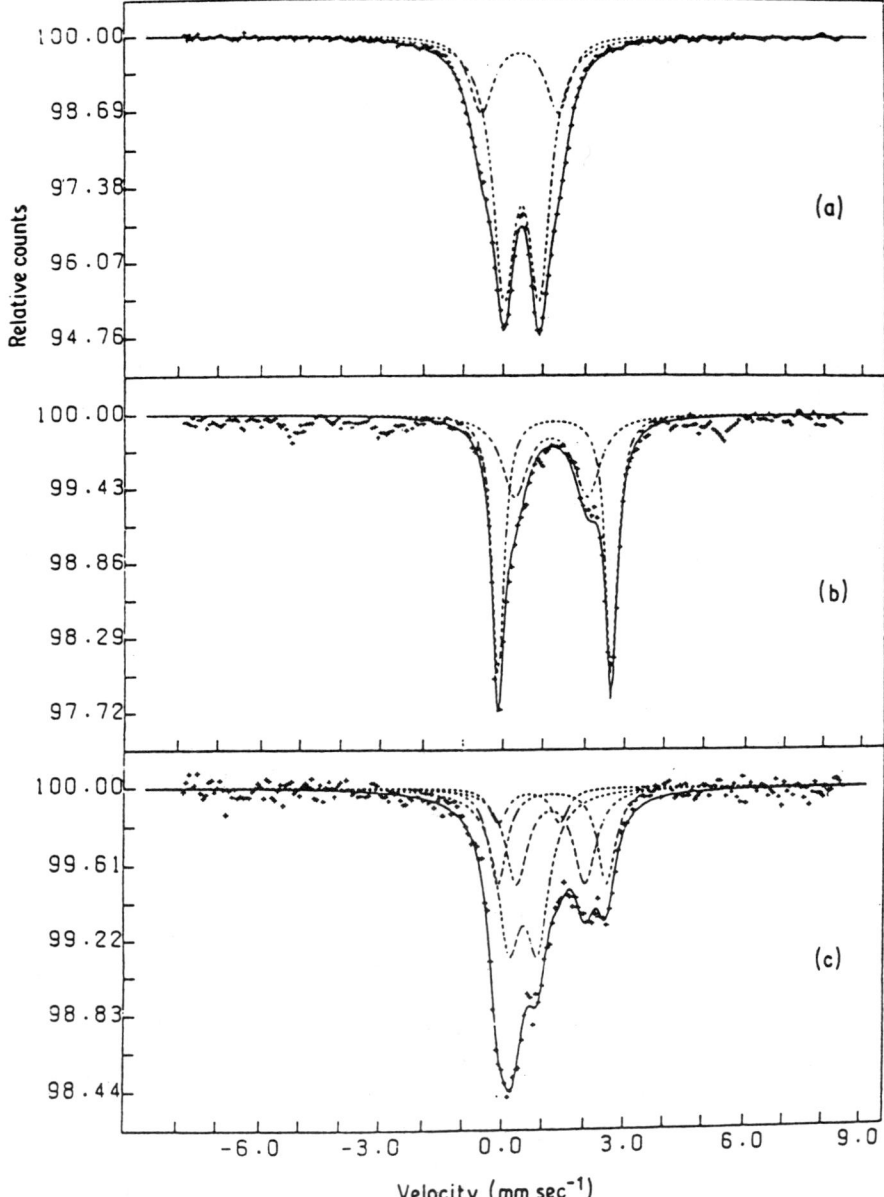

FIG. 13. Example for fitting Mössbauer spectra with up to four doublets corresponding to Fe^{3+} and Fe^{2+} in four- and sixfold coordination. (SiO_2-Fe_2O_3 glasses prepared from gels.) (Guglielmi et al., 1983.)

silicate glasses, Hanga et al. (1984) for Li-silicate glasses, Kanchan et al. (1985) for Na-vanadate glasses, and Bansal et al. (1985) for Ba-borovanadate glasses.

   c. Interpretation in terms of nonbridging oxygens.   Besides the assignment of changes of the IS to variations in the coordination number, another interpretation may be possible. Iron ions may be surrounded by bridging (BO) and nonbridging oxygens (NBO). Some authors relate these two kinds of irons to two different ISs (and QSs). The negatively charged NBOs should be closer to the oppositely charged Fe^{3+} due to increased coulombic interaction, equivalent to a transfer of electronic charge from oxygen to iron, thus increasing the s-electron density at the iron nucleus and decreasing the IS.

   This interpretation of the IS on the other hand can be easily related to the interpretation in terms of coordination numbers. Iron in OC as a network modifier is solely surrounded by six NBOs, iron as a network former in TC only by four BO. Some authors therefore use the term *coordination number* for the main changes of the ISs with the compositions of the glasses and attribute the minor changes of the ISs to changes in the amount of NBO near the irons. For example, the transition of boron from three- to fourfold coordination slightly changes the IS of iron and also of Sn (see Fig. 12), and therefore also the number of NBOs around the iron or tin ions.

   Raman et al. (1975), Raman et al. (1978), and Kishore et al. (1985) use both interpretations of IS changes in borate glasses. Shaisha et al. (1985) do the same for Sr-telluride glasses. Nishida and Takashima (1980) for K-borate glasses and Nishida et al. (1981a) for K-borosilicate glasses attribute Fe^{3+} to TC and the change in IS and QS to the borate anomaly causing a change in the number of NBOs around Fe^{3+}. The same group, Nishida et al. (1981b) and Nishida et al. (1983a) also investigated F^-, Cl^-, and Br^- in K-borate glasses and interpreted the changes in IS as being due to nonbridging F, Cl, or Br's. Nishida et al. (1985) found in glasses of the system BaF_2-ZrF_4-FeF_2 Fe^{2+} and Fe_o^{3+}. They also speculated about $[ZrF_6]$ as a result of their MB evaluations.

   A similar effect but more pronounced can be obtained with introduction of sulphur into a glass. Tomandl (1969) and Tomandl and Camara (1984) found for amber glasses (the dark yellow-brown glasses for bottles, melted with sulphur under reducing conditions) in addition to the normal Fe^{2+} doublet at least two more unusual doublets that can be attributed to Fe^{3+} but with larger ISs and QSs than usual for Fe^{3+} in oxygen environment. (See Fig. 14.) The only explanation for these two doublets is an environment of Fe^{3+} partly occupied with oxygens *and* sulphur atoms. EPR measurements also support this interpretation.

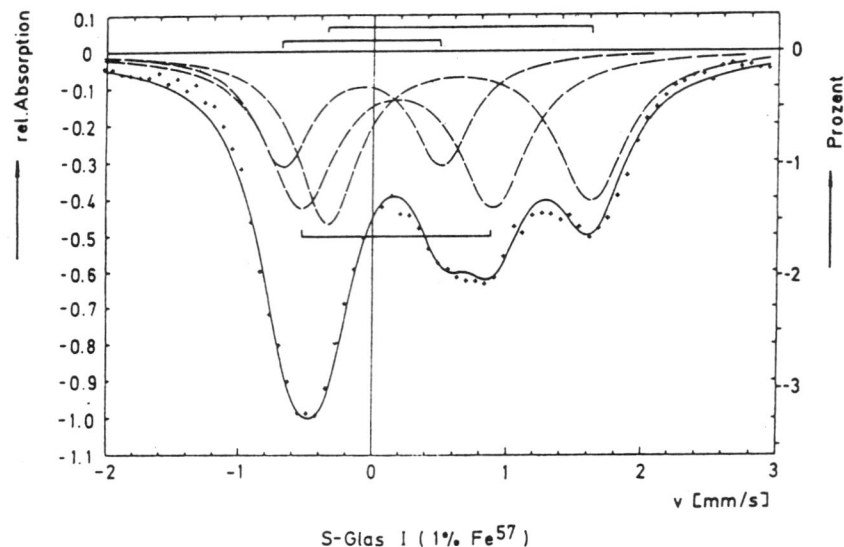

S-Glas I (1% Fe^{57})

Fig. 14.   MB spectrum of an amber glass showing one doublet due to Fe^{2+} in oxygen environment, and two doublets for Fe^{3+} surrounded partly by oxygen and sulphur. (Tomandl, 1968; Tomandl and Camara, 1984.)

d. Concluding remarks on the interpretation of the IS. It seems to the author that the interpretation of the incorporation of iron or tin into the glass structure in terms of just two coordination numbers is too oversimplified. The picture using BOs and NBOs seems to be more reliable, because there results a whole range of possibilities for the irons or tins within the glass structure giving a variation of ISs and QSs that seems to describe the real glasses in a better way. Even in the cases of fitting the spectra with one doublet for each coordination, a variation of IS and QS always resulted, indicating that there are more than just two structural parameters necessary to describe the dependence of IS and QS properly. Also because of the very disputable fits a resolution into just one doublet for the two valency states, each seems to be preferable. A detailed discussion of these problems can also be found in Müller-Warmuth and Eckert (1982).

3. f-Factor, Second-Order Doppler Effect

The f factor (or Lamb-Mössbauer factor) gives the intensity of the recoilless scattered γ-radiation. (See Section II.) It depends on the average amplitude and velocity of the MB atom within the lattice and therefore on the phonon spectrum, which is, e.g., given by the Debye model. The experimental determination of the f factor is not easy, but sometimes it is sufficient to determine the total area under the MB spectrum as a function of temperature. The quantity is proportional to the f factor. From the temperature-dependence of the f factor, the Debye temperature θ can be calculated. θ is a measure for the phonon spectrum and is also related to the specific heat. In the literature about glasses, the f factor is only determined very rarely. In most cases it is only used as a check to see whether the efficiency of the ME for different valency states is equal.

Lechtenböhmer et al. (1982) checked their evaluation of the valency ratios of tin. Bharati et al. (1983) determined the f factor of glasses up to T_g and found a strong decrease around this temperature. They explained their data in terms of a "cluster model" (structural clusters swimming in a tissue material). Coy and McEvoy (1981) found for Eu^{2+} and Eu^{3+} in fluorozirconate glasses different θ values and interpreted Eu^{2+} as being a network modifier and Eu^{3+} as a network former.

The second-order Doppler effect arises from the thermal motion of the atoms with velocity v. This gives a direction-independent contribution to the temperature-dependence of the IS proportional to v^2/c^2. Again, v depends on the phonon spectrum. (See Wegener, 1966.) Therefore, similar information as from the f factor can be deduced.

Raman et al. (1978) used the second-order Doppler effect to determine the local specific heat of Fe^{3+} in borate glasses. Gosselin et al. (1967) used both the temperature-dependence of the f factor and the IS to determine the specific heat in silicate glasses.

B. CRYSTALLIZATION

A nice feature of the ME is the possibility to distinguish between the MB isotope in glassy and crystalline environment. In some cases it is even possible to determine quantitatively the ratio of the glassy to the crystalline phase, which is often very difficult with other methods.

One way to distinguish between glassy and crystalline phases is the linewidth. Crystalline compounds most often show smaller linewidths than glassy phases. Another possibility for detecting crystalline phases is the occurrence of magnetic hyperfine patterns (e.g., sixfold splittings in the case of ^{57}Fe in a magnetic field). Precaution must be taken, however, because magnetic hyperfine patterns also can be found in paramagnetic (glassy) phases due to relaxation processes (see III.C), and, on the other hand, magnetic hyperfine patterns of crystalline compounds can be reduced up to a singulet or doublet due to superparamagnetism for very small crystallites. (See III.D.) The following papers deal with crystallization from glasses using the ME. Most often a-Fe_2O_3 or Fe_3O_4 has been found.

Gosselin et al. (1967) observed the crystallization of acmite. Tomandl (1969), and Tomandl and Camara (1984) found a crystalline phase during heat treatment of amber glasses (Fig. 15). This phase could be identified as antiferromagnetic stochiometric FeS in a somewhat distorted structure. During these investigations, another nonmagnetic crystalline phase could be detected with Fe^{3+} and an extremely large QS of $\Delta = 1.7$ mm/s which could not yet be identified. FeS, FeS_2, $NaFeS/_2$, and Fe_3C could be excluded.

Jach (1974), Tricker et al. (1974), Jach and Nabatian (1977a,b). Sekhon and Kamal (1979), Komatsu and Soga (1980), Hirao et al. (1980), Brown (1982), Prasad et al. (1982), Nishida et al. (1983b), Prasad et al. (1984), and Kanchan et al. (1985) investigated the crystallization behavior and kinetics of various glasses.

The crystallization kinetics in metallic glasses was studied by Kemeny et al. (1981), Klein et al. (1981), and Kopzewicz (1981). These authors triggered the crystallization with a high radio-frequency magnetic field, which induced crystallization via a magnetostriction effect.

C. Magnetic Effects

1. Superparamagnetism

During crystallization of magnetic phases, very small crystallites in the range of 10 to 30 nm show a collapse of the hyperfine splitting. (See II.A.3.) The transition from a six-line pattern to a doublet (or eventually singlet)

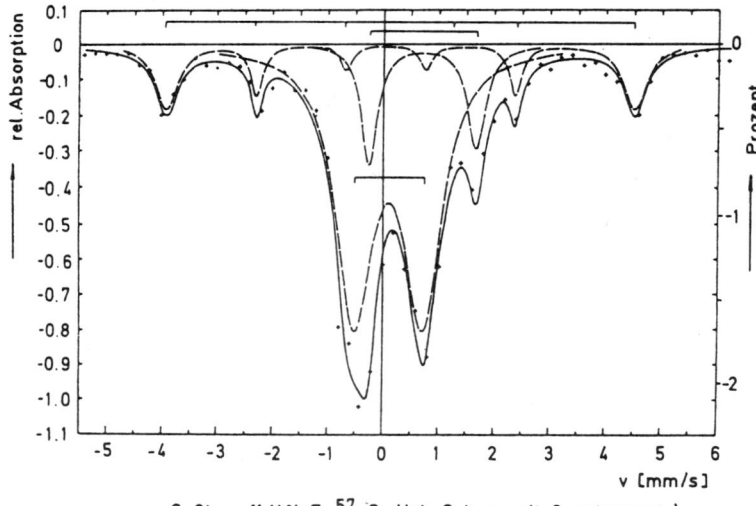

S Glas II (1% Fe^{57}, S-Ueb. Schuss, mit C, getempert)

FIG. 15.   MB spectrum of a heat-treated amber glass showing in addition to the doublets of Fig. 14 a six-line pattern due to crystallized fine FeS. (Tomandl, 1968; Tomandl and Camara, 1984.)

occurs for one crystallite size with decreasing temperature or for one temperature with decreasing crystallite size. The relaxation time τ for the fluctuations of the magnetization of the particles is

$$\tau = \frac{a}{v_L} \exp\left(\frac{KV}{kT}\right) \tag{23}$$

(a = geometric factor, v_L = frequency of precession of the magnetization, K = anisotropy constant, V = particle volume.) If τ is larger than the lifetime of the excited state of the MB niveau, then a normal magnetic HFS is observed; otherwise, the HFS gradually becomes narrower and smeared out. If τ is very much shorter than the MB lifetime, the material shows a spectrum typical for a normal paramagnetic behavior. Figure 16 shows an example for the transition from superparamagnetic to ferrimagnetic behavior for very fine a-Fe_2O_3 particles (Einmahl, 1979). The spectra are obtained at room temperature. Parameter of the single spectrum is the average particle size. Since the particles always show a distribution of sizes, each spectrum contains the information of the crystallite size distribution of the powder. Using

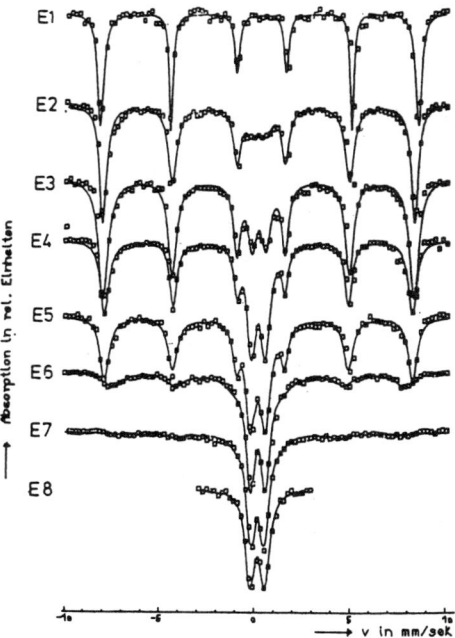

FIG. 16. Series of Mössbauer spectra of very fine -Fe_2O_3 powders exhibiting superparamagnetic behavior. (Einmahl, 1979.)

theoretical models and extensive numerical calculations, the crystallite size distribution could be determined from Mössbauer spectra (Einmahl, 1979; Einmahl and Tomandl, 1980).

In the following papers, superparamagnetism during crystallization of glasses has been detected: Bukrey et al. (1974) in the system $Na_2O \cdot Li_2O \cdot B_2O_3 \cdot Fe_2O_3$, Knorr et al. (1977) in Ca- and Al, Mn-silicate glasses, O'Horo and Levy (1978) in a Ca-Mn-silicate glass, Sekhon and Kamal (1978) in borate glasses, Hanga et al. (1981) in Li-silicate glasses containing manganese ferrite, Sanchez and Friedt (1982) in Ba-borate glasses, and Komatsu and Soga (1984) in silicate glasses. Bandyopadhyay et al. (1980b) found superparamagnetic particles of Fe_3O_4 in basalt glass. Using an external magnetic field of 50 kG at 4K, they could show that the Fe ions at the surface of the small particles have a noncollinear spin configuration, i.e., canted spins.

2. Magnetic Relaxation in Glasses

In oxide glasses at iron concentration below 1 to 2%, a significant broadening of the Fe^{3+} doublet and a very broad background can be found in the MB spectra. (See Fig. 17.) Figure 18 plots the apparent linewidth for Fe^{3+} for a phosphate glass (background neglected) as a function of the Fe_2O_3 concentration. These measurements by Frischat and Tomandl (1971) showed that even at room temperature, there is an indication for a six-line splitting for Fe^{3+} in glasses, having no crystallization at all. Kurkjian (1970) was the first who detected this effect in paramagnetic glasses. He showed that a true magnetic HFS can appear in these glasses at helium temperature and with very low Fe_2O_3 content. (See Fig. 19.)

Maksimov et al. (1979) investigated this effect in more detail for silicate, borate, germanate, and phosphate glasses. The explanation for this effect, which only can be observed for Fe^{3+}, is that with decreasing iron concen-

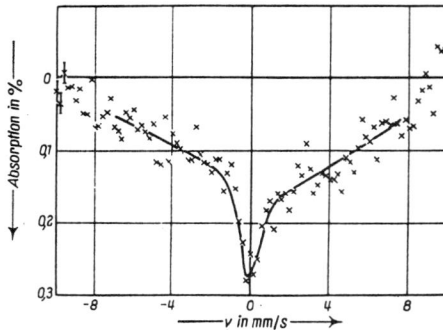

FIG. 17. MB-spectrum of a silicate glass containing 0.d1w% Fe_2O_3, melted in oxidizing atmosphere. (Frischat and Tomandl, 1971.)

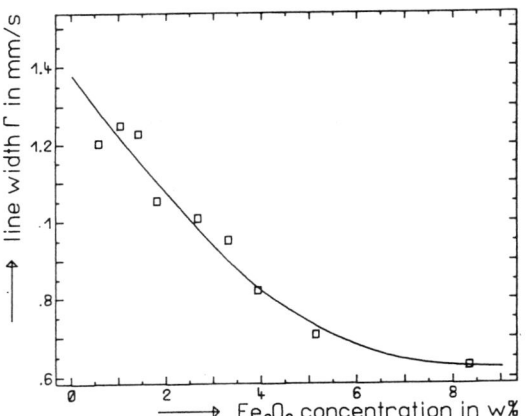

FIG. 18. Linebroadening due to long spin-lattice relaxation time in phosphate glass. (Frischat and Tomandl, 1971.)

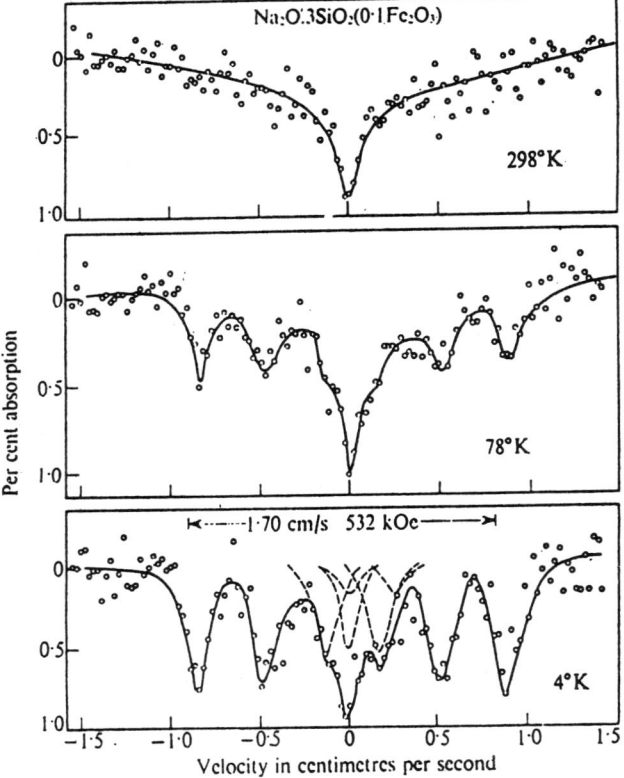

FIG. 19. Occurrence of HFS in silicate glass with low iron content at low temperature. (Kurkjian, 1970.)

tration the spin–spin interactions become more and more weak because the average distance of the iron atoms increases. At the same time, the spin-lattice relaxation time increases. Therefore, the internal magnetic field due to the paramagnetic electron does not flip fast enough to be averaged out within the lifetime of the excited state.

A very detailed study of the relaxation effects in phosphate glasses was performed by Mattern (1965). Bogomolova et al. (1983) showed relaxation for Fe_o^{3+} in Ba vanadate glasses. Varret and Naudin (1979) investigated the relaxation behavior of Fe^{3+} in silicate glasses. They separated the Fe^{2+} spectrum from the Fe^{3+} relaxation spectrum using a stripping method. Suzdalev and Imshennik (1981) reviewed magnetic relaxation effects and clustering of iron in polymers and inorganic glasses.

3. Magnetic Glasses

Besides the magnetic crystallized glasses, truly amorphous substances also exhibit magnetic behavior like the metallic glasses or spin glasses. Hooper (1973) presented an easy-to-read review about these glasses. A comprehensive review about spin glasses was given by Fischer (1983) (part I) and Fischer (1985) (part II). These very ambitious papers unfortunately do not say very much about Mössbauer applications. Only a very short chapter in the second part is devoted to ME investigations. Taneja et al. (1973) provides a survey about semiconducting glasses and metallic glasses.

The amorphous alloys consist of a metal such as Fe, Co, Ni, or Pd and a metalloid such as B, P, Si, or C and must be prepared by very rapid cooling from the melt. The MB spectra consist of more or less broadened six-line patterns (in the case of iron). Figure 20 shows an example of a MB spectra for $Fe_{82}B_{18}$ at room temperature. From (a) to (c), successive annealing results in an occurrence of a new six-line pattern with narrower linewidth due to crystallization of the glass (Ruckman et al. 1980).

Gonser et al. (1978) used a structure model for liquids by Bernal (dense random packing of atom) to explain their very complicated MB spectra of $Fe_{80}B_{20}$. They fitted these spectra with five six-line patterns according to five different coordinations.

Abd-Elmeguid et al. (1982) investigated the pressure-dependence and crystallization behavior of amorphous FeNiB alloys. Eickelman et al. (1984) performed similar studies on the same alloys but additionally using an external magnetic field.

Fujinami and Ujihira (1985) used transmission and conversion electron MB spectrometry to follow the crystallization behavior of metallic glasses in the system Fe-B-Si-C as thin ribbons. At the surface layers a lowering of the crystallization temperature connected with an increase of carbon concentration and a decrease of silicon could be observed by the conversion electron MB spectra, which are only sensitive to a few atomic layers at the surface.

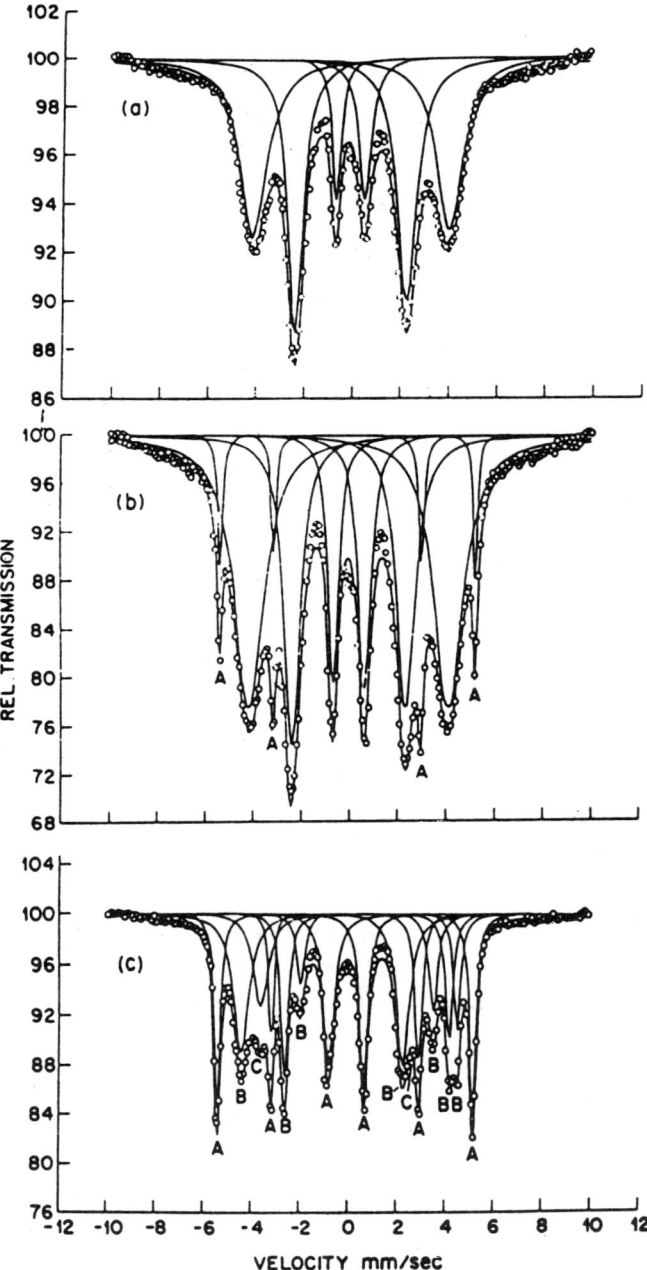

FIG. 20. Mössbauer spectrum of $Fe_{82}B_{18}$ at room temperature. (a): No annealing. (b): Annealed at 615 K for 24 hours. (c): Annealed at 615 K for 360 hours. (Ruckman et al., 1980.)

A rather unusual investigation was carried out on precipitations near freshwater springs of a ferric gel with the empirical formula $Fe(OH)_3O \cdot 9H_2O$ (Coey and Readman, 1973). Below 10K an ordered magnetic structure was found with a broadened six-line pattern. An applied external field results just in a further broadening of the lines. They explained it as due to spins within a particle statistically distributed relative to the direction of their resultant moment. They termed it sperimagnetism or speromagnetism (as by Coey, 1974).

4. Frozen Solutions

Some frozen solutions behave very much like glasses. Litterst et al. (1977) show as an example MB measurements on $FeCl_3$ in propane-(1,2)diol, frozen by quenching from 300 to 80K. Near the glass-transition temperature the QS, linewidth, and f factor show distinct anomalies. The recoil-free fraction becomes very small near the transition temperature. At higher temperatures, the signal reappears and crystallization and finally melting can be observed. The anomaly near the glass-transition temperature is correlated with structural relaxation processes. Litterst (1981) also reports about these kinds of investigations in his review about MB studies on nonmetallic amorphous materials.

D. INVESTIGATIONS OF MÖSSBAUER NUCLIDES OTHER THAN IRON

1. Tin

^{119}Sn having the same quantum numbers as ^{57}Fe is a very suitable MB isotope. Nevertheless, papers on it are not as numerous as for ^{57}Fe. Some investigations already have been reported in previous sections (especially III.A).

Tykachinskii et al. (1974) concluded from their measurements on Al-B-silicate glasses with low Sn content, finding predominantly Sn^{2+} when Si replacing by Al or B, that oxygen is bonded to Al or B more strongly than to Sn. Bartenev et al. (1976) observed the crystallization of cassiterite from glasses with help of the ME.

Mukerji and Sanyal (1982) and Sanyal and Mukerji (1985) investigated glass powder coated with SnO_2 (hot end coating with $SnCl_2$ and subsequent cold end coating with organic acids). They found solely Sn^{4+} partly bonded to the glass-oxygen, partly to the organic groups. Moreover, they found a change in coordination of Sn^{4+} in relation to the deposition temperature.

Bartenev et al. (1970) found a pronounced asymmetry of the Sn^{4+} line in fibers of silicate and borosilicate glasses. After a heat treatment, this line became symmetric as usual in Sn-containing glasses. They explained this effect with an axially symmetrical molecular field within the fibers. They also stated an asymmetry of atom oscillations that would lead to the Goldanskii-Karyagin effect. (See Section II.A.4.) The latter, however, was not suggested by them.

2. Antimony

Paul et al. (1977) reported about MB measurements of ^{121}Sb in Na_2O-B_2O_3-Sb_2O_3 glasses and also borate and silicate glasses. They found Sb^{3+} and a little Sb^{5+}. From the changes of isomer shift they deduced that the ionicity of the Sb-O bond increases with decreasing Na_2O content and also with exchange of B to Si to P.

Ruby (1972) showed in his review-like paper measurements of ^{121}Sb in crystalline Sb_2Se_3. Because of the poor resolution of these measurements, one of his conclusions about the ME was that "unique interpretations, except for iron, are unlikely." Other investigations using antimony are reported by Taneja et al. (1973).

3. Rare Earths, Actinides, and Others

Some isotopes of the rare earths show good properties for MB applications, yet not too many papers can be found in the literature. Uhrich and Barnes (1968) investigated thulium in Na-silicate glasses. They found a rather broad doublet due to Tm^{3+}. Taragin and Eisenstein (1973) studied europium in Eu_2O_3 and in a silicate glass. From the f factor they concluded that Eu^{3+} is bonded weaker in the glass than in Eu_2O_3. Also Eu was used in a MB study by Coey et al. (1981) in fluorozirconate glass. They interpreted Eu^{3+} as being a network former and Eu^{2+} a network modifier.

Czjzek (1981) describes measurements with ^{151}Eu and ^{155}Gd in amorphous materials such as $Eu_{80}Au_{20}$ and $Gd_{40}Ni_{60}$. He proposes a distribution of V_{zz} and η to explain the spectra in terms of a statistical variation of structural atom arrangements.

Friedt and Krill (1983) investigated in a similar way the systems $Eu_{80}Au_{20}$ and $Gd_{80}Au_{20}$. From the ferromagnetic interaction they assumed an EFG tensor randomly oriented with respect to the hyperfine field. Additional investigations with other systems $RE_{1-x}X_x$ (RE = rare earths, e.g., Eu; X = Mg, Zn, Cd) can be found in Friedt and Krill (1983).

Another MB-isotope, ^{125}Te, was used by Boolchand et al. (1981) in $GeSe_{2x}Te_{2-2x}$ alloy glasses. They found two QSs at He temperature and related it to the chemical order in the glasses.

A rather exotic investigation of Np in a Ba-Zr-F-glass was performed by Jove et al. (1979). They found a very broad line due to Np^{4+} in a "paramagnetic relaxation" state and a very narrow line due to Np^{2+} or Np^{3+}. (See Fig. 21.)

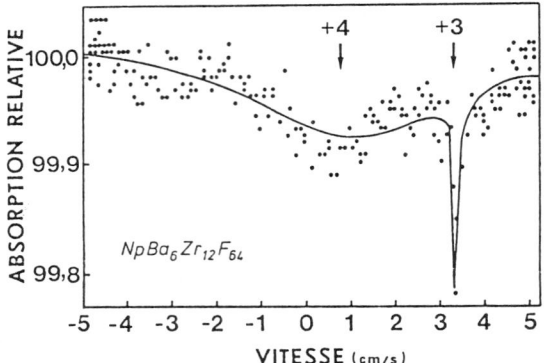

FIG. 21. Mössbauer spectrum of $NpBa_6Zr_{12}$ glass at 4.2K. (Jove et al., 1979.)

IV. Conclusions

ME has become a powerful routine method for basic research on problems of glass structure, crystallization behavior, and magnetic properties. However, it seems to the author that during the 1980s, no real progress was achieved. Many scientists still investigate other glass compositions using the Mössbauer effect in much the same way as it has been done all the years before. They interpret their results in valency states, coordination numbers, and internal magnetic fields.

The problem is that ME has very few suitable nuclides such as iron and tin and that all the other nuclides exhibit great difficulties either experimentally or for theoretical interpretation. Many of them show such poor resolution that information about the environment of the nucleus within the glass structure can hardly be obtained.

But there *are* MB isotopes that could be useful for further research, including some rare earths, ^{73}Ge, ^{129}I, ^{133}Cs, ^{157}Gd, ^{145}Nd, ^{174}Yb, ^{181}Ta, ^{197}Au, and eventually ^{67}Zn.

Furthermore, as pointed out in Section II.E, other kinds of Mössbauer experiments can be performed: selective-excitation double-Mössbauer spectroscopy (e.g., for relaxation effects, separate excitation of doublets belonging to one coordination, or especially scattering experiments). The latter are very difficult methods, but could give totally new information about the glass structure. They are available not only for Mössbauer isotopes but any kind of material.

But also the conventional MB technique can be extended to scattering experiments using scattered γ-rays or conversion electrons to get information about the surfaces of the materials. The determination of the f factor and the second-order Doppler effect is not very popular either. The application of a high external magnetic field also for nonmagnetic glasses could give valuable information, because the normal degeneration of the energy levels is lost.

Acknowledgments

I would like to thank Mr. Rödel, presently at the National Institute of Standards & Technology, Gaithersburg, Maryland, for his valuable help in the literature survey.

References

Abd-Elmeguid, M. M., Micklitz, H., and Vincze, I. (1982). *Phys. Rev.* **25**, 1–7.
Albanese, G., and Deriu, A. (1979). *Rivista del nuovo cimento* **2**, 1–40.
Albanese, G., and Ghezzi, C. (1974). *Phys. Stat. Sol.(a)* **22**, 209–216.
Archakov, Y. I., and Seregina, L. N. (1983). *Fizika i Khimiya Stekla* **9**, 410–413; *Soviet J. Glass Phys. Chem.* **9** (1983) 287–290 (transl.).
Balko, B., and Hoy, G. R. (1983). *Studies in Physical and Theoretical Chemistry*, vol. 25. Elsevier, Amsterdam.
Bandyopadhyay, A. K., Auric, P., Phallippou, J., and Zarzycki, J. (1980a). *J. Mater. Sci.* **15**, 2081–2090.
Bandyopadhyay, A. K., Zarzycki, J., Auric, P., and Chappert, J. (1980b). *J. Non-Cryst. Solids* **40**, 353–368.
Bansal, T. K., Kishore, N., and Mendiratta, R. G. (1985). *Phys. Chem. Glasses* **26**, 94–96.
Barb, D. (1980). W. Meisel (ed.). Akademie-Verlag, Berlin, DDR.
Bartenev, G. M., Suzdalev, I. P., and Tsyganov, A. D. (1970). *Phys. Stat. Sol.* **37**, 73–78.
Bartenev, G. M., Magomedov, G. M., and Tsyganov, A. D. (1976). *Neorganicheskie Materialy* **12**, 742–746.
Bharati, S., Parthasarathy, R., Rao, K. J., and Rao, C. N. R. (1983). *Solid State Comm.* **46**, 457–460.
Bogomolova, L. D., Glassova, M. P., Dubatovko, O. E., Reimann, S. I., and Spasibkina, S. N. (1983). *J. Non-Cryst. Solids* **58**, 71–89.
Boolchand, P., Bresser, W. J., Suranyi, P., and de Neufville, J. P. (1981). In "Int. Conf. on Amorphous Systems Investigated by Nuclear Methods, Proc. 1," 703, Balatonfüred, Hungary. North-Holland. Paper also in: *Nucl. Instr. Meth.* (1982).
Brown, I. W. M. (1982). *J. Non-Cryst. Solids* **50**, 233–252.
Bukrey, R. R., Kenealy, P. F., Beard, G. B., and Hooper, H. O. (1974). *Phys. Rev.* **9**, 1052–1061.
Burzo, E., and Ardelean, I. (1979a). *Phys. Chem. Glasses* **20**, 15–20.
Burzo, E., and Ardelean, I. (1979b). *Sol. State Comm.* **31**, 75–78.
Champeney, D. C., and Sedgwick, D. F. (1972). *Chem. Phys. Lett.* **15**, 377.
Champeney, D. C., and Woodhams, F. W. D. (1966). *Phys. Let.* **20**, 275–277.
Coey, J. M. D. (1974). *J. Physique* **12**, Colloque C6, 89–105.
Coey, J. M. D., and Readman, P. W. (1973). *Nature* **246**, 476–478.
Coey, J. M. D., McEvoy, A., and Shafer, M. W. (1981). *J. Non-Cryst. Sol.* **43**, 387–392.
Cremers, H., Mosel, B. D., Müller-Warmuth, W., and Frischat, G. H. (1984). *J. Non-Cryst. Solids* **63**, 329–335.
Czjzek, G. (1981). In "Int. Conf. on Amorphous Systems Investigated by Nuclear Methods, Proc. 1," 93, Balatonfüred, Hungary. North-Holland. Paper also in *Nucl. Instr. Meth.* (1982).
Dannheim, H., and Frey, T. (1978). In "Borate Glasses, Materials Science Research 12," L. D. Pye, V. D. Freschette, and N. J. Kreidl, eds. Plenum, pp. 227–238.

Dannheim, H., Oel, H. J., and Tomandl, G. (1976). *Glastechn. Ber.* **49**, 170–175.

Duncan, J. F., and Golding, R. M. (1965). *Quart. Reviews (Chem. Soc. London)* **19**, 36–56.

Eickelmann, H. J., Abd-Elmeguid, M. M., Micklitz, H., and Brand, R. A. (1984). *Phys. Rev.* **29**, 2443–2448.

Einmahl, P. (1979). Diss. thesis, Univ. Erlangen-Nürnberg, West Germany.

Einmahl, P., and Tomandl, G. (1980). *Sci. Ceram.* **10**, 691–696.

Eissa, N. A., Shaisha, E. E., and Hussiun, A. L. (1974). *J. Non-Cryst. Solids* **16**, 206–218.

Eissa, N. A., Mostafa, A. G., Sanad, A. M., and Hussein, A. L. (1981). In "Int. Conf. on Amorphous Systems Investigated by Nuclear Methods, Proc. 1," 769, Balatonfüred, Hungary. North-Holland. Paper also in *Nucl. Instr. Meth.* (1982).

Fischer, K. H. (1983). *Phys. Stat. Sol.(b)* **116**, 357–414.

Fischer, K. H. (1985). *Phys. Stat. Sol.(b)* **130**, 13–71.

Frauenfelder, H. (1962). W. A. Benjamin, New York.

Friedt, J. M., and Krill, G. (1983). *J. Phys. F. Met. Phys.* **13**, 2389–2409.

Friedt, J. M., Maurer, M., Sanchez, J. P., and Durand, J. (1982). *J. Phys. F. Met. Phys.* **12**, 821–836.

Frischat, G. H., and Tomandl, G. (1969). *Glastechn. Ber.* **42**, 182–185.

Frischat, G. H., and Tomandl, G. (1971). *Glastechn. Ber.* **44**, 173–177.

Fujinami, M., and Ujihira, Y. (1985). *J. Non-Cryst. Solids* **69**, 361–369.

Goldanskii, V. I., Makarov, E. F., and Khrapov, V. V. (1963). *Phys. Letters* **3**, 344.

Goldman, D. S., and Bewley, D. E. (1985). *J. Am. Ceram. Soc.* **68**, 691–695.

Gonser, U. (ed.) (1975). "Topics in Applied Physics," vol. 5. Springer Verlag.

Gonser, U., Ghafari, M., and Wagner, H.-G. (1978). *J. Magnetism and Magnetic Mater.* **8**, 175–177.

Gosselin, J. P., Shimony, U., Grodzins, L., and Cooper, A. R. (1967). *Phys. Chem. Glasses* **8**, 56–61.

de Grave, E., van Iseghem, P., de Batist, R., and Chambaere, D. (1980). *J. Physique* **41**, 269–270.

Greenwood, N. N., and Gibb, T. C. (1971). Chapman and Hall, London.

Guglielmi, M., Maddalena, A., and Principi, G. (1983). *J. Mater. Sci. Lett* **2**, 467–470.

Hanga, E., Tataru, E., Morariu, M., Diamandescu, L., and Popescu-Pogrion, N. (1981). *Phys. Stat. Sol.* **67**, 725–728.

Hanga, E., Tataru, E., and Morariu, M. (1984). XXIInd Congr. AMPERE on Magnetic Resonance and Related Phenomena. Proc. Zürich, pp. 92–94.

Hirao, K., Soga, N., and Kunugi, M. (1979). *J. Amer. Ceram. Soc.* **63**, 109–110.

Hirao, K., Komatsu, T., and Soga, N. (1980). *J. Non-Cryst. Solids* **40**, 315–323.

Hirayama, C., Castle J. G. Jr., and Kuriyama, M. (1968). *Phys. Chem. Glasses* **9**, 109.

Hooper, H. O. (1973). *Magnetism and Magnetic Materials. American Institute of Physics* **10**, 702–713.

Iwamoto, N., Tsunawaki, Y., Nakagawa, H., Yoshimura, T., and Wakabayashi, N. (1978). *J. Non-Cryst. Solids* **29**, 347–356.

Iwamoto, N., Tsunawaki, Y., Nakagawa, H., Yoshimura, T., Miyago, M., and Wakabayashi, N. (1979). *J. Physique* **40**, 151–152.

Jach, J. (1974). *J. Nonmetals* **2**, 89–93.

Jach, J., and Nabatian, D. (1977a). *Semiconductors and Insulators*, vol. 3. Gordon and Breach Science Publishers, New York, pp. 23–32.

Jach, J., and Nabatian, D. (1977b). *Semiconductors and Insulators*, vol. 3. Gordon and Breach Science Publishers, New York, pp. 33–41.

Jove, J., Gal, J., Potzel, W., Kalvinus, G. M., Spirlet, J. C., and Pages, M. (1979). *J. Physique* **40**, 190–191.

Kanchan, D. K., Puri, R. K., and Mendiratta, R. G. (1985). *Phys. Chem. Glasses* **26**, 217–220.

Karyagin, S. V. (1963). *Doklady Akad. Nauk. S.S.S.R.* **148**, 1102.

Kawamoto, Y., Horisaka, T., Hirao, K., and Soga, N. (1984a). *Chemistry Letters, Chem. Soc. Japan* **8**, 1441–1444.

Kawamoto, Y., Nohara, I., Hirao, K., and Soga, N. (1984b). *Solid State Commun.* **51**, 169–72.

Kemeny, T., Vincze, J., Schaafsma, A. S., van der Woude, F., and Lovas, A. (1981). In "Int. Conf. on Amorphous Systems Investigated by Nuclear Methods, Proc. 1," 355, Balatonfüred, Hungary. North-Holland. Paper also in *Nucl. Instr. Meth.* (1982).

Kishore, N., Bansal, T. K., Kamal, R., and Mendiratta, R. G. (1985). *J. Non-Cryst. Solids* **69**, 213–219.

Klein, H. P., Ghafari, M., Ackermann, M., Gonser, U., and Wagner, H. G. (1981). In "Int. Conf. on Amorphous Systems Investigated by Nuclear Methods, Proc. 1," 369, Balatonfüred, Hungary. North-Holland. Paper also in *Nucl. Instr. Meth.* (1982).

Knorr, K., Geller, R., and Prandl, W. (1977). *J. Magnetism and Magnetic Materials* **4**, 258–261.

Komatsu, T., and Soga, N. (1980). *J. Phys. Chem.* **72**, 1781–1785.

Komatsu, T., and Soga, N. (1984). *J. Mater. Sci.* **19**, 2353–2360.

Kopcewicz, M., Gonser, U., and Wagner, H. G. (1981). In "Int. Conf. on Amorphous Systems Investigated by Nuclear Methods, Proc. 1," 379, Balatonfüred, Hungary. North-Holland. Paper also in *Nucl. Instr. Meth.* (1982).

Kozhukharov, V., Nikolov, S., Marinov, M., and Troev, T. (1979). *Mat. Res. Bull.* **14**, 735–741.

Kurbanova, L. V. (1985). *The sovjet journal of glass physics and chemistry* **11**, 120–123.

Kurkjian, C. R. (1970). *J. Non-Cryst. Solids* **3**, 157–194.

Kurkjian, C. R., and Sigety, E. A. (1968). *Phys. Chem. Glasses* **9**, 73–83.

Labar, C., and Gielen, P. (1973–1974). *J. Non-Cryst. Solids* **13**, 107–119.

Lechtenböhmer, A., Mosel, B. D., Müller-Warmuth, W., and Dutz, H. (1982). *Glastechn. Ber.* **55**, 161–166.

Levitz, P., Calas, G., Bonnin, D., and Legrand, A. P. (1980). *Revue Phys. Appl.* **15**, 1169–1173.

Levy, R. A., Lupis, C. H. P., and Flinn, P. A. (1976). *Phys. Chem. Glasses* **17**, 94–103.

Lewis, G. K., and Drickamer, H. G. (1968). *J. Chem. Phys.* **49**, 3785–3789.

Litterst, F. J. (1981). In "Int. Conf. on Amorphous Systems Investigated by Nuclear Methods, Proc. 1," 223, Balatonfüred, Hungary. North-Holland. Paper also in *Nucl. Instr. Meth.* (1982).

Litterst, F. J., Ramisch, R., and Kalvius, G. M. (1977). *J. Non-Cryst. Solids* **24**, 19–28.

Lopez-Herrera, M. E., Greneche, J. M., and Varret, F. (1983). *Phys. Rev. B* **28**, 4844–4848.

Maksimov, Y. V., Suzdalev, I. P., Markvart, M., Litterst, F. I., Zhilin, A. A., and Nemilov, S. V. (1978). *Fizika i Khimiya Stekla* **4**, 529–534; *Soviet J. Glass Phys. Chem.* **5** (1979) (transl.).

Massiot, D. (1985). *J. Non-Cryst. Solids* **69**, 371–380.

Mattern, P. L. (1965). Ph.D. thesis. Univ. Microfilms, Ann Arbor, Mich.

Mössbauer, R. L. (1958a). *Z. Phys.* **151**, 124. Reprinted in Frauenfelder (1962).

Mössbauer, R. L. (1958b). *Naturwissensch.* **45**, 538. Reprinted in Frauenfelder (1962).

Mössbauer, R. L. (1959). *Z. Naturforsch.* **14a**, 211. Reprinted in Frauenfelder (1962).

Mössbauer Effect Data Center. (1967–). Mössbauer Effect Reference and Data Journal. University of North Carolina, Ashville.

Muir, A. H. Jr., Ando, K. J., and Coogan, H. M. (1966). Interscience Publishers, J. Wiley & Sons, New York.

Mukerji, J., and Sanyal, A. S. (1982). *Phys. Chem. Glasses* **23**, 79–82.

Müller-Warmuth, W., and Eckert, H. (1982). *Physics Reports* **88**, 93–149.

Nassar, A. M., Gomaa, S. S., Salman, S. M., and Mostafa, F. (1983). *Central Glass and Ceramic Bull. Calcutta* **30**, 62–65.

Nishida, T., and Takashima, Y. (1980). *J. Non-Cryst. Solids* **37**, 37–43.

Nishida, T., Hirai, T., and Takashima, Y. (1981a). *Phys. Chem. Glasses* **22**, 94–98.

Nishida, T., Kai, N., and Takashima, Y. (1981b). *Phys. Chem. Glasses* **22**, 107–109.
Nishida, T., Nonaka, T., Isobe, T., and Takashima, Y. (1983a). *Phys. Chem. Glasses* **24**, 88–91.
Nishida, T., Hirai, T., and Takashima, Y. (1983b). *Phys. Chem. Glasses* **24**, 113–116.
Nishida, T., Katada, M., and Takashima, Y. (1984). *Bull. Chem. Soc. Jpn.* **57**, 3566–3570.
Nishida, T., Nonaka, T., and Takashima, Y. (1985). *Bull. Chem. Soc. Jpn.* **58**, 2255–2259.
Nolet, D. A. (1980). *J. Non-Cryst. Solids* **37**, 99–110.
O'Horo, M. P., and Levy, R. A. (1978). *J. Appl. Phys.* **49**, 1635–1637.
Paul, A., Donaldson, J. D., and Thomas, M. J. K. (1977). *J. Mater. Sci.* **12**, 219–222.
Pollak, H., de Coster, M., and Amelinckx, S. (1962). D. M. J. Compton, A. H. Schoen, eds., Wiley, New York, p. 298.
Prasad, A., Bahadur, D., Singru, R. M., and Chakravorty, D. (1982). *J. Mater. Sci.* **17**, 2687–2692.
Prasad, A., Singru, R. M., Bahadur, D., and Chakravorty, D. (1984). *J. Mater. Sci.* **19**, 3021–3027.
Raman, T., Nagesh, V. K., Chakravorty, D., and Rao, G. N. (1975). *J. App. Physics* **46**, 972–973.
Raman, T., Rao, G. N., and Chakravorty, D. (1978). *J. Non-Cryst. Solids* **29**, 85–107.
Ruby, S. L. (1972). *J. Non-Cryst. Sol.* **8–10**, 78–84.
Ruckman, M. W., Levy, R. A., Kessler, A., and Hawegawa, R. (1980). *J. Non-Cryst. Solids* **40**, 393–406.
Sanchez, J. P., and Friedt, J. M. (1982). *J. Physique* **43**, 681–684.
Sanyal, A. S., and Mukerji, A. S. (1985). *Phys. Chem. Glasses* **26**, 135–136.
Sekhon, S. S., and Kamal, R. (1978). *J. Appl. Phys.* **49**, 3444–3445.
Sekhon, S. S., and Kamal, R. (1979). *J. Non-Cryst. Solids* **33**, 169–175.
Shaisha, E. E., Bahgat, A. A., Sabry, A. I., and Eissa, N. A. (1985). *Phys. Chem. Glasses* **26**, 91–93.
Stevens, J. G., and Stevens, V. E. (1977). *Mössbauer Effect Data Index 1966–1976.* Plenum.
Suzdalev, I. P., and Imshennik, V. K. (1981). In "Int. Conf. on Amorphous Systems Investigated by Nuclear Methods, Proc. 1," 279, Balatonfüred, Hungary. North-Holland. Paper also in *Nucl. Instr. Meth.* (1982).
Takeda, M., Sato, K., Sato, J., and Tominaga, T. (1979). *Revue Chimie Minérale* **16**, 400–401.
Taneja, S. P., Kimball, C. W., and Shaffer, J. C. (1973) (Pl. *Mössbauer Effect Methodology* **8**, 41-69 Plenum Press, New York).
Taragin, M. F., and Eisenstein, J. C. (1970). *J. Non-Cryst. Solids* **3**, 311–316.
Taragin, M. F., and Eisenstein, J. C. (1973). *J. Non-Cryst. Sol.* **11**, 395–396.
Tomandl, G. (1969). *Paper on Fachausschuß I of Deutsche Glastechn. Ges. (1968) Abstract in Glastech. Ber.* **42**, 71.
Tomandl, G. (1974). Nahordnungfelder in Gläsern, Fachausschußbericht Nr. 70 Deutsche Glastechnische Gesellschaft, Frankfurt, Germany.
Tomandl, G., and Camara, B. (1984). Abschlussbericht, DFG (Deutsche Forschungsgemeinschaft) TO 70/3.
Tomandl, G., and Oel, H. J. (1970). Kommission der europäischen Gemeinschaften Bericht Nr.28 Konferenz über die Strahlungs-und Isotopenanwendung im Bauwesen, Brüssel.
Tomandl, G., Frischat, G. H., and Oel, H. J. (1967). *Glastechn. Ber.* **40**, 293–298.
Tricker, M. J., Thomas, J. M., Omar, M. H., Osman, A., and Bishay, A. (1974). *J. Mat. Science* **9**, 1115–1122.
Tykachinskii, I. D., Fedorovskii, Y. A., Dzhakhva, N. G., Ovchinnikov, A. I., and Tsyganov, A. D. (1974). *Inorg. Mater.* **10**, 1883–1885.
Uhrich, D. L., and Barnes, R. G. (1968). *Phys. Chem. Glasses* **9**, 184–189.
Varret, F., and Naudin, F. (1979). *Rev. Phys. Appliquee* **14**, 613–617.
Vertes, A., Korecz, L., and Burger, K. (1979). "Studies in Physical and Theoretical Chemistry 5." Elsevier.
Videau, J. J., Portier, J., and Tanguy, B. (1983). *Glass Technol.* **24**, 171–172.
Walker, L. R., Wertheim, G. K., and Jaccarino, V. (1961). *Phys. Rev. Letters* **6**, 98–101.

CHAPTER 6

Chromatography*

C. R. Masson[†]

ATLANTIC RESEARCH LABORATORY
NATIONAL RESEARCH COUNCIL OF CANADA
HALIFAX, NOVA SCOTIA
CANADA

I. Introduction

Chromatographic methods have been used extensively to study the anionic constitution of phosphate glasses. In these methods the glass is dissolved in a neutral or weakly alkaline aqueous medium and the resulting solution is examined by paper, column, or thin-layer chromatography. With suitable precautions to avoid reorganization reactions in solution or during chromatographic separation, the method yields ionic distributions that reflect accurately the distributions present in the original glass.

*NRCC publication No. 25647.
[†]Deceased.

For silicate glasses such methods are less satisfactory due to the pronounced tendency for silicate ions to undergo reorganization reactions in solution and recourse has been made to derivatization procedures followed by gas-liquid chromatography. Thin layer, gel permeation, and high-performance liquid chromatography have also been used to separate the products. Although less accurate than for the phosphates, the method has proved valuable for the identification and, in some instances, quantitative determination of discrete silicate ions in glasses with leachable cations.

In general, the results obtained by chromatography are in good agreement with conclusions derived by other methods.

II. Scope and Purpose

Most silicate glasses of commercial importance have silica contents of the order of 70%, or more by weight. The three-dimensional network of $\equiv$Si—O—Si$\equiv$ linkages in these glasses is only partially broken down and this accounts largely for their inertness and chemical durability. When the silica content of such glasses is lowered by progressive addition of network-breaking oxides MO or M_2O a composition is reached at which discrete ions are eventually formed. In binary systems $MO\text{-}SiO_2$ and $M_2O\text{-}SiO_2$ ions of finite size must appear at silica contents below the metasilicate composition. It is conceivable that discrete ions could be present also in glasses of higher silica content although their proportions in this region are expected to be small.

Chromatographic methods yield information concerning the discrete ions in glasses and are therefore of value mainly for glasses more basic than the metasilicate or, in the case of the phosphates, the metaphosphate composition. For binary phosphate glasses, most studies have centered on the range 50–70 mole% metal oxide. For the silicates binary glasses with more than about 57 mole% metal oxide are difficult to prepare by normal quenching techniques, so the method has been less widely used. With the advent of fast quenching techniques similar to those used for preparing metallic glasses, however, the range of glassy silicates available for study is being extended to higher basicities and it is expected that such glasses will be subjected to increasing study by chromatographic methods in the ensuing years.

Studies of discrete ions in silicate and phosphate glasses are important for testing theories of the ionic constitution of the corresponding melts. Indeed, this has been one of the main incentives for the development of improved derivatization and chromatographic procedures. Of particular interest is the application of polymer theory to these systems. Polymer theory predicts that, below the gel point, such melts contain an array of discrete anions in thermodynamic equilibrium. The proportions of these ions in silicate melts can be inferred from the thermodynamic properties of the melts.

Ionic distributions calculated in this way, however, represent "ideal" distributions based on certain simplifying assumptions and reliable experimental data are required to assess their range of validity. The neglect of cyclic structures, for example, is an important limitation of the theory.

Chromatography has proved particularly valuable in revealing the principles that govern the constitution of phosphate glasses. The results have shown that ionic distributions in these glasses can be interpreted in terms of polymer chemistry except, as might be expected, for the equilibria between the smallest phosphate anions.

III. Phosphates

A. OUTLINE OF CHROMATOGRAPHIC METHODS

The use of paper chromatography to study the condensed phosphates was first reported independently by Ebel et al. (Ebel and Volmar, 1951; Ebel, 1952, 1953) in France, Westman et al. (Westman and Scott, 1951; Westman et al., 1952) in Canada, and Ando et al. (1952) in Japan. Our knowledge of the detailed ionic constitution of phosphate glasses is due largely to this technique. In little more than a decade after the initial publications, hundreds of articles had appeared on the constitution of crystalline and amorphous phosphates and a large measure of order had been introduced into what had previously been a complex and poorly understood field.

The voluminous literature on the chromatography of phosphates has been reviewed by Ebel (1968), Kalliney (1972), Kiso et al. (1972), Ohashi (1975), and Westman (1977) and only an outline of the various methods can be given here. For details the reader should consult the original references, including the references given for specific glasses.

1. Paper Chromatography

In a series of publications (see Table I) Westman and coworkers applied paper chromatography to study binary and ternary alkali glassy phosphates, phosphoric acids and mixed glasses in the system $Na_2O-H_2O-P_2O_5$. Descending paper chromatography was first employed but this was changed later to ascending chromatography to obtain a better separation of the longer chains (Huhti and Gartaganis, 1956). The techniques used in much of this work have been described by Smith (1959) and Crowther (1954). Linear chain ions of formula $P_nO_{3n+1}^{(n+2)-}$ with n up to 9 could be separated. Westman showed the importance of operating at 4°C to minimize hydrolysis of the chains.

TABLE I

SYSTEMATIC STUDIES OF PHOSPHATE GLASSES, FLUOROPHOSPHATE GLASSES,
AND PHOSPHORIC ACIDS BY CHROMATOGRAPHIC METHODS

Author(s)	Cation(s)
(a) *Phosphate Glasses*	
Westman and Crowther (1954)	Na
Van Wazer and Karl-Kroupa (1956)	Na
Westman and Gartaganis (1957)	Li, Na, K
Westman et al. (1959)	Na + H
Murthy et al. (1961)	Li + Na + K
Westman and Murthy (1961)	Li, Li + Na + K
Murthy (1961)	Na
Murthy and Westman (1962a)	Rb
Murthy and Westman (1962b)	Li + Rb
Meadowcroft and Richardson (1965)	Li, Na, Ca, Zn, Li + Na, Li + Ca, Zn + Na, Zn + Ca
Murthy and Westman (1966)	Li + Cs
Poch (1966)	Na
Cripps-Clark et al. (1974)	Na, Pb, Zn, Cd, Mg
(b) *Fluorophosphate Glasses*	
Murthy and Mueller (1963)	Na
Murthy (1963)	Na
Westman and Murthy (1964)	Li + Na
Stevic et al. (1982)	Na
(c) *Phosphoric Acids*	
Huhti and Gartaganis (1956)	
Ohashi and Sugatani (1957)	
Jameson (1959)	

Ascending paper chromatography was used by Ebel and coworkers (Ebel, 1968) to study the sodium phosphates and polyphosphates. Ebel introduced the technique of two-dimensional paper chromatography to separate linear and cyclic species, using mildly acidic and alkaline solvents for the respective separations. This technique was further developed by Karl-Kroupa (1956). With Ebel's acidic solvent R_f values varied from about 0.7 for the ortho- to about 0.04 for the heptaphosphate whereas, with the alkaline solvent, the R_f values for the cyclic trimer (0.53) and tetramer (0.40) were higher than for the ortho (0.33). Ebel discovered the beneficial effect of adding a small amount of ammonium hydroxide to the trichloracetic acid-containing solvent to improve the separations, a procedure adopted by most other workers.

Extensive contributions to the paper chromatography of phosphates, including studies of the relation between R_f and the degree of condensation of the polyions, were made by Thilo and coworkers (Thilo 1962). For a compilation of R_f values of phosphate ions in various solvents, see Lederer and Majani (1970).

Paper chromatographic techniques for glasses with bivalent cations were similar except that a chelating agent was used to aid in dissolving the glasses (Meadowcroft and Richardson, 1965; Cripps-Clark et al., 1974). Ions of chain length up to $n = 7$ were resolved. Best separations were achieved when the pH of the solvent was between 1.2 and 1.8.

2. Thin-Layer Chromatography

Less extensive use has been made of thin-layer chromatography (TLC) for the study of glasses although the method has its advocates. For references to this technique for the study of phosphates generally, see Kalliney (1972), Kiso et al. (1972), and Ohashi (1975). Cleaner separations of ions up to the octaphosphate, shorter times (1.5 to 2 h as compared with 16 to 24 h for paper chromatography), and more compact apparatus were among the advantages claimed. As in paper chromatography, both one- and two-dimensional development techniques have been employed.

TLC on cellulose plates was used by Minami et al. (1979) to study superionic conducting glasses in the system $AgI-Ag_2O-P_2O_5$. Fluorophosphate glasses have also been examined by this method (Stevic et al., 1982).

3. Column Chromatography

Column chromatography has been used extensively for both preparative and analytical work. Historically, ion-exchange chromatography with Dowex-1 or Amberlite IRA-400 anion-exchange resins has been the preferred method. Discontinuous elution techniques with buffered sodium and potassium chloride solutions as eluant were reviewed by Ebel (1968). Separations up to $n = 12$ (Busch et al., 1957) and $n = 13$ (Rothbart et al., 1964) were reported. Gradient elution techniques with similar resins have also been described (Ebel, 1968; Kalliney, 1972). For excellent reviews of the theory of anion exchange and gradient elution chromatography, see Kiso et al. (1972) and Ohashi (1975). Until the advent of modern, automated procedures, the most powerful of these techniques appears to have been that of Jameson (1959), who succeeded in determining quantitatively the constituents of aqueous phosphoric acids up to $n = 14$.

A combination of gel and anion-exchange chromatography was used by Kura and Ohashi (1971) to separate cyclic species up to the cyclooctaphosphate.

4. High-Performance Liquid Chromatography

Traditional column chromatography has now been largely supplanted by high-performance liquid chromatography (HPLC) as the preferred technique for the separation of oligophosphates in solution. Coupled with automatic procedures for the analysis of phosphorus in the eluant (Yagamuchi et al., 1979), these methods have eliminated much of the tedium associated with earlier work. The development of improved techniques has been spurred largely by the increasing industrial importance of the polyphosphates and their key role in biosynthesis and metabolism.

An important development has been the use of ethylenediaminetetraacetic acid (EDTA), usually as its tetrasodium salt $Na_4(EDTA)$, in the eluant to avoid hydrolytic degradation during ion-exchange chromatography (Fukuda et al., 1976). This, in turn, has allowed higher temperatures to be used in the separations, with a consequent improvement in resolution (Baba et al., 1985a).

A further development has been the use of computer-assisted procedures for the optimization of operating parameters. Both isocratic (Baba et al., 1985a) and gradient elution (Baba et al., 1985b) techniques have been improved in this way by the judicious variation of pH, eluant concentration, and column temperature. Ions of chain length up to $n = 35$ have been separated by the latter technique. This is illustrated in Fig. 1, which shows the elution profile for an aqueous solution of a sodium phosphate glass $Na_{n+2}P_nO_{3n+1}$ with an average chain length $\bar{n}$ of 10 (Baba et al., 1985b). Future work on glasses may benefit from these advanced analytical procedures.

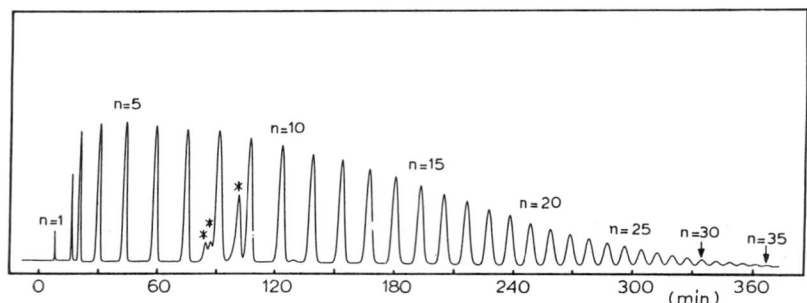

FIG. 1.   Profile obtained by gradient ion-exchange chromatography for a solution of sodium phosphate glass of composition $Na_{12}P_{10}O_{31}$ ($\bar{n} = 10$). Peaks with asterisks correspond to cyclic phosphates. (From Baba et al., 1985a.)

5. Discussion of Methods

The main source of error in the analysis of phosphate glasses by chromatography has been acid hydrolysis during dissolution of the glass and subsequent development of the chromatograms. As the ultimate product of hydrolysis is the orthophosphate ion PO_4^{3-}, the error is largest for this constituent. For the other anions the error tends to be self-corrective as random hydrolysis causes not only the disappearance of anions but also their generation from higher species. For the separation of linear chains an acidic solvents has generally been required as the neutral salts do not migrate readily on paper. With careful control of pH, however, and operation at low temperatures, such errors have been kept to a minimum. In early work, it was estimated (Hettler, 1958, 1959) that errors due to hydrolysis did not exceed about 3%. In more recent studies (Meadowcroft and Richardson, 1965; Cripps-Clark et al., 1974), such errors were estimated to be as low as 0.6%. The higher accuracy claimed by the latter authors may be due partly to the use of EDTA as a complexing agent to aid in dissolving the glasses.

In general, the results obtained for phosphate glasses by different experimenters are in good agreement and the advantages claimed for one method in preference to another appear to have been slight.

B. PHOSPHATE GLASSES AND PHOSPHORIC ACIDS

Studies of phosphate glasses by chromatographic techniques have been reviewed by Westman (1960, 1977), Gimblett (1963), Richardson (1974), and Jeffes (1981). The general picture that has emerged from these studies is that glasses more basic than the metaphosphate composition contain an array of anions of general formula $P_nO_{3n+1}^{(n+2)-}$ whose size and distribution depend on the composition and the nature of the cation(s). The concentration or ion fraction of each constituent passes through a maximum as the composition is varied, with the smaller ions becoming more abundant as the basicity is increased.

For the sodium phosphate glasses, which have been studied most extensively, it has been found that the distribution at any given composition does not depend on the nature of the starting materials, the time or temperature at which the glass is held in the molten state, the rate of quenching, or, provided the necessary precautions are taken to avoid hydrolysis, the conditions of dissolution (Van Wazer, 1958, p. 754; Westman, 1977, Meadowcroft and Richardson, 1965; Cripps-Clark et al., 1974).

Both Westman and Richardson et al. used a variety of quenching procedures to assess the possible effect of this variable. In Meadowcroft and Richardson (1965), quenching rates were estimated to vary between 300 and $400°C\,s^{-1}$. In Cripps-Clark et al. (1974), two fast-quenching techniques were employed. One was a hot-stage microscope technique similar to that described by Mercer and Miller (1963) for the preparation of beryl glass at quenching rates up to $20,000°C s^{-1}$. The other was a "splat" quenching technique similar to that described by Duwez and Williams (1963) for the fast quenching of metals at reported rates of up to $250,000°C s^{-1}$.

With these techniques, Cripps-Clark et al. found no change in the distributions for sodium phosphate glasses with Na_2O/P_2O_5 ratios R of 1.4 and 1.5, as compared with the distributions found earlier by Meadowcroft and Richardson by the slower cooling methods. Independently, also using the splat-cooling technique of Duwez and Williams, Phalippou and Zarzycki (1978) reported a slightly altered distribution for a sodium phosphate glass with $R = 1.65$, compared with a glass of the same composition that had been quenched between steel plates. The main difference was a somewhat lower percentage of dimer $P_2O_7^{4-}$ and a higher percentage of trimer $P_3O_{10}^{5-}$ in the rapidly quenched glass. In agreement with the results of Cripps-Clark et al., however, no effect was observed for a glass with $R = 1.38$.

The effect observed by Phalippou and Zarzycki for the more basic glass is probably due to the difficulty of avoiding some crystallization in a glass of so high basicity. For a glass of this composition ($\bar{n} = 3$) a higher proportion of trimer would be the expected result if some recrystallization occurred. In this respect, Cripps-Clark et al. noted that extensive devitrification can readily occur in such glasses without any obvious change in the appearance of the specimen, particularly with finely divided glasses prepared by the splat-cooling technique.

Richardson et al. concluded that the observed distributions corresponded to the fictive temperatures of the glasses, which did not vary with the cooling rate at the rates employed and could be equated with the glass transition temperatures T_g without much error.

By comparing chromatograms from solutions of phosphate glasses with those from crystalline phosphates of known structure it has been established that the ions present in the glasses are linear chains. Branched chains such as are not present in glasses more basic than the metaphosphate. Also, provided the metaphosphate composition is not closely approached, there are very few cyclic structures.

Table I shows the systems that have been studied systematically. Distributions for sodium phosphate glasses are shown in Fig. 2(a) and for sodium acid phosphate glasses and phosphoric acids in Figs. 2(b) and 2(c). In these figures the percentage by weight of phosphorus present as ortho PO_4^{3-}, pyro $P_2O_7^{4-}$, tri $P_3O_{10}^{5-}$ and higher chains, as well as cyclic and hypoly phosphates, is plotted against the average chain length $\bar{n}$ as calculated from the composition.

If R is the ratio M_2O/P_2O_5 or $(M_2O+H_2O)/P_2O_5$ and if the only anions are chain phosphates of formula $P_nO_{3n+1}^{(n+2)-}$, then $\bar{n}$ is related to R by the expression

$$\bar{n} = 2/(R - 1). \tag{1}$$

It is seen from Fig. 2 that, for glasses in the system Na_2O-H_2O-P_2O_5 distributions could be measured down to $\bar{n} = 1$, corresponding to the orthophosphate composition, $R = 3$. Such highly basic compositions could not be attained in the binary Na_2O-P_2O_5, however, due to the difficulty of quenching these melts to glasses. For this system the lowest average chain length at which data could be obtained was approximately three, corresponding to the composition $Na_5P_3O_{10}$ or $R = 5/3$.

If T represents the number of molecules per 100 atoms of phosphorus, then

$$T = 100/\bar{n}. \tag{2}$$

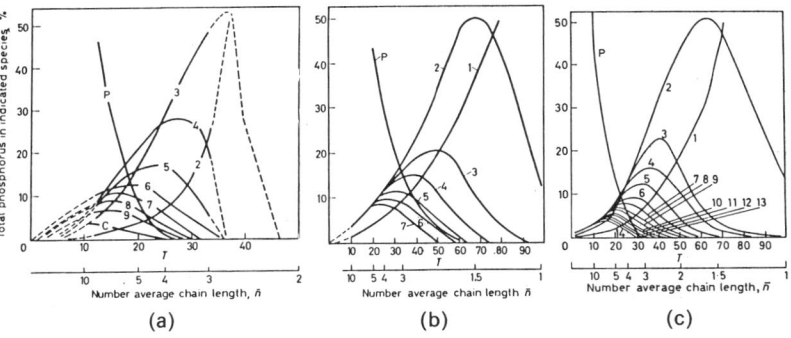

FIG. 2. Anionic constitution of (a) sodium phosphate glasses, (b) sodium acid phosphate glasses, and (c) phosphoric acids. 1 = ortho, 2 = pyro, 3 = tri, 4 = tetra, 5 = penta, 6 = hexa, 7 = hepta, P = hypoly, C = total cyclic phosphates. (From Gimblett, 1963.)

Values of T are shown also on the abscissae in Fig. 2. It is clear that $T = 0$ corresponds to the metaphosphate composition, $R = 1$. If cyclic or network structures are absent, $\bar{n} \to \infty$ at this composition.

Distributions for sodium phosphate glasses down to $\bar{n} = 2$ have been reported by Jeffes (1981). These are illustrated in Fig. 3, which shows the percentage phosphorus as x-mer in glasses of various compositions.

It is evident from Figs. 2 and 3 that little or no orthophosphate ion is present in sodium phosphate glasses with $\bar{n} \geq 2$. As shown later, this is in contrast to the silicates in which the orthosilicate ion is invariably found in glasses more basic than the metasilicate. This ion is, however, present in the acid phosphate glasses and phosphoric acids at $\bar{n} = 2$ (Figs. 2b and 2c) and its proportion increases rapidly as the orthophosphate composition ($\bar{n} = 1$) is approached. There is a corresponding decrease in the proportions of the other anions, although some pyrophosphate remains in both systems even at the ortho composition.

Distribution curves similar to those in Figs. 2 and 3 have been found for other systems. Thus the proportion of orthophosphate ion in lithium and potassium phosphate glasses with $\bar{n} \geq 4$ is very small. There is, however, a measurable effect due to the cation, with the distributions for sodium lying intermediate between those for lithium and potassium (Westman and Gartaganis, 1957). This is illustrated in Fig. 4.

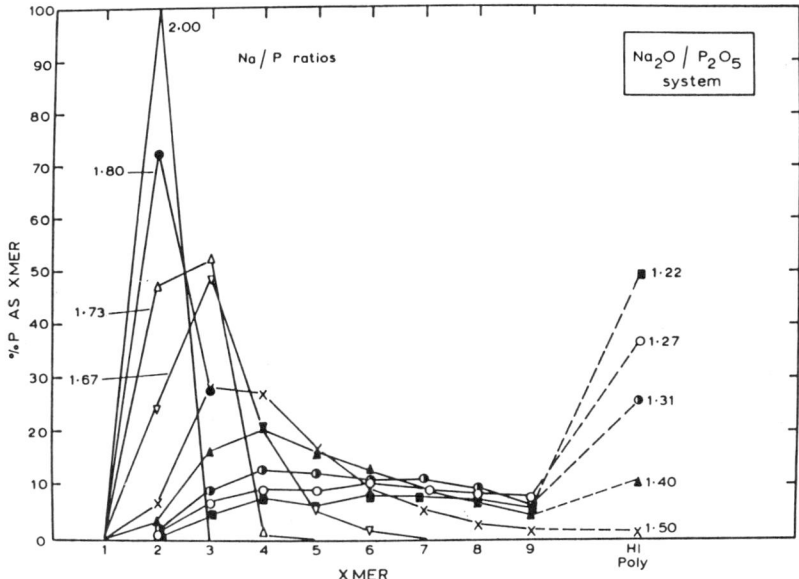

Fig. 3. Distribution of chains in sodium phosphate glasses of varying Na/P ratio. (From Jeffes, 1981.)

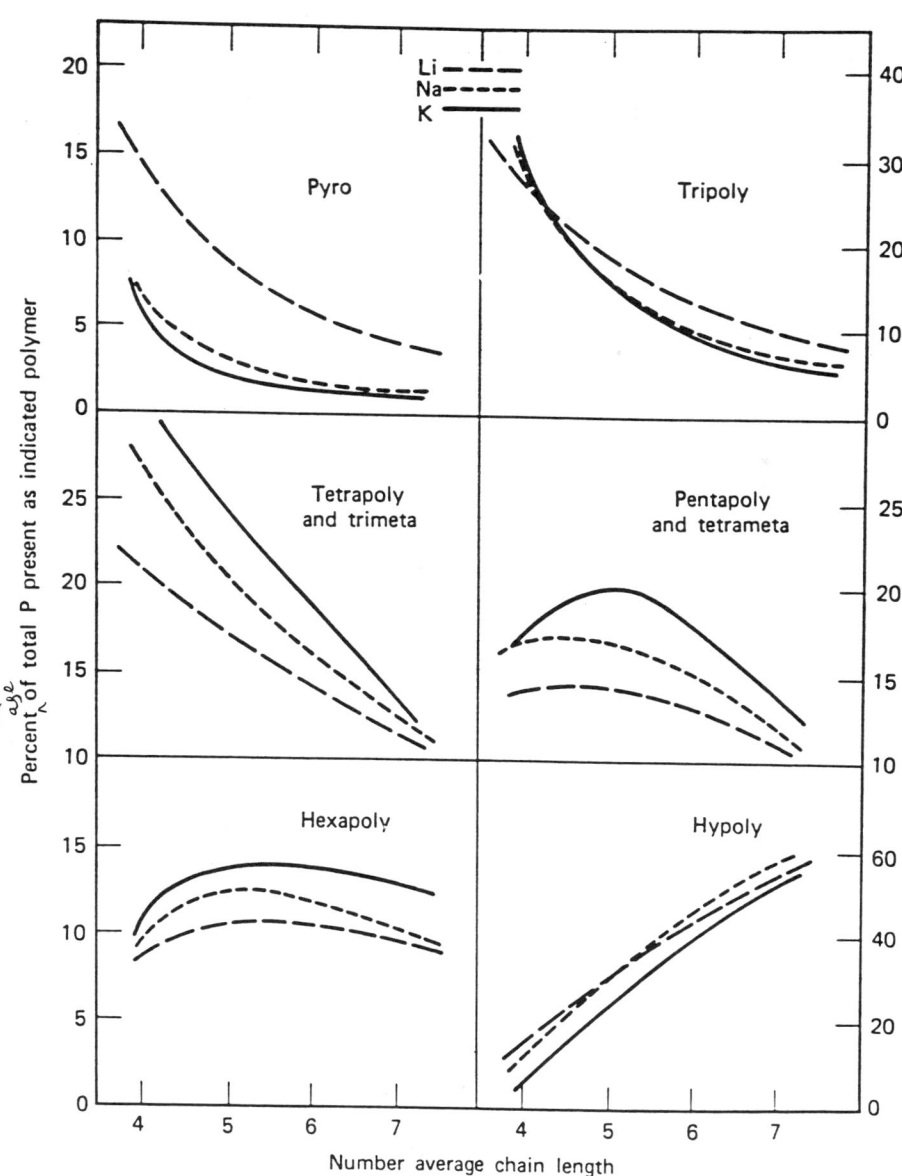

FIG. 4.   Illustration of cation effect in phosphate glasses. (From Westman and Gartaganis, 1957.)

Mixed-alkali phosphate glasses in the Li-Na-K system were studied by Murthy et al. (See Table I.) The use of mixed-alkali cations allowed glasses with $\bar{n}$ as low as 1.5 to be prepared. These contained large amounts of orthophosphate ion, the proportion of which began to increase significantly in the vicinity of $\bar{n} = 2$.

The results of Westman, Murthy, et al. for the Li and Na phosphates were confirmed by Meadowcroft and Richardson (1965) and Cripps-Clark et al. (1974) who extended these studies to glasses with bivalent cations. For glasses with Ca^{2+}, Zn^{2+}, Pb^{2+}, Cd^{2+}, and Mg^{2+} cations the proportion of orthophosphate ion, though still small, was higher than in the alkali phosphates and was significant for glasses with $\bar{n} = 2$. The distributions for glasses with bivalent cations were broader than for the alkali phosphate glasses of the same average chain length. This is illustrated in Fig. 5, which shows the distributions for Li, Pb, Zn, Cd, and Mg phosphates with $\bar{n} = 4$.

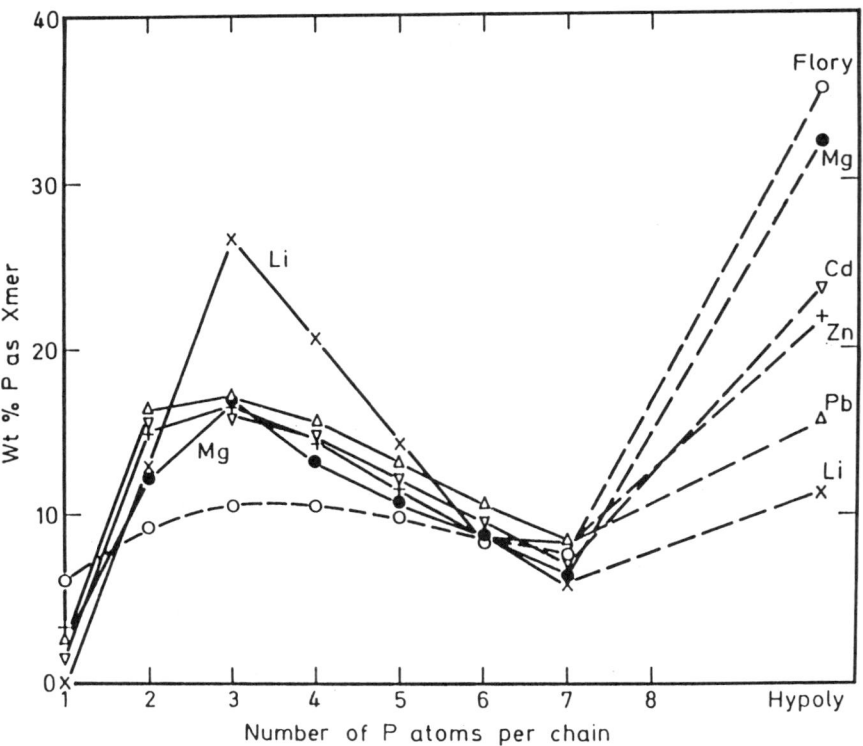

FIG. 5.   Distributions of phosphorus in phosphate glasses of average chain length $\bar{n} = 4$. x: Li; Δ: Pb; $+$: Zn; ∇: Cd; $\bullet$: Mg. The open circles O correspond to the Flory distributions. (From Cripps-Clark et al., 1974.)

The marked difference between the distributions for the sodium phosphate glasses and phosphoric acid for systems with $\bar{n} = 4$ is illustrated in Fig. 6 (Jeffes, 1981).

The proportion of cyclic ions in sodium phosphate glasses, measured by Westman and Gartaganis (1957), is shown in Fig. 7. Only the cyclic trimer $P_3O_9^{3-}$ and tetramer $P_4O_{12}^{4-}$ were observed. Their proportions were negligible in glasses of high basicity ($\bar{n} \leq 4$). At the metaphosphate composition (69.6% P_2O_5), only 7.6% of the total phosphorus was present as cyclic anions. This is in good agreement with results obtained by solubility fractionation and barium precipitation (Van Wazer, 1958).

The effect of pressure on the constitution of sodium phosphate glasses with $R = 1.457$ and 1.299 was studied by Poch (1966). For glasses prepared in the

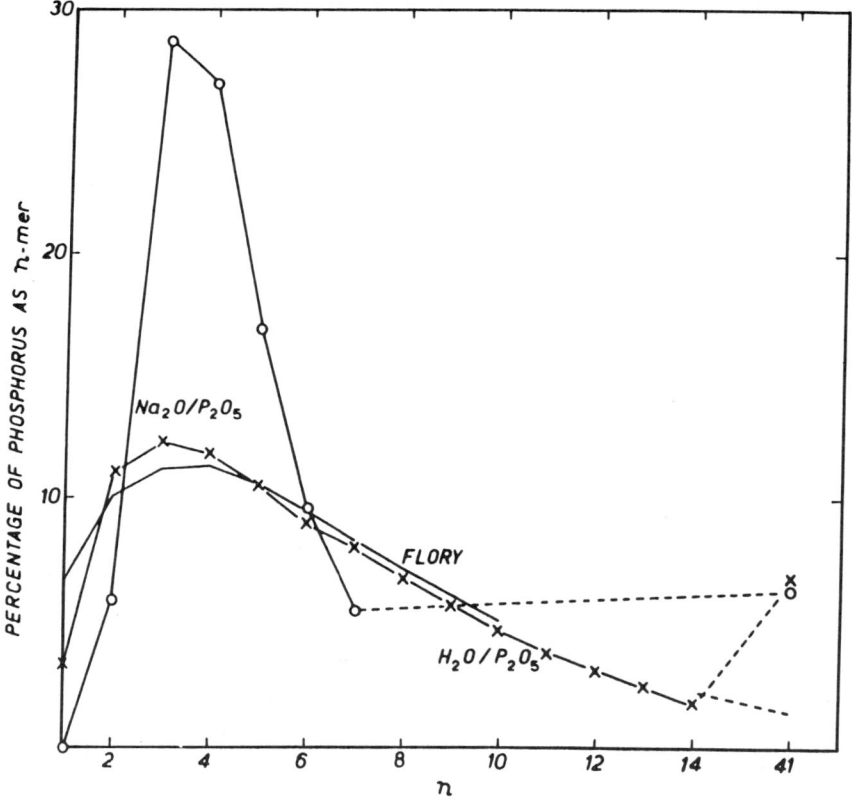

FIG. 6. Distribution of chains in a sodium phosphate glass (open circles) and phosphoric acid (crosses) of composition corresponding to the formulas $Na_6P_4O_{13}$ and $H_6P_4O_{13}$ ($\bar{n} = 4$). Shown also is the calculated Flory distribution for this composition. (From Jeffes, 1981.)

transformation range no effect on the constitution was observed for pressures of 25×10^5 and 45×10^5 kPa. For glasses prepared by cooling under pressure from above the liquidus temperatures, however, the proportion of ions with $n = 1$ to 7 was increased and that of the higher polyions ($n \geq 7$) was reduced. When these glasses were annealed, the original anionic distributions were restored.

C. DISTRIBUTION THEORIES

To interpret the chromatographic data it is assumed that the ionic distributions present in the glasses represent chemical equilibria in the corresponding melts. Two main approaches have been used to describe the equilibria in these systems: the random reorganization theory of Van Wazer et al. and conventional polymer theory.

1. Random Reorganization Theory

In general, four structural entities may be envisaged in phosphate melts. These are:

branching   middle   end   ortho

b m e o

For the phosphoric acids the equilibria between them were treated by Parks and Van Wazer (1957) in terms of the following reactions:

$$2m = b + e \qquad k_i = be/m^2 \tag{3}$$

$$2e = m + o \qquad k_{ii} = mo/e^2 \tag{4}$$

$$2o = (2e) + u \qquad k_{iii} = (2e)u/o^2 \tag{5}$$

where b, m, e, and o represent the "mole fraction" of branching, middle, end, and ortho units respectively, u is the mole fraction of unreacted H_2O, and $(2e)$ is treated as a single entity. Figure 8 shows the relative proportions of these units calculated for selected values of the equilibrium ratios of these reactions. The good agreement supports the concept that the distribution of structural units can be expressed in this way.

2. Polymer Theory

The approach of Van Wazer was significant in revealing the principles of bond interchange that govern the constitution of phosphate melts. An

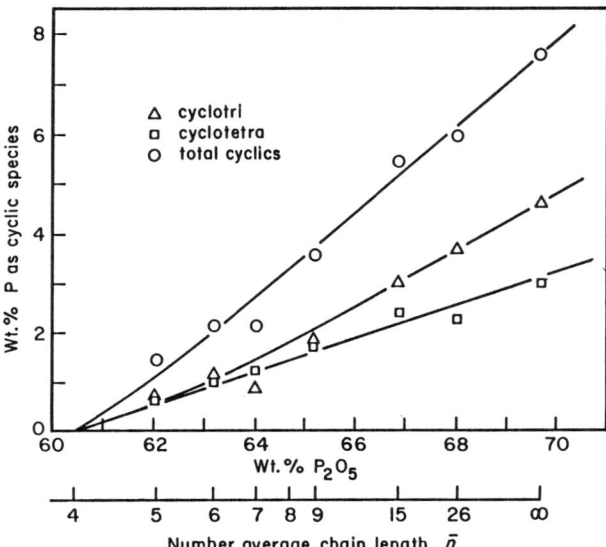

FIG. 7. Cyclic species in sodium phosphate glasses. (From Westman and Gartaganis, 1957.)

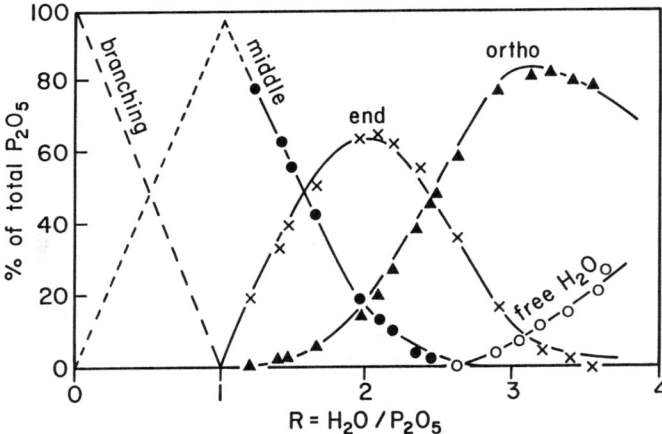

FIG. 8. Relative proportions of branching, middle, and end units as well as orthophosphate and free water as functions of composition for aqueous phosphoric acid solutions. The points are from the experimental data of Huhti and Gartaganis (1956). The smooth curves were calculated assuming $k_i = 10^{-3}$, $k_{ii} = 0.08$, and $k_{iii} = 0.02$. (From Parks and Van Wazer, 1957.)

alternative procedure, which yields more insight into the detailed ionic constitution, is to consider the data in terms of polymer theory (Meadowcroft and Richardson, 1965; Westman and Beatty, 1966; Cripps-Clark et al., 1974). In this approach the equilibria between the various polyions are expressed in terms of conventional equilibrium constants, assuming that ion or mole fractions[1] may be used in place of the activities of the individual constituents.

The set of equilibria commonly chosen to depict the data is given by the equation

$$2P_nO_{3n+1}^{(n+2)-} = P_{n-1}O_{3n-2}^{(n+1)-} + P_{n+1}O_{3n+4}^{(n+3)-} \tag{6}$$

for which

$$-\Delta G_{nn}^0 = RT \ln k_{nn} \tag{7}$$

$$k_{nn} = N_{n-1} N_{n+1} / N_n^2 \tag{8}$$

where N is the ion fraction of the indicated species. If the distribution of phosphate ions were completely random, the standard free-energy change ΔG_{nn}^0 for reaction (6) would be zero and the equilibrium ratio k_{nn} unity for all values of n greater than or equal to 2. Table II shows the mean equilibrium ratios k_{nn} for the binary systems so far investigated.

It is clear from Table II that k_{nn} rapidly converges to unity as n increases, showing that the random-mixing approximation holds for the longer chains. For the sodium phosphate glasses, values of k_{nn} close to unity were found for $n = 5$ to 10 (Cripps-Clark et al., 1974) while the data of Jameson (1959) for the phosphoric acids showed k_{nn} effectively constant and equal to unity for $n = 3$ to 14 (Meadowcroft and Richardson, 1965; Westman and Beatty, 1966).

Reaction (6) involves exchange of the segment PO_3^- between ions of the same chain length n. By analogy with the silicates it is readily shown (Masson, 1968) that if k_{nn} is unity for reaction (6), the ratios k_{mn} for all metathetical reactions of the type

$$P_mO_{3m+1}^{(m+2)-} + P_nO_{3n+1}^{(n+2)-} = P_{m-x}O_{3(m-x)+1}^{(m-x+2)-} + P_{n+x}O_{3(n+x)+1}^{(n+x+2)-} \tag{9}$$

in which the segment $(PO_3)_x^{x-}$ is exchanged are also unity. Hence the random-mixing approximation holds in the general case for ions of different size and the anions may be permuted with one another irrespective of their lengths when m and n become sufficiently large.

The value of unity for k_{nn} in equation (8) implies that

$$\frac{N_{n-1}}{N_n} = \frac{N_n}{N_{n+1}} = \cdots = r \qquad (n \geq 2), \tag{10}$$

---

[1] For a binary system $MO-P_2O_5$ or $M_2O-P_2O_5$, the ion fractions of $P_nO_{3n+1}^{(n+2)-}$ are identical with the mole fractions of the constituents $M_{n+2/2}P_nO_{3n+1}$ or $M_{n+2}P_nO_{3n+1}$.

TABLE II

EQUILIBRIUM RATIOS K_{nn} FOR BINARY PHOSPHATE GLASSES AND PHOSPHORIC ACID

Cation	$\bar{n}$	k_{22}	k_{33}	k_{44}	k_{55}	k_{66}	Source of data*
Li	3.0–7.1	very small	0.48	0.93	1.04	1.00	WG, MR
Na	3.0–9.0	very small	0.26	0.66	0.97	1.03	WG, MR
K	3.5–9.0	very small	0.265	0.68	0.94	0.82	WG
Zn	3.0–4.0	0.32	0.98	1.00	0.94	1.19	MR, CC
Cd	2.02–4.48	0.10	1.00	1.08	0.96	0.99	CC
Mg	4.01–5.32	0.38	0.98	0.94	0.98	0.99	CC
Ca	3.0	0.12	0.67	1.01	0.89	1.09	MR
	4.0	0.21	0.73	1.01	0.82	1.25	MR
	mean	0.17	0.70	1.01	0.86	1.17	MR
Pb	1.96	0.07	1.07	0.99	—	—	CC
	2.82	0.10	0.96	0.97	1.00	1.07	CC
	3.60	0.25	0.97	0.99	1.00	0.99	CC
	4.12	0.30	1.00	1.01	0.98	1.06	CC
	4.73	0.29	1.12	1.02	1.12	1.00	CC
	mean	—	1.02	0.99	1.02	1.03	CC
H	1.12	0.21	—	—	—	—	J
	1.34	0.31	0.75	—	—	—	HG
	1.56	0.20	1.15	1.28	—	—	J
	1.9	0.26	1.07	0.92	1.08	0.98	HG
	2.20	0.29	1.05	0.99	0.98	1.04	J
	2.6	0.34	1.04	0.90	0.85	1.25	HG
	2.94	0.36	1.11	0.99	0.91	1.07	J
	3.6	0.36	1.21	0.83	1.04	0.92	HG
	4.0	0.47	0.99	0.99	1.03	1.06	J
	5.1	0.49	1.44	0.78	1.01	1.01	HG
	7.8	0.77	1.56	0.70	1.05	0.62	HG
	9.1	0.92	1.00	0.92	1.09	0.96	J
	mean	—	1.12	0.93	1.00	0.99	

*WG: Westman and Gartaganis (1957); MR: Meadowcroft and Richardson (1965); CC: Cripps-Clark et al. (1974); J: Jameson (1959); HG: Huhti and Gartaganis (1956). (Source: Cripps-Clark et al., 1974.)

i.e., that at any given composition, the ion fractions N_{n-1}, N_n, N_{n+1},... decrease monotonically, with common ratio r, at sufficiently high values of n.

a. Flory distributions. For the hypothetical case in which k_{nn} is unity for all values of n greater than or equal to 2 (i.e., for the random mixing of all phosphate anions PO_4^{3-}, $P_2O_7^{4-}$, $P_3O_{10}^{5-}$,...) and if no free oxide ion is present in the system, the distributions are given by the classical Flory (1953) distributions[2] for linear chains:

$$N_n = \frac{1}{\bar{n}}\left(\frac{\bar{n}-1}{\bar{n}}\right)^{n-1} \qquad (n \geq 1) \tag{11}$$

where $\bar{n}$, the number average chain length, is determined solely by the composition (Meadowcroft and Richardson, 1965). Distributions calculated by equation (11) for systems with $\bar{n} = 4$ are shown in Figs. 5 and 6. Ion fractions are conveniently converted to percentages of phosphorus by the expression (Richardson, 1974, p. 89)

$$(\% \text{ P in glass present as } n\text{-mer}) = \frac{100n}{\bar{n}} N_n. \tag{12}$$

The Flory distributions are calculated on the assumption that all functional groups on the polymer chains are chemically equivalent, regardless of the size of the chain to which they are attached (equal reactivity of functional groups). That this assumption does not hold for the smaller anions is shown by the low percentage of orthophosphate ion PO_4^{3-} in most systems. The ion fraction of this species is always lower than predicted by the Flory distributions, reflecting a higher reactivity of the O^- groups on this ion than on the longer chains. This perturbs the distributions, which are always sharper or more "peaked" than the Flory distributions as shown in Figs. 5 and 6.

  b. *Effect of n on k_{nn}.*   Table II shows that k_{nn} reaches unity for the alkali-metal phosphates when n is equal to 5. For calcium phosphate glasses, k_{nn} becomes unity at $n = 4$, while for the other systems studied, including the phosphoric acids, k_{nn} is effectively unity for $n \geq 3$. Clearly, departures from the random-mixing approximation are associated with the nature of the cation and are more pronounced for the alkali-metal glasses than for the other systems.

  c. *Effect of $\bar{n}$ on k_{nn}.*   For all systems in Table II the mean chain length $\bar{n}$ had essentially no effect on k_{nn} for $n \geq 3$ and the values of k_{nn} shown for the alkali-metal cations and for Zn^{2+}, Cd^{2+}, and Mg^{2+} are the mean values over the ranges of $\bar{n}$ indicated in column 2. There is, however, a variation of k_{22} with $\bar{n}$, the extent of which depends on the cation. This is shown in Table II by

---

[2]Equation (11) may readily be derived as follows. For a bifunctional or pseudobifunctional monomer with two functional groups available for reaction, the fraction of functional groups that react to yield a polymer of average chain length $\bar{n}$ is $(1 - 1/\bar{n})$ and the fraction that remains unreacted (i.e., as end groups) in $1/\bar{n}$. Hence the probability that any unreacted functional group is part of a molecule with $(n - 1)$ reacted groups and one unreacted group is $(1 - 1/\bar{n}) \cdot (1/\bar{n})$, which is the ion fraction of n-mer, N_n.

the data for Ca^{2+}, Pb^{2+}, and H^+, for which the effect was most pronounced, and is illustrated in Fig. 9 where k_{22} is plotted against $\bar{n}$ for various systems.

As k_{33} and k_{44} remain constant, this variation of k_{22} with $\bar{n}$ must mean that the activity coefficient of the orthophosphate constituent varies with the composition of the melt (Cripps-Clark et al., 1974).

d. Evaluation of k_{1n}. For a complete description of the constitution it is necessary to know the values of k_{1n} for these systems. Unfortunately the chromatographic data do not allow these values to be determined as this refers to the equilibrium

$$2PO_4^{3-} = P_2O_7^{4-} + O^{2-} \tag{13}$$

and the values of $N_{O^{2-}}$ are unknown.

If the ion fraction of free oxide ion (or the mole fraction of free oxide) is large, its value can be estimated from stoichiometry. In this way it was possible to estimate the mole fraction of free H_2O in the system H_2O-P_2O_5

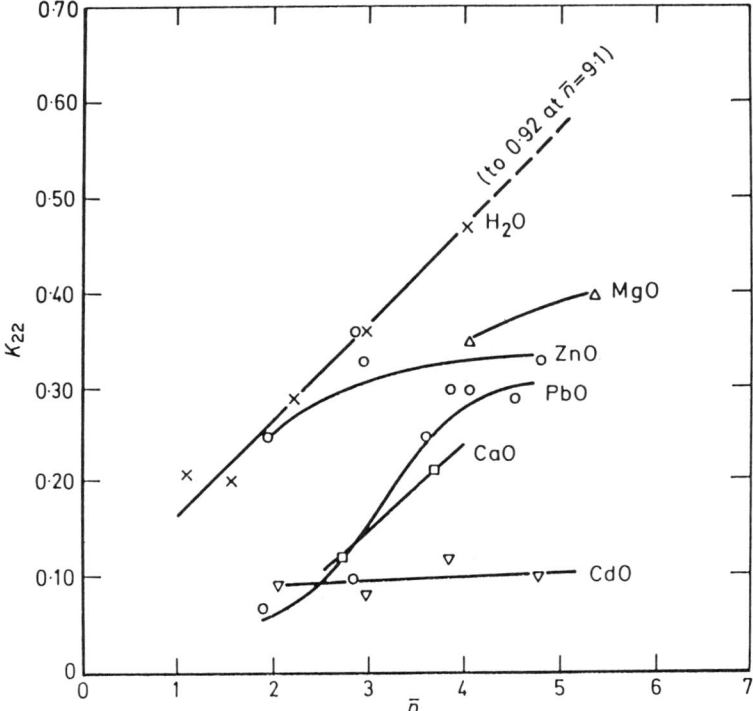

FIG. 9. Variation of k_{22} with mean chain length $\bar{n}$ for various phosphate glasses and phosphoric acid. (From Cripps-Clark et al., 1974.)

and it was shown (Cripps-Clark et al., 1974) that the value of k_{11} for this system is of the order of 10^{-2} to 10^{-3}. This is in reasonable agreement with the value of 2×10^{-2} estimated by Parks and Van Wazer from the random reorganization theory. (Cf. Fig. 8.) It also agrees with the average value of $\sim 5 \times 10^{-3}$ estimated by Huhti and Gartaganis (1956). The data suggest that k_{11} for the system H_2O-P_2O_5 decreases as $\bar{n}$ increases and this is in line with the increasing values of k_{22} with $\bar{n}$ shown in Table II.

It should be noted that although random mixing of phosphate anions requires k_{nn} to be unity for $n \geq 2$, k_{11} may have any value depending on the nature of the cation (Masson, 1965). Indeed, for silicate melts, it is the influence of the cation on k_{11} that determines the pattern of activities and, hence, ionic distributions in systems with different cations. This is un-doubtedly true also for the phosphates, but cannot be tested until values of $N_{O^{2-}}$ are known. A reliable method of determining the concentration of free oxide ion O^{2-} in phosphate glasses would represent a significant advance in this field.

Consider the array of reactions

$$PO_4^{3-} + P_nO_{3n+1}^{(n+2)-} = P_{n+1}O_{3n+4}^{(n+3)-} + O^{2-} \tag{14}$$

for which

$$k_{1n} = N_{n+1}N_{O^{2-}}/N_1N_n. \tag{15}$$

It is readily shown (Masson, 1965) that if k_{1n} is independent of n, k_{nn} must be unity for all values of n greater than or equal to 2.

Conversely, the deviations from unity of k_{nn} at low values of n in reaction (6) may be interpreted in terms of variations of k_{1n} with n for the lower members of the series in reaction (14). The two effects are equivalent, due to the interrelationships between the equilibrium ratios. Just as k_{nn} increases toward unity, so k_{1n} decreases toward a constant value as n increases. Indeed it can be shown that

$$k_{nn} = k_{1n}/k_{1(n-1)} \tag{16}$$

and

$$\lim_{n \to \infty} k_{1n} = \prod_{n=1}^{\infty} k_{nn}. \tag{17}$$

Equation (16) shows that if k_{1n} is known for all n, then k_{nn} is also known for all n and the system is completely defined.

e. Deviations from random mixing. The deviations from random mixing for the shorter chains were interpreted by Meadowcroft and Richardson (1965) in terms of endothermic enthalpy changes for reaction (6). In support of this, Cripps-Clark et al. (1974) showed there was a clear correlation between the heats of formation of the glasses and the trend of the ionic distributions for different cations, departures from the Flory distributions being smallest for those systems in which the heats of formation were small.

f. General discussion of ionic distributions. Although the ionic distributions in phosphate glasses can be conveniently discussed in terms of the Flory distributions, it should be borne in mind that the distributions calculated by Flory for the self-condensation of a bifunctional or multifunctional monomer refer to the situation in which the product of condensation (usually H_2O or NH_3 in organic systems) is removed from the system as it is formed and plays no further role in the polymerization process. In the case of silicate and phosphate melts, however, the product of condensation (i.e., free oxide or free oxide ion) is an integral part of the system and its presence must be taken into consideration in calculating the distribution. It is only when the free oxide ion is absent that the Flory distributions can be expected to pertain. This condition holds only in the limit when k_{11} is zero and the P_2O_5 or SiO_2 content is higher than the orthophosphate or orthosilicate composition. Under all other conditions, the oxide ion will be present to some extent in the melt or glass below the metasilicate composition and its presence will perturb the distributions.

A more general approach to linear chain polycondensations, which allows for the presence of the free oxide ion, has been developed for the silicates (Masson, 1965) and can readily be extended to the phosphates. To apply this approach requires a knowledge of the equilibrium ratio k_{11}. It is easily shown that this treatment yields the Flory distributions as a special case when k_{11} tends to zero and the SiO_2 or P_2O_5 content lies between the ortho and meta compositions.

When the concentration of the free oxide ion is small, as might be expected for the alkali- and alkaline-earth metal phosphates more acidic than the orthophosphate composition, two factors are responsible for departures from the Flory distributions and should be clearly recognized. The first is the variation of k_{nn} with n. This is due to departures from the principle of equal reactivity as previously discussed. It is considered that these deviations are associated with the degree of ionicity of the cation–oxygen bonds on the phosphate chains (Jeffes, 1981). The more covalent these bonds (as, for example, in the phosphoric acids), the more closely will these systems resemble "conventional" uncharged polymers and the more one would expect the principle of equal reactivity to hold.

This implies that deviations from equal reactivity will become less important as k_{11} increases, provided allowance is made for the presence of free oxide in computing the distributions. Ultimately, when $k_{11} = 1$, the principle of equal reactivity will hold for the free O^{2-} ions as well as the O^- groups on the phosphate chains. This special case corresponds to the situation in which the oxide MO or M_2O is "neutral" and acts in neither a "network-breaking" or "network-forming" capacity. An example is the oxide FeO in $FeO\text{-}SiO_2$ melts at 1600°C (Masson, 1984).

The second factor responsible for departures from the Flory distributions in systems with a low concentration of free oxide ion is the variation of k_{nn} with $\bar{n}$. This is due to nonideality in the thermodynamic sense and must be attributed to variations in the activity coefficients of the constituents with composition. As these variations are seen only in the values of k_{22} and k_{11}, they must be due to nonideal mixing of the orthophosphate and free oxide ions. This is in line with the limited thermodynamic data available for phosphate melts, which indicate a strong interaction between the PO_4^{3-} and O^{2-} ions (Jeffes, 1981).

In summary it may be stated that the pattern of ionic distributions revealed by chromatographic studies of phosphate glasses may be broadly interpreted in terms of conventional polymer theory, provided the theory is modified for systems that contain significant amounts of the free oxide ion. Allowance must be made for departures from the principle of equal reactivity for the smallest anions and, in systems containing free oxide, for nonideal mixing of the oxide and orthophosphate anions.

D. HALOPHOSPHATE GLASSES

Fluorophosphate glasses have attracted attention in recent years due to their interesting optical properties and their possible application in the fabrication of lasers.

Murphy and Mueller (1963) showed that incorporation of sodium fluoride in sodium phosphate glasses caused a breakdown in the structure of the phosphate polymers. This was attributed to the rupture of $-P-O-P-$ bridges to form $-P-ONa$ and $-P-F$ end groups:

$$
\begin{array}{cccc}
O & O & O & O \\
\parallel & \parallel & \parallel & \parallel \\
-P-O-P- & + \text{NaF} = & -P-ONa + F-P-. \\
\mid & \mid & \mid & \mid \\
O & O & O & O \\
\text{Na} & \text{Na} & \text{Na} & \text{Na}
\end{array}
\qquad (18)
$$

A band in the chromatograms between the ortho and pyro positions was identified (Murthy, 1963) as due to the monofluorophosphate ion PO_3F^{2-}. Only about half of the fluorine, however, could be accounted for in this way (Westman and Murthy, 1964) and it was concluded that the remainder was distributed among the other polymers.

This was confirmed by Stevic et al. (1982) who showed the presence of monofluorophosphate ions $P_nO_{3n}F^{(n+1)-}$ with $n = 1$ to 5 and difluorophosphate ions $P_nO_{3n-1}F_2^{n-}$ with $n = 1$ to 7 in sodium fluorophosphate and sodium aluminofluorophosphate glasses. The ratio of difluoro- to monofluorophosphate anions increased with increasing fluorine content. Increase in the fluorine content also resulted in a shortening of the chains.

These results are in line with similar observations for fluorosilicate glasses and with conclusions derived from polymer theory. (See Section IV.D.)

Glasses in the system AgI-Ag_2O-P_2O_5 were studied by Minami et al. (1979). These glasses exhibit high ionic conductivities that are determined primarily by their AgI content (Minami et al., 1977). The glasses contained 45–85 mol% AgI and had molar ratios Ag_2O/P_2O_5 (R) of 0.8 to 3.

Glasses with $R = 0.8$ and 1 contained almost no ions of low chain length, as might be expected, but discrete ions appeared as R was increased to 1.25 and 1.6. At $R = 2$, corresponding to the pyrophosphate composition, the main constituent was the pyrophosphate ion $P_2O_7^{4-}$. The orthophosphate ion was the dominant constituent at $R = 3$. No cyclic phosphates were observed. The results were consistent with conclusions from infrared spectra.

E. Phosphate Glasses with Other Oxyacid Anions

The use of paper chromatography to study silicophosphate glasses was first reported by du Plessis (1959), who examined the effect of incorporating silica in a calcium metaphosphate glass. The average chain length of the phosphate ions decreased with increasing silica content. A similar finding was reported by Ohashi and Oshima (1963) for the incorporation of silica or sodium metasilicate in sodium metaphosphate glasses.

The effect was studied more systematically by Fray and coworkers (Fray, 1974; Finn et al., 1976), who added varying amounts of sodium silicate to sodium phosphate glasses of well-defined constitution. It was established that the mean chain length of the phosphate anions was lowered by the incorporation of silica and this was interpreted in terms of silicate ions entering the phosphate chains followed by hydrolysis of the Si-O-P linkages when the glasses were dissolved in water. It was deduced that many of the silicate ions that enter the phosphate chains caused branching, and the proportion of linear and branched chains in the glass was estimated.

Kaula et al. (1981) studied the constitution of silicophosphate glasses with SrO, BaO, ZnO, PbO, Bi_2O_3, and TeO_2 as modifiers. When the ratio R of modifying oxide/P_2O_5 was unity, tripolyphosphate anions $P_3O_{10}^{5-}$ were the main structural units found in the chromatograms. When R was 1/2 and 1/3, the main units were the cyclic ions $P_3O_9^{3-}$. For glasses of varying silica content including those with MgO and MnO as modifiers, chain and cyclic phosphate ions with $n \geq 8$ were found (Kaula et al., 1982). The degree of polymerization of the phosphate portion decreased with increasing silica content. The modifiers increased the amount of cyclic phosphates in the sequence MnO, MgO, SrO, Bi_2O_3, TeO_2, PbO, and BaO.

The constitution of arsenophosphate glasses has been studied by Thilo and coworkers (Thilo, 1962). Glasses in the sodium metaphosphate-sodium meta-arsenate system can easily be prepared (Thilo and Kolditz, 1955). When dissolved in water they hydrolyze readily at the As-O-P linkages to yield a mixture of polyphosphates of varying chain length. The amounts of poly-phosphates with longer chain length increase as the P/As ratio is increased. It was concluded that anions of the type

$$\overset{\displaystyle O}{\underset{\displaystyle O^-}{\overset{\|}{O^-\!-\!P}}}\!-\!O\!-\!\overset{\displaystyle O}{\underset{\displaystyle O^-}{\overset{\|}{P}}}\!-\!O\!-\!\overset{\displaystyle O}{\underset{\displaystyle O^-}{\overset{\|}{As}}}\!-\!O\!-\!\overset{\displaystyle O}{\underset{\displaystyle O^-}{\overset{\|}{P}}}\!-\!O\!-\!\overset{\displaystyle O}{\underset{\displaystyle O^-}{\overset{\|}{As}}}\!-\!O\!-\!\overset{\displaystyle O}{\underset{\displaystyle O^-}{\overset{\|}{P}}}\!-\!O\!-\!\overset{\displaystyle O}{\underset{\displaystyle O^-}{\overset{\|}{P}}}\!-\!O$$

were present, with AsO_4 tetrahedra incorporated statistically in the chains.

Polyarsenophosphate glasses devitrify on heating to yield a variety of crystalline compounds with As-O-P linkages. These have been well characterized by X-ray diffraction (Thilo, 1962; Corbridge, 1974). In contrast, few crystalline materials with Si-O-P linkages are known (Dent Glasser and Smith, 1978).

Sodium vanadophosphate glasses with P/V ratios of 1 to 200 have been prepared and their constitution examined by paper chromatography (Ohashi and Matsumura, 1962). The glasses were prepared by melting appropriate mixtures of sodium metaphosphate glass, V_2O_5, and Na_2CO_3 and quenching between copper plates. During this process some of the vanadium (V) was reduced to vanadium (IV), with loss of oxygen. Aqueous solutions of glasses with P/V ratios of 2 to 10 contained larger amounts of pyrophosphate ion $P_2O_7^{5-}$ than expected by comparison with other short chain phosphates and it was suggested that these glasses contain vanadophosphate rings of the type

With increasing P/V ratio, increasing amounts of longer chains were present. Aqueous solutions of glasses with P/V ratios of 50, 100, and 200 could not be analyzed quantitatively as almost all the phosphorus was present as chains with $n \geq 7$.

Glasses of the system $NaBO_2 + NaPO_3$ with P/B ratios of 1 to 98.9 were studied chromatographically by Nakamura and Ohashi (1967). It was concluded that glasses with P/B = 1 are composed mainly of chains with P-O-P, P-O-B, and B-O-B linkages. With increasing P/B, the proportion of B-O-B linkages decreased and branched BO_3 units appeared. When P/B exceeded 4.0, all B atoms were present as branched units.

In a series of publications Watanabe and coworkers studied the constitution of glasses prepared by incorporating the oxides of antimony (Watanabe et al., 1971; Watanabe and Kato, 1972; Watanabe, 1973), molybdenum and tungsten (Watanabe, 1974), aluminum (Watanabe et al., 1975), chromium (Watanabe et al., 1976), and germanium (Watanabe et al., 1977) in sodium metaphosphate glass. It was concluded that they possess Sb-O-P, Mo-O-P, W-O-P, Al-O-P, Cr-O-P, and Ge-O-P linkages, respectively. In all cases the average chain length of the phosphate portion increased as the P/M ratio increased. In the molybdo- and tungstophosphate glasses, the presence of Mo-O-P and W-O-P bonds was confirmed by infrared absorption measurements.

IV. Silicates

The successful application of chromatographic methods to phosphate glasses is due to the circumstance that the ions extracted from the glass undergo little or no reorganization in solution, as established by experiments with crystalline phosphates of known structure. Hence the ionic distributions found in solution correspond closely to those in the glass.

Similar methods have met with only partial success for the silicates due to the highly labile nature of the silicate ions in aqueous solution and their pronounced tendency to undergo rapid reorganization to yield a distribution characteristic of the solution itself.

A paper chromatographic method for silicates was described by Wieker and Hoebbel (1969) in which reorganization reactions were kept to a minimum by maintaining the pH of both the solution and chromatographic solvent between 2 and 3, where the rates of condensation and hydrolysis are lowest. In addition, the concentration of silicate in solution was kept small to minimize condensation reactions and the time required for chromatographic separation was kept as short as possible. With these precautions it was possible to separate ortho-, pyro, poly- and a variety of cyclosilicate ions in solutions containing mixtures of these species. Hydrolytic reactions were undetectable for the cyclooctamer $Si_8O_{24}^{16-}$, but were about 20% for the pyrosilicate ion $Si_2O_7^{6-}$ and 50% for the cyclotrimer $Si_3O_9^{6-}$, which dissociated into the ortho and pyro species. The technique was applied successfully to a large number of crystalline silicates (Hoebbel and Wieker, 1971, 1972, 1974; Wieker and Hahn, 1973; Götz et al., 1975a, b; Hoebbel et al., 1975, 1980, 1984; Winkler et al., 1983) and to the study of changes in the constitution of silicic acid solutions during condensation (Hoebbel and Wieker, 1973).

A. METHOD OF TRIMETHYLSILYLATION

As of the late 1980s, the most successful technique developed for the determination of discrete silicate ions in amorphous materials is that of trimethylsilylation, introduced by Lentz (1964), in which the ions are simultaneously leached from the solid under investigation and allowed to react with a trimethylsilylating agent, usually trimethylchlorosilane $(CH_3)_3SiCl$, under carefully selected conditions. The function of the trimethylsilylating agent is to block the reactive sites on the silicate ions during dissolution, thereby hindering their participation in reorganization reactions.

The overall reaction is illustrated in Eq. (19) for the mineral fayalite Fe_2SiO_4, which consists of isolated SiO_4 groups:

$$
\begin{array}{c}
Fe \\
| \\
O \\
| \\
-Fe-O-Si-O-Fe- \ + \ 4(CH_3)_3SiCl \\
| \\
O \\
| \\
Fe
\end{array}
$$

fayalite   trimethylchlorosilane

$$
\rightarrow (CH_3)_3Si-O-
\begin{array}{c}
Si(CH_3)_3 \\
| \\
O \\
| \\
Si \\
| \\
O \\
| \\
Si(CH_3)_3
\end{array}
-O-Si(CH_3)_3 \ + \ 2FeCl_2. \qquad (19)
$$

trimethylsilyl derivative
of the orthosilicate anion

The products of this reaction, i.e., the trimethylsilyl (TMS) derivatives of the silicate ions, are chemically inert, soluble in organic solvents, and thermally stable. Equally important, the derivatives of low molecular weight are sufficiently volatile to be capable of separation and quantitative determination by gas-liquid chromatography (GLC) at temperatures up to about 350°C. Separation by TLC, gel permeation chromatography (GPC), and HPLC have also been employed.

Typical compounds that have been identified in extracts of silicate glasses and aqueous silicate solutions are illustrated in Fig. 10.

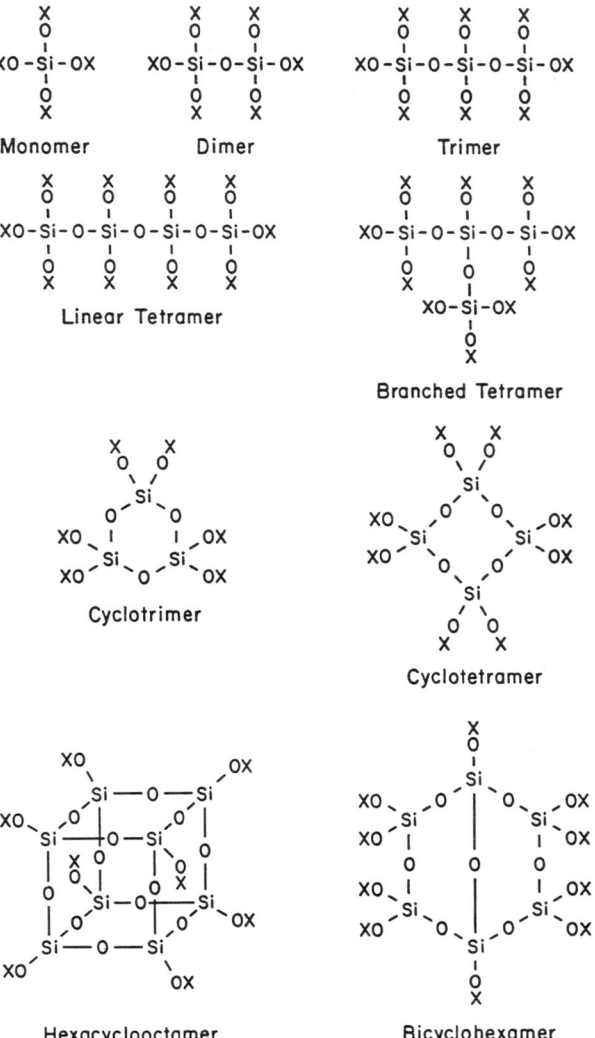

FIG. 10.   Some compounds identified in extracts of silicate glasses and aqueous silicate solutions after trimethylsilylation. X represents a trimethylsilyl group -Si(CH$_3$)$_3$.

In conjunction with techniques such as infrared, nuclear magnetic resonance, and high-resolution mass spectrometry to identify the products, the method offers, in principle, a powerful tool for the determination of silicate structures.

Although increasing use has been made of this technique since the 1960s, its full potential was only being attained in the late 1980s. The main difficulty encountered in almost all studies has been that of arranging conditions so that the trimethylsilylation occurs sufficiently rapidly to suppress reorganization reactions completely. As with most chemical methods, yields of the desired or expected product rarely amount to 100%. Side reactions can occur that lower the yield and lead to the formation of silicate structures not present in the original materials. These may trap the unwary into false or misleading conclusions. Furthermore, discrete ions capable of detection may not be leached from the solid under the conditions employed. In spite of these limitations, considerable progress has been made and improvements in the technique are constantly being effected.

The technique of trimethylsilylation has been applied to the study of crystalline silicates, soils, glasses, cementitious products, aqueous silicate solutions, and metallurgical slags. A comprehensive review has been published elsewhere (Calhoun and Masson, 1981).

1. Scope of the Method

When GLC is used to identify the products, the range of ions that can be detected is limited by the volatility of their TMS derivatives at temperatures up to 350°C, the limit of stability of the liquid phase in most chromatographic columns. Figure 11 shows a chromatogram of an extract from 65 PbO · 35 SiO$_2$ glass (Masson et al., 1986). The derivatives emerge from the column in order of increasing charge of the parent ion. Derivatives with up to 14 TMS groups per molecule (i.e., of ions up to charge 14^-) have been identified in this way.

Derivatives with up to 16 TMS groups per molecule have been identified by combined HPLC and mass spectrometry (Dent Glasser et al., 1981a, b). Derivatives up to that of the linear octamer $Si_8O_{25}^{20-}$ have been distinguished by GPC (Hirljac et al., 1983).

To illustrate the variety of structures that can, in principle, be detected, consider the ions of charge 10^-. These are the linear and branched chain isomers $Si_4O_{13}^{10-}$, the monocyclic ions $Si_5O_{15}^{10-}$ (including chain-cyclic combinations), bicyclic ions of formula $Si_6O_{17}^{10-}$, ions with three ring closures $Si_7O_{19}^{10-}$, and so forth. If we confine attention to the first three groups, the number of possibilities is illustrated in Fig. 12. For simplicity, we use the QM notation where Q and M refer to the entities $Si(O_{1/2})_4$ and $(CH_3)_3SiO_{1/2}$, respectively. It is seen that in addition to the two isomers of the tetrameric ion

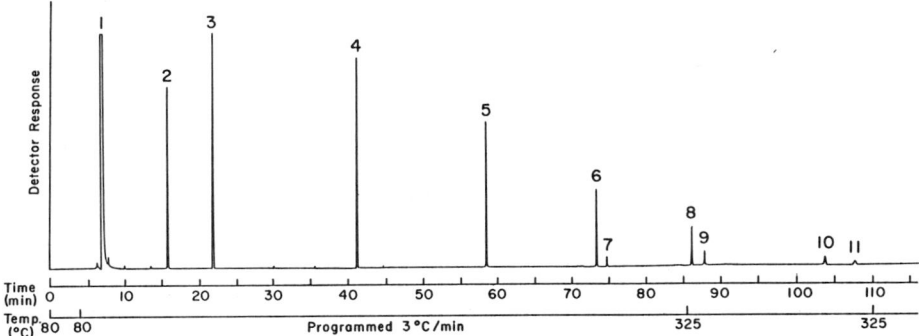

FIG. 11. Chromatogram of extract from $65PbO$ $35SiO_2$ glass, after trimethylsilylation by the DMF method. Fused silica capillary column 60 m long by 0.25 mm i.d. coated with DURABOND DB-5 (polymethyl 5% phenyl siloxane) 0.25μ thick. The temperature of the column was held at 80°C for 4 min, programmed from 80 to 325°C at 3°C/min, then held at 325°C for 25 min. Helium flow rate was 2 ml/min. Volume of extract injected was $0.8\,\mu l$. The instrument was equipped with a flame ionization detector. The peaks are due to 1 hexamethyldisiloxane, 2 reagent blank, and TMS derivatives of 3 SiO_4^{4-}, 4 $Si_2O_7^{6-}$, 5 $Si_3O_{10}^{8-}$, 6 $Si_4O_{13}^{10-}$ (linear), 7 $Si_4O_{13}^{10-}$ (branched), 8 and 9 isomers of $Si_5O_{16}^{12-}$, 10 and 11 isomers of $Si_6O_{19}^{14-}$. (Masson et al., 1986.)

$Si_4O_{13}^{10-}$ (structures 1 and 2, Q_4M_{10}), there are four isomers of the cyclopentamer $Si_5O_{15}^{10-}$ (structures 3 to 6, Q_5M_{10}) and fifteen isomers of the bicyclohexamer $Si_6O_{17}^{10-}$ (Q_6M_{10}). Garzo et al. (1978) estimated that for ions up to charge 12^-, more than 250 structures are sterically possible.

Fluorosilicate ions have also been identified as their TMS derivatives (Calhoun et al., 1979). If these are included, the number of possibilities is increased considerably.

Further refinement of methods such as TLC, GPC, and HPLC, in which the volatility of the derivative is not an important consideration, and direct examination of derivatives by mass spectrometry without prior chromatographic separation may further enhance the scope of the method. The use of supercritical fluid chromatography (SFC) for more efficient separation of the high-molecular-weight derivatives (Jackson et al., 1984) would also appear attractive.

2. Techniques of Derivatization

No universal method of trimethylsilylation has been developed that is applicable to all silicates. The yield is determined by the leachability of the cations as well as the nature of the anions in the solid and a variety of techniques have been described for different materials. The careful investigator will assess the results obtained by more than one method before selecting the method most suitable for the system under investigation.

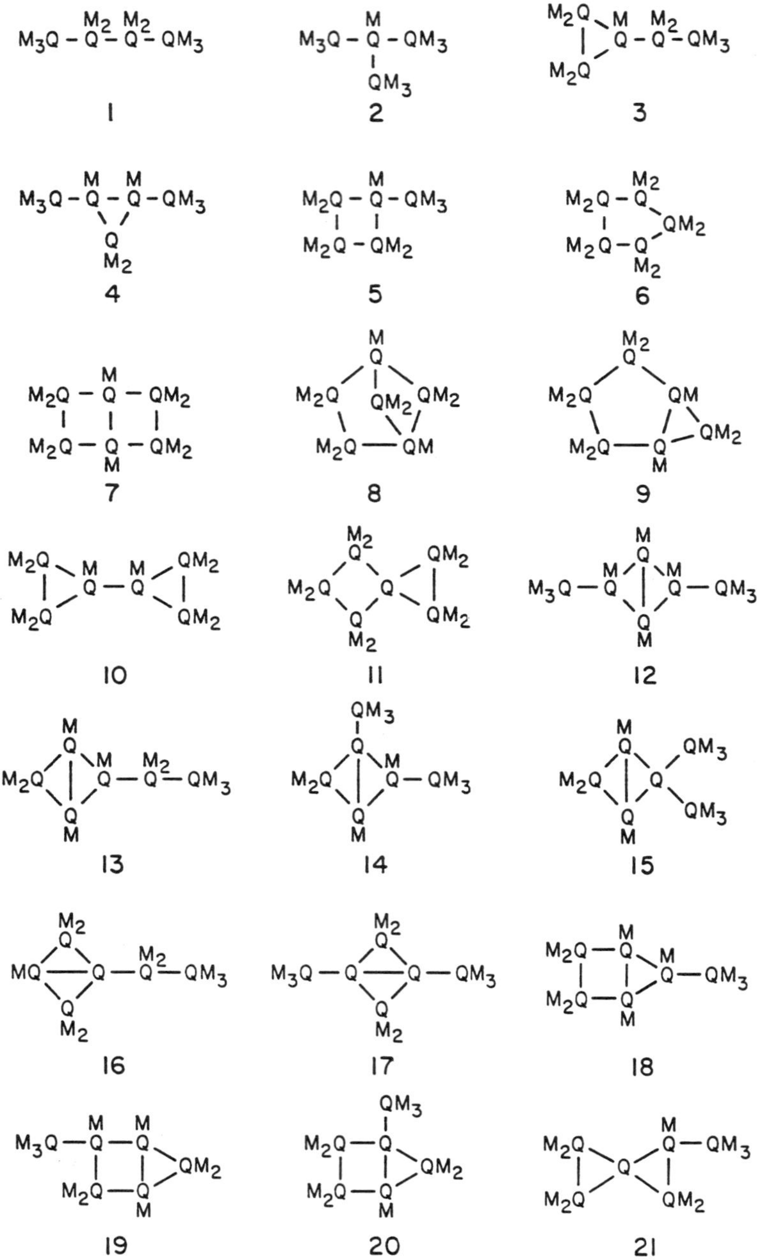

　　　C. R. MASSON

FIG. 12. TMS derivatives of some ions of charge 10^-.

a. The method of Lentz. In the original technique as described by Lentz (1964) for the mineral olivine $(Fe, Mg)_2SiO_4$ the finely divided solid (≤ 100 mesh) was slurried with water and added to a mixture of ice, concentrated HCl, isopropyl alcohol, and hexamethyldisiloxane $(CH_3)_3Si\text{-}O\text{-}Si(CH_3)_3$ that had been stirring at room temperature for 1 h. Stirring was continued for 48 h and the mixture was filtered to remove unreacted solids. The upper (siloxane) layer was separated, washed with water, treated to remove dissolved or suspended water, and concentrated by distillation. The residue was analyzed by GLC. The main product was the TMS derivative of the orthosilicate ion, $SiO_4[Si(CH_3)_3]_4$ or $SiO_4(TMS)_4$, along with a smaller amount of the pyrosilicate derivative $Si_2O_7(TMS)_6$. Similar techniques were employed for hemimorphite $Zn_4(Si_2O_7)(OH)_2 \cdot H_2O$, natrolite $Na_2(Al_2Si_3O_{10}) \cdot 2H_2O$, and laumontite $CaAl_2Si_4O_{12} \cdot 4H_2O$. For the mineral sodalite $Na_8(Al, SiO_4)_6Cl_2$, trimethylchlorosilane $(CH_3)_3SiCl$ was also employed as derivatizing agent.

The products of some of these reactions contained -OH as well as -O(TMS) groups on the silicon atoms. Thus the products from hemimorphite contained, in addition to $Si_2O_7(TMS)_6$, about 20% of the partially trimethyl-silylated derivative $Si_2O_7(TMS)_5H$. In these cases a solution of the crude product in hexamethyldisiloxane was stirred with a small quantity of Amberlyst 15, a sulfonic acid functional ion-exchange resin, to complete the trimethylsilylation.

Yields are shown in Table III. In each case the main product was the TMS derivative of the ion present in the parent mineral.

TABLE III

YIELDS OF VARIOUS DERIVATIVES BY THE LENTZ METHOD

Mineral	% Si recovered as TMS derivative of				
	SiO_4^{4-}	$Si_2O_7^{6-}$	$Si_3O_{10}^{8-}$	$Si_4O_{12}^{8-}$	Total
Olivine $(Mg, Fe)SiO_4$	70.0	11.1	—	—	81.1
Hemimorphite $Zn_4(Si_2O_7)(OH)_2 \cdot H_2O$	22.0	77.6	2.4	—	102.0
Sodalite $Na_8(AlSiO_4)_6Cl_2$	76.0	8.7	2.5	—	87.2
Natrolite $Na_2(Al_2Si_3O_{10}) \cdot 2H_2O$	10.0	13.1	67.5	—	90.6
Laumontite $CaAl_2Si_4O_{12} \cdot 4H_2O$	—	—	—	80.9	80.9

Source: Lentz (1964).

For minerals with Si-O-Al-O-Si linkages, the structure was disrupted at the O-Al-O bonds and only the silicate portion was recovered intact as its TMS derivative. Thus sodalite, which has alternating AlO_4 and SiO_4 tetrahedra, yielded $SiO_4(TMS)_4$ as the main derivative. Laumontite, a zeolite made of Si_4O_{12} rings linked through aluminum, yielded $Si_4O_{12}(TMS)_8$. For natrolite, whose structure may be represented as

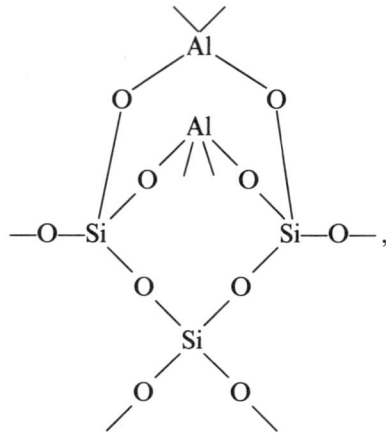

the main product was the derivative of the chain ion $Si_3O_{10}^{8-}$.

Table III shows that although the major product is, in each case, the derivative of the parent ion, other derivatives are formed as well. Thus, for olivine, only 70% of the silicon was recovered as the orthosilicate derivative $SiO_4(TMS)_4$; 11.1% was obtained as the pyrosilicate derivative $Si_2O_7(TMS)_6$. Similarly, for hemimorphite, only 77.6% was recovered as the pyrosilicate derivative, the remainder being distributed between the orthosilicate (22.0%) and trisilicate (2.4%) derivatives.

It was established that these additional products were not due to impurities in the starting materials but to side reactions during trimethylsilylation.

The sequence of reactions for olivine may be illustrated by the equations

$$(Fe, Mg)SiO_4 + 4HCl = H_4SiO_4 + FeCl_2 + MgCl_2 \qquad (20)$$

and

$$H_4SiO_4 + 4(CH_3)_3SiCl = SiO_4[Si(CH_3)_3] + 4HCl. \qquad (21)$$

The $(CH_3)_3SiCl$ for the trimethylsilylation in reaction (21) is derived from the reaction of HCl with hexamethyldisiloxane:

$$(CH_3)_3SiOSi(CH_3)_3 + HCl = (CH_3)_3SiCl + (CH_3)_3SiOH. \qquad (22)$$

Reaction (22) shows that trimethylsilanol is also a product. This compound may also act as a trimethylsilylating agent.

Reactions (20) and (21) are believed to proceed stepwise rather than in the concerted manner indicated. Thus replacement of cations by protons in reaction (20) to yield silicic acid groups $\equiv$Si—OH probably occurs in a stepwise manner and these groups react with trimethylchlorosilane or trimethylsilanol as they are formed. The concentration of free silicic acid groups at any time thus depends on the relative rates of reactions (20) and (21) and may be expected to vary with the concentration of HCl and $(CH_3)_3SiCl$ and with the extent of reaction.

Side reactions are of two main types: (a) condensation of silicic acid groups before trimethylsilylation is complete to yield products of higher molecular weight and (b) hydrolytic cleavage of siloxane linkages in the parent material. It should be noted that such cleavage probably occurs *after* dissolution of the partially trimethylsilylated derivatives because (a) the final products are stable and (b) experiments with long-chain silicates such as augite and enstatite, or with SiO_2 itself, do not yield derivatives of discrete ions.

Götz and Masson (1970) showed that side reactions in the Lentz method are markedly influenced by the proportion of HCl in the reagents. This is shown in Fig. 13 for hemimorphite. The ratio of peak areas $Si_2:Si_1$ for the dimeric and monomeric derivatives exhibits a maximum as the concentration of HCl is increased. This is consistent with the dual role of HCl in the Lentz method. It acts as a leaching agent to solubilize the mineral by reaction (20)

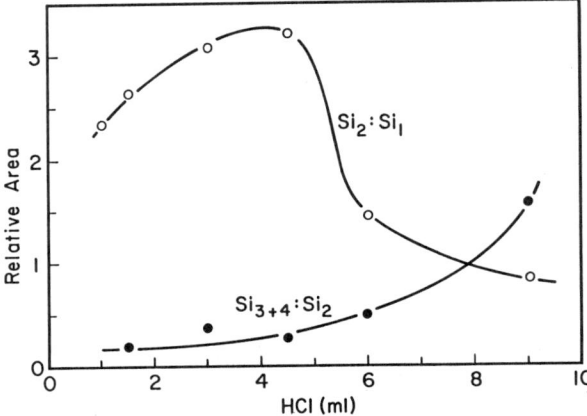

FIG. 13. Effect of HCl on the ratio of peak areas Si_2/Si and Si_{3+4}/Si_2 in the trimethylsilylation of hemimorphite by the Lentz method. (From Götz and Masson, 1970.)

and determines the concentration of trimethylsilylating agent(s) by reaction (21). The decrease in this ratio at high acidities (Fig. 13) is obviously due to rupture of the Si-O-Si linkage in the mineral. That this is accompanied by further side reactions is shown by the simultaneous increase in the proportions of the Si_3 and Si_4 derivatives, the combined yield of which exceeds that of the Si_2 derivative at high acidities.

b. The direct method.   To avoid these difficulties Götz and Masson (1970, 1971a, b) introduced a "direct" method of trimethylsilylation in which HCl was excluded from the reagents and derivatization was effected directly by $(CH_3)_3SiCl$ in the presence of water. As in the Lentz method, hexamethyldisiloxane was used to provide an organic phase to act as a solvent for the products and remove them from the aqueous medium as they are formed. The presence of isopropyl alcohol was necessary to obtain adequate yields.

Of critical importance in the direct method is the proportion of water in the reagents. Figure 14 shows the ratio of peak areas Si_2/Si_1 and Si_1/Si_2 in the products from hemimorphite and olivine, respectively, plotted against the volume of water employed. In the final method, sufficient water (0.2 ml in Fig. 14) was used to provide an adequate yield of monomeric derivative from olivine ($Si_1/Si_2 \approx 6.5$) without causing excessive hydrolysis of the pyrosilicate structure in hemimorphite ($Si_2/Si_1 \approx 32$).

Table IV shows relative yields obtained in the trimethylsilylation of various crystalline materials by the direct method. Absolute yields for some systems are shown in Table V.

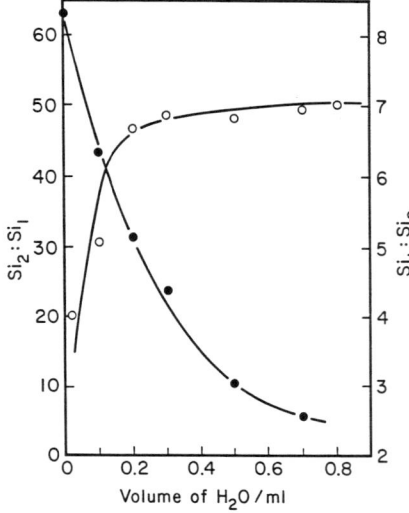

FIG. 14.   Effect of water on the ratio of peak areas Si_1/Si_2 in the trimethylsilylation of olivine (○) and on the ratio Si_2/Si_1 in the trimethylsilylation of hemimorphite (●) by the direct method. (From Götz and Masson, 1971a,b.)

TABLE IV

RELATIVE YIELDS IN THE TRIMETHYLSILYLATION OF VARIOUS CRYSTALLINE SILICATES BY THE
DIRECT METHOD

Substance	Wt % Si in the products as TMS derivative of			
	SiO_4^{4-}	$Si_2O_7^{6-}$	$Si_3O_{10}^{8-}$	$Si_4O_{12}^{8-}$
Olivine, $(Mg, Fe)_2SiO_4$	88.0	9.7	2.3	—
Forsterite, Mg_2SiO_4	81.9	15.2	2.9	—
Tephroite, $MnSiO_4$	87.2	11.3	1.5	—
Fayalite, Fe_2SiO_4	84.9	14.1	1.0	—
Andradite, $Ca_3Fe_2(SiO_4)_3$	83.0	11.6	5.4	—
Dicalcium silicate, Ca_2SiO_4	85.0	13.7	1.0	—
Hemimorphite, $Zn_4(Si_2O_7)(OH)_2 \cdot H_2O$	1.1	93.6	—	5.3
Rankinite, $Ca_3Si_2O_7$	—	96.7	3.3	—
Barysilite, $Pb_3Si_2O_7$	1.4	94.2	1.2	3.2
Potassium hydrogen metasilicate, $K_4H_4(Si_4O_{12})$	14.7	0.9	0.4	84.0
Laumontite, $CaAl_2Si_4O_{12} \cdot 4H_2O$	3.8	—	—	96.2

Source: Götz and Masson (1971), Smart and Glasser (1978), Calhoun and Masson (1980).
(From Calhoun and Masson, 1981.)

Table IV shows that hydrolytic reactions were almost entirely suppressed for hemimorphite, rankinite, barysilite, and laumontite, for which 93.6–96.7% of the silicon in the products was recovered as the parent derivative. For pseudowollastonite, whose structure is that of a cyclic trimer $Si_3O_9^{6-}$, the yield was much lower (Table V) and the product was almost exclusively the derivative of the chain ion $Si_3O_{10}^{8-}$. This is in line with the results of paper

TABLE V

ABSOLUTE YIELDS IN THE TRIMETHYLSILYLATION OF VARIOUS MINERALS BY THE DIRECT
METHOD

Mineral	Wt % Si recovered as TMS derivative of			
	SiO_4^{4-}	$Si_2O_7^{6-}$	$Si_3O_{10}^{8-}$	$Si_4O_{12}^{8-}$
Dicalcium silicate, Ca_2SiO_4	83	13	—	—
Rankinite, $Ca_3Si_2O_7$	—	96	—	—
Pseudowollastonite, $Ca_3Si_3O_9$	—	—	24	—
Hemimorphite, $Zn_4(Si_2O_7)(OH)_2 \cdot H_2O$	—	83	—	—
Barysilite, $Pb_3Si_2O_7$	0.8	52.9	0.7	1.8

Source: Götz and Masson (1978), Smart and Glasser (1978), Calhoun and Masson (1980).
(From Calhoun and Masson, 1981.)

chromatographic studies quoted earlier that showed that the ion $Si_3O_9^{6-}$ is unstable in acid aqueous media. This ring-opening reaction is also catalyzed by Amberlyst 15 exchange resin. (See the following Section C.)

Tables IV and V show that although the yield of monomer derivative from the orthosilicate minerals is considerably improved by this method, quantitative recovery of the highly labile orthosilicate ion is difficult. For the orthosilicate minerals, approximately 10–15% of the silicon in the products was always present as the dimer derivative, with smaller amounts of trimer.

Sharma et al. (1973) made a detailed study of the crystalline sodium silicate hydrates $Na_2H_2SiO_4 \cdot xH_2O$ ($x = 4$, 5, and 8) by the direct method. These are orthosilicates that contain the anion $H_2SiO_4^{2-}$ or $SiO_2(OH)_2^{2-}$. With these highly soluble materials, the probability of side reactions due to a high concentration of the orthosilicate ion in solution is increased. It was shown that an important parameter is the order of addition of the reagents. With these materials it was necessary to add the trimethylchlorosilane last, with continuous stirring, to obtain high yields. In addition, the yield depended on the weight of solid, approaching 100% only in the limit as small amounts of solid (ca. 2 mg) were employed. Quantitative recovery of silicon as the orthosilicate derivative from these materials is a severe test of the method.

Trimethylsilylation of dioptase $Cu_6Si_6O_{18} \cdot 6H_2O$ by the direct method (Calhoun and Masson, 1978) yielded three isomers of the expected cyclohexamic derivative $Si_6O_{18}(TMS)_{12}$ and three isomers of the bicyclo hexameric derivative $Si_6O_{17}(TMS)_{10}$. The side reaction in this case must involve condensation of O^- groups on the same silicate ion:

(a)

or

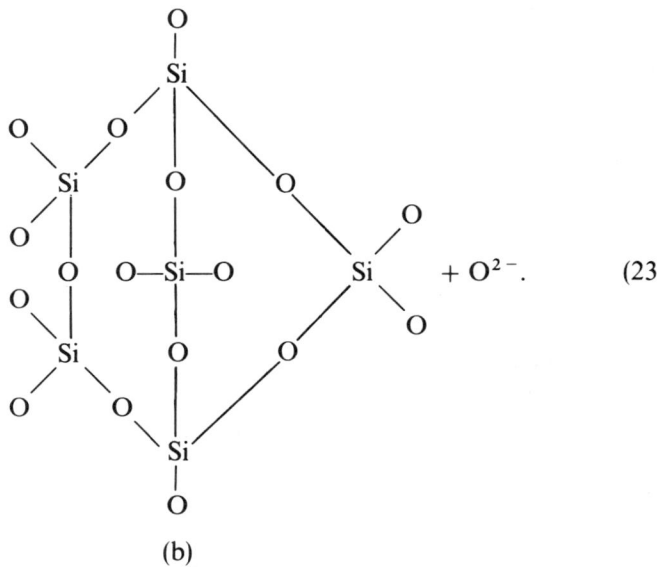

$$+ \, O^{2-}. \qquad (23)$$

(b)

When the method of Lentz was used, only traces of the parent cyclohexameric derivative were obtained, the main products being the bicyclohexameric derivatives, and it was concluded that reaction (23) is strongly catalyzed by HCl.

Studies of the products from dioptase by ^{29}Si NMR spectroscopy (Hoebbel et al., 1976) showed that both structures (a) and (b) mentioned previously were formed. The structure of the third isomer and of the three isomers of the monocyclic derivative $C_6H_{18}(TMS)_{12}$ are unknown.

Garzo et al. (1980) showed that reaction (23) is catalyzed by Amberlyst 15 exchange resin and that this resin also catalyzes interconversion of the isomers of $C_6H_{18}(TMS)_{12}$.

c. The DMF method. An important advance in this field was made by Tamas et al. (1976), who studied the trimethylsilylation of the orthosilicates β-Ca_2SiO_4 and Ca_3SiO_5 and the pyrosilicates $Ca_3Si_2O_7$ and $Zn_4Si_2O_7(OH)_2 \cdot H_2O$ (hemimorphite) by $(CH_3)_3SiCl$ in the presence of hexamethyldisiloxane and a wide variety of other reagents in the absence of water. The reagents included isopropyl alcohol, dioxane, formamide, dimethylformamide, pyridine, ethylenediamine, decane, 1,4 butanediol, 1,2 dichlorethane, nitrobenzene, carbon tetrachloride, benzene, and xylene. Dimethylformamide (DMF) gave the best results. It suppressed side reactions to a large extent and proved to be a good solvent for the by-products of the reactions, $CaCl_2$ and $ZnCl_2$.

Excellent yields (91–94%) of monomeric derivative were obtained for the orthosilicate minerals, with only 6–8% of the products appearing as the dimeric and about 1% as the trimeric derivatives. Side reactions with the pyrosilicate minerals were even smaller and the presence of water in hemimorphite did not cause side reactions to a large extent.

In this procedure the finely ground solid was added with stirring to a mixture of DMF, hexamethyldisiloxane, and trimethylchlorosilane that had been stirring previously at room temperature for 15 min. When reaction was complete, as shown by the dissolution of the solid, the solution was washed with water, dried over anhydrous $CaCl_2$, and concentrated by distillation. Treatment with Amberlyst has been shown (Garzo et al., 1980) to be unnecessary.

The DMF method has been used extensively to study the silicate structures in cement pastes and has been reported to be the most reliable method available for this purpose (Lachowski, 1979; Sarkhar and Roy, 1979). As pointed out by the authors, it is not suitable for all acid-soluble silicates, as shown by experiments with metakaolinite and natrolite.

Currell et al. (1985) compared the results for cement pastes by the DMF and a modified version of the Lentz method. They concluded that the yield of TMS derivatives by the DMF method was sometimes low and consisted mainly of monomer and dimer derivatives, possibly because this method only solubilized the low-molecular-weight species. The Lentz method gave a higher yield of products that contained more complex mixtures of derivatives and species of higher molecular weight.

Pseudowollastonite, α-$CaSiO_3$, resists trimethylsilylation by the DMF method (Masson, 1986). When a mixture of Ca_2SiO_4, $Ca_3Si_2O_7$, and α-$CaSiO_3$ was examined by the DMF method, the derivatives of the ions SiO_4^{4-} and $Si_2O_7^{6-}$ were observed in the products and an insoluble residue remained. When this residue was examined by the anhydrous direct method (see Section d, next) the derivative of the ion $Si_3O_{10}^{8-}$ was obtained.

d. The anhydrous direct method. A further significant development was made by Nakamura and Suginohara (1980) who showed that crystalline calcium silicates could be trimethylsilylated in high yield by an anhydrous direct method in which methyl alcohol was employed. Trimethylsilylation of Ca_2SiO_4 and $Ca_3Si_2O_7$ yielded 94.8 and 100% of $SiO_4(TMS)_4$ and $Si_2O_7(TMS)_6$, respectively, provided water was absent and sufficient methyl alcohol was present in the reagents. Only a single product, reported as $Si_4O_{12}(TMS)_8$ but probably $Si_3O_{10}(TMS)_8$ (to be discussed shortly), was obtained from crystalline α-$CaSiO_3$.

In this method a mixture of methyl alcohol and trimethylchlorosilane was stirred for 2 min. Hexamethyldisiloxane was added, stirring was continued for a further 5 min, and the solid was introduced. Reaction was allowed to proceed for 20 min. The upper layer was separated and treated twice with Amberlyst 15 exchange resin for periods of 12 h, with an intermediate washing between the treatments, to complete the trimethylsilylation. The resulting solution was examined by GLC.

In a later study, Okusu et al. (1982) extended this work to other silicates. Yields are shown in Table VI. Recovery of silicon from Zn_2SiO_4, Sr_2SiO_4, Ca_2SiO_4, and $Ca_3Si_2O_7$ was essentially quantitative, the major product in each case being the derivative of the ion present in the parent material. Crystalline Li_2SiO_3 was considered to be composed of infinite chains because no TMS derivatives were detected.

For α-$CaSiO_3$ (pseudowollastonite), known to have the cyclic structure $Si_3O_9^{6-}$, only 47.8% of the silicon was recovered as TMS derivatives, the remaining 52.2% being attributed to polysilicate, undetected by GLC. Of the total silicon, 40.9% was recovered as the derivative of the linear chain ion $Si_3O_{10}^{8-}$. The low-temperature form of calcium metasilicate, β-$CaSiO_3$, known to consist of infinite chains, was highly resistant to silylation.

The low yield of trimeric derivative from pseudowollastonite and the appearance of the product as the chain rather than the cyclic derivative is in line with the results of Calhoun and Masson (1980) by a similar method.

The products of trimethylsilylation before treatment with Amberlyst were shown to be mixed TMS-methyl derivatives.

For pseudowollastonite, a considerable amount of the fully silylated cyclic derivative $Si_3O_9(TMS)_6$ was found in the products before treatment with Amberlyst, confirming that conversion of $Si_3O_9(TMS)_6$ to $Si_3O_{10}(TMS)_8$ is

TABLE VI

YIELDS OF VARIOUS DERIVATIVES BY THE ANHYDROUS DIRECT METHOD

Crystalline silicate	Wt % Si recovered as TMS derivative of			Total % Si recovered % Si in by GLC	polysilicate
	SiO_4^{4-}	$Si_2O_7^{6-}$	$Si_3O_{10}^{8-}$		
Zn_2SiO_4	88.46	11.27	2.60	102.74	—
Sr_2SiO_4	98.84	3.64	1.28	103.75	—
Ca_2SiO_4	95.89	8.65	1.22	106.30	—
$Ca_3Si_2O_7$	2.60	94.00	2.00	98.60	—
α-$CaSiO_3$	0.20	0.10	40.90	47.80	52.20
$SrSiO_3$	5.60	—	60.40	74.20	25.80
Li_2SiO_3	—	—	—	0	100.00

Source: Okusu et al. (1982).

catalyzed by Amberlyst, as reported earlier by Hoebbel et al. (1979), and Calhoun and Masson (1980).

The anhydrous method of Suginohara and coworkers has been used successfully to study calcium and lead silicate glasses and appears to offer considerable promise for future work in this field.

Calhoun and Masson (1980) showed that dicalcium silicate Ca_2SiO_4 and rankinite $Ca_3Si_2O_7$ could be trimethylsilylated in high yield by an anhydrous direct method in which ethyl or isopropyl alcohol was employed. Highest yields (86–88% as the parent derivative) were obtained with isopropanol. Mixed isopropyl-TMS derivatives of the type $SiO_4(TMS)_x(Pr^i)_{4-x}$ ($x = 1$ to 4) and $Si_2O_7(TMS)_x(Pr^i)_{6-x}$ ($x = 2$ to 6) were formed and the trimethylsilylation was completed by treatment with Amberlyst.

The reaction is considered to proceed by the following mechanism:

$$(CH_3)_3SiCl + Pr^iOH \rightleftharpoons (CH_3)_3SiOPr^i + HCl \qquad (24)$$

$$HCl + Pr^iOH \rightleftharpoons Pr^iCl + H_2O \qquad (25)$$

$$\equiv Si-O^- + HCl \xrightarrow{H_2O} \equiv Si-OH + Cl^- \qquad (26)$$

$$\equiv Si-OH + (CH_3)_3SiOPr^i \longrightarrow \equiv Si-OPr^i + (CH_3)_3SiOH \qquad (27)$$

$$\equiv Si-OH + (CH_3)_3SiCl \longrightarrow \equiv Si-O-Si(CH_3)_3 + HCl \qquad (28)$$

$$(CH_3)_3SiCl + H_2O \longrightarrow (CH_3)_3SiOH + HCl. \qquad (29)$$

Side reactions that yield derivatives of higher anions may be ascribed to the reaction

$$\equiv Si-OH + HO-Si\equiv \rightarrow \equiv Si-O-Si\equiv + H_2O \qquad (30)$$

in competition with reactions (27) and (28). These are not eliminated by employing an excess of $(CH_3)_3SiCl$ because this increases the steady-state concentration of HCl.

Other reactions may be envisaged in this system but the preceding scheme accounts for the wide array of mixed derivatives and the observation that the presence of both ROH (R = Et, Pr^i, H) and $(CH_3)_3SiCl$ was necessary for reaction to occur.

Pseudowollastonite $Ca_3(Si_3O_9)$ gave predominantly $Si_3O_9(TMS)_x^-$ $(Pr^i)_{6-x}$ ($x = 2$ to 6) and $Si_3O_{10}(TMS)_x(Pr^i)_{8-x}$ ($x = 3$ to 8) by this method. These were converted to the fully trimethylsilylated chain derivative $Si_3O_{10}(TMS)_8$ by treatment with Amberlyst. The absolute yield, however, was only 24%. The reason for this is unknown as no significant amounts of higher derivatives were found in the products. See Section IV.B.4 for a further discussion.

Mixed alkyl-TMS derivatives were found in the trimethylsilylation of hemimorphite (Kuroda and Kato, 1979) and olivine (Kuroda et al., 1981) by similar methods when the trimethylsilylation was performed in the presence of an alcohol ROH (R = Me, Et, Pr^i, Bu^n). Fully silylated derivatives were obtained when acetone, tetrahydrofuran, or butane-2-one were employed.

e. The BSA method. This method, in which the silylating agent is weakly acidified N, O-bis-(trimethylsilyl)-acetamide, $CH_3C[OSi(CH_3)_3] = NSi(CH_3)_3(BSA)$, was developed by Garzo et al. (1978) to study silicic acid solutions. Prior to use, the BSA was acidified with gaseous HCl to give a 0.075 M H^+ solution. The silicic acid solution was added, with stirring, to a mixture of hexamethyldisiloxane, acetone, and acidified BSA at 15°C. After stirring for 15 min, the mixture was allowed to stand for 3 to 4 h, and the organic phase was separated, washed with water, and examined by GLC. No treatment with Amberlyst was required.

When a freshly prepared solution of orthosilicic acid was examined, the method gave higher yields of the expected ortho derivative than either the Lentz or direct methods. For silicic acid, solutions that had been allowed to stand for 24 h to attain a distribution of ions the results of the three methods were more closely comparable.

f. The nitric acid method. This method was developed by Calhoun et al. (1980) for the trimethylsilylation of $Ag_{10}Si_4O_{13}$. This compound, prepared by Jansen and Keller (1979), was the first linear chain tetrasilicate to be synthesized. Trimethylsilylation by the Lentz method yielded the derivative of the cyclic tetramer $Si_4O_{12}^{8-}$ as the main product, showing that the linear chain ion $Si_4O_{13}^{10-}$ readily undergoes cyclization in acidic aqueous media.

In this method the finely ground solid was slurried with hexamethyldisiloxane and absolute ethanol at $-20°C$. Trimethylchlorosilane followed by 0.2 N HNO_3 were added and stirring was continued for 5 h at $-20°C$, followed by 15 h at room temperature. The upper layer was separated, clarified by centrifugation, concentrated by distillation, and treated with Amberlyst 15.

With this technique, 32% of the silicon in $Ag_{10}Si_4O_{13}$ was recovered as TMS derivatives, the main product being the derivative of the linear chain ion $Si_4O_{13}^{10-}$. This may be compared with a 35% yield of the cyclic tetramer by the Lentz method.

The yield of $Si_2O_7(TMS)_6$ from barysilite $Pb_3Si_2O_7$ by this method was 83%, with only trace amounts of other products ($\leq 1\%$). This may be compared with 56.2% by the direct method (Smart and Glasser, 1978).

The higher yield obtained by the nitric acid method may be due to the solubility of the lead salt in the aqueous medium. In the absence of this reagent, the lead precipitates as the chloride, which may hinder derivatization.

It should be noted that the DMF method of Tamas et al. (1976) also preserves intact the structure of the linear chain ion $Si_4O_{13}^{10-}$ (Dent Glasser and Lachowski, 1980).

g. The DMF-pyridine method. Kalmychkov (1982) described a modification of the DMF method in which pyridine was used as solvent for the trimethylsilylation of orthosilicate minerals. The presence of pyridine was reported to (a) lower the rate of condensation of silanol groups and thereby inhibit side reactions leading to formation of oligomers and (b) accelerate the rate of silylation. The effect of pyridine was attributed to its polar nature and its ability to form a salt with HCl. The yield of orthosilicate from Li_4SiO_4 by this method was reported to be quantitative. The use of Amberlyst to complete the trimethylsilylation was not required.

For the trimethylsilylation of larger anions, up to 30% dioxane was added to increase their solubility in the medium. The time required for trimethylsilylation was 10–15 min for hemimorphite and 35–40 min for melilite, $Ca_3Si_2O_7$.

This method would appear to be attractive for determination of the orthosilicate ion in some systems. The detailed procedure however was not fully described in the work just mentioned. The method was subsequently used (Kalmychkov and Almukhamedov, 1984) to study lead silicate glasses and glasses of geological interest.

h. Other methods. Milestone (1977) reported that natrolite, pseudowollastonite, laumontite, and dioptase could be trimethylsilylated to give linear trisilicate, cyclic trisilicate, cyclic tetrasilicate, and cyclic hexasilicate derivatives, respectively, by a modified Lentz method in which isopropanol was replaced by t-butanol and the aqueous phase was saturated with sodium chloride. The t-butanol and salt were considered to cause a preferential salting out of the silicic acids into the organic phase. Yields were not reported.

An anhydrous method that gave 60–80% yields of the parent derivatives from hemimorphite, laumontite, and natrolite, with only a few percent of side products, was reported by Sharma and Hoering (1977). The finely ground mineral was reacted with a large excess of trimethylchlorosilane in hexamethyldisiloxane as solvent. Anhydrous methanol saturated with dry hydrogen chloride and Amberlyst 15 ion-exchange resin were added. Derivatization was completed by final treatment with hexamethyldisiloxane in the presence of the ion-exchange resin.

These authors also reported a method in which the minerals were stirred with a large excess of trimethylchlorosilane and hexamethyldisilazane $(CH_3)_3Si-N-Si(CH_3)_3$ in the solvent hexamethyldisiloxane. Water was added slowly over a period of several hours by a piston-driven syringe. Thus the concentration of water in the reaction mixture was always low. Yields, however, were lower than by the anhydrous method.

i. Discussion of methods of derivatization. Understanding the factors that govern side reactions in the trimethylsilylation of silicates requires a detailed knowledge of the constitution of silicate solutions. This is a subject of intensive study and no general principles have yet emerged as of the late 1980s. It is clear, however, that the constitution depends markedly on the pH, concentration of silicate, and nature of the cations. It may also depend on the presence of other materials, organic or inorganic, in the solution.

Unlike phosphate solutions, which are stable for long periods in neutral or nearly neutral solution at low temperatures, silicate solutions are most stable when the pH is approximately two (Iler, 1979). According to Alexander (1954), 0.1 M orthosilicic acid prepared from the sodium salt by ion exchange polymerizes almost instantly at pH 6, even at 2°C.

Side reactions during trimethylsilylation can now be recognized as being of five distinct types:

(1) *Condensation of Si-OH groups on different anions to yield larger species.* These can be minimized by working at low concentrations of silicate and keeping the pH of the aqueous phase between 2 and 3.

(2) *Ring-opening reactions.* The main side reaction observed in this category is conversion of the cyclic trimer $Si_3O_9^{6-}$ to the chain ion $Si_3O_{10}^{8-}$. This appears to occur in almost all methods of silylation and is probably due to strain in the Si-O-Si linkage (Hoebbel et al., 1979). As this is an isolated case, the occurrence of this reaction is not considered to be a serious limitation of the method.

(3) *Isomerization reactions.* These are known to be catalyzed by Amberlyst exchange resin and involve rupture and reforming of Si-O-Si linkages (Garzo et al., 1980). Such reactions usually occur slowly and can be recognized by following the change in chromatographic pattern during treatment with Amberlyst. They can be avoided in the DMF method.

(4) *Hydrolytic cleavage of Si-O-Si linkages to yield smaller anions.* This can be minimized by avoiding the use of water for some materials or by using only the minimum amount of water to obtain adequate yields from others. This conflicts to some extent with requirement (1). This reaction is not considered to be a serious limitation because it appears to occur only after solubilization of the mineral and before derivatization is complete. It is not a source of discrete ions from the rupture of insoluble chains.

(5) *Condensation of Si-OH groups on the same silicate ion.* This is the most important and troublesome side reaction encountered in the trimethylsilylation of silicates. It leads to rapid cyclization of the chain ion $Si_4O_{13}^{10-}$ in both

the Lentz and direct methods. It leads also to the formation of double-ring structures as, for example, in the trimethylsilylation of dioptase and is probably responsible for a host of other minor products with multiple-ring structures often observed in TMS extracts (Masson et al., 1974).

Fortunately, side reactions of type (5) can be minimized for silicates with certain cations (e.g., Ca^{2+}, Pb^{2+}) that do not require the use of water for trimethylsilylation. Some minerals, however (e.g., the olivines and aluminosilicates), cannot as of the late 1980s be trimethylsilylated by anhydrous techniques and further work is required to develop improved methods for these substances.

The method of trimethylsilylation is limited to silicates that can be solubilized by leaching with acids. However, some minerals (e.g., thorite, uranothorite) which gelatinize with acid according to Murata (1943) are resistant to trimethylsilylation (Götz and Masson, 1971a,b). Acid-resistant minerals such as zircon, garnet, and beryl, although they contain discrete silicate ions, resist trimethylsilylation.

A disadvantage of the method is the complexity of the derivative that must be formed to stabilize the silicate structure. Development of a procedure that would yield the alkyl derivatives as stable products without disrupting the structure would be a significant advance (Jeffes, 1981). Derivatives of this nature (the organic siloxanes) are well-known commercial products and are readily separated by GLC at moderate temperatures.

B. SILICATE GLASSES

Chromatographic methods have not yet been used extensively for silicate glasses. This has been due largely to the difficulties in developing a method of derivatization suitable for quantitative studies and free of side reactions. With the improved techniques now available, however, it is expected the method will find increasing use, particularly for systems that can be studied by anhydrous methods.

A further reason for the paucity of chromatographic studies of silicate glasses has been the limited number of glasses available in regions of composition where discrete ions are considered to predominate. For most binary silicates, glasses with more than about 57 mole% metal oxide are difficult to prepare by normal quenching techniques. An exception is the system PbO-SiO_2, for which glasses with up to 80 mol% PbO can readily be prepared, and this system has been studied most extensively.

1. Lead Silicate Glasses

Early studies of lead silicate glass by the method of Lentz (Götz et al., 1971) revealed a distribution of silicate ions but the chromatographic pattern varied with the conditions of trimethylsilylation and reliable conclusions could not be drawn. More consistent results were obtained by the direct method and a comparison of chromatograms from crystalline and glassy Pb_2SiO_4 (Götz and Masson, 1971a, b) yielded firm evidence for an array of discrete ions in glass of the orthosilicate composition.

This was supported by the work of Götz et al. (1976a) who studied crystalline and glassy Pb_2SiO_4 by a combined technique of trimethylsilylation and paper chromatography. The latter method was used to determine the orthosilicate ion SiO_4^{4-} while the relative proportions of $Si_2O_7^{6-}$ and $Si_3O_{10}^{8-}$ were determined by derivatization followed by GLC.

Götz et al. (1976b) extended this work to glasses with molar PbO/SiO_2 ratios R between 1 and 4.14. Glasses with $R \geq 2.5$ contained mainly SiO_4^{4-}, $Si_2O_7^{6-}$, and $Si_3O_{10}^{8-}$ ions. Increasing the silica content promoted the formation of more complex species, as indicated by paper chromatography. No discrete ions were detected at the metasilicate composition.

Nakamura et al. (1977) found that the chromatographic pattern for Pb_2SiO_4 glass by the direct method varied to some extent with the proportion of water in the reagents and the yield of TMS derivatives sharply declined when increasing volumes of water were employed.

A study of ionic distributions in $3PbO \cdot SiO_2$ glass by the direct method was reported by Ezikov et al. (1983).

In a detailed study of $PbO-SiO_2$ glasses with 15 to 45 mole % SiO_2, Smart and Glasser (1978) showed that, on a molar basis, the monomeric ion was the most abundant single species at all compositions. This is in marked contrast with the situation for phosphate glasses, outlined in Section III.B.

The general pattern of ionic distributions emerging from these early studies was in qualitative agreement with expectations from polymer theory (Calhoun and Masson, 1981).

In the work of Smart and Glasser, studies of crystalline lead silicates with well-defined structure generally yielded the expected derivative as the dominant product, but a troublesome feature in all studies was the difficulty of accounting quantitatively for the total silicon as TMS derivatives. Thus, although confidence could be placed in the chromatographic patterns, absolute yields were uncertain.

Calhoun et al. (1980) showed that the derivative of the cyclic tetramer $Si_4O_{12}^{8-}$, found in all previous studies, arose by cyclization of the linear chain ion $Si_4O_{13}^{10-}$ during derivatization. This could be avoided by use of the nitric acid method. Dent Glasser and Lachowski (1980) confirmed the absence of the cyclic tetramer in lead orthosilicate glass and showed that cyclization of the chain ion $Si_4O_{13}^{10-}$ could be avoided by the DMF method.

A detailed study of lead silicate glasses by the DMF method was made by Hoebbel et al. (1984). The results were in good agreement with the conclusions from earlier studies except that the derivative of the cyclic tetramer was replaced mainly with that of the linear chain ion $Si_4O_{13}^{10-}$. The derivative of the branched chain ion $Si_4O_{13}^{10-}$ was also reported along with higher derivatives attributed to the linear and branched isomers of $Si_5O_{16}^{12-}$ and $Si_6O_{19}^{14-}$. The chromatogram for glassy $3PbO \cdot SiO_2$ is illustrated in Fig. 15.

The data of Hoebbel et al. are shown in Table VII. The results are plotted in Fig. 16, where the curve for $x = 4$ refers to the combined yields of the linear and branched chain isomers. For comparison, distributions calculated by polymer theory are presented in Section IV.C.

Table VII shows that all the silicon in glasses with $R = 3.08$ and 4.14 could be accounted for as TMS derivatives. This is a marked improvement over

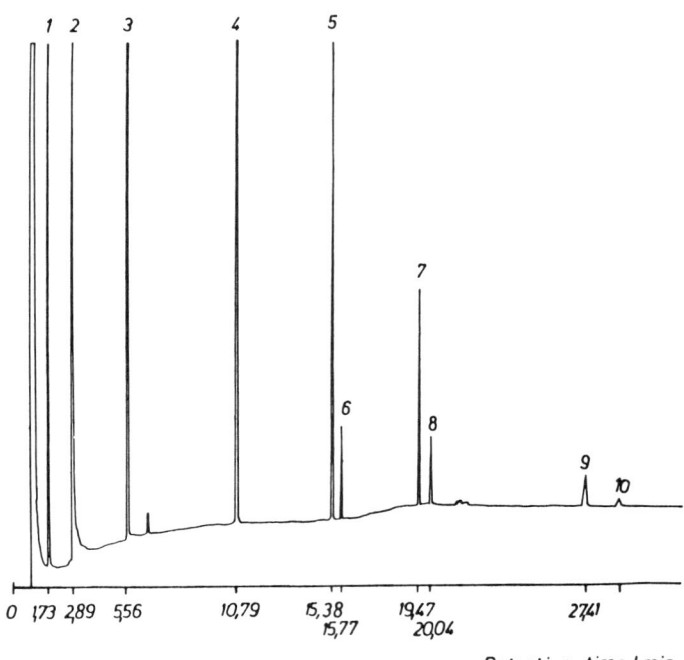

FIG. 15. Chromatogram of TMS derivatives from $3PbO \cdot SiO_2$ glass by the DMF method. Pyrex glass capillary column 20 m long $\times$ 0.25 mm diameter coated with OV-1 as stationary phase. A flame ionization detector was employed. Peak 2 is due to the internal standard used for calibration. The other peaks were attributed to the TMS derivatives of the following anions: 1 SiO_4^{4-}, 3 $Si_2O_7^{6-}$, 4 $Si_3O_{10}^{8-}$, 5 $Si_4O_{13}^{10-}$ (linear chain), 6 $Si_4O_{13}^{10-}$ (branched chain), 7 $Si_5O_{16}^{12-}$ (linear chain), 8 $Si_5O_{16}^{12-}$ (branched chain), 9 $Si_6O_{19}^{14-}$ (linear chain?), 10 $Si_6O_{19}^{14-}$ (branched chain)? (From Hoebbel et al., 1984.)

TABLE VII

YIELDS OF VARIOUS DERIVATIVES IN THE TRIMETHYLSILYLATION OF PbO-SiO$_2$ GLASSES BY THE DMF METHOD

TMS derivative of (ion)	Wt % Si in glass for molar PbO/SiO$_2$ ratio				
	4.14	3.08	2.01	1.50	1.0
SiO$_4^{4-}$	57	38	19	9	—
Si$_2$O$_7^{6-}$	26	29	14	7	—
Si$_3$O$_{10}^{8-}$	11	19	11	7	—
[Si$_4$O$_{13}$]$_1^{10-}$	4	8	10	4	—
[Si$_4$O$_{13}$]$_2^{10-}$	1	1	1	1	—
[Si$_5$O$_{16}$]$_1^{12-}$	1	3	5	3	—
[Si$_5$O$_{16}$]$_2^{12-}$	—	1	2	1	—
Unknown	—	1	3	2	1
Higher silicates	—	—	35	66	99

Source: Hoebbel et al. (1984).

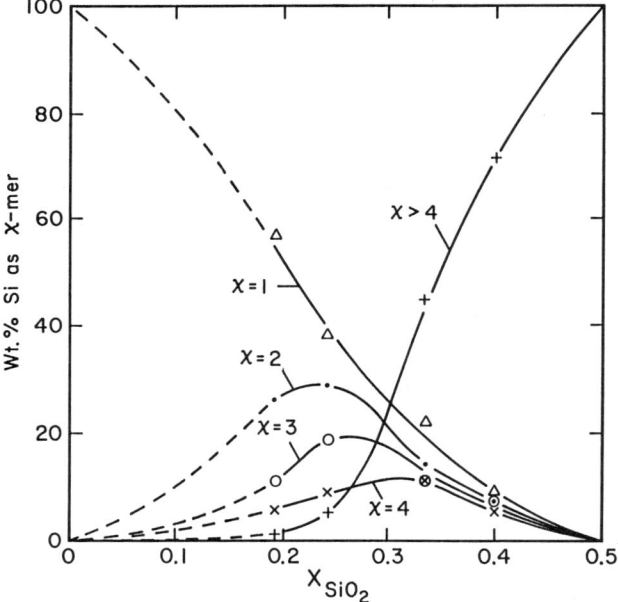

FIG. 16. Anionic distributions in PbO-SiO$_2$ glasses as determined by Hoebbel et al. (1984) by the DMF method. Wt % Si in the glass present as monomer SiO$_4^{4-}$ (Δ), dimer Si$_2$O$_7^{6-}$ ($\cdot$), trimer Si$_3$O$_{10}^{8-}$ (O), sum of linear and branched tetramers Si$_4$O$_{13}^{10-}$ ($\times$), and sum of polyions with $x > 4$ (+).

previous studies and lends confidence that the recovery of silicon as discrete ions from the glasses of higher silica content is also quantitative. The data in Table VII are considered to be the most reliable available at this time for lead silicate glasses.

Quantitative yields of products from lead silicate glasses by the DMF-pyridine and DMF-dioxane methods have been claimed (Kalmychkov, 1982; Kalmychkov and Almukhamedov, 1984), but sufficient details are not available to assess these results.

2. Crystallization of Lead Silicate Glasses

The method of trimethylsilylation is valuable for studying changes in the anionic constitution of materials with the same chemical composition. Differences are revealed and data can be obtained that are sometimes unattainable by other means. Examples are changes in the constitution of silicic acid and silicate solutions on polymerization, the aging of cement pastes, and the conversion of minerals from one form to another by thermal treatment (Calhoun and Masson, 1981). Another useful application is to study the devitrification of glasses.

Götz et al. (1972) studied the devitrification of lead orthosiliate glass by the direct method. The results are summarized in Fig. 17 which shows the change in chromatographic pattern with time of crystallization at 500°C. Clearly the kinetics involve initial dimerization of the orthosilicate ion SiO_4^{4-} to yield the pyrosilicate structure $Si_2O_7^{6-}$ accompanied by self-condensation of this ion to yield the cyclic tetramer $Si_4O_{12}^{8-}$, which was found to be the dominant structure in the fully crystalline material.

The unexpected finding that crystalline Pb_2SiO_4 has the structure characteristic of a cyclic tetramer has been confirmed by other workers by both trimethylsilylation (Götz et al., 1975a, 1976a; Smart and Glasser, 1978; Dent Glasser and Lachowski, 1980) and crystal structure analysis (Dent Glasser et al., 1981a). Crystalline lead "orthosilicate" is thus more correctly described by the formula $Pb_8O_4(Si_4O_{12})$ and is one of the few "orthosilicates" that does not contain SiO_4^{4-} groups. This probably accounts for the relative ease with which liquid $2PbO \cdot SiO_2$ and $PbO\text{-}SiO_2$ melts in general can be quenched to glasses at high basicities.

The crystallization of Pb_2SiO_4 glass at various temperatures was studied by Okusu et al. (1982, 1983) by the anhydrous direct method. The distribution of anions in crystalline Pb_2SiO_4 varied widely with the temperature of crystallization, the $Si_4O_{12}^{8-}$ structure being most stable at 530 to 540°C.

Devitrification of glassy $PbSiO_3$ was studied by Hoebbel et al. (1984). Crystallization at 550°C occurred in two stages: (a) rupture of polysilicate chains to yield discrete ions, mainly $Si_4O_{13}^{10-}$ and an ion tentatively identified as $Si_7O_{22}^{16-}$ followed by (b) recombination of these ions to yield higher species undetectable by GLC.

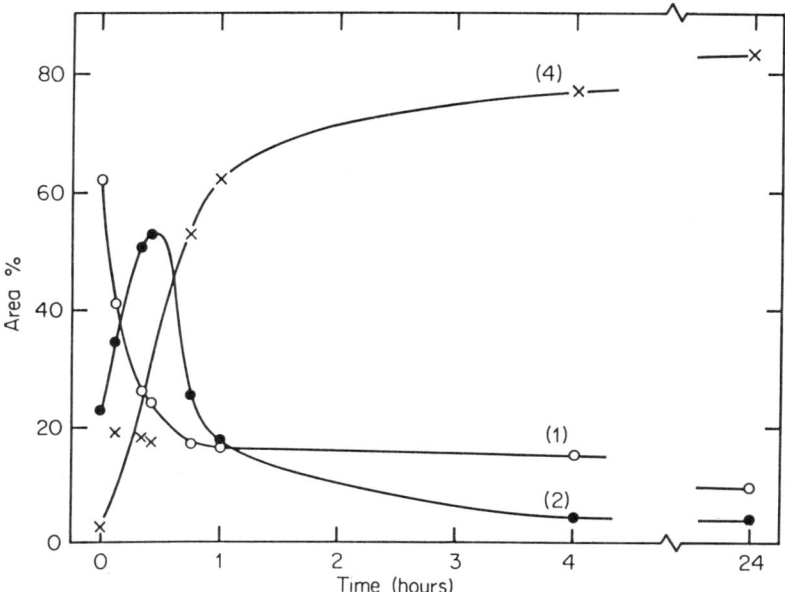

FIG. 17.   Change in anionic constitution of $2PbO \cdot SiO_2$ glass on crystallization at 500°C. Percentage of total area in the chromatograms plotted against time of devitrification for SiO_4^{4-} (O), $Si_2O_7^{6-}$ (●), and $Si_4O_{12}^{8-}$ (×) derivatives. (From Götz et al., 1972.)

3. Calcium Silicate Glasses

The presence of a distribution of anions in calcium silicate glass was shown elegantly by Nakamura and Suginohara (1980) by the anhydrous direct method. Trimethylsilylation of crystalline calcium silicate containing 43.3 mole% SiO_2 yielded mainly the dimeric $Si_2O_7^{6-}$ and trimeric $Si_3O_{10}^{8-}$ derivatives, as shown in Fig. 18. When this crystalline material was melted and quenched to a glass, the chromatogram in Fig. 19 was obtained.

The relative yields of various derivatives from this glass ($R = 1.309$) and from a glass of the metasilicate composition ($R = 1$) are shown in Table VIII. Yields for the metasilicate glass were calculated assuming the same calibration factor for both glasses. The derivatives are listed in Table VIII in the order in which they emerge from the chromatographic column. In the case of unresolved peaks, the derivative of the main ion, as determined by mass spectrometry, is underlined. Derivatives of ions shown in parentheses were not confirmed by mass spectrometry; these were tentative assignments. These results are discussed in Section IV.C in terms of conventional polymer theory.

4. Crystallization of Calcium Silicate Glasses

Changes in constitution during devitrification of calcium silicate glasses with $R = 1.309$ and 1.0 were studied by Okusu et al. (1982).

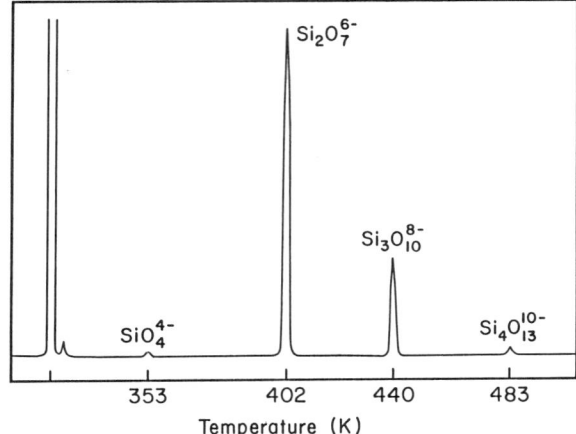

FIG. 18. Chromatogram of products from the trimethylsilylation of crystalline 56.7 CaO 43.3 SiO_2 by the anhydrous direct method. (From Nakamura and Suginohara, 1980.)

For the metasilicate glass, crystallization occurred at 1173 K, yielding β-$CaSiO_3$, as shown by DTA and X-ray diffractometry. No significant change in the chromatographic pattern was observed on thermal treatment up to 1373 K.

Between 1373 and 1570 K, a sharp increase in the $Si_3O_{10}^{8-}$ derivative occurred, corresponding to the appearance of the pseudowollastonite structure $Si_3O_9^{6-}$. (That this compound yields the $Si_3O_{10}^{8-}$ derivative on

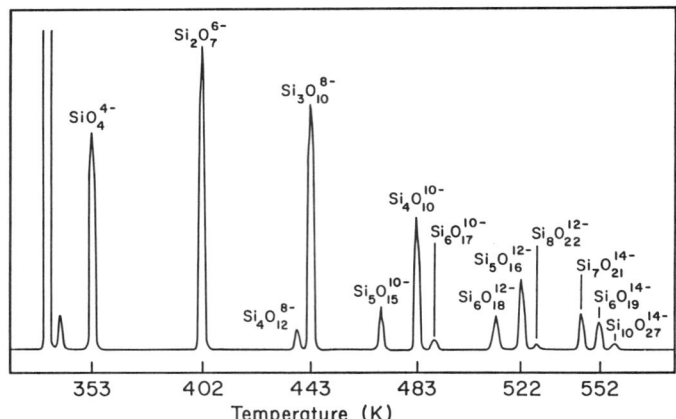

FIG. 19. Chromatogram of products from the trimethylsilylation of glassy 56.7 CaO 43.3 SiO_2 by the anhydrous direct method. (From Nakamura and Suginohara, 1980.)

TABLE VIII

TRIMETHYLSILYLATION OF CaO-SiO$_2$ GLASSES BY THE ANHYDROUS DIRECT METHOD

Derivative of (ion)	Column temperature (°C)	% total area for R =		% Si in products[1] for R =	
		1.309	1.0	1.309	1.0
SiO_4^{4-}	80	18.7	22.8	31.6	38.5
$Si_2O_7^{6-}$	129	26.0	20.6	29.4	23.5
$Si_4O_{12}^{8-}$	167	1.8	2.0	1.6	1.8
$Si_3O_{10}^{8-}$	170	20.8	18.2	17.6	15.4
$Si_5O_{15}^{10-}$	200	3.7	3.6	2.5	2.4
$Si_4O_{13}^{10-}$	210	11.4	11.1	7.8	7.6
$\left.\begin{array}{l}Si_4O_{13}^{10-}\\Si_6O_{17}^{10-}\\ \rule{1cm}{0.4pt}\\Si_7O_{19}^{10-}\\Si_8O_{21}^{10-}\end{array}\right\}$ b)	214	1.0	2.1	0.7	1.4
$Si_6O_{18}^{12-}$	243	3.8	3.8	2.1	2.1
$Si_5O_{16}^{12-}$	249	6.1	7.1	3.5	4.1
$\left.\begin{array}{l}Si_5O_{16}^{12-}\\Si_7O_{20}^{12-}\\Si_8O_{22}^{12-}\end{array}\right\}$ b)	253	0.4	—	0.2	—
(Si_7O^{14-}) a)	271	3.3	3.8	1.6	1.8
$(Si_6O_{19}^{14-})$ a)	279	2.5	4.3	1.2	2.1
$(Si_{10}O_{27}^{14-})$ a)	282	0.5	0.6	0.2	0.2

Source: Nakamura and Suginohara (1980).

Note: a) indicates tentative assignment; b) indicates main constituent in unresolved group.

[1] Values for $R = 1$ were calculated assuming the same sensitivity factor as for $R = 1.309$.

trimethylsilylation has already been noted.) Not all the glass was converted to this structure, however. About 55% of the silicon remained as polysilicate derivative. As pointed out by the authors, these results, and those of other workers, suggest that the structure of pseudowollastonite should be reinvestigated.

For the glass with $R = 1.309$, the percentage silicon found as $Si_2O_7^{6-}$ derivative after crystallization at 1573 K was 62%, compared with 61.8% calculated from the phase diagram. (Cf. Fig. 18.)

The infrared absorption spectra of the crystallized glass in the range 1173–1473 K showed the presence of both Ca_2SiO_4 and α-CaSiO$_3$ in the products of crystallization. This was in line with the chromatographic studies which showed significant proportions of SiO_4^{4-} and $Si_3O_{10}^{8-}$, although these results are not in accord with the accepted phase diagram.

The reason for this is not known as of the late 1980s, but the results show that the method of trimethylsilylation is valuable in providing direct evidence of silicate ions in synthetic silicate crystals and glasses.

5. Borosilicate Glasses

The distribution of silicate ions in borosilicate and lithium borosilicate glasses was studied by Kolb and Hansen (1965) by the Lentz method. The results are presented in Table IX. In both systems the order of abundance was monomer > dimer > trimer > tetramer. For a B_2O_3/SiO_2 glass of composition 90/10, the distribution was unchanged when the melt was heated for 4 h and 6 h. There was a tendency to form more monomer in the boric oxide glasses.

6. Other Glasses

Glasses with 40–50 mole % SiO_2 in the system ZnO-SiO_2 were studied by Nakamura et al. (1977) by the direct method. The main products were the derivatives of SiO_4^{4-}, $Si_2O_7^{6-}$, and $Si_3O_{10}^{8-}$, with the monomer always in greatest abundance. Yields of discrete ions decreased, as expected, with increasing silica content.

Discrete ions in basalt glass containing 44.3 wt. % SiO_2, 12.4% Al_2O_3, and a variety of other components were studied by Yang et al. (1985) by the Lentz method, with ultrasonic agitation. Glass melted in a reducing atmosphere contained more SiO_4^{4-} derivative due, presumably, to reduction of Fe_2O_3 in the glass and the stronger "network-breaking" tendency of the ferrous ion.

TABLE IX

IONIC DISTRIBUTIONS IN BOROSILICATE AND LITHIUM BOROSILICATE GLASSES BY THE METHOD OF LENTZ

Composition (mole %)			Melting temp. (°C)	% Si in products as derivative of			
SiO_2	B_2O_3	$LiBO_2$		SiO_4^{4-}	$Si_2O_7^{6-}$	$Si_3O_{10}^{8-}$	$Si_4O_{12}^{8-}$
5	95	—	1400	66.4	21.8	10.9	3.3
10	90	—	1400	52.6	27.0	15.0	7.3
10	90	—	1400^a	53.8	25.3	13.3	3.9
10	90	—	1200	36.1	14.5	10.6	5.6
15	85	—	1400	37.2	18.0	12.6	7.2
5	—	95	1200	35.1	21.1	7.3	4.7
10	—	90	1200	34.8	23.4	9.7	6.3
15	—	85	1200	29.6	24.8	13.7	9.3

Source: Kolb and Hansen (1965).
^{a}Melted for 6 h with stirring; all other samples melted for 4 h.

A fragment of vesicular glass from lunar fines (Apollo 11) was examined by Masson et al. (1971) by the direct method. The orthosilicate ion was present in greatest abundance although its yield was low, only 0.34% of the yield from olivine. The dimer and trimer were also found. The ratio of peak areas due to monomer, dimer, and trimer was 100:32:16. The chromatogram for Apollo 11 fines closely resembled that for the glass and was attributed to the high proportion of olivine in the lunar regolith.

Pasishnik et al. (1985) studied the anionic composition of Pb_4SiO_6 and $Na_6Si_2O_7$ glasses prepared by rapid quenching of the corresponding melts. The Pb_4SiO_6 glass contained mainly SiO_4^{4-} and some $Si_2O_7^{6-}$ whereas $Na_6Si_2O_7$ contained SiO_4^{4-}, $Si_2O_7^{6-}$, $Si_3O_9^{6-}$, and $Si_4O_{12}^{8-}$ groups.

C. POLYMER THEORY

Application of polymer theory to silicate melts and glasses has been discussed in detail elsewhere (Masson, 1977). Figure 20 shows the distributions calculated for a binary melt $MO-SiO_2$ or M_2O-SiO_2 for which $k_{1n} = 0$,

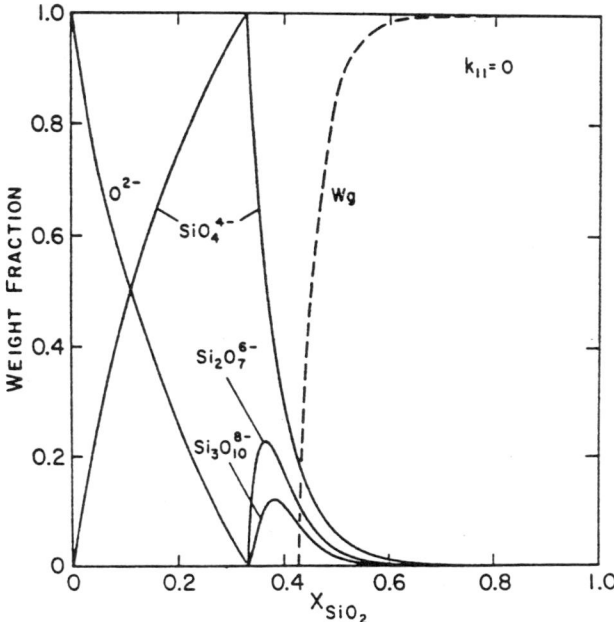

FIG. 20. Calculated weight fractions of ortho-, pyro-, and tripolysilicate ions and free oxide ion O^{2-} in a binary silicate melt for which $k_{11} = 0$, corresponding closely to a typical glass-forming system with highly basic cations. The broken line is the weight fraction of gel, w_g, considered as an "infinite" network. (From Masson, 1977.)

where k_{1n} is the equilibrium ratio for the reaction

$$SiO_4^{4-} + Si_nO_{3n+1}^{(2n+2)-} = Si_{n+1}O_{3n+4}^{(2n+4)-} + O^{2-}. \qquad (31)$$

A value of $k_{1n} = 0$ corresponds closely to that expected when M is a highly basic cation, with a strong "network-breaking" tendency, such as a cation of the alkali or alkaline-earth groups (Masson, 1984).

The curves in Fig. 20 are identical to the classical Flory distributions for the self-condensation of a tetrafunctional monomer in the region $\frac{1}{3} \leqq X_{SiO_2} \leqq 1$ (i.e., where the free oxide ion is absent).

Below the gel point ($X_{SiO_2} = 0.428$), the theory predicts that the melt will consist entirely of discrete ions.

At $X_{SiO_2} = 0.433$, approximately 19% of the Si will, in theory, be present as gel, undetectable chromatographically, while 19% will be present as monomer SiO_4^{4-}, 10% as dimer $Si_2O_7^{6-}$, and 6% as trimer $Si_3O_{10}^{8-}$, the remainder being discrete ions of higher molecular weight, each in decreasing abundance.

At the metasilicate composition ($X_{SiO_2} = 0.5$), Fig. 20 predicts that 85% of the Si will be present as gel, of infinite molecular weight. Only 5.8% will be present as monomer, 2.9% as dimer, and 1.5% as trimer.

If we confine our attention to these three ions, which can be measured accurately chromatographically, it is evident for the glass with $X_{SiO_2} = 0.433$ that, of the silicon extracted as (monomer + dimer + trimer), 54.3% will be present as monomer, 28.6% as dimer, and 17.1% as trimer. For the metasilicate glass, the corresponding figures are 56.9, 28.4, and 14.7%, respectively.

In Table X, these values are compared with those calculated from the relative yields reported by Nakamura and Suginohara (1980) for CaO-SiO$_2$ glasses and listed in Table VIII. For both glasses, the observed values are slightly lower than predicted for the monomer and slightly higher than predicted for the dimer and trimer but the agreement is remarkably good considering the simplifying assumptions in the theory.

TABLE X

Relative Amounts of Monomer, Dimer, and Trimer in CaO-SiO$_2$ Glasses Compared with Values Calculated from Polymer Theory

% Si in (monomer + dimer + trimer) present as	$R = 1.309$		$R = 1.0$	
	Obs.	Theor.	Obs.	Theor.
SiO_4^{4-}	40.2	54.3	49.7	56.9
$Si_2O_7^{6-}$	37.4	28.6	30.4	28.4
$Si_3O_{10}^{8-}$	22.4	17.1	19.9	14.7

Source: Nakamura and Suginohara (1980).

At the metasilicate composition, the theory predicts that only 10.2% of the Si will be present as (monomer + dimer + trimer). This is in good agreement with the observation of Okusu et al. (1982) that at this composition the sum of the silicate ions detected by GLC accounted for only 10% of the Si in the original glass.

It is clear from Fig. 20 that, as the metasilicate composition is approached, the percentage of discrete ions in the glass decreases markedly and this is accompanied by a correspondingly rapid increase in the proportion of gel. This is in line with the observations of many workers. For example, Nakamura et al. (1977) reported that for CaO-SiO$_2$ glasses, 96.7% of the Si was soluble as TMS derivatives at $X_{SiO_2} = 0.433$ whereas only 17.0% was soluble at $X_{SiO_2} = 0.5$.

It may thus be stated that for CaO-SiO$_2$ glasses, both the chromatographic patterns and the yields are in good agreement with predictions of polymer theory.

For PbO-SiO$_2$ glasses, the Flory theory cannot be used because there is a considerable proportion of free oxide in basic melts of this system. Ionic distributions calculated for this case by the more general treatment referred to in Section III.C, with $k_{1n} = 0.196$, are illustrated in Fig. 21 (Calhoun and Masson, 1981).

These may be compared with the experimental distributions of Hoebbel et al. (1984), shown in Fig. 16. The general features are similar. Again, the observed values for monomer are lower than predicted by the theory due, presumably, to the higher reactivity of the O$^-$ groups on the SiO$_4^{4-}$ ions. The ion fractions of Si$_2$O$_7^{6-}$ and larger species are correspondingly higher.

D. FLUOROSILICATE GLASS

Application of polymer theory to fluorosilicate melts (Masson and Caley, 1978) predicts that such melts will contain an array of silicate, monofluorosilicate, difluorosilicate... ions in thermodynamic equilibrium. On a molar basis, theory predicts that the silicate ions will be more abundant than the corresponding monofluorosilicate ions, which will, in turn, be more abundant than the corresponding difluorosilicate ions, and so forth. For a summary the reader is referred elsewhere (Masson, 1984).

Figure 22 shows the chromatogram obtained for a 3PbO · PbF$_2$ · SiO$_2$ glass after trimethylsilylation by the direct method (Calhoun and Masson, 1978). The identities of the peaks are given in Table XI. Of the total area in the chromatogram, 16.5% was due to derivatives of fluorosilicate ions. Of these, the monofluorosilicate ions SiO$_3$F^{3-}, Si$_2$O$_6$F^{5-}, Si$_3$O$_9$F^{7-}, and Si$_4$O$_{11}$F^{7-} were the most abundant, together totalling 14.2% of the total area. These were followed by the difluorosilicate derivatives. Only traces of trifluorosilicate derivatives were observed.

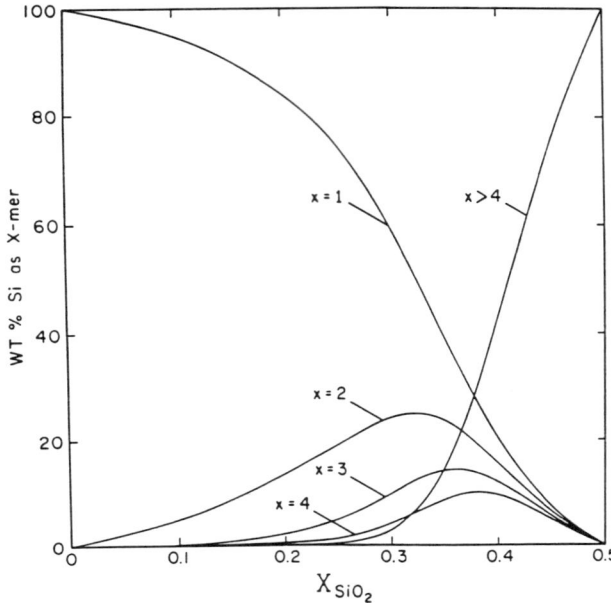

FIG. 21.   Wt % Si as $Si_xO_{3x+1}^{(2x+2)-}$ ($x = 1$ to 4) and higher anions ($x > 4$) plotted against mole fraction of SiO_2 for PbO-SiO_2 glasses calculated from polymer theory. (From Calhoun and Masson, 1981.)

The identities of the peaks in Fig. 22 were established by mass spectrometry. It was shown that peak 18 is due to the derivative of the linear chain ion $Si_4O_{13}^{10-}$ by comparison with the product of trimethylsilylation of $Ag_{10}Si_4O_{13}$. Peak 19, of the same molecular weight, is thus due to the branched chain ion $Si_4O_{13}^{10-}$. The mass spectra of the material corresponding to peaks 15c, 15d, and 16 showed that these substances had molecular weights corresponding to derivatives of the ion $Si_4O_{02}F^{9-}$, but had different cracking patterns. These peaks were thus attributed to isomers of this ion.

E. CHARACTERIZATION OF TMS DERIVATIVES

Conditions for the gas-chromatographic separation of TMS derivatives of silicates have been described in detail by Wu et al. (1970), Garzo and Alexander (1971), Garzo and Hoebbel (1976), Hoebbel et al. (1976), and Garzo et al. (1978).

Both packed and open-tubular (capillary) columns have been employed. Sharper separations are obtained with the latter and the amounts injected are smaller (< 1 μl). For preparative work, packed columns are required. A vent suitable for collecting small quantities of pure derivatives in capillary tubes

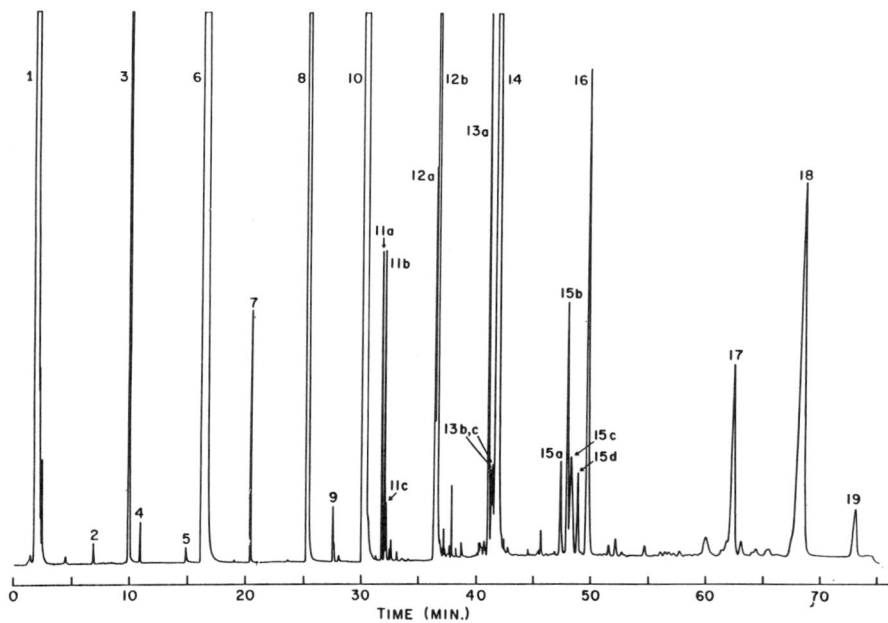

FIG. 22. Chromatogram of TMS derivatives from $3PbO \cdot PbF_2 \cdot SiO_2$ glass by the direct method. Glass capillary column, 25 m long by 0.27 mm i.d., coated with OV-101. Temperature program: 70°C for 6 min then 5°C/min to 240°C; held at 240°C for 30 min. For identity of peaks, see Table XI. (From Calhoun et al., 1979.)

for the preparation of standard solutions, or for transfer to a mass spectrometer, has been described by Wu et al. (1970). TLC has also been used for preparative purposes (Garzo and Hoebbel, 1976).

Columns with either a polar or nonpolar stationary phase may be employed, but retention times are longer with a polar phase and nonpolar phases such as OV-1, SE-30, and OV-101 (dimethylpolysiloxane) have generally been employed. Retention indices for phases of different polarity have been measured by Garzo and Alexander (1971) and Garzo and Hoebbel (1976).

The maximum temperature for continuous operation with the just-mentioned columns is about 300°C, although temperatures up to 350°C can be employed with intermittent use. At these temperatures, however, phase bleeding becomes important. This results in the appearance of spurious peaks and limits the life of the column. In the late 1980s, columns with a cross-linked stationary phase that also adheres strongly to the wall of the column or to the packing material became available. "Bonded" columns of this nature, such as DB-1 (methylpolysiloxane) and DB-5 (5% phenylmethylpoly-

TABLE XI

PRODUCTS OF TRIMETHYLSILYLATION OF $3PbO \cdot PbF_2 \cdot SiO_2$ GLASS BY THE DIRECT METHOD

Peak number[1]	TMS derivative of	Percentage of total area (excluding solvent)
1	(solvent)	
2	(unidentified)	0.06
3	SiO_3F^{3-}	3.84
4	(unidentified)	0.05
5	(unidentified)	0.04
6	SiO_4^{4-}	45.05
7	$Si_2O_5F_2^{4-}$	0.24
8	$Si_2O_6F^{5-}$	5.80
9	$Si_3O_7F_3^{5-}$	0.05
10	$Si_2O_7^{6-}$	24.25
11a	$Si_4O_{10}F_2^{6-}$ ⎫	
11b	$Si_3O_8F_2^{6-}$ ⎬	0.24
11c	$Si_3O_9^{6-}$ ⎭	
12a	$Si_4O_{11}F^{7-}$ ⎫	
12b	$Si_3O_9F^{7-}$ ⎬	4.52
13a	$Si_4O_{12}^{8-}$ ⎫	
13b	$Si_5O_{13}F_2^{8-}$ ⎬	0.93
13c	$Si_4O_{11}F_2^{8-}$ ⎭	
14	$Si_3O_{10}^{8-}$	9.46
15a	$Si_6O_{16}F^{9-}$ ⎫	
15b	$Si_5O_{14}F^{9-}$ ⎬	1.01
15c	$Si_4O_{12}F^{9-}$	
15d	$Si_4O_{12}F^{9-}$ ⎭	
16	$Si_4O_{12}F^{9-}$	0.83
17	$Si_5O_{15}^{10-}$	0.75
18	$Si_4O_{13}^{10-}$ (linear)	2.74
19	$Si_4O_{13}^{10-}$ (branched)	0.14

Source: Calhoun et al. (1979).

[1] Peak numbers refer to Fig. 22.

siloxane), have greater stability than the earlier ones and retain their characteristics even when used regularly up to 320 or 325°C. Columns of this type are recommended for future work.

The retention of TMS derivatives of silicates depends mainly on the number of TMS groups per molecule. This results in similar retentions for derivatives of ions with widely differing molecular weights and structures. This has been discussed in detail by Garzo and Hoebbel (1976) and Hoebbel et al. (1976), who have shown how retention indices may be used to identify unknown derivatives, based on well-known GC homologue rules.

Due to the higher resolution attainable, the elution time of a given compound can be considerably shorter in a capillary than a packed column. The use of the former is thus advantageous if species of high molecular weight have to be determined (Garzo et al., 1978).

Flame ionization has been the preferred method of detection in all studies by GLC. For quantitative work, calibration of the detector response with standard solutions of the pure derivatives is necessary. When standard solutions are not available, as is often the case for the less common derivatives, response factors relative to an internal standard may be employed. Garzo et al. (1978) showed there was a linear relation between the relative response factor and the C/Si ratio in the TMS derivative.

Mass spectrometry (MS) is the most powerful method available for identifying TMS derivatives of silicate ions. Samples can be collected in glass capillary tubes and transferred to the mass spectrometer for direct sublimation into the ionization chamber (Wu et al., 1970). Combined GC-MS can also be employed. The major ion of structural significance in electron impact spectra is $(M-15)^+$, due to loss of a methyl group from the molecular ion. Mass spectra of silicate derivatives have been reported by Wu et al. (1970), Götz and Masson (1971b), Masson et al. (1974), Calhoun and Masson (1978), and Calhoun et al. (1979).

Characterization of TMS derivatives of silicate ions by ^{29}Si NMR spectroscopy has been described by Hoebbel et al. (1976), Harris and Newman (1977), and Dent Glasser et al. (1979).

V. Conclusions

Until about 1980, chromatographic methods for silicate glasses have, in general, been less reliable than for the phosphates. However, considerable progress has been made during the 1980s in understanding the factors responsible for side reactions in the derivatization of silicates and reliable methods are now available for glasses with certain cations. It is anticipated that this work will now be extended to other systems.

With the improved techniques now available we are rapidly gaining a more detailed knowledge of the constitution of basic silicate glasses. One of the goals in this work should be to determine equilibrium ratios k_{nn} for silicate glasses, as has already been done for the phosphates. Such data will provide a valuable test of the ionic distributions calculated from polymer theory. Indications are that, for lead and calcium silicate glasses, the proportion of orthosilicate ion predicted by the theory is somewhat too large and that this perturbs the distribution of the higher species. The effect, however, would appear to be less pronounced than for phosphate glasses, in which the orthophosphate ion is often a minor constituent.

Chromatography is the only method available for the direct determination of discrete ions in amorphous materials, although the presence of such ions can often be inferred by other means. Considerable scope exists for the study of silicate glasses by chromatographic methods. It is hoped that this chapter will stimulate further interest in this expanding field.

References

Alexander, G. B. (1954). *J. Am. Chem. Soc.* **76**, 2094–2096.

Ando, T., Ito, J., Ishii, S. H., and Soda, T. (1952). *Bull. Chem. Soc. Japan* **25**, 78–79.

Baba, Y., Yoza, N., and Ohashi, S. (1985a). *J. Chromatog.* **348**, 27–37.

Baba, Y., Yoza, N., and Ohashi, S. (1985b). *J. Chromatog.* **350**, 119–125.

Busch, N., Ebel, J. P., and Blanck, M. (1957). *Bull. Soc. Chim.* (Fr.), 486–487.

Canic, V. D., Turic, M. N., Petrovic, S. M., and Petrovic, S. E. (1965). *Anal. Chem.* **37**, 1576–1577.

Calhoun, H. P., and Masson, C. R. (1978). *J. Chem. Soc. Dalton Trans.*, 1342–1349.

Calhoun, H. P., and Masson, C. R. (1980). *J. Chem. Soc. Dalton Trans.*, 1282–1291.

Calhoun, H. P., and Masson, C. R. (1981). In "Reviews on Silicon, Germanium, Tin and Lead Compounds," vol. V, M. Gielen, ed. Freund Publishing House, Tel-Aviv, Israel, pp. 153–243.

Calhoun, H. P., Jamieson, W. D., and Masson, C. R. (1979). *J. Chem. Soc. Dalton Trans.*, 454–459.

Calhoun, H. P., Masson, C. R., and Jansen, M. (1980). *J. Chem. Soc. Chem. Commun.*, 576–577.

Corbridge, D. E. C. (1974). "The Structural Chemistry of Phosphorus." Elsevier, Amsterdam.

Cripps-Clark, C. J., Sridhar, R., Jeffes, J. H. E., and Richardson, F. D. (1974). In "Physical Chemistry of Process Metallurgy: The Richardson Conference," J. H. E. Jeffes and R. J. Tait, eds. Institution of Mining and Metallurgy, London, pp. 233–245.

Crowther, J. P. (1954). *Anal. Chem.* **26**, 1383–1386.

Currell, B. R., Midgley, H. G., Montecinos, M., and Parsonage, J. R. (1985). *Cem. Concr. Res.* **15**, 889–900.

Dent Glasser, L. S., and Lachowski, E. E. (1980). *J. Chem. Soc. Chem. Commun.*, 973–974.

Dent Glasser, L. S., and Smith, D. N. (1978). *Nature* **274**, 878–879.

Dent Glasser, L. S., Lachowski, E. E., Harris, R. K., and Jones, J. (1979). *J. Mol. Str.* **51**, 239–245.

Dent Glasser, L. S., Howie, R. A., and Smart, R. M. (1981a). *Acta Cryst.* **B37**, 303–306.

Dent Glasser, L. S., Lachowski, E. E., Qureshi, M. Y., Calhoun, H. P., Embree, D. J., Jamieson, W. D., and Masson, C. R. (1981b). *Cem. Concr. Res.* **11**, 775–780.

Du Plessis, D. J. (1959). *Angew. Chem.* **71**, 697–700.

Duwez, P., and Williams, R. H. (1963). *Trans. A.I.M.E.* **227**, 362–365.

Ebel, J. P. (1952). *Compt. Rend. Acad. Sci.*, Paris **234**, 621–623, 732–733.

Ebel, J. P. (1953). *Bull. Soc. Chim.* (Fr.) 991–998, 1089–1095, 1096–1099.

Ebel, J. P. (1968). *Bull. Soc. Chim.* (Fr.), 1663–1670.

Ebel, J. P., and Volmar, Y. (1951). *Compt. Rend. Acad. Sci.* (Paris) **233**, 415–417.

Ezikov, V. I., Pasishnik, S. V., Chuchmarev, S. K., and Oleksa, V. S. (1983). *Visn. L'viv. Politekh. Inst.* **171**, 17–20. [*Chem. Abst.* **99**, 127126.]

Finn, C. W. F., Fray, D. J., King, T. B., and Ward, J. G. (1976). *Phys. Chem. Glass.* **17**, 70–76.

Flory, P. J. (1953). "Principles of Polymer Chemistry." Cornell University Press, Ithaca, New York.

Fray, D. J. (1974). In "Physical Chemistry of Process Metallurgy: The Richardson Conference," J. H. E. Jeffes and R. J. Tait, eds. Institution of Mining and Metallurgy, London, pp. 215–221.

Fukuda, T., Nakamura, T., and Ohashi, S. (1976). *J. Chromatog.* **128**, 212–217.

Garzo, G., and Alexander, G. (1971). *Chromatographia* **4**, 554–560.
Garzo, G., and Hoebbel, D. (1976). *J. Chromatog.* **119**, 173–179.
Garzo, G., Hoebbel, D., Ecsery, Z. J., and Ujszaszi, K. (1978). *J. Chromatog.* **167**, 321–336.
Garzo, G., Vargha, A., Székely, T., and Hoebbel, D. (1980). *J. Chem. Soc. Dalton*, 2068–2074.
Gimblett, F. G. R. (1963). "Inorganic Polymer Chemistry," Butterworths, London.
Götz, J., and Masson, C. R. (1970). *J. Chem. Soc. (A)*, 2683–2686.
Götz, J., and Masson, C. R. (1971a). *J. Chem. Soc. (A)*, 686–688.
Götz, J., and Masson, C. R. (1971b). In "Int. Congr. Glass, Sci. Tech. Commun., 9th," vol. 1. Institut du Verre, Paris, pp. 261–276.
Götz, J., Jamieson, W. D., and Masson, C. R. (1971). In "Proc. Ann. Meeting Int. Comm. on Glass, 1969," S. Bateson and A. G. Sadler, eds. Can. Ceramic Society, Toronto, pp. 69–74.
Götz, J., Masson, C. R., and Castelliz, L. M. (1972). In "Amorphous Materials," R. W. Douglas and B. Ellis, eds. Wiley, New York, pp. 317–325.
Götz, J., Hoebbel, D., and Wieker, W. (1975a). *Z. anorg. allg. Chem.* **416**, 163–168.
Götz, J., Hoebbel, D., and Wieker, W. (1975b). *Z. anorg. allg. Chem.* **418**, 29–34.
Götz, J., Hoebbel, D., and Wieker, W. (1976a). *J. Non-Cryst. Solids* **20**, 413–425.
Götz, J., Hoebbel, D., and Wieker, W. (1976b). *J. Non-Cryst. Solids* **22**, 391–398.
Harris, R. K., and Newman, R. H. (1977). *Org. Magn. Reson.* **9**, 426–431.
Hettler, H. (1958). *J. Chromatog.* **1**, 389–410.
Hettler, H. (1959). *Chromatog. Rev.* **1**, 225–245.
Hirljac, J., Wu, Z.-Q., and Young, J. F. (1983). *Cem. Concr. Res.* **13**, 877–886.
Hoebbel, D., and Wieker, W. (1971). *Z. anorg. allg. Chem.* **384**, 43–52.
Hoebbel, D., and Wieker, W. (1972). *Z. Chem.* **12**, 295–297.
Hoebbel, D., and Wieker, W. (1974). *Z. anorg. allg. Chem.* **405**, 267–274.
Hoebbel, D., Wieker, W., Franke, P., and Otto, A. (1975). *Z. anorg. allg. Chem.* **418**, 35–44.
Hoebbel, D., Garzo, G., Engelhardt, G., Jancke, H., Franke, P., and Wieker, W. (1976). *Z. anorg. allg. Chem.* **424**, 115–127.
Hoebbel, D., Garzo, G., Engelhardt, G., and Till, A. (1979). *Z. anorg. allg. Chem.* **450**, 5–20.
Hoebbel, D., Garzo, G., Engelhardt, G., Ebert, R., Lippmaa, E., and Alla, M. (1980). *Z. anorg. allg. Chem.* **465**, 15–33.
Hoebbel, D., Garzo, G., Ujszasji, K., Engelhardt, G., Fahlke, B., and Vargha, A. (1982). *Z. anorg. allg. Chem.* **484**, 7–21.
Hoebbel, D., Götz, J., Vargha, A., and Wieker, W. (1984). *J. Non-Cryst. Solids* **69**, 149–159.
Huhti, A.-L., and Gartaganis, P. A. (1956). *Can. J. Chem.* **34**, 785–797.
Ida, T., and Yamabe, T. (1971). *J. Chromatog.* **54**, 413–421.
Iler, R. K. (1979). "The Chemistry of Silica." Wiley, New York.
Jackson, W. P., Richter, B. E., Fjeldsted, J. C., Kong, R. C., and Lee, M. L. (1984). In "Ultrahigh Resolution Chromatography," S. Ahuja, ed. A.C.S. Symposium Series 250. American Chemical Society, Washington, D.C., pp. 121–133.
Jameson, R. F. (1959). *J. Chem. Soc.* **752**.
Jansen, M., and Keller, H. L. (1979). *Angew Chem. Int. Ed. Engl.* **18**, 464.
Jeffes, J. H. E. (1981). *Can. Met. Quart.* **20**, 37–50.
Kalliney, S. Y. (1972). In "Topics in Phosphorus Chemistry," vol. 7. Interscience, New York, pp. 255–309.
Kalmychkov, G. V. (1982). *Zh. Anal. Khim.* **37**, 1247–1250. *J. Anal. Chem. USSR* **37**, 953–956.
Kalmychkov, G. V., and Almukhamedov, A. I. (1984). *Fiz.-Khim. Modeli Petrog. Rudoobraz.*, 169–176. [*Chem. Abst.* **103**, 16065 (1985).]
Karl-Kroupa, E. (1956). *Anal. Chem.* **28**, 1091–1097.
Kaula, A., Sedmalis, U., Vitina, I., and Igaune, S. (1981). *Neorg. Stekla, Pokrytiya Mater.* **5**, 44–51. [*Chem. Abst.* **96**, 108939 (1982).]

Kaula, A., Vitina, I., Sulce, A., Sedmalis, U., and Mezhinskii, G. K. (1982). *Latv. PSR Zinat. Akad. Vestis, Kim. Ser.* 547–550. [*Chem. Abst.* **98**, 21048 (1983).]

Kiso, Y., Kobayashi, M., and Kitaoka, Y. (1972). In "Analytical Chemistry of Phosphorus Compounds," M. Halmann, ed. Wiley-Interscience, New York, pp. 93–147.

Kolb, K. E., and Hansen, K. W. (1965). *J. Am. Ceram. Soc.* **48**, 439–440.

Kura, G., and Ohashi, S. (1971). *J. Chromatog.* **56**, 111–120.

Kuroda, K., and Kato, C. (1979). *J. Inorg. Nucl. Chem.* **41**, 947–951.

Kuroda, K., Koike, T., and Kato, C. (1981). *J. Chem. Soc. Dalton Trans.*, 1957–1960.

Lachowski, E. E. (1979). *Cem. Concr. Res.* **9**, 111–114.

Lederer, M., and Majani, C. (1970). *Chromatog. Rev.* **12**, 239–426.

Lentz, C. W. (1964). *Inorg. Chem.* **3**, 574–579.

Masson, C. R. (1965). *Proc. Roy. Soc.* **A287**, 201–221.

Masson, C. R. (1968). *J. Am. Ceram. Soc.* **51**, 134–143.

Masson, C. R. (1977). *J. Non-Cryst. Solids* **25**, 1–41.

Masson, C. R. (1984). In "Second International Symposium on Metallurgical Slags and Fluxes," H. A. Fine and D. R. Gaskell, eds. Metallurgical Society of AIME, Warrendale, Penn., p. 3–44.

Masson, C. R. (1986). (unpublished).

Masson, C. R., and Caley, W. F. (1978). *J. Chem. Soc. Faraday Trans.I*, **74**, 2942–2967.

Masson, C. R., Götz, J., Jamieson, W. D., McLachlan, J. L., and Volborth, A. (1971). "Proc. Second Lunar Science Conference," vol. 1. MIT Press, Cambridge, Mass., pp. 957–971.

Masson, C. R., Jamieson, W. D., and Mason, F. G. (1974). In "Physical Chemistry of Process Metallurgy: The Richardson Conference," J. H. E. Jeffes and R. J. Tait, eds. Institution of Mining and Metallurgy, London, pp. 223–231.

Masson, C. R., Jamieson, W. D., Sim, P. G., Caines, G. W., and Embree, D. J. (1986). (unpublished)

Meadowcroft, T. R., and Richardson, F. D. (1965). *Trans. Faraday Soc.* **61**, 54–70.

Mercer, R. A., and Miller, R. D. (1963). *J. Sci. Instrum.* **40**, 352–354.

Milestone, N. B. (1977). *Cem. Concr. Res.* **7**, 345–346.

Minami, T., Takuma, Y., and Tanaka, M. (1977). *J. Electrochem. Soc.* **124**, 1659–1662.

Minami, T., Katsuda, T., and Tanaka, M. (1979). *J. Phys. Chem.* **83**, 1306–1309.

Murata, K. T. (1943). *Am. Mineral.* **28**, 545–562.

Murthy, M. K. (1961). *J. Am. Ceram. Soc.* **44**, 412–417.

Murthy, M. K. (1963). *J. Am. Ceram. Soc.* **46**, 558–559.

Murthy, M. K., and Mueller, A. (1963). *J. Am. Ceram. Soc.* **46**, 530–535.

Murthy, M. K., and Westman, A. E. R. (1962a). *Inorg. Chem.* **1**, 712–714.

Murthy, M. K., and Westman, A. E. R. (1962b). *J. Am. Ceram. Soc.* **45**, 401–407.

Murthy, M. K., and Westman, A. E. R. (1966). *J. Am. Ceram. Soc.* **49**, 310–311.

Murthy, M. K., Smith, M. J., and Westman, A. E. R. (1961). *J. Am. Ceram. Soc.* **44**, 97–105.

Nakamura, R., and Suginohara, Y. (1980). *J. Jap. Inst. Metals* **44**, 352–358.

Nakamura, R., Arikata, K., Suginohara, Y., and Yanagase, T. (1977). *Kyushu Daigaku Kogaku Shuho* **50**, 635–641.

Nakamura, T., and Ohashi, S. (1967). *Bull. Chem. Soc. Japan* **40**, 110–115.

Nakamura, T., Yano, T., Fujita, A., and Ohashi, S. (1977). *J. Chromatog.* **130**, 384–386.

Ohashi, S. (1975). *Pure and Applied Chemistry* **44**, 415–438.

Ohashi, S., and Matsumura, T. (1962). *Bull. Chem. Soc. Japan* **35**, 501.

Ohashi, S., and Oshima, F. (1963). *Bull. Chem. Soc. Japan* **36**, 1489–1494.

Ohashi, S., and Sugatani, H. (1957). *Bull. Chem. Soc. Japan* **30**, 864–867.

Okusu, H., Masuda, G., Wakita, M., and Suginohara, Y. (1982). *Trans. Jap. Inst. Metals* **23**, 250–258.

Okusu, H., Takeshita, S., Mizoguchi, K., and Suginohara, Y. (1983). *J. Jap. Inst. Metals* **47**, 956–963.

Parks, J. R., and Van Wazer, J. R. (1957). *J. Am. Chem. Soc.* **79**, 4890–4897.

Pasishnik, S. V., Ezikov, V. I., Chuchmarev, S. K., and Voznyuk, V. S. (1985). *Vestn. L'vov. Politekh. Inst.* **191**, 18–20. [*Chem. Abst.* **104**, 60938 (1986).]

Phalippou, J., and Zarzycki, J. (1978). *Ann. Chim. Fr.* **3**, 369–377.

Poch, W. (1966). *Glastech. Ber.* **39**, 45–50.

Pollard, F. H., Nickless, G., Rogers, D. E., and Rothell, M. T. (1965). *J. Chromatog.* **17**, 157–167.

Pollard, F. H., Nickless, G., and Murray, J. D. (1966). *J. Chromatog.* **22**, 139–142.

Richardson, F. D. (1974). "Physical Chemistry of Melts in Metallurgy," vol. 1. Academic Press, London.

Rothbart, H. L., Weymouth, H. W., and Rieman, W. III (1964). *Talanta* **11**, 33–41.

Sarkhar, A. K., and Roy, D. M. (1979). *Cem. Concr. Res.* **9**, 343–352.

Sharma, S. K., and Hoering, T. C. (1978). In "Carnegie Institute of Washington Year Book 77." Port City Press, Baltimore, Md., pp. 662–665.

Sharma, S. K., Dent Glasser, L. S., and Masson, C. R. (1973). *J. Chem. Soc. Dalton Trans.*, 1324–1328.

Smart, R. M., and Glasser, F. P. (1978). *Phys. Chem. Glasses* **19**, 95–102.

Smith, M. J. (1959). *Anal. Chem.* **31**, 1023–1025.

Stevic, S., Radosavljevic, S., and Poleti, D. (1982). *Rev. Chim. Miner.* **19**, 192–198.

Tamas, F. D., Sarkar, A. K., and Roy, D. M. (1976). In "Hydraulic Cement Pastes: Their Structure and Properties," P. Cook and V. Maxwell, eds. Cement and Concrete Association, London, pp. 55–72.

Thilo, E. (1962). *Adv. Inorg. Chem. Radiochem.* **4**, 1–75.

Thilo, E., and Kolditz, L. (1955). *Z. anorg. allgem. Chem.* **278**, 122–135.

Van Wazer, J. R. (1958). "Phosphorus and Its Compounds." Interscience, New York.

Van Wazer, J. R., and Karl-Kroupa, E. (1956). *J. Am. Chem. Soc.* **78**, 1772.

Wada, H., Araki, S., Kuroda, K., and Kato, C. (1985). *Polyhedron* **4**, 653–656.

Watanabe, M. (1973). *Bull. Chem. Soc. Japan* **46**, 2468–2469.

Watanabe, M. (1974). *Nippon Kagaku Kaishi*, 1407–1416.

Watanabe, M., and Kato, M. (1972). *Bull. Chem. Soc. Japan* **45**, 1058–1060.

Watanabe, M., Tanabe, K., Takahara, T., and Yamada, T. (1971). *Bull. Chem. Soc. Japan* **44**, 712–715.

Watanabe, M., Sato, S., and Saito, H. (1975). *Bull. Chem. Soc. Japan* **48**, 893–895.

Watanabe, M., Sato, S., and Saito, H. (1976). *Bull. Chem. Soc. Japan* **49**, 3265–3268.

Watanabe, M., Ito, M., Sato, S., and Yamada, T. (1977). *Bull. Chem. Soc. Japan* **50**, 3251–3254.

Wieker, W., and Hahn, U. (1973). *Z. Chem.* **13**, 195–196.

Wieker, W., and Hoebbel, D. (1969). *Z. anorg. allgem. Chem.* **366**, 139–151.

Winkler, A., Hoebbel, D., Grimmer, A. R., Wieker, W., Ujszaszy, K., and Magi, M. (1983). *Rev. Chim. Miner.* **20**, 801–806.

Westman, A. E. R. (1960). In "Modern Aspects of the Vitreous State," vol. I, J. D. MacKenzie, ed. Butterworths, Washington, D.C., pp. 63–91.

Westman, A. E. R. (1977). In "Topics in Phosphorus Chemistry," vol. 9, E. J. Griffith, ed. Wiley, New York, pp. 231–405.

Westman, A. E. R., and Beatty, R. (1966). *J. Am. Ceram. Soc.* **49**, 63–67.

Westman, A. E. R., and Crowther, J. (1954). *J. Am. Ceram. Soc.* **37**, 420–427.

Westman, A. E. R., and Gartaganis, P. A. (1957). *J. Am. Ceram. Soc.* **40**, 293–299.

Westman, A. E. R., and Murthy, M. K. (1961). *J. Am. Ceram. Soc.* **44**, 97–105, 475–480.

Westman, A. E. R., and Murthy, M. K. (1964). *J. Am. Ceram. Soc.* **47**, 375–378.

Westman, A. E. R., and Scott, A. E. (1951). *Nature* **168**, 740.

Westman, A. E. R., Scott, A. E., and Pedley, J. T. (1952). *Chem. in Canada* **4**, 189–194.
Westman, A. E. R., Gartaganis, P. A., and Smith, M. J. (1959). *Can. J. Chem.* **37**, 1764–1775.
Wu, F. F. H., Götz, J., Jamieson, W. D., and Masson, C. R. (1970). *J. Chromatog.* **48**, 515–520.
Yagamuchi, H., Nakamura, T., Hirai, Y., and Ohashi, S. (1979). *J. Chromatog.* **172**, 131–140.
Yang, H., Ye, D., Shi, Y. and Qiu, X. (1985). *Kexue Tongbao* (foreign lang. ed.) **30**, 224–228.

INDEX

4d and 5d ions, 191

A

Absorber, 283
  thickness, 276, 280
Absorption
  recoil-free gamma-resonance, 273
  resonance, 275, 276
Abundance, natural, 283
Accelerator, particle, 282, 283
Acmite, 299
Activation-energy (E_A), 87, 93
a-Fe_2O_3, 301
$Ag^{\pm}$, 241
AgI-Ag_2O-P_2O_5, 335
Alkali-associated electron center (AEC), 232
Alkali borate glasses, 185, 190, 191,
    229, 230, 231
Alkali silicate glasses, 223, 225, 228
Alkaline-earth metal phosphates, 333
Alloy, FeNiB, amorphous, 304
Aluminophosphate glass, 181
Alumino silicate glasses, 364
Aluminum, 337
  ^{27}Al NMR, 130, 140
Amorphous system, 274
Andradite, 347
Anisotropy constant, 280
Anomaly, borate, 293, 297
Antimony, 237–241, 337

Arrangement, measuring, 283
Arsenic, 23, 237–241
  oxide glass, As_2O_3, 29
Arsenophosphate glasses, 336
Asymmetry, 306
Atomic hydrogen, 206
$Au°$, 242

B

B_2O_3, 210
Background, 281
Barium aluminoborate glasses, 192
Barysilite, 347
Beryllium, ^9Be NMR, 143
Bivalent cations, 324
Bonding
  homopolar v. heteropolar, 37
  mesomeric, 23, 34
  random, 23, 30
Borate glasses, 96, 101, 179, 229–233
  silver diborate, $Ag_2O(B_2O_3)_2$, 60
  sodium diborate, $Na_2O(B_2O_3)_2$, 44
Boron
  ^{10}B NMR, 98
  ^{11}B NMR, 97, 102, 129
    in borosilicate glasses, 123
    in commercial glasses, 129
    in modified borate glasses, 101
    in sulfide glasses, 144
Boron electron center (BEC), 212, 215, 232

377

N